Developments in fiber-reinforced polymer (FRP) composites for civil engineering

© Woodhead Publishing Limited, 2013

Related titles:

Advanced fibre-reinforced polymer (FRP) composites for structural applications
(ISBN 978-0-85709-418-6)

Eco-efficient concrete
(ISBN 978-0-85709-424-7)

Nanotechnology in eco-efficient construction: Materials, processes and applications
(ISBN 978-0-85709-544-2)

Details of these books and a complete list of titles from Woodhead Publishing can be obtained by:

- visiting our web site at www.woodheadpublishing.com
- contacting Customer Services (e-mail: sales@woodheadpublishing.com; fax: +44 (0) 1223 832819; tel.: +44 (0) 1223 499140 ext. 130; address: Woodhead Publishing Limited, 80 High Street, Sawston, Cambridge CB22 3HJ, UK)
- in North America, contacting our US office (e-mail: usmarketing@woodhead-publishing.com; tel.: (215) 928 9112; address: Woodhead Publishing, 1518 Walnut Street, Suite 1100, Philadelphia, PA 19102–3406, USA)

If you would like e-versions of our content, please visit our online platform: www.woodheadpublishingonline.com. Please recommend it to your librarian so that everyone in your institution can benefit from the wealth of content on the site.

We are always happy to receive suggestions for new books from potential editors. To enquire about contributing to our Materials series, please send your name, contact address and details of the topic/s you are interested in to gwen.jones@woodheadpublishing.com. We look forward to hearing from you.

The Woodhead team responsible for publishing this book:
Commissioning Editor: Jess Rowley
Publications Coordinator: Lucy Beg
Project Editor: Cathryn Freear
Editorial and Production Manager: Mary Campbell
Production Editor: Richard Fairclough
Cover Designer: Terry Callanan

© Woodhead Publishing Limited, 2013

Woodhead Publishing Series in Civil and
Structural Engineering: Number 45

Developments in fiber-reinforced polymer (FRP) composites for civil engineering

Edited by
Nasim Uddin

WOODHEAD PUBLISHING

Oxford   Cambridge   Philadelphia   New Delhi

© Woodhead Publishing Limited, 2013

Published by Woodhead Publishing Limited,
80 High Street, Sawston, Cambridge CB22 3HJ, UK
www.woodheadpublishing.com
www.woodheadpublishingonline.com

Woodhead Publishing, 1518 Walnut Street, Suite 1100, Philadelphia,
PA 19102-3406, USA

Woodhead Publishing India Private Limited, G-2, Vardaan House, 7/28 Ansari Road, Daryaganj, New Delhi – 110002, India
www.woodheadpublishingindia.com

First published 2013, Woodhead Publishing Limited
© Woodhead Publishing Limited, 2013. Note: the publisher has made every effort to ensure that permission for copyright material has been obtained by authors wishing to use such material. The authors and the publisher will be glad to hear from any copyright holder it has not been possible to contact.
The authors have asserted their moral rights.

This book contains information obtained from authentic and highly regarded sources. Reprinted material is quoted with permission, and sources are indicated. Reasonable efforts have been made to publish reliable data and information, but the authors and the publisher cannot assume responsibility for the validity of all materials. Neither the authors nor the publisher, nor anyone else associated with this publication, shall be liable for any loss, damage or liability directly or indirectly caused or alleged to be caused by this book.

Neither this book nor any part may be reproduced or transmitted in any form or by any means, electronic or mechanical, including photocopying, microfilming and recording, or by any information storage or retrieval system, without permission in writing from Woodhead Publishing Limited.

The consent of Woodhead Publishing Limited does not extend to copying for general distribution, for promotion, for creating new works, or for resale. Specific permission must be obtained in writing from Woodhead Publishing Limited for such copying.

Trademark notice: Product or corporate names may be trademarks or registered trademarks, and are used only for identification and explanation, without intent to infringe.

British Library Cataloguing in Publication Data
A catalogue record for this book is available from the British Library.

Library of Congress Control Number: 2013934746

ISBN 978-0-85709-234-2 (print)
ISBN 978-0-85709-895-5 (online)
ISSN 2052-4714 Woodhead Publishing Series in Civil and Structural Engineering (print)
ISSN 2052-4722 Woodhead Publishing Series in Civil and Structural Engineering (online)

The publisher's policy is to use permanent paper from mills that operate a sustainable forestry policy, and which has been manufactured from pulp which is processed using acid-free and elemental chlorine-free practices. Furthermore, the publisher ensures that the text paper and cover board used have met acceptable environmental accreditation standards.

Typeset by Toppan Best-set Premedia Limited, Hong Kong
Printed by MPG Printgroup, UK

Contents

Contributor contact details		*xiii*
Woodhead Publishing Series in Civil and Structural Engineering		*xix*
Introduction		*xxiii*

Part I General developments — 1

1 Types of fiber and fiber arrangement in fiber-reinforced polymer (FRP) composites — 3
Y. GOWAYED, Auburn University, USA

1.1	Introduction	3
1.2	Fibers	5
1.3	Fabrics	10
1.4	Composites	14
1.5	Future trends	15
1.6	Sources of further information and advice	16
1.7	References	16

2 Biofiber reinforced polymer composites for structural applications — 18
O. FARUK and M. SAIN, University of Toronto, Canada

2.1	Introduction	18
2.2	Reinforcing fibers	19
2.3	Drawbacks of biofibers	22
2.4	Modification of natural fibers	24
2.5	Matrices for biocomposites	26
2.6	Processing of biofiber-reinforced plastic composites	31
2.7	Performance of biocomposites	36
2.8	Future trends	43
2.9	Conclusion	45
2.10	References	46

3	Advanced processing techniques for composite materials for structural applications	54
	R. EL-HAJJAR, University of Wisconsin-Milwaukee, USA, H. TAN, Hewlett-Packard Company, USA and K. M. PILLAI, University of Wisconsin-Milwaukee, USA	
3.1	Introduction	54
3.2	Manual layup	54
3.3	Plate bonding	55
3.4	Preforming	56
3.5	Vacuum assisted resin transfer molding (VARTM)	57
3.6	Pultruded composites	65
3.7	Automated fiber placement	69
3.8	Future trends	71
3.9	Sources of further information	72
3.10	References	72
4	Vacuum assisted resin transfer molding (VARTM) for external strengthening of structures	77
	N. UDDIN, S. CAUTHEN, L. RAMOS and U. K. VAIDYA, The University of Alabama at Birmingham, USA	
4.1	Introduction	77
4.2	The limitations of hand layup techniques	79
4.3	Comparing hand layup and vacuum assisted resin transfer molding (VARTM)	81
4.4	Analyzing load, strain, deflections, and failure modes	83
4.5	Flexural fiber-reinforced polymer (FRP) wrapped beams	86
4.6	Shear and flexural fiber-reinforced polymer (FRP) wrapped beams	90
4.7	Comparing hand layup and vacuum assisted resin transfer molding (VARTM): results and discussion	94
4.8	Case study: I-565 Highway bridge girder	97
4.9	Conclusion and future trends	111
4.10	Acknowledgment	113
4.11	References	113
5	Failure modes in structural applications of fiber-reinforced polymer (FRP) composites and their prevention	115
	O. GUNES, Cankaya University, Turkey	
5.1	Introduction	115
5.2	Failures in structural engineering applications of fiber-reinforced polymer (FRP) composites	116

5.3	Strategies for failure prevention	123
5.4	Non-destructive testing (NDT) and structural health monitoring (SHM) for inspection and monitoring	129
5.5	Future trends	140
5.6	Conclusion	141
5.7	Acknowledgment	141
5.8	Sources of further information	142
5.9	References	143
6	Assessing the durability of the interface between fiber-reinforced polymer (FRP) composites and concrete in the rehabilitation of reinforced concrete structures J. Wang, The University of Alabama, USA	148
6.1	Introduction	148
6.2	Interface stress analysis of the fiber-reinforced polymer (FRP)-to-concrete interface	149
6.3	Fracture analysis of the fiber-reinforced polymer (FRP)-to-concrete interface	155
6.4	Durability of the fiber-reinforced polymer (FRP)–concrete interface	163
6.5	References and further reading	171

Part II Particular types and applications   175

7	Advanced fiber-reinforced polymer (FRP) composites for civil engineering applications S. Moy, University of Southampton, UK	177
7.1	Introduction	177
7.2	The use of fiber-reinforced polymer (FRP) materials in construction	178
7.3	Practical applications in buildings	181
7.4	Future trends	202
7.5	Sources of further information	203
7.6	References	204
8	Hybrid fiber-reinforced polymer (FRP) composites for structural applications D. Lau, City University of Hong Kong, P. R. China	205
8.1	Introduction	205
8.2	Hybrid fiber-reinforced polymer (FRP) reinforced concrete beams: internal reinforcement	207

8.3	Hybrid fiber-reinforced polymer (FRP) composites in bridge construction	218
8.4	Future trends	221
8.5	Sources of further information	222
8.6	References	223
9	**Design of hybrid fiber-reinforced polymer (FRP)/autoclave aerated concrete (AAC) panels for structural applications** N. Uddin, M. A. Mousa, U. Vaidya and F. H. Fouad, The University of Alabama at Birmingham, USA	**226**
9.1	Introduction	226
9.2	Performance issues with fiber-reinforced polymer (FRP)/autoclave aerated concrete (AAC) panels	227
9.3	Materials, processing, and methods of investigation	229
9.4	Comparing different panel designs	233
9.5	Analytical modeling of fiber-reinforced polymer (FRP)/autoclave aerated concrete (AAC) panels	237
9.6	Design graphs for fiber-reinforced polymer (FRP)/autoclave aerated concrete (AAC) panels	239
9.7	Conclusion	244
9.8	Acknowledgment	244
9.9	References	244
9.10	Appendix A: λ calculations for fiber-reinforced polymer (FRP)/autoclave aerated concrete (AAC) using unidirectional fiber-reinforced polymer (FRP) facesheets (UFFS)	245
9.11	Appendix B: symbols	246
10	**Impact behavior of hybrid fiber-reinforced polymer (FRP)/autoclave aerated concrete (AAC) panels for structural applications** N. Uddin, M. A. Mousa and F. H. Fouad, The University of Alabama at Birmingham, USA	**247**
10.1	Introduction	247
10.2	Low velocity impact (LVI) and sandwich structures	249
10.3	Materials and processing	250
10.4	Analyzing sandwich structures using the energy balance model (EBM)	253
10.5	Low velocity impact (LVI) testing	255
10.6	Results of impact testing	258
10.7	Analysis using the energy balance model (EBM)	266
10.8	Conclusion	269

10.9	Acknowledgment	269
10.10	References	270
10.11	Appendix: symbols	271

11	Innovative fiber-reinforced polymer (FRP) composites for disaster-resistant buildings N. UDDIN and M. A. MOUSA, The University of Alabama at Birmingham, USA	272
11.1	Introduction	272
11.2	Traditional and advanced panelized construction	273
11.3	Innovative composite structural insulated panels (CSIPs)	274
11.4	Designing composite structural insulated panels (CSIPs) for building applications under static loading	279
11.5	Composite structural insulated panels (CSIPs) as a disaster-resistant building panel	288
11.6	Conclusion	299
11.7	Acknowledgment	299
11.8	References	299

12	Thermoplastic composite structural insulated panels (CSIPs) for modular panelized construction N. UDDIN, A. VAIDYA, U. VAIDYA and S. PILLAY, The University of Alabama at Birmingham, USA	302
12.1	Introduction	302
12.2	Traditional structural insulated panel (SIP) construction	304
12.3	Joining of precast panels in modular buildings	305
12.4	Manufacturing of composite structural insulated panels (CSIPs)	307
12.5	Connections for composite structural insulated panels (CSIPs)	311
12.6	Conclusion	315
12.7	Acknowledgment	315
12.8	References	315

13	Thermoplastic composites for bridge structures N. UDDIN, A. M. ABRO, J. D. PURDUE and U. VAIDYA, The University of Alabama at Birmingham, USA	317
13.1	Introduction	317
13.2	Manufacturing process for thermoplastic composites	318
13.3	Bridge deck designs	320
13.4	Design case studies	323
13.5	Comparing bridge deck designs	329

x   Contents

13.6	Prefabricated wraps for bridge columns	332
13.7	Compression loading of bridge columns	333
13.8	Impact loading of bridge columns	338
13.9	Conclusion	343
13.10	Acknowledgment	345
13.11	References	345

14 Fiber-reinforced polymer (FRP) composites for bridge superstructures — 347
Y. Kitane, Nagoya University, Japan and A. J. Aref, University at Buffalo – The State University of New York, USA

14.1	Introduction	347
14.2	Fiber-reinforced polymer (FRP) applications in bridge structures	351
14.3	Hybrid fiber-reinforced polymer (FRP)–concrete bridge superstructure	356
14.4	Conclusion	378
14.5	References	379

15 Fiber-reinforced polymer (FRP) composites for strengthening steel structures — 382
M. Dawood, University of Houston, USA

15.1	Introduction	382
15.2	Conventional repair techniques and advantages of fiber-reinforced polymer (FRP) composites	383
15.3	Flexural rehabilitation of steel and steel–concrete composite beams	386
15.4	Bond behavior	394
15.5	Repair of cracked steel members	399
15.6	Stabilizing slender steel members	400
15.7	Case studies and field applications	401
15.8	Future trends	402
15.9	Sources of further information	404
15.10	References	405

16 Fiber-reinforced polymer (FRP) composites in environmental engineering applications — 410
R. Liang and G. Hota, West Virginia University, USA

16.1	Introduction	410
16.2	Advantages and environmental benefits of fiber-reinforced polymer (FRP) composites	412

16.3	Fiber-reinforced polymer (FRP) composites in chemical environmental applications	414
16.4	Fiber-reinforced polymer (FRP) composites in sea-water environment	418
16.5	Fiber-reinforced polymer (FRP) composites in coal-fired plants	423
16.6	Fiber-reinforced polymer (FRP) composites in mining environments	429
16.7	Fiber-reinforced polymer (FRP) composites for modular building of environmental durability	435
16.8	Fiber-reinforced polymer (FRP) wraps	437
16.9	Recycling composites	441
16.10	Green composites	447
16.11	Durability of composites	455
16.12	Design codes and specifications	458
16.13	Future trends	461
16.14	Acknowledgment	462
16.15	References	463
17	**Design of all-composite structures using fiber-reinforced polymer (FRP) composites** P. QIAO, Washington State University, USA and J. F. DAVALOS, The City College of New York, USA	**469**
17.1	Introduction	469
17.2	Review on analysis	470
17.3	Systematic analysis and design methodology	473
17.4	Structural members	485
17.5	Structural systems	502
17.6	Design guidelines	503
17.7	Conclusion	504
17.8	References	505
	Index	*509*

Contributor contact details

(* = main contact)

Editor

Professor Nasim Uddin
Department of Civil, Construction
  and Environmental Engineering
The University of Alabama at
  Birmingham
1075 13th Street South
Birmingham, AL 35294-4440
USA

E-mail: nuddin@uab.edu

Chapter 1

Professor Yasser Gowayed
Department of Polymer and Fiber
  Engineering
311 W Magnolia Ave
Auburn University, AL 36849-5327
USA

E-mail: gowayed@auburn.edu

Chapter 2

Dr Omar Faruk* and Professor
  Mohini Sain
Centre for Biocomposites and
  Biomaterials Processing
University of Toronto
33 Willcocks Street
Toronto
ON M5S 3B3
Canada

E-mail: o.faruk@utoronto.ca

Chapter 3

Dr R. El-Hajjar*
Department of Civil Engineering
  and Mechanics
College of Engineering and
  Applied Science
University of Wisconsin-Milwaukee
3200 N Cramer Street
Milwaukee, WI 53211
USA

E-mail: elhajjar@uwm.edu

Dr H. Tan
Engineering Modeling and
    Analysis Group
Hewlett-Packard Company
1000 NE Circle Blvd
Corvallis, OR 97330
USA

Dr K. M. Pillai
Laboratory for Flow and Transport
    Studies in Porous Media
Department of Mechanical
    Engineering
College of Engineering and
    Applied Science
University of Wisconsin-Milwaukee
3200 N Cramer Street
Milwaukee, WI 53211
USA

E-mail: krishna@uwm.edu

Chapter 4

Professor Nasim Uddin,*
    S. Cauthen and L. Ramos
Department of Civil, Construction
    and Environmental Engineering
The University of Alabama at
    Birmingham
1075 13th Street South
Birmingham, AL 35294-4440
USA

E-mail: nuddin@uab.edu

Dr Uday K. Vaidya
Department of Materials Science
    and Engineering
The University of Alabama at
    Birmingham
1075 13th Street South
Birmingham, AL 35294-4461
USA

Chapter 5

Dr Oguz Gunes
Department of Civil Engineering
Cankaya University
Eskisehir yolu 29. km
Ankara 06810
Turkey

E-mail: ogunes@alum.mit.edu

Chapter 6

Dr Jialai Wang
Department of Civil, Construction,
    and Environmental Engineering
The University of Alabama
Tuscaloosa, AL 35487-0205
USA

E-mail: jwang@eng.ua.edu

Chapter 7

Professor Stuart Moy
School of Civil Engineering and
    the Environment
University of Southampton
Southampton SO17 1BJ
UK

E-mail: ssjm@soton.ac.uk

Chapter 8

Dr Denvid Lau
Department of Civil and
    Architectural Engineering
City University of Hong Kong
Tat Chee Avenue
Kowloon
Hong Kong
P.R. China

E-mail: denvid.lau@cityu.edu.hk

Chapter 9

Professor Nasim Uddin,* Dr Mohammed A. Mousa and Fouad H. Fouad
Department of Civil, Construction and Environmental Engineering
The University of Alabama at Birmingham
1075 13th Street South
Birmingham, AL 35294-4440
USA

E-mail: nuddin@uab.edu; mmousa@uab.edu

Professor Uday Vaidya
Department of Materials Science and Engineering
The University of Alabama at Birmingham
1075 13th Street South
Birmingham, AL 35294-4461
USA

Chapter 10

Professor Nasim Uddin,* Dr Mohammed A. Mousa and Professor Fouad H. Fouad
Department of Civil, Construction and Environmental Engineering
The University of Alabama at Birmingham
1075 13th Street South
Birmingham, AL 35294-4440
USA

E-mail: nuddin@uab.edu; mmousa@uab.edu

Chapter 11

Professor Nasim Uddin* and Dr Mohammed A. Mousa
Department of Civil, Construction and Environmental Engineering
The University of Alabama at Birmingham
1075 13th Street South
Birmingham, AL 35294-4440
USA

E-mail: nuddin@uab.edu; mmousa@uab.edu

Chapter 12

Professor Nasim Uddin*
Department of Civil, Construction and Environmental Engineering
The University of Alabama at Birmingham
1075 13th Street South
Birmingham, AL 35294-4440
USA

E-mail: nuddin@uab.edu

Dr Uday Vaidya and Dr Selvum Pillay
Department of Materials Science and Engineering
The University of Alabama at Birmingham
1075 13th Street South
Birmingham, AL 35294-4461
USA

Chapter 13

Professor Nasim Uddin,* Abdul Moeed Abro and John D. Purdue
Department of Civil, Construction and Environmental Engineering
The University of Alabama at Birmingham
1075 13th Street South
Birmingham, AL 35294-4440
USA

E-mail: nuddin@uab.edu

Professor Uday Vaidya
Department of Materials Science and Engineering
The University of Alabama at Birmingham
1075 13th Street South
Birmingham, AL 35294-4461
USA

Chapter 14

Dr Yasuo Kitane
Department of Civil Engineering
Nagoya University
Furo-cho, Chikusa-ku
Nagoya
Aichi 464-8603
Japan

E-mail: ykitane@civil.nagoya-u.ac.jp

Professor Amjad J. Aref*
Department of Civil, Structural and Environmental Engineering
University at Buffalo – The State University of New York
235 Ketter Hall
Buffalo, NY 14260
USA

E-mail: aaref@buffalo.edu

Chapter 15

Dr Mina Dawood
Department of Civil and Environmental Engineering
University of Houston
N107 Engineering Bldg. 1
Houston, TX 77204-4003
USA

E-mail: mmdawood@uh.edu

Chapter 16

Dr Ruifeng (Ray) Liang*
Constructed Facilities Center
NSF I/UCRC Center for Integration of Composites into Infrastructure
West Virginia University
Morgantown, WV 26506-6103
USA

E-mail: rliang@mail.wvu.edu

Professor Gangarao Hota
Constructed Facilities Center
Center for Integration of
 Composites into Infrastructure
West Virginia University
Morgantown, WV 26506
USA

E-mail: ghota@mail.wvu.edu

Chapter 17

Professor Pizhong Qiao*
Department of Civil and
 Environmental Engineering
Washington State University
Sloan 120
Spokane Street
Pullman, WA 99164-2910
USA

E-mail: qiao@wsu.edu

Professor Julio F. Davalos
Department of Civil Engineering
Grove School of Engineering
The City College of New York
Steinman Hall 139
160 Convent Avenue at 140th
 Street-West
New York, NY 10031
USA

E-mail: jdavalos@ccny.cuny.edu

Woodhead Publishing Series in Civil and Structural Engineering

1. **Finite element techniques in structural mechanics**
   C. T. F. Ross
2. **Finite element programs in structural engineering and continuum mechanics**
   C. T. F. Ross
3. **Macro-engineering**
   F. P. Davidson, E. G. Frankl and C. L. Meador
4. **Macro-engineering and the earth**
   U. W. Kitzinger and E. G. Frankel
5. **Strengthening of reinforced concrete structures**
   Edited by L. C. Hollaway and M. Leeming
6. **Analysis of engineering structures**
   B. Bedenik and C. B. Besant
7. **Mechanics of solids**
   C. T. F. Ross
8. **Plasticity for engineers**
   C. R. Calladine
9. **Elastic beams and frames**
   J. D. Renton
10. **Introduction to structures**
    W. R. Spillers
11. **Applied elasticity**
    J. D. Renton
12. **Durability of engineering structures**
    J. Bijen
13. **Advanced polymer composites for structural applications in construction**
    Edited by L. C. Hollaway
14. **Corrosion in reinforced concrete structures**
    Edited by H. Böhni

15 **The deformation and processing of structural materials**
   Edited by Z. X. Guo
16 **Inspection and monitoring techniques for bridges and civil structures**
   Edited by G. Fu
17 **Advanced civil infrastructure materials**
   Edited by H. Wu
18 **Analysis and design of plated structures Volume 1: Stability**
   Edited by E. Shanmugam and C. M. Wang
19 **Analysis and design of plated structures Volume 2: Dynamics**
   Edited by E. Shanmugam and C. M. Wang
20 **Multiscale materials modelling**
   Edited by Z. X. Guo
21 **Durability of concrete and cement composites**
   Edited by C. L. Page and M. M. Page
22 **Durability of composites for civil structural applications**
   Edited by V. M. Karbhari
23 **Design and optimization of metal structures**
   J. Farkas and K. Jarmai
24 **Developments in the formulation and reinforcement of concrete**
   Edited by S. Mindess
25 **Strengthening and rehabilitation of civil infrastructures using fibre-reinforced polymer (FRP) composites**
   Edited by L. C. Hollaway and J. C. Teng
26 **Condition assessment of aged structures**
   Edited by J. K. Paik and R. M. Melchers
27 **Sustainability of construction materials**
   J. Khatib
28 **Structural dynamics of earthquake engineering**
   S. Rajasekaran
29 **Geopolymers: Structures, processing, properties and industrial applications**
   Edited by J. L. Provis and J. S. J. van Deventer
30 **Structural health monitoring of civil infrastructure systems**
   Edited by V. M. Karbhari and F. Ansari
31 **Architectural glass to resist seismic and extreme climatic events**
   Edited by R. A. Behr
32 **Failure, distress and repair of concrete structures**
   Edited by N. Delatte
33 **Blast protection of civil infrastructures and vehicles using composites**
   Edited by N. Uddin
34 **Non-destructive evaluation of reinforced concrete structures Volume 1: Deterioration processes**
   Edited by C. Maierhofer, H.-W. Reinhardt and G. Dobmann

35 **Non-destructive evaluation of reinforced concrete structures Volume 2: Non-destructive testing methods**
Edited by C. Maierhofer, H.-W. Reinhardt and G. Dobmann
36 **Service life estimation and extension of civil engineering structures**
Edited by V. M. Karbhari and L. S. Lee
37 **Building decorative materials**
Edited by Y. Li and S. Ren
38 **Building materials in civil engineering**
Edited by H. Zhang
39 **Polymer modified bitumen**
Edited by T. McNally
40 **Understanding the rheology of concrete**
Edited by N. Roussel
41 **Toxicity of building materials**
Edited by F. Pacheco-Torgal, S. Jalali and A. Fucic
42 **Eco-efficient concrete**
Edited by F. Pacheco-Torgal, S. Jalali, J. Labrincha and V. M. John
43 **Nanotechnology in eco-efficient construction**
Edited by F. Pacheco-Torgal, M. V. Diamanti, A. Nazari and C.-G. Granqvist
44 **Handbook of seismic risk analysis and management of civil infrastructure systems**
Edited by F. Tesfamariam and K. Goda
45 **Developments in fiber-reinforced polymer (FRP) composites for civil engineering**
Edited by N. Uddin
46 **Advanced fibre-reinforced polymer (FRP) composites for structural applications**
Edited by J. Bai
47 **Handbook of recycled concrete and demolition waste**
Edited by F. Pacheco-Torgal, V. W. Y. Tam, J. A. Labrincha, Y. Ding and J. de Brito
48 **Understanding the tensile properties of concrete**
Edited by J. Weerheijm
49 **Eco-efficient construction and building materials**
Edited by F. Pacheco-Torgal, J. Labrincha, G. Baldo and A. de Magalhaes

Introduction

A composite material combines two or more materials (e.g., fiber reinforcements and binder or matrix resins). These combinations are designed to obtain a material that achieves targeted performance objectives and properties. As an example, fiber-reinforced polymer (FRP) composites are made of thermosetting or thermoplastic resins and glass and/or carbon fibers. The fiber network provides the load-bearing component of the composite while the resin contributes towards transferring loads to the fiber network and maintains fiber orientation. The resin controls the manufacturing process and processing variables. Resins also protect the fabrics from environmental factors such as humidity, high temperature, and chemical attack.

There has been significant research on the development of FRP composite materials and their innovative applications. Many of these R&D efforts have successfully demonstrated materials with improved structural performances. FRP composites are being promoted as twenty-first-century materials because of their superior corrosion resistance, excellent thermo-mechanical properties, and high strength-to-weight ratio. FRPs composites are also 'greener' in terms of embodied energy than conventional materials such as concrete and steel. The use of FRP composites in civil and military infrastructure can improve innovation, increase productivity, enhance performance, and provide longer service lives, i.e. reduced life-cycle costs. It is obvious from these efforts that the use of innovative composite materials and designs has significant potential to reduce vulnerability of infrastructure. The new generation of composite materials has great potential for numerous infrastructure applications, hence the focus of the book.

The composites industry had achieved great strides in making a product that is light, stiff, has excellent strength, and is environmentally stable. As it continues to press forward with a product that has such unique characteristics, the industry is realizing the need to advance in different directions, such as fiber types, fabric geometry, and composite applications. There is a strong need for a high strength and environmentally stable polymer fiber that can be stacked and stitched without the fear of breaking like ceramic fibers.

FRP composite materials have increasingly been used in civil engineering applications in the past three decades. They are ideal for structural applications where high strength-to-weight and stiffness-to-weight ratios are required. Their widespread use, however, has still not been realized because of a number of fundamental issues including high material costs, a relatively short history of applications, and gaps in the development of established standards. While a good number of technical publications exist, there are relatively few sources providing engineers with the information required to design structures using FRPs. In an effort to fill this apparent gap, we have asked for contributions from experts in the field to cover some significant aspects of this broad topic. These contributions have raised critical issues for existing theories and practices relating to civil engineering structural applications. They also show how innovative composites materials can help meet these challenges.

Part I discusses general technical issues with chapters on recent developments in fiber and fiber arrangements, biofibers, processing techniques, external strengthening of structures, failure modes and durability assessment. Chapter 1 by Dr Gowayed introduces some of the fibers typically used in composite materials and their formation into fabric layers as well as other types of FRP. As the discussion goes into different material configurations, the importance of anisotropy in the selection of fibers and the formation of fabrics is also highlighted.

In Chapter 2 by Dr Faruk and Dr Sain, the development of biocomposites reinforced with natural fibres is discussed. These composites have developed significantly over recent years because of their significant processing advantages, biodegradability, low cost, low relative density, high specific strength, and renewable nature. This chapter gives an overview of the most common biofibers in biocomposites covering their sources, types, structure, composition, and properties. Drawbacks of biofibers such as dimensional instability, moisture absorption, biological, ultraviolet, and fire resistance are discussed. The chapter also discusses fiber modification and processing of biofiber reinforced plastic composites. The mechanical and physical properties of biocomposites are also reviewed.

Chapter 3 by Drs El-Hajjar, Tan and Pillai introduces the reader to some of the recent developments in advanced processing of composite materials for manufacturing large composite structures. The goal is to introduce the reader to the methods coupling advanced numerical and analytical approaches for developing predictive capabilities especially useful for larger projects. This chapter also discusses recent developments related to the manufacturing of composites using vacuum assisted resin transfer molding (VARTM), pultrusion, and automated fiber placement (AFP) that have been successfully used to create large composite structures for civilian engineering applications.

High quality and expedient repair methods are necessary to address deterioration that can occur in concrete bridge structures. Most infrastructure-related applications of fiber-reinforced plastics (FRPs) use hand lay-up methods. Hand lay-up is tedious, labor-intensive and results depend on the level of skill of personnel. Chapter 4 by Dr Uddin et al. introduces this alternative for the application of FRP composites. VARTM uses single-sided molding technology to infuse resin over fabrics which can wrap large structures such as bridge girders and columns. This chapter investigates the shear and flexural strength gains of a concrete beam reinforced with FRP composites using the VARTM method. Tests were conducted to determine and document the gains of FRP rehabilitated beams applied by the VARTM method compared to the hand lay-up method of application. This newly introduced technique was used, for example, to repair and retrofit a simple span I-565 prestressed concrete bridge girder in Huntsville, Alabama (USA).

Design safety requires that all possible modes and mechanisms of failure are identified, characterized, and accounted for in the design procedures. Chapter 5 by Dr Gunes provides a review of the failure types encountered in structural engineering applications of FRP and the preventive methods and strategies that have been developed to eliminate or delay such failures. As part of these preventive measures, various non-destructive testing (NDT) and structural health monitoring (SHM) methods used for monitoring FRP applications are discussed with illustrative examples.

Strengthening reinforced concrete (RC) members using FRP composites through external bonding has emerged as a viable technique to retrofit/repair deteriorated infrastructure. The interface between the FRP and concrete plays a critical role in this technique. In Chapter 6, Dr Wang discusses the analytical and experimental methods used to examine the integrity and long-term durability of this interface. Interface stress models, including the commonly adopted two-parameter elastic foundation model and a novel three-parameter elastic foundation model (3PEF) are first presented, which can be used as general tools to analyze and evaluate the design of the FRP strengthening system. The two interface fracture model, linear elastic fracture mechanics and cohesive zone models are used to analyze the potential and full debonding process of the FRP–concrete interface. A novel experimental method, environment-assisted subcritical debonding testing, is then introduced to evaluate this deterioration process. A series of subcritical cracking wedge tests are then used to help accurately predict the long-term durability of the FRP–concrete interface.

Part II reviews innovative FRP applications related to civil engineering. The author of Chapter 7 is Dr Moy who has extensive experience of the use of FRP products in construction in the USA, Canada, Europe, Asia and Australasia. In this chapter, he describes the uses of advanced FRP composites in construction applications, the fabrication techniques which have driven those uses, and the advice available to someone thinking of using these materials. Examples are given of typical products and how they are used. The emphasis is on practicalities rather than theory.

In Chapter 8 by Dr Lau, the advantages of hybrid structural systems including the cost effectiveness and the ability to optimize the cross section based on material properties of each constituent material are presented. In this chapter, two major applications of hybrid FRP composites are discussed: (1) the internal reinforcement in RC structures, and (2) the cable in long-span cable-stayed bridges. In order to improve the flexural ductility of FRP-reinforced concrete (FRPRC) beam, the addition of steel longitudinal reinforcement is proposed such that the hybrid FRPRC beam contains both FRP and steel reinforcement. In order to improve the vibrational problem in pure FRP cable used in bridge construction, an innovative hybrid FRP cable which can inherently incorporate a smart damper is proposed. The objective of this chapter is to deliver the up-to-date developments in hybrid FRP composite structures including both the industrial practice and the research in academia.

Chapter 9 by Dr Uddin *et al.* presents, for the first time, a finite element (FE) modeling for fiber-reinforced polymer (FRP)/autoclave aerated concrete (AAC) sandwich panels. The finite element analysis(FEA) results are compared with the experimental ones and show an acceptable agreement. The validation of both results, FE and experimental, concludes that nonlinear FEA modeling could predict accurately the deflections of the FRP/AAC panel. Analytical models are presented to predict the deflection and strength of the FRP/AAC panels. Design graphs have been developed to help in designing the floor and wall panels made from FRP/AAC panels. Also, those panels have been compared by the commercially used reinforced AAC panels and the results showed that FRP/AAC panels offer a relatively cost-effective solution for longer life cycles. Therefore, they have the potential to be implemented in high wind and seismic load areas. This research would be a major step toward the FE modeling and design of FRP/AAC panels.

The structural sandwich panels composed of a FRP/AAC combination have shown excellent characteristics in terms of high strength and high stiffness-to-weight ratios. In addition to adequate flexural and shear properties, the behavior of FRP/AAC sandwich panels needs to be investigated when subjected to impact loading. During service, the structural members in the building structures are subjected to impact loading that varies from object-caused impact, blast due to explosions to high velocity impact of debris during tornados, hurricanes, or storms. Low velocity impact (LVI) testing serves as a mean to quantify the allowable impact energy that the structure is able to withstand and to assess the typical failure modes encountered during this type of loading. The low velocity impact response of plain AAC and FRP/AAC sandwich panels is investigated in Chapter 10 by Dr Uddin et al. The chapter includes results from the study on the response of plain AAC and CFRP/AAC sandwich structures to low velocity impact and to assess the damage performance of the panels, to study the effect of FRP laminates on the impact response of CFRP/AAC panels, to study the effect of the processing method (hand lay-up versus VARTM) and panels' stiffness on the impact response of the hybrid panels. Impact testing was conducted using an Instron drop-tower testing machine. Experimental results show a significant influence of CFRP laminates on the energy absorbed and peak load of the CFRP/AAC panels. Further, a theoretical analysis is conducted to predict the energy absorbed of CFRP/AAC sandwich panel using the energy balance model, and results are in good accordance with the experimental ones.

Today, fiber-reinforced composite materials are used in a wide range of civil infrastructure applications. Most of these applications utilize prepreg thermosetting composites, the most common of which is carbon fiber-reinforced polymer. Thermoplastic composites are relatively new materials in civil engineering applications and lack the history of use in civil infrastructure. Limited time has been spent investigating the usage of thermoplastic materials. These materials offer comparable material characteristics to thermosetting composites. The ability to readily form these materials using epoxy resins makes them much more desirable. Thermoplastic polymers have several advantages over thermosets: they can be reshaped by reheating, are recyclable, are cost-effective, and possess superior impact properties. They have comparable mechanical properties, higher notched impact strength, reduced creep tendency, and very good stability at elevated temperatures in humid conditions. Long fiber reinforcement thermoplastics have significantly higher heat deflection temperature and better heat aging properties than the corresponding short fiber-reinforced matrix materials. Thermoplastic composites typically comprise a commodity matrix such as polypropylene (PP), polyethylene (PE), or polyamide (PA) reinforced with glass, carbon, or aramid fibers. Progress in low cost thermoplastic materials and fabrication technologies offer new solutions for very lightweight, cost efficient composite structures with enhanced damage resistance and sustainable designs.

In Chapter 11 by Dr Uddin and Dr Mousa, a new type of sandwich panel called composite structural insulated panels (CSIPs) is developed for structural floor and wall applications. This new hybrid panel is intended to replace the traditional SIPs that are made of wood–based materials. CSIPs are made of bi-directional glass-PP facesheets and EPS foam core resulting in a very stiff panel with very high face/core moduli ratio. A comprehensive explanation of the CSIP concept, materials characteristics, and manufacturing techniques is provided. A detailed analytical modeling procedure is developed in order to determine the global buckling, interfacial tensile stress at facesheet/core debonding, critical wrinkling stress at facesheet/core debonding, equivalent stiffness, and deflection for CSIPs. The proposed models are validated using experimental results that have been conducted on full-scale CSIP wall and floor panels. The good correlation between the analytical and experimental results demonstrates the accuracy of the developed formulas for modeling the behavior of CSIPs under different types of loading. In order to be used as a hazard resistance material, a detailed section was presented to show the resistance of CSIP elements to the different types of hazard effects that a building can experience during its lifetime. The hazards included impact loading, floodwater effect, fire effect, and windstorm loading. As noticed from the impact testing, CSIPs have excellent performance in terms of strength and stiffness compared to traditional SIPs. Further, their impact resistance is much stronger than SIPs which recommends CSIPs for areas that are prone to windstorm debris. Further, flood testing showed that CSIPs had insignificant strength and stiffness degradations and accordingly proved to be a hazard resistant building material that can survive during a flood event. Finally, the windstorm testing on full-scale CSIPs proved that they can withstand high wind loading (up to Category 5) which therefore recommends their use as a building material in severe windstorm locations.

The primary objective of Chapter 12 by Dr Uddin *et al.* is to introduce and demonstrate the application of thermoplastic (woven glass reinforced polypropylene) in the design of modular panelized housing construction. Modular panelized construction is a modern form of construction technique in which precast multifunctional structural panels are used. In this technique, precast panels are fabricated in the manufacturing facility and are transported to the construction site. Traditional structural insulated panels (SIPs) consist of oriented strand boards (OSB) as facesheets and expanded polystyrene (EPS) foam as the core. These panels are highly energy efficient but have issues in terms of poor impact resistance and higher life cycle costs. Proposed panels consist of E-glass/PP laminates as facesheets and EPS foam as core and are termed 'composite structural insulated panels' (CSIPs). Proposed CSIPs overcome the issues of traditional SIPs and retain all the energy saving benefits of the traditional SIPs. This chapter covers manufacturing techniques developed for CSIPs and connection details for bonding CSIPs on the construction site. Based on experimental investigation, ultrasonic welding was found to be the most suitable technique for joining the proposed CSIPs.

The primary objective of Chapter 13 by Dr Uddin *et al.* is to introduce and demonstrate the application of thermoplastic (woven glass reinforced polypropylene) first, in the design of modular fiber-reinforced bridge decks, and next on the development of jackets for confining concrete columns against compression and impact loading. The design concept and manufacturing processes of thermoplastic bridge deck composite structural systems are presented by recognizing the structural demands required to support highway traffic. The deck system is carefully engineered by considering the structural efficiency and manufacturing ease of deck components. Glass/PP woven tape material is used based on its effective utilization to produce structural deck components with flat geometries and gradual radii/curvatures.

The structural system presented possesses several special features that contribute to its effectiveness, including the use of curved panels (sine ribs) which provide the nonplanar core configurations to increase the performance of the bridge deck system. The proposed design is compared to two published composite bridge concepts. On the other hand, bridge columns are expected to sustain breaching and large inelastic rotation in plastic hinges during impact loading, a prime concern for the retrofit design to enhance the breaching and ductility capacity. Ductility will normally be provided by column plastic hinges. It is the plastic rotation of potential plastic hinge that is of greatest interest. The available plastic rotation capacity, and hence the ductility capacity, depends on the distribution of transverse reinforcement within the plastic hinge region. Transverse reinforcement provides the dual function of confining the core concrete, thus enhancing its breaching strength and enabling it to sustain higher compression strains, and restraining the longitudinal compression reinforcement against buckling. Most of the current bridge retrofitting application utilizes wet lay-up thermosetting composites, the most common of which is carbon fiber-reinforced polymer. However, FRPs possess a limited strain capacity relative to conventional material such as steel. Finally FRP materials are relatively expensive. This chapter discusses the results of the small-scale static cylinder tests and the impact tests of concrete columns. As summarized in the following, thermoplastic reinforcement jackets act to restrain the lateral expansion of the concrete that accompanies the onset of crushing, maintaining the integrity of the core concrete, and enabling much higher compression strains (compare to CFRP composites wrap) to be sustained by the compression zone before failure occurs. The impact tests were conducted to assess the energy absorption capacity of three concrete columns strengthened by PP confinement, a carbon/epoxy confined, and one unconfined control specimen. All the results conclusively demonstrated the superior impact resistant properties of PP wrapped specimens over the CFRP.

Chapter 14 by Drs Kitane and Aref first reviews current structural applications of FRP composites in bridge structures, and describes the advantages of FRP in bridge applications. This chapter then introduces the design of a hybrid FRP–concrete bridge superstructure, which has been developed at the University at Buffalo for the past ten years, and discusses the structural performance of the superstructure based on extensive experimental and analytical studies. The concept of the hybrid design is a very prudent way to design a structure because different materials can be used efficiently where they perform best. In this chapter, the hybrid FRP–concrete bridge superstructure is introduced, where the novel concept of the hybrid FRP–concrete design is applied. The trial design of the bridge superstructure was proposed, and its structural performance was investigated through detailed FEA and a series of static and fatigue loading tests. In addition, simple methods of analysis for this hybrid bridge superstructure were also proposed. Results from the research to date showed that the proposed hybrid bridge superstructure has excellent structural performance, and the hybrid FRP–concrete system for the bridge superstructure is highly feasible from a structural engineering point of view.

Dr Dawood in Chapter 15 summarizes the recent advances in the use of FRP materials for repair, rehabilitation, and strengthening of steel structures. Conventional methods of strengthening and repairing steel structures are presented. The advantages and limitations of using FRP materials are summarized. Topics presented include strengthening of flexural members, strengthening with prestressed FRP materials, stress-based and fracture mechanics-based approaches to evaluating bond behavior, repair of cracked steel members, and strengthening of slender members subjected to compression forces. The chapter concludes with a brief discussion of future trends in this field and a summary of other resources for further information.

Chapter 16 by Drs Liang and Hota deals with FRP composites for environmental engineering applications. Environmental engineering covers a wide range of applications from applying science and engineering principles to improve the natural environment (air, water, and land resources), to providing healthy water, air, and land for human habitation (house or home) and for other organisms, and to remediating environmental pollution issues. It directly deals with environmental sustainability issue that our planet Earth is currently facing. It would be impossible for this chapter to provide a complete coverage on the applications of FRP composites in environmental engineering. Instead, this chapter will introduce a number of select FRP field applications related to environmental engineering, that the authors have recently been involved with or are currently researching, including: 1) gas and oil storage tank; 2) decking for ocean environment; 3) cold water pipe for ocean thermal energy conversion power generation;

4) cooling water tower; 5) chimney liners; 6) environmentally-friendly utility poles; 7) corrosion resistant pipelines; 8) rock bolt for underground mining; 9) modular buildings of environmental durability; 10) engineered recycled rail-road tie; and 11) green composites. Finally, design codes and specifications to promote and advance the applications of FRP composites along with future research directions on composite materials are presented with emphasis on durability and sustainability.

Structures produced from all fiber-reinforced polymer or plastic (FRP) composites have shown to provide efficient and economic applications in industrial and public works, termed civil infrastructure. Chapter 17 by Dr Qiao and Dr Davalos presents a systematic analysis and design methodology for all-FRP composite structures. Research on analysis of FRP composites in civil infrastructure is first reviewed, and a 'bottom-up' analysis concept is introduced based on a systematic approach for material characterization, analysis, and design of all-FRP composite structures. Concepts of micro/macromechanics and member/system properties and characterizations are introduced for the analyses of: (1) constituent (fiber and matrix) materials and plies, (2) laminated panels, (3) structural members and components, (4) honeycomb cores, (5) thin-walled cellular and honeycomb sandwich panels, and (6) structural systems. Step-by-step design guidelines are presented for: (1) FRP structural shapes, accounting for bending, shear, local/global buckling, and material failure; and (2) FRP deck-and-stringer systems, based on equivalent properties for cellular or honeycomb decks and using a first-order shear-deformation macro-flexibility analysis. The 'bottom-up' analysis concept and systematic design methodology described in this chapter can be used in practice by structural engineers concerned with design of FRP composite structures, and the guidelines provided for analysis and design can be used to develop new efficient FRP sections and to design FRP structural systems.

The book should appeal to all those concerned with the innovative application of FRP for civil engineering structures in both military and civil sectors. With its team of expert contributors who reflects many years of specialized experience, including the private, governmental, and academic perspectives, the book will be a standard reference in many fields of engineering such as fiber-reinforced polymer composites, engineering mechanics, numerical analysis, materials science and engineering, and structural engineering. It can serve as the text for advanced materials course on civil engineering, and materials science and engineering.

Part I
General developments

1
Types of fiber and fiber arrangement in fiber-reinforced polymer (FRP) composites

Y. GOWAYED, Auburn University, USA

DOI: 10.1533/9780857098955.1.3

Abstract: In fiber reinforced plastics (FRP), as a special type of polymer matrix composite, fibers provide the stiffness and strength while the surrounding plastic matrix transfers the stress between fibers and protects them. In this chapter, the role of fibers in FRP is delineated, their types and properties are discussed and the fabric forms in which they can be formed and used to reinforce FRP are presented. A special focus is given to the effect of the chemical structure of fibers on the stability and the level of anisotropy of their mechanical response. Furthermore, the effect of assembling these fibers into yarns and fabrics on the response of the FRP is presented as basis for further readings.

Key words: matrix, strength, stiffness, composite, carbon fibers, Kevlar® fibers, glass fibers, unidirectional fabrics, woven fabrics, stitched fabrics.

1.1 Introduction

Fiber-reinforced polymer (FRP) composites are made of a mixture of two solid materials, a material with high strength and stiffness surrounded by a homogeneous material that protects it and keeps it in place. The stiff material or reinforcement is typically made of a directional component such as fibers, rods, or sheets, while the surrounding material is typically isotropic and is called the matrix. This chapter will focus on the reinforcement.

Fibers are long solid objects with an extremely high aspect ratio and a unique set of directional properties. They are also flexible with a very low bending stiffness. Despite such combination of directionality, high aspect ratio, and flexibility, it is almost impossible to find a durable, high strength application for fibers without protecting them from the surrounding environment. They can easily be damaged by friction or even simple handling. Coating a group of fibers with a polymeric film or any other material will not only protect them but also help transfer the stresses between them, making them behave as if they are a single body, rather than a loose group of fibers.

The role of fibers in FRP composites is typically limited to providing strength and stiffness while the matrix keeps the fibers in their intended location, protects them, transfers stresses between the fibers and provides the rigidity for the FRP composite as a whole. This means the properties of the interface between the fibers and the matrix play a major role in how the stress is transferred between them, affecting the way the FRP behaves under sustained loads. A strong interface ensures that the entire load is transferred causing the composite to have high stiffness and strength. On the other hand, a weak interface may not allow such full load transfer, lowering the strength and stiffness of the composite.

The toughness of the FRP composite is also affected by the interfacial strength between the fibers and the matrix. Defining toughness as the energy needed to break the composite, it is important that a fracture-causing crack be forced to take a long and tortuous path consuming as much of its energy as possible to achieve a tough composite. This can only be achieved if the crack path is faced by different material phases with varying levels of strength and stiffness. A weak interface next to a stiff fiber would help deviate the crack path and accordingly enhance the toughness of the composite.

FRP composites are usually anisotropic with different material properties in different directions. Such anisotropy is caused by the alignment of the fibers in the composite. For example, if the fibers are only aligned in one direction, the composite is called unidirectional with properties along the direction of the fibers different than those in any other directions. If the fibers are aligned in two perpendicular directions, then the properties in those two directions are different than the properties in all other directions. Depending on the volume of fibers in each direction, the properties in both directions may not be the same.

The concept of anisotropy and, in particular, the ability to design for anisotropy has made composite materials what they are today; not only is it easy to achieve an optimum design, but they are also an inexpensive answer to critical loading scenarios. From an optimum design perspective, an anisotropic material design can provide certain properties only in required directions. Since a property such as stiffness can be qualitatively equated to cost, such optimum design will also reduce the cost. For isotropic materials, stiffness is provided in all directions of the material causing additional unnecessary cost, but anisotropic design strategies can meet target stiffness and strength in required directions without waste.

Fibrous composites are typically manufactured using either the process of stacking lamina or fabrics. In some cases, these fiber layers are pre-impregnated (typically called prepregs) with a resin and kept at a sub-zero temperature to slow down the chemical reaction for at least 6 months or even a year. These layers are typically placed in the expected load directions of the target product for an optimum design. After stacking, if the fibers are not 'prepreged', resin is introduced to the composite and cured, or, in the case of prepregs, heat is applied to initiate the curing process.

The ensuing sections introduce some of the fibers typically used in composite materials and their formation into fabric layers as well as other types of FRP. As the discussion goes into different material configurations, the reader has to keep in mind the importance of anisotropy in the selection of fibers and the formation of fabrics.

1.2 Fibers

High strength and modulus fibers that are commonly used in composite materials can be categorized based on their molecular conformation in three groups:

- polymeric fibers
- carbon fibers
- other inorganic fibers

The first group has one-dimensional primary bonds that are somewhat aligned with the longitudinal axis of the fibers. This one-dimensional bond can be as simple as in polyethylene with a linear chain of carbon atoms attached to side groups of a single hydrogen atom or a much more complex chain configuration with rings of atoms and more complex side groups. Most of these fibers are either anisotropic or transversely isotropic with the plane of isotropy perpendicular to the fiber longitudinal axis.

Carbon fibers have two-dimensional (2D) graphite sheets in a hexagonal planar network of primary bonds that are aligned parallel to the fiber axis with secondary bonds connecting the sheets in the radial direction of the fiber. Further connection between the sheets is due to some disorder and imperfection in the alignment of the atoms in the sheets. Graphite/carbon fibers are expected to be transversely isotropic. The last group of fibers has a three-dimensional (3D) network of primary bonds that can provide stiffness and strength and, additionally, a good thermal stability at higher temperatures. These networks can be random or crystalline and typically exist in ceramic fibers such as glass. Most of the fibers in this type are expected to be isotropic or pseudo-isotropic. The next sections will provide some examples of the most common of the above-mentioned types of fibers. This by no means represents all that the market has of these types of fibers; to the contrary, there are many other fibers that are used in making composite materials and there are new fibers that are being developed.

1.2.1 Para-aramid fibers such as Kevlar® and Twaron®

The generic term 'aramid' designates a long chain synthetic polyamide molecule (-CO-NH-) and para-aramids, such as Kevlar®, are wholly aromatic polyamides. DuPont started commercialization of Kevlar® as the first high strength aromatic polyimide fiber in 1971. Another fiber with a similar chemical structure is Twaron® produced by Akzo Nobel since the late 1970s.

Kevlar® fibers are manufactured by solution spinning of 1,4-phenylenediamine and terephthaloyl chloride in a condensation reaction yielding hydrochloric acid as a by-product. The spinning solution is extruded through the holes of a spinneret and subjected to mechanical stretch to elongate and orient the molecules in the axial direction of the fibers. The higher the level of mechanical stretch, the higher the strength and modulus of the fiber and the lower its strain to failure. The resulting polymer (p-phenylene terephthalamide), shown in Fig. 1.1, exhibits a liquid-crystalline behavior in solution. The long chain molecules act as if they are rigid rods and when the solution is sheared their crystalline regions tend to reorient themselves in the direction of the flow, providing high modulus and strength in the axial direction of the fiber.

Kevlar® fibers are known for their ultraviolet (UV) degradation in the presence of oxygen. Upon degradation, the fiber changes in color from a lustrous golden hue to dark yellow or brown. Zhang et al. studied the effects of simulated solar UV irradiation on the mechanical properties of Twaron®. They showed a decrease in mechanical properties and reported that UV irradiation deteriorated the surface and defect areas of the fiber severely by photo-induced chain scission, while the crystalline structure remained almost unchanged (Zhang et al., 2006). Wang et al. reported a reduced compressive strength and modulus of Kevlar®/epoxy fabric composites as compared to its tensile strength (Wang et al., 1995). In support of this finding, Fidan et al. reported that the compressive strength of Kevlar® fibers was reduced from their tensile strength due to the kink band formation in the fiber from microbuckling of separated microfibrils due to elastic instability in the fibers (Fidan et al., 1993). Furthermore, the weak bonding between the chains caused the fibers to split into smaller fibrils and microfibrils. This phenomenon is typically witnessed during the failure of Kevlar® reinforced composites, especially those subjected to compressive stresses (Hull and Clyne, 1996). Different types of Kevlar® fibers are available in the market, covering a large range of strengths and moduli. Table 1.1 lists

1.1 The chemical structure of Kevlar®.

Table 1.1 Typical properties of some Kevlar® fibers (manufacturer data)

	Kevlar® 29	Kevlar® 49
Tensile modulus (GPa)	70.5	112.4
Poisson's ratio	–	0.36
Strength (GPa)	2.92	3
Strain to failure (%)	3.6	2.4
Density (g/cm^3)	1.44	1.44
Thermal conductivity (W/m.°K)	0.04	0.04
Coefficient of thermal expansion ($\times 10^{-6}$/°K)	–4	–4.9

the properties of some of the most popular types of Kevlar® fibers used in composite reinforcement.

Kevlar® fibers are transversely isotropic with the tensile modulus in the axial direction around 52 times the transverse compressive modulus of Kevlar® 49 and 38 times for Kevlar® 29. Additionally, the longitudinal tensile strength is about 42 times the transverse compressive strength of Kevlar® 49 and 54 times for Kevlar® 29 (Kawabata, 1990).

1.2.2 Carbon fibers

Carbon fibers have been known since their development by Thomas Edison in 1870 and have been under continuous development for the past 60 years (Hearle, 2001; Peebles, 1995). The current interest in carbon fibers stems from their excellent mechanical properties and thermal stability. Carbon can be found in nature in many forms. Diamond, graphite, and ash are all made from pure carbon atoms with different atomic arrangements. Diamonds are made from covalent 3D bonds between carbon atoms. Graphite and carbon fibers are made from sheets of covalent-bonded carbon atoms. These sheets are connected to each other via a weak secondary bond.

Carbon fibers can be made by pyrolysis of a hydrocarbon precursor. Rayon was one of the first precursors used to make carbon fibers. During the processing of Rayon fibers into carbon fibers, only 25% of the fiber mass is retained. This made carbon fibers manufactured from Rayon precursors very expensive. Another precursor that has proved to be economical is the polyacrylontrile (PAN) fiber with a conversion yield of around 50–55%. Carbon fibers made from a PAN precursor generally have higher strength than fibers made from other precursors. This is due to the lack of surface defects, which act as stress concentrators, and hence reduce tensile strength. Another commonly used precursor is the Pitch precursor which is a by-product of petroleum refining. Pitches are relatively low in cost and high in carbon yield. Their most significant drawbacks are the irregular surfaces of the fiber, which reduces the fiber tensile and compressive strengths (Hearle, 2001). Currently, fibers are manufactured from either a PAN or a Pitch precursor. Rayon precursors are used in less than 1% of the production of carbon fibers (Department of Defense MIL-HDBK-17-5, 2002). Physical properties of some carbon fibers used by the industry are listed in Table 1.2 (Hearle, 2001).

There are several processes that a PAN precursor has to go through to be converted to a carbon fiber. Typically, the precursor-to-carbon-fiber conversion process follows the following sequence: stabilization, carbonization, graphitization, surface treatment, and application of sizing and spooling. Stabilization is carried out at temperatures <400°C in various atmospheres. The fiber is held under tension during this stage to enhance molecular orientation, which increases fiber modulus and strength. Carbonization is accomplished at temperatures from 800 to 1200°C in an inert atmosphere. Fiber tensioning is still maintained during this process. Graphitization is an additional process at a temperature of >2000°C in an inert environment. This process reduces the level of impurities and stimulates crystal growth. Various materials can be applied to the surface of carbonized/graphitized fiber during the surface treatment process. These materials will help control the interaction between the fibers and the matrix materials in a composite. Sizing is applied to fiber tows (yarns) to enhance their handling characteristics in further textile forming operations (e.g., weaving and braiding). After these processes, the fiber is spooled on a carrier tube to form a stable package. For a pitch precursor, the petroleum mixture is heated above 350°C to allow a condensation reaction to occur, which produces large flat molecules, and then the mixture is extruded through a spinneret to produce a 'green fiber' aligning the molecules in the fiber axial direction. The fibers are further heated without tension at around 2000°C to produce a carbon fiber.

Table 1.2 Mechanical properties of some carbon fibers (Hearle, 2001)

Precursor type	Product name	Young's modulus (GPa)	Tensile strength (GPa)	Strain to failure (%)
PAN	T300	230	3.53	1.5
	T1000	294	7.06	2.0
	M55J	540	3.92	0.7
	IM7	276	5.30	1.8
Pitch	KCF200	42	0.85	2.1
	Thornel P25	140	1.40	1.0
	Thornel P75	500	2.00	0.4
	Thornel P120	820	2.20	0.2

Most carbon fibers are transversely isotropic with the value of anisotropy scattered for both PAN and Pitch fibers. Kawabata reported that the ratio of the longitudinal tensile modulus divided by the transverse compressive modulus is 51.1 for T300 which is a PAN fiber, 19.1 for Thornel P25, 108.9 for Thornel P75, and 247.4 for Thornel P120 with all the Thornel fibers manufactured from a Pitch base. On the other hand, the strength tells a different story. The longitudinal tensile strength of a PAN-based fiber divided by its transverse compressive strength is around 1.1, while for a Pitch-based fiber like Thornel P25 it is 3.8, for Thornel P75 it is 21.5, and for Thornel P120 it is 43.4 (Kawabata, 1990).

1.2.3 Glass fibers

Drawing glass into fibers is an ancient art. These early fibers were used to reinforce clay vessels as well as clay statues. Recent development for glass fibers started in the early years of the twentieth century with the work by Griffith (1920), in which he used glass fibers as a model for his theories on fracture mechanics. Glass fibers are manufactured by melting various raw materials including silica and other salts followed by extruding the melt to form glass fibers. The primary component of glass fiber is sand, but it also includes varying quantities of limestone, soda ash, borax, sodium sulfate, boric acid, etc. Raw materials are heated in a furnace to temperatures ranging from 1500 to 1700°C and are refined and transformed through a sequence of chemical reactions to molten glass. The molten glass is forced through a heated platinum spinneret containing very small holes. The fibers emerging from the spinneret are immediately coated with a water-soluble sizing or a coupling agent. The coat protects the fiber from dust which can scratch its surface and reduce the fiber strength, while the coupling agent can enhance the interfacial adhesion of the fiber to matrix material in a composite. The coat also improves the yarn handlability and protects it during further fabric forming processes.

Fiber glass typically used to reinforce composite materials include E-glass (E for electrical) which is an alumino-borosilicate glass, C-glass (C for corrosion) which has a close chemical structure to that of E-glass but with a better resistance to corrosion, and S-glass (S for strength) which is an alumino silicate glass with a high tensile strength. Some of the typical fiber properties are listed in Table 1.3.

The basic building block of the glass fiber, as shown in Fig. 1.2, is a tetrahedral structural unit of silica (SiO_4), where the Si atom shows tetrahedral coordination, with four oxygen atoms surrounding a central Si atom. The presence of Ca, Na, and K breaks up the silica network lowering the stiffness and strength of the fiber but improving the formability of the melt to form the fibers. The structure of the fiber is typically amorphous,

10 Developments in FRP composites for civil engineering

Table 1.3 Mechanical properties of some glass fibers

	Tensile modulus (GPa)	Tensile strength (GPa)	Poisson's ratio	Density (g/cm^3)	Strain to failure (%)
E-glass	72	3.45	0.22	2.55	1.8–3.2
C-glass	69	3.3	–	2.49	–
S-glass	87	3.5	0.23	2.5	4

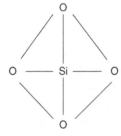

1.2 The chemical structure of glass fibers.

although crystalline regions may form after prolonged heating at high temperatures. The bonds are randomly oriented causing the fiber to be isotropic or pseudo-isotropic.

1.3 Fabrics

Polymer matrix composite materials are made by mixing fibers with a resin material (matrix). These fibers can either have a non-specific arrangement like in the case of chopped fiber (nonwovens) or long fibers grouped together, called tows or yarns, and assembled into fabrics such as wovens, braids, or knits. Since the fibers are the main source for the stiffness and strength of polymer matrix composites, it could be correctly inferred that the direction of these fiber will play a major role in the mechanical and thermal behavior of the composite. The more the fibers are aligned in the load direction(s), the higher the stiffness(s) and strength(s) of the composite in these direction(s). Hence, a chopped nonwoven glass composite with fibers randomly oriented in many directions will have stiffness inferior to that of a woven fabric composite with fibers aligned in the load direction, and so on.

A logical conclusion from the above can be to align the yarns exactly in the load direction without involving any weaving or braiding which typically moves the yarns away from the load direction to achieve fabric integrity. This is typically done by using layers of unidirectional yarns in which each group of yarns is aligned in one of the load directions. This alignment of the yarns will definitely achieve the highest possible level of stiffness and strength in the direction of the aligned yarns.

The yarn arrangement utilized in unidirectional composites comes at the expense of the toughness of the composite; its ability to resist failure and crack propagation. Unidirectional layers of yarns have a clean matrix interface between planes of yarns. If a crack initiates in this matrix area, it will have a clear path to follow and break the composites. Besides having high moduli and strengths in the plane of the yarns, composites made from unidirectional lamina are known to have low toughness, low interlaminar shear strength as well as low out-of-plane modulus and strength. On the other hand, composites made from nonwoven fibers, and woven, braided, or knitted fabrics have the advantage of fibers and yarns moving from one plane into another. This complex arrangement increases the energy required for the crack to propagate, increasing the toughness of the composite. So they are known to have in-plane moduli and strength lower than those of composites made from unidirectional layers, but have higher toughness, interlaminar shear strength and out-of-plane moduli and strengths.

The above-mentioned yarn and fiber forms can come either in a 'dry' form or impregnated with a resin. Prepreg(ing) is a process where fibers and yarns, regardless of their form, are impregnated with a resin with a known fiber volume and yarn distribution in the form of a sheet. The chemical reaction of the resin is halted by placing the prepregs in a freezer for a specific period of time. The user typically takes these sheets of prepregs, aligns them in the proposed directions, stacks them, and then heats and presses them to initiate the chemical reaction and cure the composite. The following sections give some details of some typical fiber arrangements.

1.3.1 Unidirectional laminates

Most unidirectional composites are manufactured from prepreged layers of yarns. Each layer is called a lamina and a group of lamina is called a laminate. Each lamina in the laminate can have a specific direction based on the design requirements starting from an arbitrary direction. The only important matter to remember is that these laminates must have a mirror symmetry around their neutral axis; otherwise they will warp immediately after manufacture and cannot be straightened. This happens due to the difference in the thermal expansion coefficients between the fibers and the matrix and between different lamina causing one side to permanently expand more than the other side and eventually warp. Figure 1.3 shows a schematic of a symmetrical laminate. Note the mirror symmetry in the drawing. The

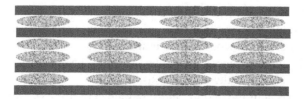

1.3 A (0,90,0,90)$_s$ laminate composite.

nomenclature of these composites follows the stacking procedure. For example, the laminate in Fig. 1.3 is called (0,90,0,90)$_s$, where the subscript (s) entails the symmetry. Any other angle beside the 0 and 90 is possible.

1.3.2 Nonwovens (chopped fibers)

Nonwoven fabrics are formed directly from short fibers that are assembled in sheets. The fibers are connected together to enhance the handlability of such sheet of otherwise loose fibers. Rows of needles are sometimes used to punch through the sheet of fibers and reorient some of the fibers to achieve mechanical adhesion between the fibers (needle punched fabrics). This process, although very successful with polymer fibers such as para-aramids, has less success with brittle ceramic fibers such as glass and carbon because fibers can easily break. Another method used to connect the short fibers is by using an adhesive to glue the short fibers to one another. In the case of short fiber composites, this glue will typically have a chemical affinity to the resins.

1.3.3 2D woven fabrics

2D woven fabrics are made by interlacing yarns in a weaving loom. Yarns are divided into two components: one called the warp, running along the length of the loom, and the other is the weft, running in the cross direction. The warp yarns are also divided into multiple parts and each part is passed through a group of eyelets held in a harness frame. Each of the frames is moved up or down and the weft yarns are inserted at each step following a defined pattern of movement, creating a specific fabric structure. Warp and weft yarns cross at cross-over points and sometimes they bend (called crimp) in order to pass over or under one another. A beat-up comb is used to stack the yarns and pack them into the fabric. The angle between the warp and the weft yarns is typically 90°. The fabric structure is characterized by the movement of its yarns as they relate to one another. For example, the plan weave shown in Fig. 1.4, has a 1/1 arrangement, where a warp yarn alternates being under or over a weft yarn at each crossover point.

Types of fiber and fiber arrangement in FRP composites 13

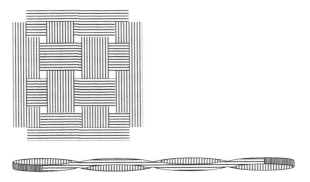

1.4 Drawing of a plain weave plan and side views as created by pcGINA© software (Gowayed and Barrowski, 2004).

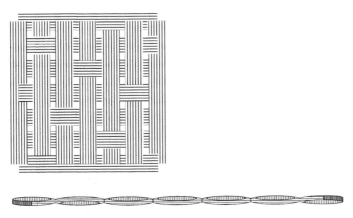

1.5 Drawing of a 5-HS weave plan and side view as created by pcGINA© software (Gowayed and Barrowski, 2004).

A 5-harness satin (5-HS) weave will have its warp yarns running over four weft yarns and under one weft yarn as shown in Fig. 1.5. Other n-HS weaves used in composite materials include 8-HS, where the warp yarn passes over seven weft yarns and under one, and 12-HS where the warp yarn passes over eleven weft yarns and under one. It can easily be understood that the more yarn passes over the other yarn, the straighter and less crimped will be the yarns present in the fabric. The straighter the segments, the more the fabric composite behaves as a lamina in a laminated composite, increasing the in-plane properties at the expense of out-of-plane properties and interlaminar shear strength; and of course vice versa.

1.3.4 Stitched fabrics

Stitched fabrics represent one of the possible forms of 3D fabrics. These dry fabrics are manufactured by simple stitching of layers of fabrics together to allow for better handlability, and enhance the out-of-plane properties of the composite. The yarn used for stitching has to be ductile enough to sustain the stitching forces and the very small radius of curvature at the end of the stitch path. Typically a polymeric yarn such as para-aramid, nylon or high density polyethylene is used to stitch fabrics.

It is important to notice that stitching typically damage the yarns of the fabric, especially if they are brittle fibers such as glass or carbon. The needle going from one side of the stack of fabrics to the other side encounters and damages various yarns and fibers. Accordingly, the density of the stitch repeat has to be low enough to limit the damage but high enough to achieve some advantage to the composite.

1.4 Composites

Composite materials are typically made using the above-mentioned fibers and fabrics, or other types, as well as a matrix material. A resin, such as epoxy, is used to impregnate the fabric and then the composite is heated to cure and harden. Sometimes pressure is applied to form the composite into a specific shape with exact target dimensions. In the case of unidirectional or fabric prepregs, the layers are stacked and then heated under pressure to form the composite.

Composite can be manufactured *in-situ* like in the case of strengthening a column or a beam by wrapping the structural element with layers of the fabric, then impregnating them with a resin using hand lay-up or a vacuum bag. In other scenarios, the composite is manufactured in a separate facility then shipped to the construction site where it is fastened to the structural member, or adhesive is used to integrate it with the structural element. Composites manufactured outside the site come in different shapes. The most common type is in the form of a sheet of material with specific dimensions. These sheets can be manufactured in a non-continuous form such as using hand lay-up, vacuum bagging or compression molding. They can also be manufactured in a continuous form using pultrusion, where the fibers, continuous or chopped, are fed through a resin bath and then pulled through a fixed orifice, with the final cross-section shape of the part, then immediately cured.

Many other forms can be manufactured and used in different civil engineering applications. One of these forms is rebars that are used to replace steel in order to prolong the life of the structure. Some of these rebars are made from unidirectional fibers at the core, mostly glass or para-aramid fibers, which are wrapped by a helical wrap or a braid and then infused with resin. These rebars are used to reinforce the concrete and must have a high resistance to alkaline environments (Abbasi and Hogg, 2005).

1.5 Future trends

The composite industry has achieved great strides in making a product that is light, stiff, has excellent strength and is environmentally stable. As it continues to press forward with a product that has such unique characteristics, the industry realizes the need to advance in different directions, such as those of the fibers, the fabric geometry and the composite application. There is a strong need for a high strength and environmentally stable polymer fiber that can be stacked and stitched without the fear of breaking like the case of ceramic fibers. The recent work on PBO fibers such as Zylon (Toyobo, 2001), which is a thermoset liquid crystalline polyoxazole, shows promise of such development, although reports suggest that these fibers are susceptible to strength degradation in high humidity and direct sunlight. Ultra high molecular polyethylene fibers such as Spectra and Dyneema are environmentally stable, have very low density and excellent stiffness and strength. Further work is needed to enhance the interfacial strength of these fibers with the matrix material. Recent developments in M5 fibers by Akzo Nobel prepared by condensation polymerization of tetraaminopyridine and dihydroxyterephthalic acid using diphosphourus pentoxide as a dehydrating agent show great promise (Sikkema, 1998). It has properties similar to those of carbon fibers.

The composite fabric manufacturing industry has focused its efforts on producing 2D fabrics that have consistent and stable dimensions and with minimum damage. Work on 3D fabric structures has produced orthogonal fabrics (Bogdanovich and Mohamed, 2009) and angle interlock fabrics as shown in Figs 1.6 and 1.7, as well as other new fabric structures to limit the need for stitched layers and the associated damage to the yarns.

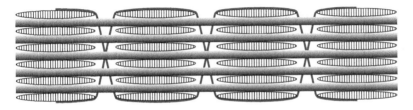

1.6 Orthogonal weave structure.

1.7 Angle interlock weave.

1.6 Sources of further information and advice

The reader is encouraged to read more on the above mentioned subjects. On the subject of fibers, the book by J.W.S. Hearle (ed.), *High performance fibers* (CRC Press, Boca Raton, 2001) provides an excellent reference. Peter R. Lord and Mansour H. Mohamed wrote *Weaving: Conversion of Yarn to Fabric* (Merrow Publishing Co., Darlington, UK, 1976), which is considered one of the most important books in the field. Two and three-dimensional fabrics are covered in the book by A. Miravete, *3-D Textile Reinforcements in Composite Materials* (Woodhead Publishing Limited, CRC Press, Boca Raton, FL, 1999). Two of the most concise composite mechanics of composite material books are S. Tsai and H. Hahn, *Introduction to Composite Materials* (Technomic Publishers, Lancaster, PA, 1989) and Robert M. Jones, *Mechanics of Composite Materials* (McGraw-Hill, New York, 1975). A. Bogdanovich and C. Pastore wrote a treatise on the mechanics of fabric composites in their *Mechanics of Textile and Laminated Composites* (Chapman and Hall, London, 1996). Finally on composite manufacturing, Timothy G. Gutowski edited the book *Advanced Composites Manufacturing* (Wiley-Interscience, New York, 1997).

1.7 References

Abbasi, A. and Hogg, P.J., 'Temperature and environmental effects on glass fibre rebar: modulus, strength and interfacial bond strength with concrete,' *Composites: Part B* 36 (2005) 394–404.

Bogdanovich, A.E. and Mohamed, M.H., 'Three-dimensional reinforcements for composites,' *SAMPE Journal* 45 (2009) 8–28.

Department of Defense MIL-HDBK-17-5, '*Handbook of Composite Materials*,' Vol. 5, *Ceramic Matrix Composites*, Department of Defense, Washington, DC (2002).

Fidan, S., Palazotto, A., Tsai, C.T. and Kumar, S., 'Compressive properties of high-performance polymeric fibers,' *Composites Science and Technology* 49 (1993) 291–297.

Gowayed, Y. and Barrowski, L., pcGINA©, A pc-based Graphical Integrated Numerical Analysis of composite materials to calculate the properties of fabric based composites, version 10.2004.

Griffith, A., 'The phenomena of rupture and flow in solids,' *Philosophical Transactions of the Royal Society, London* A221 (1920) 163.

Hearle, J.W.S. (ed), *High Performance Fibers*, CRC Press, Boca Raton (2001).

Hull, D. and Clyne, T., *An Introduction to Composite Materials*, 2nd edn, Cambridge University Press, Cambridge (1996).

Kawabata, S., 'Measurement of the transverse mechanical properties of high performance fibers,' *Journal of the Textile Institute* 81 (1990) 432–447.

Peebles, L., *Carbon Fibers: Formation, Structure and Properties*, CRC Press, Boca Raton (1995).

Sikkema, D.J., 'Design, synthesis and properties of a novel rigid rod polymer, PIPD or "M5": high modulus and tenacity fibres with substantial compressive strength,' *Polymer* 39 (1998) 5981–5986.

Toyobo Technical information, PBO Fiber Zylon, Toyobo Co. Ltd (revised 2001.9).

Wang, Y., Li, J. and Zhao, D., 'Mechanical properties of fiber glass and Kevlar woven fabric reinforced composites,' *Composites Engineering* 5 (1995) 1159–1175.

Zhang, H., Zhang, J., Chen, J., Hao, X., Wang, S., Feng, X. and Guo, Y., 'Effects of solar UV irradiation on the tensile properties and structure of PPTA fiber,' *Polymer Degradation and Stability* 91 (2006) 2761–2767.

2
Biofiber reinforced polymer composites for structural applications

O. FARUK and M. SAIN, University of Toronto, Canada

DOI: 10.1533/9780857098955.1.18

Abstract: Biofibers are emerging as a low cost, lightweight and environmentally superior alternative in composites. Generally, different fibers exhibit different properties that are fundamentally important to the resultant composites. This chapter gives an overview of the most common biofibers in biocomposites, covering their sources, types, structure, composition, and properties. Drawbacks of biofibers, such as dimensional instability, moisture absorption, biological, ultraviolet and fire resistance, will be discussed. The chapter will focus on their modifications (physical and chemical methods), matrices based on their petrochemical resources and bio-based, processing of biofiber reinforced plastic composites covering the factors influencing processing (humidity, additives, machinery, processing parameter, fiber content and length), and processing techniques (compounding, compression molding, extrusion, injection molding, pultrusion and others) will be discussed. The properties of the biocomposites based on their mechanical, physical, and biological behavior will also be covered. Lastly, this chapter concludes with recent developments and trends of biocomposites in the near future in civil engineering.

Key words: biofiber, dimensional instability, moisture absorption, biopolymer, modification, compression molding, extrusion, injection molding, pultrusion, mechanical properties, physical properties, biological properties.

2.1 Introduction

Recent advances within the worldwide biocomposites research community are just beginning to lead to the early stages of a fundamental understanding of the relationships between materials, process, and composites performance properties. Biocomposites products have created substantial commercial markets for value-added products and have become popular for building and construction, automotive, interiors and internal finishes, garden and outdoor products, industrial and infrastructure and other low-volume, niche applications. The list of areas of application is seemingly endless and continues with products such as decking, railing, fencing, roofing, automotive interior panels, automotive door and head liners, skirting boards, etc. A recent market study found that the global natural fiber composites market reached US$2.1 billion in 2010. Current indicators are that demand for natural fibers will continue to grow rapidly worldwide. The use of natural fiber composites has increased substantially in the construction and automotive sectors, with electronic, electrical, and consumer goods as emerging market segments over the last five years. According to estimates, over the next 5 years (2011–2016), the natural fiber composites market is expected to grow 10% globally. Newest survey results show that the use of wood plastic composites in Europe is increasing by 10% annually. The demand for wood plastic composites is expected to advance more than 10% per year to $5.6 billion by 2013 in the USA. Consumer acceptance of these products will be driven as they come to be seen replacements for more traditional materials. According to the report, decking will remain the leading application, and windows and doors will be among the fastest growing types of application.

Wood plastic composite (WPC) application is even more extensive in China [1]. In recent years, windows, doors, thermal insulating systems, park benches, garden sheds, and sun screens for tower buildings, have all been made from WPC in China. The growth of WPC production in China is at a level of 30% per year and is predicted to increase by up to 5 million tons per year between now and 2015.

The development of wood and natural fiber composites depends on their performance and sustainability. By the time that biocomposite materials and associated design methods are sufficiently mature to allow their widespread use, issues related to civil engineering construction material sustainability are likely to have become paramount in material choice. The development of methods, systems, and standards could see biocomposite materials at a distinct advantage over traditional materials. There is a significant research effort underway to develop biocomposite materials and explore their use in civil engineering applications. This research needs to continue in conjunction with development of conventional composite materials in order to provide a solution in the future which will allow wider use of the biocomposite materials in civil engineering applications.

2.2 Reinforcing fibers

The twenty-first century could be called the cellulosic century, because more and more renewable plant resources for products have been discovered. It has generally been claimed that natural fibers are renewable and sustainable, but they are in fact neither. The living plants are renewable and sustainable from which the natural fibers are taken, but not the fibers themselves. Among the most used natural fibers, flax, jute, hemp, sisal, ramie, and kenaf fibers have been extensively researched and employed in different applications. But nowadays, abaca, pineapple leaf, coir, oil plam, bagasse, and rice husk fibers are gaining interest and importance in both research and applications due to their specific properties and availability.

It should be mentioned that there are also shortcomings: lack of consistency of fiber qualities, high level of variability in fiber properties related to location and time of harvest, processing conditions, and their sensitivity to temperature, moisture, and UV radiation. A multi-step manufacturing process is required in order to produce high-quality natural fibers, which contributes to the cost of high-performance natural fibers.

2.2.1 Fiber source

The plants that produce natural fibers are classified as primary or secondary depending on their utilization. Primary plants are those grown for their fiber content, while secondary plants are plants in which the fibers are produced as a by-product. Jute, hemp, kenaf, and sisal are examples of primary plants. Pineapple, oil palm, and coir are examples of secondary plants. Table 2.1 shows the main fiber types used commercially in composites, which are now produced throughout the world [2, 3].

2.2.2 Fiber types

There are six basic types of natural fibers according to their botanical origin. They are classified as follows:

- bast fibers (such as jute, flax, hemp, ramie, and kenaf)
- leaf fibers (such as abaca, sisal, agave, and pineapple)

Table 2.1 Commercial major fiber sources

Fiber source	World production (10^3 tonnes)
Bamboo	30,000
Jute	2,300
Kenaf	970
Flax	830
Sisal	378
Hemp	214
Coir	100
Ramie	100
Abaca	70
Sugar cane bagasse	75,000
Grass	700
Wood	1,750,000
Satks (corn, cotton)	1,145,000

- seed fibers (such as coir, cotton, oil palm, rice hulls, and kapok)
- core fibers (such as kenaf, hemp, and jute)
- grass and reed fibers (such as bagasse, bamboo, Johnson grass, wheat, corn, and rice)
- other types (such as wood and roots).

2.2.3 Structure and chemical composition

Climatic conditions, age, and the degradation process influence not only the structure of fibers, but also the chemical composition. The major chemical component of a living tree is water. However, on a dry basis, all plant cell walls consist mainly of sugar-based polymers (cellulose, hemi-cellulose) that are combined with lignin with lesser amounts of extractives, protein, starch, and inorganics. The chemical components are distributed throughout the cell wall, which is composed of primary and secondary wall layers. The chemical composition varies from plant to plant, and within different parts of the same plant. Table 2.2 [4–7] shows the range of the average chemical constituents for a wide variety of plant types.

2.2.4 Properties

The properties of natural fibers differ among cited works for various reasons: different fibers were used, there were different moisture conditions,

Table 2.2 Chemical composition of some common natural fibers

Fiber	Cellulose (wt%)	Hemicellulose (wt%)	Lignin (wt%)	Waxes (wt%)
Bagasse	55.2	16.8	25.3	–
Bamboo	26–43	30	21–31	–
Flax	71	18.6–20.6	2.2	1.5
Kenaf	72	20.3	9	–
Jute	61–71	14–20	12–13	0.5
Hemp	68	15	10	0.8
Ramie	68.6–76.2	13–16	0.6–0.7	0.3
Abaca	56–63	20–25	7–9	3
Sisal	65	12	9.9	2
Coir	32–43	0.15–0.25	40–45	–
Oil palm	65	–	29	–
Pineapple	81	–	12.7	–
Curaua	73.6	9.9	7.5	–
Wheat straw	38–45	15–31	12–20	–
Rice husk	35–45	19–25	20	14–17
Rice straw	41–57	33	8–19	8–38
Wood	45–50	20–30	22–30	0–10

Table 2.3 Physico-mechanical properties of natural fibers and synthetic fibers

Fiber	Tensile strength (MPa)	Young's modulus (GPa)	Elongation at break (%)	Density (g/cm^3)
Abaca	400–760	12	3–10	1.5
Bagasse	290	17	–	1.25
Bamboo	140–230	11–17	–	0.6–1.1
Flax	345–1,500	27.6	2.7–3.2	1.5
Hemp	690	70	1.6	1.48
Jute	393–800	26.5	1.5–1.8	1.3
Kenaf	930	53	1.6	–
Sisal	468–700	9.4–22	3–7	1.5
Ramie	400–938	61.4–128	2.5	1.5
Oil palm	248	3.2	25	0.7–1.55
Pineapple	413–1627	34.5–82.5	1.6	0.8–1.6
Coir	131–220	4–6	15–40	1.2
Curaua	500–1,150	11.8	3.7–4.3	1.4
Soft wood	1,000	40	1.5	1.5
E-glass	3,400	73	2.5	2.55
Kevlar	3,000	60	2.5–3.7	1.44
Carbon	3,400–4,800	240–425	1.4–1.8	1.78

and different testing methods were employed. A single natural fiber is a three-dimensional, biopolymer composite composed mainly of cellulose, hemicelluloses, and lignin, with minor amounts of free sugars, starch, protein, extractives, and inorganics. The performance of a given fiber used in a given application depends on several factors, including the chemical composition, cell dimensions, microfibrillar angle, defects, structure, physical properties, mechanical properties, the interaction of a fiber with the composite matrix, and how that fiber or fiber/matrix performs under a given set of environmental conditions. Table 2.3 [4, 5] illustrates the important physico-mechanical properties of commonly used natural fibers.

In order to expand the use of natural fibers for composites, the availability of information about the fiber characteristics and the factors which affect performance of that fiber are essential. It is also necessary to know the factors which affect the performance of a given fiber in a given application.

2.3 Drawbacks of biofibers

The using of biofibers in building materials has some disadvantages such as low modulus elasticity, high moisture absorption, decomposition in alkaline environments or in biological attack, and variability in mechanical and physical properties [8]. It has been observed that dimensional stability, flammability, biodegradability, and degradation are attributed to acids bases and UV radiation that alters the biocomposites back into their basic building blocks (carbon dioxide and water).

2.3.1 Dimensional instability

Dimensional changes are one of the main problems of using natural fibers in composites, in particular expansion in thickness. Linear expansion is the result of reversible and irreversible swelling, which is caused by the release of residual compressive stresses imparted to the composite material during the composite pressing process. Therefore, dimensional instability of natural fibers restricts their further use in composites applications. However, the dimensional stability of natural fibers could be improved by bulking the fiber cell wall with simple bonded chemicals, and by impregnation of the cell wall with water-soluble polymers. For example, acetylation of the cell wall polymers by using acetic anhydride could improve the dimensional stability and biological resistance of fibers in composites.

2.3.2 Moisture absorption

Another problem in building components such as decking, fencing, etc., manufactured from biocomposites is absorbing moisture from the environment. Due to the presence of hydroxyl and oxygen-containing groups in biofibers, moisture is attracted through hydrogen bonding, and the end effect is the dimensional changes in products. The noncrystalline cellulose, lignin, hemicelluloses in the biofibers, and the surface of crystalline cellulose have effects on the moisture absorption. Table 2.4 [3] shows the equilibrium moisture content of some natural fibers.

Table 2.4 The equilibrium moisture content of different natural fibers at 65% relative humidity (RH) and 21°C

Fiber	Equilibrium moisture content (%)
Sisal	11
Hemp	9.0
Jute	12
Flax	7
Abaca	15
Ramie	9
Pineapple	13
Coir	10
Bagasse	8.8
Bamboo	8.9

2.3.3 Biological resistance

Biodegradability of biocomposites is attributed to the organisms that have specific enzyme systems to hydrolyze the carbohydrate polymers in the cell wall into digestible units. By cellulose degradation through oxidation, hydrolyses, and dehydration reactions, the strength reduces. There are several methods for improving the biological resistance of natural fibers in biocomposites. First, by bonding chemicals to the cell wall polymers, biological resistance increases because of the lowering of the moisture content below that required for microorganism attack. Another solution is to use toxic chemicals in composites to avoid biological attack.

2.3.4 Ultraviolet resistance

The problem for biocomposite products that are exposed outdoors, such as decking and fencing, is photochemical degradation because of UV light. The main reason for UV degradation is the presence of lignin which is responsible for color alteration. Whenever lignin degrades, the content of the cellulose, which is less susceptible to the UV light degradation, increases in the surface. Therefore, by lignin degradation, the weakly bonded carbohydrate-rich fibers erode from the surface and new lignin is exposed to extra degradative reactions. This 'weathering' process results in the rough surface of biocomposite products. The following methods are available to overcome this problem: bonding chemicals to the cell wall polymers that decrease the lignin degradation, and adding polymers to the cell matrix to keep the structure of the degraded fibers together.

2.3.5 Fire resistance

Fire resistance of biocomposite products could be improved by introducing flame retardants. The function of flame retardants is to decrease/avoid combustibility of the biocomposites. The flame retardants act physically by cooling (endothermic process), thinning (addition of fillers which thins the flammable materials in solid or gaseous phase) and by formation of a protective film coating which isolates the material from the fire source.

2.4 Modification of natural fibers

Natural fiber-based composites are being developed that could benefit from a thorough and fundamental understanding of the fiber surface. These products may require new adhesive systems to reach their full commercial potential. The extent of the fiber–matrix interface is significant for the application of natural fibers as reinforcement fibers for plastics.

An exemplary strength and stiffness could be achieved with a strong interface, which is very brittle in nature with easy crack propagation through the matrix and fiber. The efficiency of stress transfer from the matrix to the fiber could be reduced with a weaker interface.

2.4.1 Physical methods

The reinforcement of fibers can be modified using physical methods, such as stretching, calendaring, thermotreatment, and the production of hybrid yarns. Physical treatments change the structural and surface properties of the fiber and thereby influence the mechanical bonding of polymers. Physical treatments do not extensively change the chemical composition of the fibers. Therefore the interface is generally enhanced via an increased mechanical bonding between the fiber and the matrix. Corona treatment is one of the most interesting techniques for surface oxidation activation. This process changes the surface energy of the cellulose fibers. Plasma treatment is another physical treatment method and is similar to corona treatment. The property of plasma is exploited by the method to induce changes on the surface of a material. A variety of surface modifications can be achieved depending on the type and nature of the gases used.

2.4.2 Chemical methods

Cellulose fibers, which are strongly polarized, are inherently incompatible with hydrophobic polymers due to their hydrophilic nature. In many cases, it is possible to induce compatibility in two incompatible materials by introducing a third material that has properties intermediate between those of the other two. The development of a definite theory for the mechanism of bonding using coupling agents in composites is a complex problem. The main chemical bonding theory alone is not sufficient. So consideration of other concepts appears to be necessary. These include the morphology of the interphase, the acid-base reactions in the interface, surface energy, and the wetting phenomena.

Chemical modifications of natural fibers aimed at improving the adhesion within the polymer matrix using different chemicals have been investigated. The surface energy of fibers is closely related to the hydrophilic nature of the fiber. Some investigations are concerned with methods to decrease hydrophility. Silane coupling agents may contribute hydrophilic properties to the interface, especially when amino-functional silanes, such as epoxies and urethane silanes, are used as primers for reactive polymers. An old method of cellulose fiber modification is alkaline treatment or mercerization; it has been widely used on cotton textiles. Mercerization is an alkali treatment of cellulose fibers which is dependent on the type and concentration of the solution, its temperature, time of treatment, tension of the material as well as on the additives. Acetylation is another method of modifying the surface of natural fibers and making them more hydrophobic. It describes the introduction of an acetyl functional group into an organic compound. The main idea of acetylation is to coat the OH groups of fibers which are responsible for their hydrophilic character with molecules that have a more hydrophobic nature. The most important and popular grafting method is the application of maleic anhydride modified polymers as compatibilizers in natural fiber composites. There are numerous published studies in which the effect of maleic anhydride grafting on the mechanical properties of natural fibers has been investigated and it is impossible to list all of them here.

2.5 Matrices for biocomposites

The shape, surface appearance, environmental tolerance, and overall durability of composites are dominated by the matrix while the fibrous reinforcement carries most of the structural loads, thus providing macroscopic stiffness and strength. The polymer market is dominated by commodity plastics with 80% based on non-renewable petroleum resources. Governments, companies, and scientists are striving to find an alternative matrix to the conventional petroleum-based matrix given public awareness of the environment, climate change, and limited fossil fuel resources. Therefore biobased plastics, which consist of renewable resources, have experienced a renaissance in the past decades. Figure 2.1 shows that the matrices currently used in biofiber composites depend on biobased or petroleum-based plastics and also on their biodegradability [9].

2.5.1 Petrochemical-based

The effects of the incorporation of natural fibers in petrochemical-based thermoplastics and thermoset matrices were extensively studied. Polypropylene (PP), polyethylene (PE), polystyrene (PS), and PVC (polyvinyl chloride) were used for the thermoplastic matrices. Polyester, epoxy resin, phenol formaldehyde, and vinyl ester were used for the thermoset matrices and are reportedly the most widely used matrices for natural fiber-reinforced polymer composites.

Thermoplastic

The mechanical properties, deformation and fracture, thermal diffusivity, thermal conductivity, and specific heat of flax fiber/HDPE biocomposites were evaluated [10, 11]. Sisal fiber-reinforced PE [12] and HDPE [13, 14]

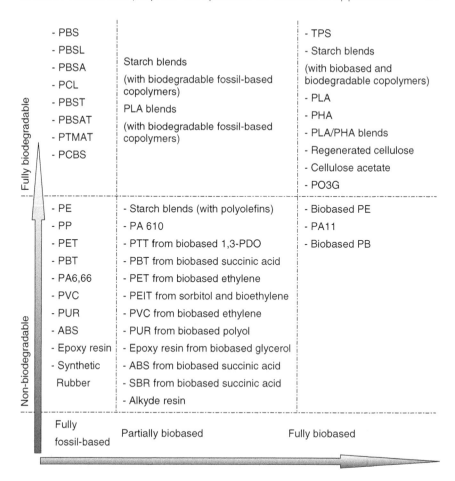

2.1 Current and emerging plastics and their biodegradability.

composites were examined regarding their interfacial properties, isothermal crystallization behavior and mechanical properties.

Traditionally the mechanism of moisture absorption is defined by diffusion theory; but the relationship between the microscopic structure-infinite 3D-network and the moisture absorption could not be explained. Wang *et al.* [15] introduced the percolation theory and a percolation model was developed to estimate the critical accessible fiber ratio, and ultimately, the moisture absorption and electrical conduction behavior of composites. At high fiber loading when fibers are highly connected, the diffusion process is the dominant mechanism; while at low fiber loading close to and below the percolation threshold, the formation of a continuous network is key and hence percolation is the dominant mechanism. The model can be used to estimate the threshold value which can in turn be used to explain moisture absorption and electrical conduction behavior. Figure 2.2 shows the increment of electrical conductivity with the increasing moisture content of composite with 65% rice hull fiber loading. The composite started showing conductivity after it absorbed approximately 50% of maximum moisture. After this, conductivity increased quickly with further moisture absorption. The pattern of the increment of electrical conductivity suggests a diffusion process of moisture absorption.

Bledzki *et al.* [16] examined the mechanical properties of abaca fiber-reinforced PP composites regarding different fiber lengths (5, 25, and 40 mm) and different compounding processes (mixer-injection molding, mixer-compression molding, and direct compression molding). It was observed that, with increasing fiber length (5 mm to 40 mm), the tensile and flexural properties showed an increasing tendency though not a significant one. Among the three different compounding processes compared, the mixer-injection molding process displayed a better mechanical performance (tensile strength is around 90% higher) than the other processes.

The environmental performance of hemp fiber-reinforced PP composites based on natural fiber mat thermoplastic (NMT) were evaluated by quantifying carbon storage potential and CO_2 emissions and comparing the results with commercially available glass fiber composites [17].

Figure 2.3 shows CO_2 emissions in t/t of composite for both natural and glass fibers. These values are estimated by converting energy into CO_2 emissions using standard conversion factors for different fossil fuels. The heat energy liberated by incineration of hemp fiber and PP is also added

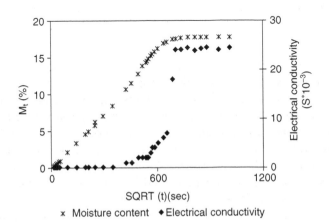

2.2 The moisture absorption and electrical conductivity for HDPE–rice hull composite with 65% fiber content.

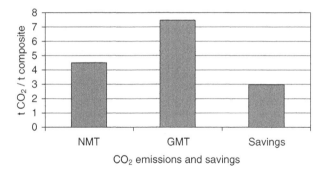

2.3 CO_2 emissions per ton composite and reduction in emissions by substituting glass fiber with hemp fibers.

in non-renewable energy requirements to make calculations simpler. The results demonstrate a net reduction in emissions of 3 t CO_2/t of product if glass fibers are substituted by hemp fibers.

Only limited studies were reported regarding the usage of PS and PVC as matrices for natural fibers. The mechanical behavior [18] and the rheological properties [19] of PS composites reinforced with sisal fibres were studied. The impact properties of PVC composites reinforced with bamboo fibers [20] and the interfacial modification of bagasse fiber [21] reinforced PVC composites were also evaluated.

Thermosets

The hybrid effect on the mechanical properties of abaca and sisal fiber-reinforced polyester composites was evaluated [22]. A positive hybrid effect was observed for the flexural properties. The tensile strength was found to be increased when the volume fraction of banana increased. A negative effect was observed for the impact properties.

The effect of fiber treatment on the mechanical properties of unidirectional sisal/epoxy composites was reported [23]. Ganan *et al.* [24] evaluated the mechanical and thermal properties of sisal/epoxy composites as a function of fiber modification.

Phenolic resins show superior fire resistance to other thermosetting resins. Sreekala *et al.* [25–29] extensively investigated oil palm fiber reinforcement in phenolic resins. The fiber surface modification [25], a comparison of mechanical properties to glass fiber/phenolic resin composites [26], the dynamic mechanical properties regarding the fiber content and hybrid fiber ratio [27], the water absorption [28] and stress–relaxation behavior [29] of oil palm/phenolic resin composites were evaluated.

2.5.2 Biobased

Public concerns about the environment, climate change and limited fossil fuel resources are important driving forces, which motivate researchers to find alternatives to crude oil. Biobased plastics may offer important contributions by reducing the dependence on fossil fuels and, in turn, the related environmental impacts. Biopolymers have experienced a renaissance in recent years. Many new polymers have been developed from renewable resources, such as starch, which is a naturally occurring polymer that was rediscovered as plastic material. Others are polylactic acid (PLA) that can be produced via lactic acid from fermentable sugar and polyhydroxyalkanoate (PHA), which can be produced from vegetable oils next to other biobased feed stocks.

Bledzki et al. [30] investigated PLA biocomposites with abaca and man-made cellulose fibers and compared these to PP composites. The composites were processed using combined molding technology: first a two-step extrusion coating process was carried out and consecutively an injection molding was completed. With man-made cellulose of 30 wt%, the tensile strength and modulus increased by factors of 1.45 and 1.75 times in comparison to neat PLA. Reinforcing with abaca fibers (30 wt%) enhanced both the E-modulus and the tensile strength by factors of 2.40 and 1.20, respectively.

Polyhydroxybutyrate-co-valerate (PHBV) bioplastic was reinforced with flax [31], hemp [32], and bamboo [33] fibers. The mechanical, thermomechanical and morphological properties were evaluated. A soy-based biodegradable resin matrix was applied to bamboo [34], and pineapple leaf [35] fiber-reinforced composites.

The petroleum-derived thermoplastics PP and PE are the two most commonly employed thermoplastics in natural fiber-reinforced composites. There is increasing interest in developing biocomposites with a thermoplastic rather than thermoset matrix, mainly due to their recyclability. Also the choice of a thermoplastic matrix fits well within the eco-theme of biocomposites, but there are some important limitations on the recyclability and mechanical performance of thermoplastics. Generally, the mechanical properties of thermosets are higher than the thermoplastic (lower modulus and strength). In addition, a dramatic loss in properties is observed above the glass transition temperature, which leads to a decrease in other thermally sensitive properties such as creep resistance. On the contrary, thermoplastics show greater fracture toughness than thermosets and thus are more useful in resisting impact loads. Another remarkable change to happen in recent years has been the increased introduction of biopolymers with the aim of decreasing reliance on petroleum-based thermoplastics. The availability and outstanding mechanical properties of biopolymer PLA have led to this matrix system being one of the most thoroughly investigated in the biocomposites research area.

2.6 Processing of biofiber-reinforced plastic composites

The application of natural fibers in plastic technologies required an adjustment to the available processing methods. New materials establish new application fields. The aim is to get high-strength engineering composites. In order to achieve this, innovative technologies and process solutions need to be intensively researched, in addition to new material combinations. New processing aids are only one possible way. New combined molding techniques are the trend of the future. A combination of extrusion and injection molding machines within one unit, the so-called IMC (Injection Moulding Compounder by Krauss-Maffei), improves composite properties and leads to cost reduction in manufacture.

Drying fibers before processing is important, because water on the surface acts like a separating agent in the fiber–matrix interface. The moisture content at a given relative humidity can have a great effect on the biological performance of a composite made from cellulosic fibers. A composite made from pennywort fibers would have a much greater moisture content (57%) at 90% RH humidity than would a composite made from bamboo fibers (15%). The pennywort product would be much more prone to decay compared to the bamboo product.

Another influencing factor is the correct choice of the right fiber type and content, which is generally essential for furthering the sustainability of the composite. Besides fiber type and content, the fiber length and its geometry play a decisive role in composites. Usually, most mechanical properties of a fiber can be enhanced by increasing the aspect ratio [36]. The main differences between organic fillers, wood flour, and natural fibers are their geometrical form and chemical composition. For example, wood flour contains particles that are mostly cubic or spherical, while natural fibers are longish (fibrous), with a high aspect ratio (length/diameter). Organic fillers, such as rice, rye, and wheat hull, contain much more silicates than natural fibers. The addition of natural fibers or other organic fillers can influence processing; thus some negative effects such as corrosion or abrasion of the screws, the barrel, and the mold can appear. It is also important to consider that in some cases selected processing parameters need to be changed. Furthermore, fiber composite compounding and processing methods determine the fiber length. The compounding process significantly influences the shortening, fibrillation, as well as the thermal deterioration of the fibers in the early stages; the final properties of the product are already determined at the beginning of the production process [37]. In general an increasing content of organic fillers or natural fibers decreases mold viscosity drastically. For this reason, some of the processing parameters need to be changed.

Fiber contents also affect the odor concentrations of the natural fiber composites. Odor concentrations of abaca fiber–PP composites with fiber content 20 wt% to 50 wt% have also been measured [38]. It is seen that odor concentration increased significantly with increasing fiber content of the abaca–PP composites, as illustrated in Fig. 2.4. Compression molding showed relatively lower odor concentrations, which is favorable for the automotive sector. It seems that injection molding decomposes the composite materials more than compression molding which results in higher odor concentration.

2.6.1 Compounding

Compounding is of key importance for obtaining materials with appropriate processing characteristics. Before biofiber-reinforced composites can be processed into a final product, they usually have to undergo compounding. Not only the base materials, but also the choice of additives determines the properties of the finished product. Compounding is the process of imparting the desired distribution to two or more components that may be present in solid or liquid form. The fibers and additives are dispersed in the molten polymer to produce a homogeneous blend. The developed available processes for compounding are to use either batch (e.g., internal and thermo-kinetic mixers) or continuous mixers (e.g., extruders, kneaders). The processing parameters (e.g., residence time, shear, and temperature) are easier to control in batch systems, whereas the continuous compounding systems do not have the problem of variations in batch-to-batch quality [39].

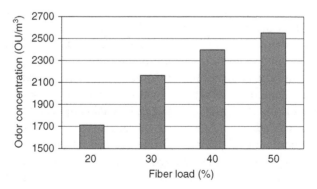

2.4 Influence of fiber load on the odor concentration of abaca fiber–PP composites.

2.6.2 Compression molding

Thermoplastic biofiber-reinforced composites are distinguished from thermoset-reinforced composites primarily by a high elongation at break, short cycle times and the possibility of recycling. The compression molding process has proved suitable for the production of profiles with any thermoplastic prepreg. Compression molding forms the thermoplastic prepreg gently into the required shape without overcompressing the material. The different layer orientations are thus retained after molding.

Whole and split wheat straws with lengths up to 10 cm were used with PP to make lightweight composites by means of the compression molding process [40].

2.6.3 Extrusion

The extrusion process is required so as to be able to melt the polymer and mix the molten polymer with the wood fiber, thus producing a wood plastic composite. By doing so, a homogeneous melt is achieved. The remaining moisture can be removed through vacuum venting. The extruder then compresses the blend, passing it through the die. Processing should not be detrimental to the wood fiber. Single-screw and twin-screw extruders that run co- or counter-rotating, conical and co-extrusion are used for wood plastic composites. Single-screw extruders are used when the mixing effect does not have to be very high. As a rule, co-rotating twin-screw extruders are used in the production of granules (compounding) or in the processing of wood fiber-reinforced plastics. Due to the excellent mixing effect of the twin-screw extruder, the wood fiber material can be homogeneously distributed and wetted in the thermoplastic melt.

HDPE composites with bagasse [41] and curaua [42] fiber-reinforced HDPE composites were obtained by the extrusion process. Curaua fibers were also reinforced with PA-6 by using a co-rotating twin-screw extruder [43].

2.6.4 Injection molding

Injection molding makes it possible to produce complex geometric components with functional elements quickly and in great numbers. Injection molding requires a polymer with a low molecular weight, so as to maintain a low viscosity. By contrast, extrusion requires a polymer with a higher molecular weight for better melt strength. Injection molding offers a number of advantages compared to compression moulding [44], including economics of scale, minimal warping and shrinkage, possibility of high function integration, the possibility of employing recycling material, and the fact that hardly any finishing is needed. In injection molding the raw material is usually added as granules to the injection molding machine and melted into a fluid mass. The plasticized thermoplastic material is then injected into the form under high pressure. The reinforcing fibers influence the injection molding process when fiber-reinforced granules are used.

Injection-molded flax, hemp, core hemp, bleached kraft pulp (BKP), and wood flour-reinforced PP composites were prepared and the effect of MAPP on the mechanical properties was investigated [45] (see Table 2.5). Bledzki et al. [46] investigated the different separation processes (mechanical, refiner, and enzymatic separation) with injection molded hemp and partially with flax and wheat straw-reinforced PP composites. It was found that thermomechanical processed hemp fiber–PP composites possessed better mechanical properties compared to other processes and composites.

2.6.5 Resin transfer molding

Resin transfer molding (RTM) is a method for the production of component parts made of fiber–plastic composites. During the RTM procedure, dry semi-finished fiber parts are streamed and subsequently soaked with reaction resin by a pressure gradient within a closed vessel. The component hardens within the vessel. The pressure gradient can be produced by evacuation of the vessel or by admission of the resin with high pressure. The following methods can be distinguished with regard to the admission by the pressure gradient: high pressure injection, twin wall injection, vacuum injection, differential pressure injection.

Hemp fiber–unsaturated polyester composites were manufactured using a resin transfer molding (RTM) process [47]. The hemp fiber composites manufactured with the RTM process were found to have a very homogeneous structure with no noticeable defects [48]. The tensile, flexural, and

Table 2.5 Effect of filler type on mechanical strength performance of filled PP composites (40% filler, 2% compatibilizer)

Fiber type	BKP	Flax	Milled hemp	Core hemp	Wood flour
Tensile (MPa)	50	42	42	29	35
Tensile modulus (GPa)	3.0	3.2	3.0	2.3	2.9
Flexural (MPa)	78	67	70	52	59
Flexural modulus (GPa)	3.3	3.4	3.5	2.6	3.0
Notched Izod (J/m)	40	44	42	20	28
U-notched Izod (J/m)	205	150	145	100	105

impact properties of these materials were found to increase linearly with increasing fiber content (Table 2.6). It was observed that the optimum properties were not reached in this study and that fiber content higher than 35 vol% should yield better mechanical properties. When compared to a glass fiber composite, however, these natural fiber composites had much lower performances.

2.6.6 Pultrusion

Fibers are pulled from a creel through a resin bath and then on through a heated die. The impregnation of the fiber controls of the resin content and curing of the materials into their final shape is completed using the die. Although pultrusion is a continuous process, which produces a profile of constant cross sections, a variant known as pulforming allows for some variation to be introduced into the cross sections.

Flax fiber-reinforced PP composites were developed by means of the thermoplastic pultrusion process and their physical-mechanical properties were evaluated [49]. Water absorption behavior of pultruded jute fiber-reinforced unsaturated polyester composites was examined [50].

2.6.7 Other processes

Thermosets compression molding

The mat compression process uses mats made of natural fibers. The mats are just sprayed, not moistened, with resin and compressed into their final contour in a hot tool; due to the air permeability, the parts can be covered easily in a vacuum covering process.

The thermal conductivity, diffusivity, and specific heat of thermoset compression molded polyester/natural fiber (abaca/sisal) composites were investigated for several fiber surface treatments as functions of the filler concentration [51].

Table 2.6 Mechanical properties of 20 vol% glass fiber, 20 vol% hemp fiber and 35 vol% hemp fiber reinforced unsaturated polyester composites

	20 vol% glass fibers	20 vol% hemp fibers	35 vol% hemp fibers
Tensile strength (MPa)	85.0	32.9	60.2
Tensile modulus (GPa)	1.719	1.421	1.736
Flexural strength (MPa)	175.9	54.0	112.9
Flexural modulus (GPa)	7.74	5.02	6.38
Impact strength (kJ/m^2)	60.8	4.8	14.2

Thermoforming

Thermoforming is a manufacturing process in which a composite sheet is heated to a pliable forming temperature, formed to a specific shape in a mold, and trimmed to create a usable product. The sheet, or 'film' when referring to thinner gauges and certain material types, is heated in an oven to a temperature that makes it possible to be stretched into or onto a mold and cooled to a finished shape.

Bhattacharyya *et al.* processed wood fiber–PP composites sheets by the thermoforming process [52] and wood fiber–biopole composites studying the thermoforming performance and biodegradability of the composites [53]. Umer *et al.* [54, 55] investigated the liquid composites molding process for wood plastic composite materials.

2.7 Performance of biocomposites

2.7.1 Mechanical properties

It is important to be knowledgeable about certain mechanical properties of each natural fiber, in order to be able to exploit the highest potential of the fiber. Among these properties are the tensile, flexural, impact, dynamic mechanical, and creep properties. In general, natural fibers are suitable for reinforcing plastics, due to their relatively high strength, stiffness, and low density.

Tensile properties

The tensile properties are among the most widely tested properties of natural fiber-reinforced composites. The fiber strength can be an important factor regarding the selection of a specific natural fiber for a specific application. A tensile test reflects the average property through the thickness, whereas a flexural test is strongly influenced by the properties of the specimen closest to the top and bottom surfaces. The stresses in a tensile test are uniform throughout the specimen cross section, whereas the stresses in flexure vary from zero in the middle to maximum in the top and bottom surfaces.

Comparison of the tensile properties of the HDPE/hemp fiber composites showed that the silane treatment and the matrix-resin pre-impregnation of the fiber produced a significant increase in tensile strength, while the tensile modulus remained relatively unaffected [56, 57]. Rice husk reinforced PP composites with filler loadings of 10 wt%, 20 wt%, 30 wt% and 40 wt% were evaluated [58]. A study on the effect of alkaline treatment on tensile properties of sugar palm fiber-reinforced epoxy composites was carried out [59]. Composites consisting of aliphatic polyester (Bionolle) with flax fibers were prepared via batch mixing [60]. The following investigations were completed with thermoplastic matrices; the effect of moisture absorption of sisal fiber/PP composites [61], the influence of surface treatment (NaOH solution) of coir fiber/PP composites [62], the surface esterification of bagasse/LDPE composites [63], the processing conditions of flax fiber/HDPE composites [10], the influence of MAH treatment of jute and hemp fiber/PP composites [64], and the effects of surface treatments of luffa fiber/PP composites [65]. It was observed that the tensile properties were influenced by all the mentioned factors.

The tensile properties of natural fiber-reinforced thermosets were investigated regarding the siloxane treatment of jute fiber/polyester and epoxy composites [66], the temperature and loading rate effects of kenaf fiber/epoxy composites [67], the effects of a differing geometry of abaca fiber/epoxy composites [68], the influence of moisture absorption of bamboo/vinylester composites [69], the fiber loading of oil palm empty fruit bunch/epoxy composites [70], and the influence of the fiber orientation and the volume fraction of alfa fiber/polyester composites [71].

The effects of hybridization and the chemical modification of oil palm/sisal fiber-reinforced natural rubber composites [72, 73], the effects of high temperature on ramie fiber/biodegradable resin composites [74], the effect of biodegradable matrix type (PLA, PHBV, PBS) on regenerated cellulose fiber/biopolymer composites [75], the influence of biobased coupling agent on bamboo fiber/PLA and PBS composites [76], the compounding effects of hemp/PHBV composites [32], and the influence of different thermal treatments of flax fiber/PLA composites [77] on the tensile properties were evaluated.

Flexural properties

The flexural stiffness is a criterion of measuring deformability. The flexural stiffness of a structure is a function based upon two essential properties: the elastic modulus (stress per unit strain) of the material that composes it, and the moment of inertia, a function of the cross-sectional geometry.

Zampaloni *et al.* [78] focused on the fabrication of kenaf fiber-reinforced PP sheets that could be thermoformed for a wide variety of applications with properties that are comparable to existing synthetic composites. Composites of PP and HDPE reinforced with 20 wt% curaua fibers were prepared and the effect of screw rotation speed was evaluated by measuring the flexural properties of the composites [79]. The effect of acetylation [80] on the flexural properties of bagasse/PP composites (properties decreased due to acetylation), and the effect of different maleated coupling agents (flexural properties increased above 60% with optimum loading of coupling agent) on jute and flax fiber-reinforced PP composites [81] have been evaluated. Composites were fabricated using abaca fiber with varying fiber lengths and fiber loading [82]. The flexural properties of coir fiber/polyester composites were evaluated [83].

Impact properties

Impact strength is the ability of a material to resist fracture under stress applied at high speed. Biofiber-reinforced plastic composites have properties that can compete with the properties of glass fiber thermoplastic composites, especially concerning specific properties. However, one property, namely the impact strength, is often listed among the major disadvantages of biofiber-reinforced composites. In recent years, the development of new fiber manufacturing techniques and improved composite processing methods along with enhancement of fiber/matrix adhesion has improved the current situation somewhat.

Damping indexes of abaca fiber–PP composites in different processes are presented in Fig. 2.5 [16]. Mixer-injection molding showed comparatively lower damping index in comparison to other processes. MAH–PP reduced the damping index significantly, regardless of the processing conditions, and it showed a maximum reduction of 50% in the mixer-compression molding process.

Composite panels consisting of virgin and recycled HDPE and four types of rice straw components, including rice husk, rice straw leaf, rice straw stem, and whole rice straw, were made via melt compounding and compression molding [84]. Recycled HDPE composites were reinforced with bagasse fibers and the influence of coupling agent types/concentrations on the composite impact properties was studied [85].

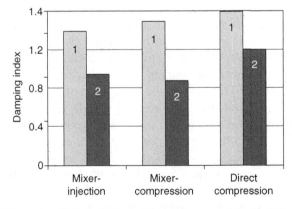

2.5 Damping index of abaca fiber–PP composites in different processes (fiber content: 30 wt%, fibre length 5 mm): 1 – without MAH–PP, 2 – with 5 wt% of MAH–PP.

A comparative study was performed between jute, abaca, and flax fiber-reinforced PP composites focusing on their mechanical properties [38]. Figure 2.6 illustrates that jute–PP composites showed better tensile strength than abaca–PP and flax–PP composites for both cases (with and without MAH–PP). MAH–PP has significant effect for all types of composites and strength properties improved 20–40%. It is also observed that abaca–PP composites showed better flexural strength than jute–PP and flax–PP composites (both cases). MAH–PP has a significant effect on flexural strength for all types of composites and the improvement range was 20–35%.

The moduli of abaca–PP, flax–PP and jute–PP composites are illustrated in Fig. 2.7. It is seen that the tensile modulus shows a discrete effect, where

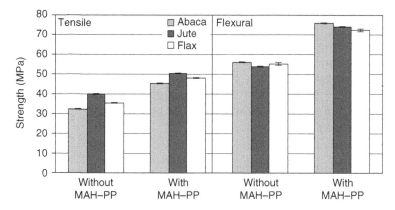

2.6 Comparison of tensile and flexural strength of abaca/jute/flax fiber–PP composites with and without MAH–PP.

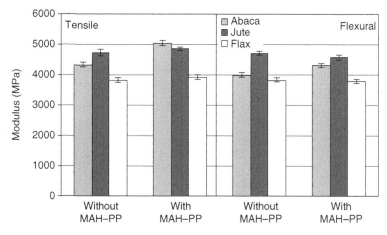

2.7 Comparison of tensile and flexural modulus of abaca/jute/flax fiber–PP composites with and without MAH–PP.

jute–PP composites showed better tensile modulus than abaca–PP and flax–PP composites without MAH–PP. The abaca–PP composites showed better tensile modulus than jute–PP and flax–PP composites with using a coupling agent MAH–PP. MAH–PP has a significant effect on the modulus properties of abaca–PP composite, but very little effect was observed on the modulus properties of jute–PP and flax–PP composites. It may be that the relatively hard and tough abaca fiber bonded with the matrix by MAH–PP and improved the modulus properties. Jute–PP composites showed somewhat improved flexural modulus than abaca–PP and flax–PP composites with and without the coupling agent MAH–PP. When abaca fiber–PP composites were compared with jute and flax fiber–PP composites, abaca fiber composites had the best notched Charpy (Fig. 2.8) and falling weight impact properties.

The toughness of the short fiber-reinforced composites can be influenced by a number of factors, such as the intrinsic properties of the matrix, the fiber volume fraction, and interfacial bond strength. Therefore, strong interactions between the hydroxyl groups of biofibers and the coupling agents are needed to overcome the incompatibility problem. In doing so, the impact, tensile, and flexural strengths of biofiber-reinforced composites can be increased.

2.7.2 Physical properties

When dry, biofiber has unique properties. Knowledge of water–polymer interactions in polymeric composite materials is critical in order to be able to predict their behavior in applications in which they are exposed to water or humid environments. The application of biofiber reinforcement is limited mainly because of the changes in geometry caused by swelling.

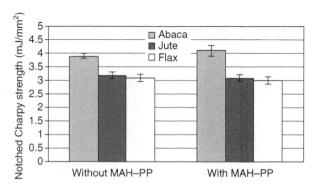

2.8 Comparison of notched Charpy strength of abaca/jute/flax fiber–PP composites with and without MAH–PP.

Water absorption

Drying fibers before processing is of importance, because water on the surface acts like a separating agent in the fiber–matrix interface.

Bledzki et al. [86] investigated the water absorption (static and cyclic) abilities of soft wood fiber–PP composites at two different temperatures (23 and 50°C). Wood flour plastic composites based on PP, PE, and UPVC were found to be as strong as medium fiber board and superior to wooden materials due to their lower water absorption upon exposure to water [87]. Water absorption tests on maleic anhydride modified wood fiber-reinforced composites indicated that they were more hydrophobic than the unmodified ones [88].

Panthapulakkal and Sain investigated the effect of compatibilizer on the water absorption behavior of HDPE composites with different fillers (wheat straw, cornstalk, and corncob). This is shown in Fig. 2.9 [89]. The presence of compatibilizer had no effect on the water absorption of corncob-filled composites. Wheat straw and cornstalk-filled composites showed a decrease in the water uptake with the incorporation of compatibilzers. This indicates that the flaws and gaps at the interface of the filler and HDPE are the main

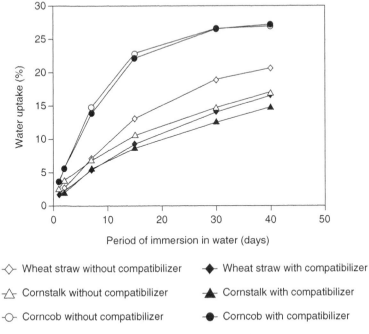

2.9 Water absorption of HDPE/biofiber composites with and without compatibilizers.

factors for moisture diffusion in the composites without compatibilizer. As observed in mechanical properties, the observed reduction in water uptake is higher in wheat straw-filled composites compared to cornstalk-filled composites. After 40 days of water absorption, the percentage reduction in the water uptake of composites is 19%, 12%, and 0% respectively for wheat straw, corn stalk, and corncob-filled composites.

Swelling

The hydroxyl groups (–OH) in cellulose, hemicelluloses, and lignin build a large amount of hydrogen bonds between the macromolecules of the wood polymers. Exposing the wood to humidity causes these bonds to be broken. The hydroxyl groups then form new hydrogen bonds with water molecules which induce the swelling.

The effects of an ambient environment, temperature, and relative humidity on the hygroscopic thickness swelling rate of wood plastics composites were investigated [90]. Hygroscopic thickness swelling rate of wood fiber–PP composites was investigated regarding different recycled plastics (HDPE and PP) [91] and the influence of temperature [92].

Moisture content

Moisture content at a given relative humidity can have a great effect on the biological performance of a composite made from wood fibers. The effects of hydrothermal environment on moisture diffusion [93], moisture content [94] and dependence of the mechanical properties on moisture content [95] of wood plastic composites were evaluated.

2.7.3 Biological properties

The growing markets for biofiber-reinforced composites are in outdoor applications, where they are exposed to moisture, light, temperature changes, freezing, thawing, and biological attacks by fungi and bacteria.

The biological properties of biofiber–plastic composites depend on the load of biofiber material in the matrix and on the surface conditions of the samples produced. These properties are significantly different from the biological properties of the natural components used. The improved spectrum of properties is the basis for success in the market. Biofiber-reinforced composites such as wood plastic composites are sold for exterior application as decay-resistant materials that require no maintenance. The plastic matrix is presumed to protect the fiber or wood particles against biological attack. However, wood particles remain susceptible to fungal degradation since certain amounts of water can be absorbed [96]. In order to prevent potential decaying, borates, such as zinc borate, can be incorporated as effective and leach-resistant preservatives [97].

Fungal

Wood-rotting fungi require a suitable temperature, oxygen, and water supply for their growth. Water is the key parameter in decay mechanisms and in controlling the durability of decay [98].

Wood plastic composites are relatively immune to fungal attacks since the plastic matrix largely encapsulates the wood particles. There are reports describing fungal attacks on wood fiber-filled materials [99]. The decay happens at far lower rates than those found for natural wood. Traditional methods of evaluating biological durability such as weight loss and visual inspection seem to be insufficient when attempting to adequately describe the extent of decay. Inconsistency in the different reports suggests that methods of manufacturing may be the reason for different fungal decay results. Decay susceptibility increases with wood loading [100]. The biological resistance regarding fungal attack of wood plastic composites was also investigated with the influence of chemical modification [101, 102], and moisture dynamics and impact on fungal testing [103].

Bacterial

Bacteria tend to colonize wood fibers with high moisture content. They can affect wood permeability, attack the structure, and work together with other bacteria or fungi. Lignocellulose degrades in a very slow process [104]. While the bacterial degradation for wood fibers has been thoroughly investigated, the literature on the bacterial biodegradation of plastic composites is limited.

2.8 Future trends

The advanced biofiber-reinforced plastic composite contributes to enhancing the development of biocomposites in regards to performance and sustainability. Biocomposites have created substantial commercial markets for value-added products especially in the automotive sector.

Biocomposites are currently the subject of extensive research, specifically in the construction and building industries due to their many advantages such as lower weight and lower manufacturing costs. Green building is a movement that has gained global attention over the past few years. Green buildings are planned to be environmentally responsible, economically viable, and healthy places to live and work. One of the main materials currently used in green buildings is biocomposite. Biocomposites may be classified, with respect to their applications in the building industry, into two main groups: structural (roof structure and bridge) and nonstructural biocomposites (exterior construction, window, door frame, and composite panels).

By the time that biocomposite materials and associated design methods are sufficiently mature to allow their widespread use, issues related to construction materials are likely to have become paramount in material choice. The development of methods, systems, and standards could see biocomposite materials at a distinct advantage over traditional materials. There is a significant research effort underway to develop biocomposite materials and explore their use as construction materials, especially for load-bearing applications.

Burgueno et al. have manufactured cellular beams and plates as load-bearing structural components from hemp, jute, and flax fibers with unsaturated polyester resin [105–108] for housing panel applications. The material and structural performance were experimentally assessed and compared with results from short-fiber composite micro-mechanics models and sandwich analyses. It was demonstrated that cellular biocomposite components can be used for load-bearing components by improving their structural efficiency through cellular material arrangements. Furthermore, it was verified that they can compete with components made from conventional materials.

Natural fibers (such as sisal, jute, and coconut) cement composites are produced in the form of short filament fibers [109, 110]. Short filament geometry composites presented a tension softening behavior with low tensile strength, resulting in products which are more suitable for non-structural applications. Pulp fibers derived from wood, bamboo, and sisal have also been used as reinforcement in biofiber cement composites [111–115].

Uddin et al. [116–118] studied the manufacturing and structural feasibility of natural fiber-reinforced polymeric structural insulated panels for panelized construction, mainly focusing on the manufacturing feasibility and structural characterization of natural fiber-reinforced structural insulated panels (NSIPs) using jute fiber-reinforced polypropylene (NFRP) laminates as skin. The natural fibers were bleached before their use as reinforcement.

Mathur presented an overview of building materials from local resources (India) with particular attention on natural fibers-based composites [119]. The performance of polymer composites made from jute and sisal fibers and unsaturated polyester/epoxy resin was evaluated under various humidity, hygrothermal, and weathering conditions and consequently various composite products (laminates/panels, doors, roofing sheets, shuttering, and dough molding compounds) have been prepared. The process know-how for the manufacture of natural fiber composite panels/door shutters has also been commercialized.

Thomson *et al.* [120] illustrated the opportunities for wood plastic composite (WPC) products in the US highway construction sector, focusing on the market potential of WPC to replace non-renewable materials (e.g., virgin plastic, steel) and preservative-based products (treated wood).

Biobased structural composites for housing and infrastructure applications are of significant importance in the building materials of the next generation of construction in fencing, decking, siding, doors, windows, bridges, fiber cement and so on. This research needs to continue in conjunction with development of conventional composite materials in order to provide a solution in the future which will allow wider use of biocomposite materials by civil engineering applications.

In the future, these biocomposites will see increased use in structural and various other applications depending on their further improvements. Several drawbacks of natural fiber composites which would be even more pronounced in their use in infrastructure include their higher moisture absorption, inferior fire resistance, non-linearity in mechanical properties and durability, variation in quality and price, and difficulty using established manufacturing processes when compared to synthetic composites.

2.9 Conclusion

Biocomposites reinforced with natural fibers have developed significantly over the past years because of their significant processing advantages, biodegradability, low cost, low relative density, high specific strength and renewable nature. These composites are predestined to find more and more applications in the near future, especially in Europe, where pressure from legislation and the public is rising. Interfacial adhesion between the biofiber and the matrix will remain the key issue in terms of overall performance, since it dictates the final properties of the composites.

Further research is still required to overcome obstacles such as moisture absorption, inadequate toughness, and reduced long-term stability for outdoor applications. Especially for outdoor applications, changing weathering conditions such as temperature, humidity, and UV radiation affect the service life of the product. The major detrimental effects of hygrothermal and UV exposure are property deterioration, discoloration and deformation under constant stress for a longer duration.

Significant research is underway around the world to address and overcome these obstacles. Based on their positive economic and environmental outlook, as well as their ability to uniquely meet human needs worldwide, biofiber-reinforced composites are showing a good potential for use in infrastructure applications.

2.10 References

[1] Carus M. Market growth WPC. Press release from the Nova-Institut 2009.
[2] Staiger MP, Tucker N. Natural-fibre composites in structural applications. In Pickering KL (ed.), *Properties and Performance of Natural-Fibre Composites*, Woodhead Publishing, Cambridge, 2008.
[3] Rowell RM. Natural fibres: types and properties. In Pickering KL (ed.), *Properties and Performance of Natural-Fibre Composites*, Woodhead Publishing, Cambridge, 2008.
[4] Hattallia S, Benaboura A, Ham-Pichavant F, Nourmamode A, Castellan A. Adding value to Alfa grass (*Stipa tenacissima* L.) soda lignin as phenolic resins 1. Lignin characterization. *Polym. Degrad. Stab* 2002; 76(2): 259–264.
[5] Hoareau W, Trindade WG, Siegmund B, Castellan A, Frollini E. Sugar cane bagasse and curaua lignins oxidized by chlorine dioxide and reacted with furfuryl alcohol: characterization and stability. *Polym. Degrad. Stab.* 2004; 86(3): 567–576.
[6] Marti-Ferrer F, Vilaplana F, Ribes-Greus A, Benedito-Borras A, Sanz-Box C. Flour rice husk as filler in block copolymer polypropylene: effect of different coupling agents. *J. Appl. Polym. Sci.* 2006; 99: 1823–1831.
[7] Bledzki AK, Sperber VE, Faruk O. Natural and wood fibre reinforcements in polymers, Report 152. RAPRA Technology, 2002.
[8] Swamy RN. Vegetable fibre reinforced cement composites – a false dream or a potential reality? In Sobral HS (ed.), *Proceedings, 2nd International Symposium on Vegetable Plants and their Fibres as Building Materials*. Rilem Proceedings 7. Chapman and Hall, London, 1990, 3–8.
[9] Shen L, Haufe J, Patel MK. Product overview and market projection of emerging bio-based plastics. PRO-BIP 2009, Final Report, Utrecht, The Netherlands.
[10] Singleton CAN, Baillie CA, Beaumont PWR, Peijs T. On the mechanical properties, deformation and fracture of a natural fibre/recycled polymer composite. *Composites Part A: Applied Science and Manufacturing* 2003; 34: 519–526.
[11] Li X, Tabil LG, Oguocha IN, Panigrahi S. Thermal diffusivity, thermal conductivity, and specific heat of flax fiber–HDPE biocomposites at processing temperatures. *Composites Science and Technology* 2008; 68: 1753–1758.
[12] Torres FG, Cubillas ML. Study of the interfacial properties of natural fibre reinforced polyethylene. *Polymer Testing* 2005; 24: 694–698.
[13] Li Y, Hu C, Yu Y. Interfacial studies of sisal fiber reinforced high density polyethylene (HDPE) composites. *Composites Part A: Applied Science and Manufacturing* 2008; 39: 570–578.
[14] Choudhury A. Isothermal crystallization and mechanical behavior of ionomer treated sisal/HDPE composites. *Materials Science and Engineering* 2008; 491: 492–500.
[15] Wang W, Sain M, Cooper PA. Study of moisture absorption in natural fiber plastic composites. *Composites Science and Technology* 2006; 66: 379–386.
[16] Bledzki AK, Faruk O, Mamun AA. Influence of compounding processes and fibre length on the mechanical properties of abaca fibre-polypropylene composites. *Polimery* 2008; 53(2): 35–40.

[17] Pervaiz M, Sain MM. Carbon storage potential in natural fiber composites. *Resources, Conservation and Recycling* 2003; 39: 325–340.
[18] Nair KCM, Thomas S, Groeninckx G. Thermal and dynamic mechanical analysis of polystyrene composites reinforced with short sisal fibres. *Composites Science and Technology* 2001; 61: 2519–2529.
[19] Nair KCM, Kumar RP, Thomas S, Schit SC, Ramamurthy K. Rheological behavior of short sisal fiber-reinforced polystyrene composites. *Composites* 2000; 31: 1231–1240.
[20] Wang H, Chang R, Sheng KC, Adl M, Qain XQ. Impact response of bamboo–plastic composites with the properties of bamboo and polyvinylchloride (PVC). *Journal of Bionic Engineering* Suppl 2008: 28–33.
[21] Zheng YT, Cao DR, Wang DS, Chen JJ. Study on the interface modification of bagasse fibre and the mechanical properties of its composite with PVC. *Composites* 2007; 38: 20–25.
[22] Idicula M, Neelakantan NR, Oommen Z, Joseph K, Thomas S. A study of the mechanical properties of randomly oriented short banana and sisal hybrid fiber reinforced polyester composites. *Journal of Applied Polymer Science* 2005; 96: 1699–1709.
[23] Rong MZ, Zhang MQ, Liu Y, Yang GC, Zeng HM. The effect of fiber treatment on the mechanical properties of unidirectional sisal-reinforced epoxy composites. *Composites Science and Technology* 2001; 61: 1437–1447.
[24] Ganan P, Garbizu S, Llano-Ponte R, Mondragon I. Surface modification of sisal fibers: effects on the mechanical and thermal properties of their epoxy composites. *Polymer Composites* 2005; 121–127.
[25] Sreekala MS, Kumaran MG, Joseph S, Jacob M, Thomas S. Oil palm fibre reinforced phenol formaldehyde composites: influence of fibre surface modifications on the mechanical performance. *Applied Composite Materials* 2000; 7: 295–329.
[26] Sreekala MS, George J, Kumaran MG, Thomas S. The mechanical performance of hybrid phenol-formaldehyde-based composites reinforced with glass and oil palm fibres. *Composites Science and Technology* 2002; 62: 339–353.
[27] Sreekala MS, Thomas S, Groeninckx G. Dynamic Mechanical Properties of Oil Palm Fiber/Phenol Formaldehyde and Oil Palm Fiber/Glass Hybrid Phenol Formaldehyde Composites. *Polymer Composites* 2005; 26(3): 388–400.
[28] Sreekala MS, Kumaran MG, Thomas S. Water sorption in oil palm fiber reinforced phenol formaldehyde composites. *Composites* 2002; 33: 763–777.
[29] Sreekala MS, Kumaran MG, Joseph R, Thomas S. Stress-relaxation behavior in composites based on short oil-palm fibres and phenol formaldehyde resin. *Composites Science and Technology* 2001; 61: 1175–1188.
[30] Bledzki AK, Jaszkiewicz A, Scherzer D. Mechanical properties of PLA composites with man-made cellulose and abaca fibres. *Composites Part A: Applied Science and Manufacturing* 2009; 40: 404–412.
[31] Barkoula NM, Garkhail Sk, Peijs T. Biodegradable composites based on flax/polyhydroxybutyrate and its copolymer with hydroxyvalerate. *Industrial Crops and Products* 2010; 31: 34–42.
[32] Keller A. Compounding and mechanical properties of biodegradable hemp fibre composites. *Composites Science and Technology* 2003; 63: 1307–1316.

[33] Singh S, Mohanty AK, Sugie T, Takai Y, Hamada H. Renewable resource based biocomposites from natural fiber and polyhydroxybutyrate-*co*-valerate (PHBV) bioplastic. *Composites* 2008; 39: 875–886.

[34] Huang X, Netravali A. Biodegradable green composites made using bamboo micro/nano-fibrils and chemically modified soy protein resin. *Composites Science and Technology* 2009; 69: 1009–1015.

[35] Liu W, Misra M, Askeland P, Drzal LT, Mohanty AK. 'Green' composites from soy based plastic and pineapple leaf fiber: fabrication and properties evaluation. *Polymer* 2005; 46: 2710–2721.

[36] Liu W, Drzal LT, Mohanty AK, Misra M. Influence of processing methods and fiber length on physical properties of kenaf fiber reinforced soy based biocomposites. *Composites* 2007; 38: 352–359.

[37] Bledzki AK, Specht K, Cescutti G, Müssig M. Comparison of different compounding processes by an analysis of fibres degradation. In *Proceedings EcoComp*, Stockholm, 2005.

[38] Bledzki AK, Mamun AA, Faruk O. Abaca fibre reinforced PP composites and comparison with jute and flax fibre PP composites. *eXPRESS Polymer Letters* 2007; 1(11): 755–762.

[39] Youngquist JA. Wood-based composites and panel products. Available at: www.woodweb.com/Resources/wood_eng_handbook/Ch10.pdf

[40] Zou Y, Huda S, Yang Y. Lightweight composites from long wheat straw and polypropylene web. *Bioresource Technology* 2010; 101: 2026–2033.

[41] Mulinari DR, Voorwald HJC, Cioffi MOH, da Silva MLCP, da Cruz TG, Saron C. Sugarcane bagasse cellulose/HDPE composites obtained by extrusion. *Composites Science and Technology* 2009; 69: 214–219.

[42] Araujo JR, Waldman WR, De Paoli MA. Thermal properties of high density polyethylene composites with natural fibres: coupling agent effect. *Polymer Degradation and Stability* 2008; 93: 1770–1775.

[43] Santos PA, Spinacé MAS, Fermoselli KKG, De Paoli MA. Polyamide-6/vegetal fiber composite prepared by extrusion and injection molding. *Composites* 2007; 38: 2404–2411.

[44] Scheruebl B. Application of natural fibre reinforced plastics for automotive exterior parts, with a focus on underfloor systems. In: *Proceedings of the 8th International AVK-TV Conference* 2005; p. D5/1–8.

[45] Haijun L, Sain M. High stiffness natural fiber-reinforced hybrid polypropylene composites. *Polymer Plastics Technology and Engineering* 2003; 42(5): 853–862.

[46] Bledzki AK, Faruk O, Specht K. Influence of separation and processing systems on morphology and mechanical properties of hemp and wood fibre reinforced polypropylene composites. *Journal of Natural Fibers* 2007; 4(3): 37–56.

[47] Rouison D, Sain M, Couturier M. Resin transfer molding of natural fiber reinforced composites: cure simulation. *Composites Science and Technology* 2004; 64: 629–644.

[48] Rouison D, Sain M, Couturier M. Resin transfer molding of hemp fiber composites: optimization of the process and mechanical properties of the materials. *Composites Science and Technology* 2006; 66: 895–906.

[49] Van de Velde K, Kiekens P. Thermoplastic pultrusion of natural fibre reinforced composites. *Composite Structures* 2001; 54: 355–360.

[50] Akil HM, Cheng LW, Ishak ZAM, Abu Bakar A, Abd Rahman MA. Water absorption study on pultruded jute fibre reinforced unsaturated polyester composites. *Composites Science and Technology* 2009; 69: 1942–1948.

[51] Idicula M, Boudenne A, Umadevi L, Ibos L, Candau Y, Thomas S. Thermophysical properties of natural fibre reinforced polyester composites. *Composites Science and Technology* 2006; 66: 2719–2725.

[52] Bhattacharyya D, Bowis M, Jayaraman K. Thermoforming woodfibre-polypropylene composite sheets. *Composites Science and Technology* 2003; 63: 353–365.

[53] Peterson S, Jayaraman K, Bhattacharyya D. Forming performance and biodegradability of woodfibre-Biopol™ composites. *Composites A* 2002; 33: 1123–1134.

[54] Umer R, Bickerton S, Fernyhough A. Modelling the application of wood fibre reinforcements within liquid composite moulding processes. *Composites A* 2008; 39: 624–639.

[55] Umer R, Bickerton S, Fernyhough A. Characterising wood fibre mats as reinforcements for liquid composite moulding processes. *Composites A* 2007; 38: 434–448.

[56] Herrera-Franco PJ, Valadez-Gonzalez A. Mechanical properties of continuous natural fibre-reinforced polymer composites. *Composites* 2004; 35: 339–345.

[57] Herrera-Franco PJ, Valadez-Gonzalez A. A study of the mechanical properties of short natural-fiber reinforced composites. *Composites* 2005; 36: 597–608.

[58] Yang HS, Kim HJ, Son J, Park HJ, Lee BJ, Hwang TS. Rice-husk flour filled polypropylene composites; mechanical and morphological study. *Composite Structures* 2004; 63: 305–312.

[59] Bachtiar D, Sapuan SM, Hamdan MM. The effect of alkaline treatment on tensile properties of sugar palm fibre reinforced epoxy composites. *Materials and Design* 2008; 29: 1285–1290.

[60] Baiardo M, Zini E, Scandola M. Flax fibre-polyester composites. *Composites* 2004; 35: 703–710.

[61] Chow CPL, Xing XS, Li RKY. Moisture absorption studies of sisal fibre reinforced polypropylene composites. *Composites Science and Technology* 2007; 67: 306–313.

[62] Gu H. Tensile behaviours of the coir fibre and related composites after NaOH treatment. *Materials and Design* 2009; 30: 3931–3934.

[63] Pasquini D, de Morais Teixeira E, da Silva Curvelo AA, Belgacem MN, Dufresne A. Surface esterification of cellulose fibres: processing and characterisation of low-density polyethylene/cellulose fibres composites. *Composites Science and Technology* 2008; 68: 193–201.

[64] Park JM, Quang ST, Hwang BS, DeVries KL. Interfacial evaluation of modified jute and hemp fibers/polypropylene (PP)–maleic anhydride polypropylene copolymers (PP-MAPP) composites using micromechanical technique and nondestructive acoustic emission. *Composites Science and Technology* 2006; 66: 2686–2699.

[65] Demir H, Atikler U, Balköse D, Tıhmınlıoglu F. The effect of fiber surface treatments on the tensile and water sorption properties of polypropylene–luffa fiber composites. *Composites* 2006; 37: 447–456.

[66] Seki Y. Innovative multifunctional siloxane treatment of jute fiber surface and its effect on the mechanical properties of jute/thermoset composites. *Materials Science and Engineering* 2009; 508: 247–252.
[67] Xue Y, Du Y, Elder S, Wang K, Zhang J. Temperature and loading rate effects on tensile properties of kenaf bast fiber bundles and composites. *Composites* 2009; 40: 189–196.
[68] Sapuan SM, Leenie A, Harimi M, Beng YK. Mechanical properties of woven banana fibre reinforced epoxy composites. *Materials and Design* 2006; 27: 689–693.
[69] Chen H, Miao M, Ding X. Influence of moisture absorption on the interfacial strength of bamboo/vinylester composites. *Composites* 2009; 40: 2013–2019.
[70] Kalam A, Sahari BB, Khalid YA, Wong SV. Fatigue behaviour of oil palm fruit bunch fibre/epoxy and carbon fibre/epoxy composites. *Composite Structures* 2005; 71: 34–44.
[71] Ben Brahim S, Ben Cheikh R. Influence of fibre orientation and volume fraction on the tensile properties of unidirectional alfa-polyester composite. *Composites Science and Technology* 2007; 67: 140–147.
[72] Jacob M, Thomas S, Varughese KT. Mechanical properties of sisal/oil palm hybrid fiber reinforced natural rubber composites. *Composites Science and Technology* 2004; 64: 955–965.
[73] John MJ, Francis B, Varughese KT, Thomas S. Effect of chemical modification on properties of hybrid fiber biocomposites. *Composites* 2008; 39: 352–363.
[74] Nakamura R, Goda K, Noda J, Ohgi J. High temperature tensile properties and deep drawing of fully green composites. *eXPRESS Polymer Letters* 2009; 3(1): 19–24.
[75] Shibata M, Oyamada S, Kobayashi SI, Yaginuma D. Mechanical properties and biodegradability of green composites based on biodegradable polyesters and lyocell fabric. *Journal of Applied Polymer Science* 2004; 92: 3857–3863.
[76] Lee SH, Wang S. Biodegradable polymers/bamboo fiber biocomposite with bio-based coupling agent. *Composites* 2006; 37: 80–91.
[77] Le Duigou A, Davies P, Baley C. Interfacial bonding of flax fibre/poly (L-lactide) bio-composites. *Composites Science and Technology* 2010; 70: 231–239.
[78] Zampaloni M, Pourboghrat F, Yankovich SA, Rodgers BN, Moore J, Drzal LT, Mohanty AK, Misra M. Kenaf natural fiber reinforced polypropylene composites: a discussion on manufacturing problems and solutions. *Composites* 2007; 38: 1569–1580.
[79] Mano B, Araújo JR, Spinacé MAS, De Paoli MA. Polyolefin composites with curaua fibres: effect of the processing conditions on mechanical properties, morphology and fibres dimensions. *Composites Science and Technology* 2010; 70: 29–35.
[80] Luz SM, Del Tio J, Rocha GJM, Goncalves AR, Del'Arco Jr AP. Cellulose and cellulignin from sugarcane bagasse reinforced polypropylene composites: effect of acetylation on mechanical and thermal properties. *Composites* 2008; 39: 1362–1369.
[81] Keener TJ, Stuart RK, Brown TK. Maleated coupling agents for natural fibre composites. *Composites* 2004; 35: 357–362.
[82] Joseph S, Sreekala MS, Oommen Z, Koshy P, Thomas S. A comparison of the mechanical properties of phenol formaldehyde composites reinforced with

banana fibres and glass fibres. *Composites Science and Technology* 2002; 62: 1857–1868.
[83] Monteiro SN, Terrones LAH, D'Almeida JRM. Mechanical performance of coir fiber/polyester composites. *Polymer Testing* 2008; 27: 591–595.
[84] Yao F, Wu Q, Lei Y, Xu Y. Rice straw fiber-reinforced high-density polyethylene composite: effect of fiber type and loading. *Industrial Crops and Products* 2008; 28: 63–72.
[85] Lei Y, Wu Q, Yao F, Xu Y. Preparation and properties of recycled HDPE/natural fiber composites. *Composites* 2007; 38: 1664–1674.
[86] Bledzki AK, Letman-Sakiewicz M, Murr M. Influence of static and cyclic climate condition on bending properties of wood plastic composites (WPC). *eXPRESS Polymer Letters* 2010; 4(6): 364–372.
[87] Bledzki AK, Sperber V, Theis St, Gassan J, Nishibori S. Wood filled thermoplastics as an alternative to natural wood. *International Polymer Science and Technology* 1999; 26(6): T/1–T/4.
[88] Marcovich NE, Aranguren MI, Reborredo MM. Modified woodflour as thermoset fillers Part I Effect of the chemical modification and percentage of filler on the mechanical properties. *Polymer* 2001; 42(2): 815–825.
[89] Panthapulakkal S, Sain M. Agro-residue reinforced high-density polyethylene composites: fiber characterization and analysis of composite properties. *Composites Part A: Applied Science and Manufacturing* 2007; 38: 1445–1454.
[90] Shi SQ, Gardner DJ. Hygroscopic thickness swelling rate of compression molded wood fiberboard and wood fiber/polymer composites. *Composites* 2006; 37: 1276–1285.
[91] Najafi SK, Kiaeifar A, Tajvidi M, Hamidinia E. Hygroscopic thickness swelling rate of composites from sawdust and recycled plastics. *Wood Science Technology* 2008; 42: 161–168.
[92] Najafi A, Najafi SK. Effect of temperature on hygroscopic thickness swelling rate of composites from lignocellulosic fillers and HDPE. *Polymer Composites* 2009; 30: 1570–1575.
[93] Marcovich NE, Reboredo MM, Aranguren MI. Moisture diffusion in polyester–woodflour composites. *Polymer* 1999; 40(26): 7313–7320.
[94] Lin Q, Zhou X, Dai G. Effect of hydrothermal environment on moisture absorption and mechanical properties of wood flour–filled polypropylene composites. *Journal of Applied Polymer Science* 2002; 85: 2824–2832.
[95] Marcovich NE, Reboredo MM, Aranguren MI. Dependence of the mechanical properties of woodflour-polymer composites on the moisture content. *Journal of Applied Polymer Science* 1998; 68: 2069–2076.
[96] Li R. Environmental degradation of wood-HDPE composite. *Polymer Degradation and Stability* 2000; 70(2): 135–145.
[97] Prasad SV, Pavithran C, Rahatgi PK. Alkali treatment of coir fibres for coir-polyester composites. *Journal of Material Science* 1983; 18: 1443–1454.
[98] Zabel RA, Morell JJ. *Wood Microbiology: Decay and its Prevention*, Academic Press, San Diego, CA (1992).
[99] English BW, Falk RH. Factors that affect the application of woodfiber-plastic composites. *Proceeding of 2nd International Conference on Woodfiber-Plastic Composites* (1996).

[100] Verhey SA, Laks PE, Richter DL. The effect of composition on the decay resistance of woodfiber-thermoplastic composites. *Proceeding of 6th International Conference on Woodfiber-Plastic Composites*, 79 (2001).

[101] Timar MC, Pitman A, Mihai MD. Biological resistance of chemically modified aspen composites. *International Biodeterioration & Biodegradation* 1999; 43: 181–187.

[102] Baysal E, Yalinkilic MK, Altinok M, Sonmez A, Peter H, Colak M. Some physical, biological, mechanical, and fire properties of wood polymer composite (WPC) pretreated with boric acid and borax mixture. *Construction and Building Materials* 2007; 21: 1879–1885.

[103] Defoirdt N, Gardin S, Van den Bulcke J, Van Acker J. Moisture dynamics of WPC and the impact on fungal testing. *International Biodeterioration & Biodegradation* 2010; 64: 65–72.

[104] Erikson KEL, Blanchette RA, Ander P. *Microbiol and Enzymatic Degradation of Wood and Wood Components*. Springer-Verlag, New York (1990).

[105] Burgueno R, Quagliata MJ, Mohanty AK, Mehta G, Drzal LT, Misra M. Load-bearing natural fiber composite cellular beams and panels. *Composites Part A* 2004; 35: 645–656.

[106] Burgueno R, Quagliata MJ, Mohanty AK, Mehta G, Drzal LT, Misra M. Hierarchical cellular designs for load-bearing biocomposite beams and plates. *Materials Science and Engineering A* 2005; 390: 178–187.

[107] Burgueno R, Quagliata MJ, Mohanty AK, Mehta G, Drzal LT, Misra M. Hybrid biofiber-based composites for structural cellular plates. *Composites: Part A* 2005; 36: 581–593.

[108] Burgueno R, Quagliata MJ, Mehta G, Mohanty AK, Misra M, Drzal LT. Sustainable cellular biocomposites from natural fibers and unsaturated polyester resin for housing panel applications. *Journal of Polymers and the Environment* 2005; 13: 139–149.

[109] Toledo FRD, Scrivener K, England GL, Ghavami K. Durability of alkali sensitive sisal and coconut fibers in cement mortar composites. *Cement and Concrete Composites* 2000; 22: 127–143.

[110] Ramakrishna G, Sundararajan T. Impact strength of a few natural fibre reinforced cement mortar slabs: a comparative study. *Cement and Concrete Composites* 2005; 27: 547–553.

[111] Silva FA, Ghavami K, D'Almeida JRM. Bamboo-wollastonite hybrid cementitious composites: toughness evaluation. In *Joint ASME/ASCE/SES Conference on Mechanics and Materials*, Baton Rouge, 2005.

[112] Silva FA, Ghavami K, D'Almeida JRM. Toughness of cementitious composites reinforced by randomly distributed sisal pulps. In *Eleventh International Conference on Composites/Nano Engineering (icce-11)*, Hilton-Head Island, 2004.

[113] Silva FA, Ghavami K, D'Almeida JRM. Behavior of CRBP-AL subjected to impact and static loading. In *17th ASCE Engineering Mechanics Conference (EM 2004)*, Delaware, 2004.

[114] Mohr BJ, Nanko H, Kurtis KE. Durability of Kraft pulp fiber-cement composites to wet/dry cycling. *Cement and Concrete Composites* 2005; 27: 435–448.

[115] Roma JR, Martello LC, Savastano JRH. Evaluation of mechanical, physical and thermal performance of cement-based tiles reinforced with vegetable fibers. *Construction & Building Materials* 2008; 22: 668–674.

[116] Uddin N, Kalyankar RR. Manufacturing and structural feasibility of natural fiber reinforced polymeric structural insulated panels for panelized construction. *International Journal of Polymer Science* 2011; 963549.
[117] Kalyankar RR, Uddin N. Structural characterization of natural fiber-reinforced polymeric structural insulated panels for panelized construction. *Journal of Reinforced Plastics and Composites* 2011; 30(11): 988–993.
[118] Kalyankar RR, Uddin N. Structural characterization of natural fiber reinforced polymeric (NFRP) laminates for building construction. *Journal of Polymers and the Environment* 2012; 20(1): 224–229.
[119] Mathur VK. Composite materials from local resources. *Construction and Building Materials* 2006; 20: 470–477.
[120] Thomson DW, Hansen EN, Knowles C, Muszynski L. Opportunities of wood plastic composite products in the U.S. highway constructor sector. *Bioresources* 2010; 5(3): 1336–1352.

3
Advanced processing techniques for composite materials for structural applications

R. EL-HAJJAR, University of Wisconsin-Milwaukee, USA, H. TAN, Hewlett-Packard Company, USA and K. M. PILLAI, University of Wisconsin-Milwaukee, USA

DOI: 10.1533/9780857098955.1.54

Abstract: Modern structural applications of composite materials are dictated by the processing methods available. In this chapter, we introduce recent developments related to the manufacturing of composites in civil engineering applications using vacuum assisted resin transfer molding, pultrusion, and automated fiber placement.

Key words: vacuum assisted resin transfer molding, pultrusion, automated fiber placement, processing.

3.1 Introduction

This chapter introduces the reader to some of the recent developments in advanced processing of composite materials for manufacturing large composite structures. The goal is to introduce the reader to the methods coupling advanced numerical and analytical approaches for developing predictive capabilities useful especially for larger projects. Traditional methods such as plate bonding, filament winding, and hand layup have been covered to large extent in other resources (Miracle and Donaldson, 2001; Strong, 2008) and will not be covered in detail here. This chapter will focus more on the recent developments related to the manufacturing of composites using vacuum assisted resin transfer molding (VARTM), pultrusion, and automated fiber placement (AFP). These manufacturing methods have been successfully used to create large composite structures for civilian engineering applications.

3.2 Manual layup

The process of manual layup for retrofit or repair focuses on creating a 'shell' of composite around large civilian structures in order to boost their structural strength. The manual layup or the hand layup falls under the category of open molding processes which has seen significant application in the composites industry (Strong, 2008). The advantages of the hand layup process include low investment costs, no need for special training, and ability to create complicated structures (Miracle and Donaldson, 2001). In the area of civil engineering, the hand layup process has seen significant applications in retrofitting old bridges (Uddin et al., 2004).

The first step in the hand layup process is the cleaning and preparing of the surface of the structure. An appropriate thermosetting resin is mixed with its initiator in a suitable ratio in order to start the cross-linking (curing) reaction. A layer of such a resin is applied to the surface followed by draping of the surface with the reinforcement material. These reinforcements are usually fiber mats, woven fabrics or knits. (Typical reinforcement materials are made from glass fibers because of their low cost.) This is followed by application of another layer of resin. Small hand-held rollers, which are used for rolling the resin into the reinforcement, are used to ensure proper wet-out of the reinforcement material.

The resin of choice for general composites manufacturing has been the unsaturated polyesters (acting as the thermosetting resins) because of their low cost with respect to the other resins. However, use of styrene as a solvent in such polyester resins leads to the problem of the environmental management of the harmful styrene vapors at the work sites. The use of polyesters requires the use of catalysts or some kind of heating to promote the curing reaction. This again is cumbersome for many civil engineering applications, since it means taking additional equipment for heating the layup. In order to overcome these two problems associated with the polyester resins, epoxy resin has emerged as the resin of choice for civil engineering applications. Epoxy resins can be cured at room temperature with no harmful styrene emissions. However, epoxies suffer from higher costs compared with the polyester resins.

Despite having the advantages listed above, the hand layup process suffers from some difficulties in the area of civil engineering applications (Uddin et al., 2004): (1) need for special workers for handling fabrics, (2) improper wetting of the fabrics, and (3) problems in controlling the processes on long bridge girders and columns. Some of these problems can be ameliorated by the emerging technology of VARTM described in Section 3.5.

3.3 Plate bonding

Strengthening of civil engineering infrastructure using plate bonding involves adhesive and/or mechanical bonding of fiber-reinforced plastic materials to the original structure. Typically this process is implemented to strengthen or retrofit an existing structure such as steel reinforced concrete. The most common fiber materials are fiberglass or carbon fibers that offer a high specific stiffness and strength but possess generally low values of coefficient of thermal expansion in the fiber direction. The low weight of the repair panels confers a significant advantage during these repairs since they can be more easily transported to the field compared to steel plates. The fiber systems are available in a variety of forms from the uniaxial to multidirectional fabrics, in addition to the 'off-the-shelf' pultruded plates that are available in various thicknesses and properties (see Section 3.6). The properties of the panel/structure interface become of primary importance and efforts to understand the effects of reduced interfacial properties on the structural response have been documented (Cheng et al., 1997).

Typically the bonded plates can be analyzed as an integrated structure in the absence of any delaminations or failures. This allows using structural analysis methodologies familiar to civil engineers. Typical failures have less to do with failure of the actual plates than of failure at the interface due to peeling stresses. These failure stresses are generally classified into flexural, shear, or axial peeling stresses. The most common failures are associated with adhesive shear stresses that peak at the ends of the plate, especially those with low shear span/depth ratios (Garden et al., 1998). Anchor plates are used to resist the peeling stresses occurring at the ends of the composite and the superstructure it is bonded to. Inverse analysis can be used to determine the best processing parameters for reducing undesirable peeling stresses (Kim et al., 2008). With careful design practices, the flexural, shear, and axial peeling failures can be prevented (Oehlers, 2001).

3.4 Preforming

Preforming refers to the process of fabricating a reinforcement package, for making the polymer composites through one of the several available technologies, into a shape and size that is very close to the desired shape and size of the final composite part (Strong, 2008). The final part can be made from the preform by adding resin and then curing. The use of preforms allows one to make parts very close to the net shape required and thus minimize the need for subsequent cutting, machining, and polishing operations. The use of preforms also allows one to have a good control over the density and orientation of fibers in the final part as they prevent relative movement of fabric layers during composites processing. The application of preforms in composites manufacturing is dictated by several factors such as the need for reduction in cycle time of the process or the need for maintaining consistent quality from part to part. It is also dependent on automated techniques for fiber handling and forming. Of the processes used to make composites described in the present chapter, preforming can be applied usefully to pultrusion, and to a lesser extent to VARTM.

Two-dimensional preforms are created by stacking into layers the traditional two-dimensional materials such as fiber mats and woven or knitted fabrics, and then cutting them to the final part-shape required. The layers are held together through the use of lightly applied adhesive or light stitching or pinning. The two-dimensional preforms are also created by thermoforming the layers with slight binder coating into the final shape, and then trimming to remove the excess material. Such preforms can also be created through the use of a chopper gun to deposit fibers and binder resins into a perforated mold designed to be similar to the desired net-shaped part. The strong vacuum suction behind the perforations holds the fibers in place till the binder cures and creates the preform.

Three-dimensional preforms are created using the 3D weaving processes employed for making textiles. Hence machines using processes such as 3D weaving and knitting are used to create such preforms. Since fiber yarns are continually directed across the plane of fabrics during the weaving or knitting of the preform, the mechanical properties such as inter-laminar strength are much higher in the final part, thereby ensuring less chance of failure through delamination. Three-dimensional I-beams, T-stiffened panels, and similar shapes can be created automatically by the machines employing the 3D weaving or knitting techniques. Such preforms can be useful in processes such as pultrusion (Strong, 2008). As the final composite part becomes thicker, the labor required to manage the reinforcement, prior to wetting by the resin, increases dramatically; in such situations, the use of three-dimensional preforms is eminently justifiable.

3.5 Vacuum assisted resin transfer molding (VARTM)

Vacuum assisted resin transfer molding (VARTM), a member of the family of liquid composite molding (LCM) processes, is a closed-mold process for making fiber-reinforced polymer composites (Parnas, 2000). VARTM has been developed as a variant of the traditional resin transfer molding (RTM) process to reduce the cost and design difficulties associated with large mold tools. VARTM is a single-sided molding process, where the other side of the mold is a flexible vacuum bag that provides the compaction and sealing during resin infusion (Fig. 3.1).

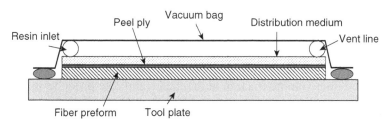

3.1 A typical cross section of VARTM mold.

Since one side of the mold in VARTM is replaced by the vacuum bag, this eliminates the need for making a precise matched mold as required by RTM. As a result, the tooling costs of VARTM are reduced significantly because of such one-sided molds. VARTM is very cost-effective for the manufacture of large composite structures such as boat hulls, car bodies, and wind-turbine blades (Parnas, 2000). VARTM is currently implemented in numerous fields such as:

- shipbuilding industry: manufacture of naval structural components (i.e., masts, hulls, and bridge decks);
- automotive industry: manufacture of chassis and body components;
- aerospace industry: manufacture of fuselage and wing components.

Recently, there have been some important applications of this technology in civil engineering areas as well. VARTM has been used in making strong, light, corrosion-free bridges as well as retrofitting old, damaged structures including bridges. For example, VARTM was recently used in the strengthening of a highway bridge girder in Huntsville, Alabama (see Fig. 3.2). For containing the stress- and environmental-induced cracks in the old girder, it was observed that repairs implemented through VARTM are superior to repairs achieved through the traditional method of injecting polyurethane foam into the cracks. VARTM has also been applied to retrofit glass or fiber composites to steel, concrete, and masonry structures for blast protection (Buchan and Chen, 2007).

A typical VARTM process involves four steps as shown in Fig. 3.3. First, the reinforcing fibers in the form of fiber preforms are placed over a mold surface. Then a highly porous distribution medium, often in the form of a fabric, is placed on top. A layer of peel ply, a thin layer facilitating the separation of the distribution medium from the main composite after the cure, separates the two layers. Plastic tubes for the distribution of resin as well as pulling of vacuum are then placed strategically in the mold to ensure the complete wetting of the preform. Finally a sheet of some flexible transparent material (vacuum bag) such as nylon or polyvinyl acetate is then placed over the sandwich layered structure on the mold in order to form a vacuum-tight seal. In the second step, a vacuum is applied between the bag and the preform through vents to draw the resin to the preform through various tubes. During infiltration of the fiber preform by the resin, pressure difference due to the atmospheric pressure outside the vacuum bag and vacuum inside the bag tends to compact the preform against the hard mold side or tooling surface. The resultant pressure difference also helps in achieving the desired fiber volume fraction of the final composite products by forcing out excess resin. The vacuum inside the bag also helps in reducing the formation of bubbles due to the entrapment of air inside the preform, thus reducing the presence of porosity in the final part. In the third step, the matrix

Advanced processing techniques for composite materials

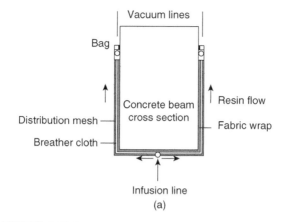

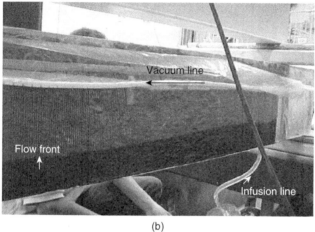

3.2 (a) A schematic of the VARTM setup to put a composites wrap around a damaged concrete beam. (b) VARTM processing on a 6 ft RC beam (Uddin *et al.*, 2004).

material is allowed to undergo a cross-linking reaction (also called the curing reaction) that leads to its solidification. The part may be cured at room temperature or in an oven depending on the resin used. Finally, the composite part is removed from the mold after the resin is completely cured.

The driving force for resin flow in VARTM is the pressure difference between the injection port (atmospheric pressure) and the vent (vacuum pressure). Due to the relatively low pressure difference (~1 atm) and low permeability of preform arising from its compaction under the vacuum bag, as mentioned before, a resin distribution medium with high permeability is often incorporated into the vacuum bag layup on top of the dense preform

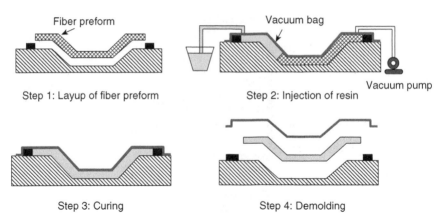

3.3 Manufacturing steps involved in VARTM.

to facilitate resin flow. Because of the low flow resistance in the distribution medium, the resin flows preferentially across the distribution medium while simultaneously seeping into the preform through the thickness. Use of the distribution medium reduces the processing time significantly, and ensures complete wet-out of the preform.

The main materials used in VARTM are resins as matrix and fibers as reinforcements. Although a wide range of resins are available for different applications, the common requirements are: low viscosity and long gel time to permit complete impregnation and fiber wetting; appropriate curing characteristics to provide acceptable cycle times; and adequate mechanical properties and physical characteristics to meet the performance specifications. The most significant practical limitation on the suitability of a resin system is imposed by its viscosity. Because of these requirements, most resins used in VARTM are polyesters, vinyl esters, epoxies, and bismaleimides.

Fibers used in VARTM can be in a number of forms ranging from individual filaments to intermediate products such as chopped strand mat and fabrics made from rovings, yarns, strands and tows, each consisting of thousands of filaments. A number of preforming processes are based on the use of either rovings or yarns, often utilizing techniques originally developed for the textile industry. These yarns or tows are stitched, woven, braided, or knitted into one-, two-, or three-dimensional (1D, 2D, 3D) fabrics to create a textile preform as shown in Fig. 3.4. These fabrics are usually in the form of flat sheets (often called mats), which are then rolled up for transport to the composites manufacturing facility. The fibrous preforms are the skeletons of the VARTM composites, which not only provide a mechanism for the structural toughening of composites, but also facilitate the processing of composites into net or near-net shape structural parts. A

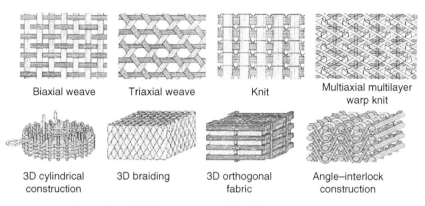

3.4 Examples of different types of braiding, weaving, and knitting patterns employed to create preforms.

2D fabric consists of the planar interlaced and interloped weave pattern. Due to a lack of through-thickness fiber reinforcement between fabric layers, the interlaminar strength is limited by the fracture toughness of the polymer matrix. In contrast, the fibers of a 3D fibrous preform are oriented in various in-plane and out-of-plane directions, providing the additional reinforcement in the through-thickness direction. The 2D fabrics are usually formed into shapes by molding or stitching, while 3D preforms are more suitable for creating the near net-shape structural parts with more complex geometries. Architecture of the fibrous preforms plays a key role in composite manufacturing as it affects various processing steps including forming and resin infiltration.

Typical fibers used in making composites through VARTM are made from glass, carbon (graphite), or aramid. The choice of the fiber type used in a particular application depends mainly on the intended cost and performance. Glass fibers have been widely used in automotive and shipbuilding industries due to their relatively low cost. Higher performance applications, including components for the aerospace and auto-racing industries, often use carbon or aramid fibers, where the increased cost is justified by the associated improvement in mechanical properties. Carbon fibers offer high strength and stiffness whereas aramid fibers offer important advantages of toughness and impact resistance.

Compared to other composite manufacturing techniques, VARTM has several advantages. The pressure used in VARTM is relatively lower than that required for the compression and injection molding processes, which means that the tooling costs and operating expenses are low. It also allows for a fairly mobile setup that can be taken to the field (e.g., for use in repair applications). VARTM can make complex parts at intermediate volume rates, which allows limited production runs in a cost-effective manner.

Fabrication of large-scale composite structures is easy and affordable as well. A high fiber volume fraction with controlled fiber directions can be achieved by VARTM. Continuous fibers used in VARTM lead to the production of near net-shape parts, so material wastage and machining cost are reduced. Because the closed molding processes offer low volatile emission during processing, VARTM processes are more environmentally friendly. However, since one side of the mold consists of the flexible vacuum bag, surface quality and dimensional tolerance are issues for the VARTM parts.

The quality of VARTM product and the efficacy of the process depend strongly on complete wetting or impregnation of fiber preforms inside the mold. Such impregnation is affected by several parameters including the location of resin-inlet tubes and air vents, change in resin viscosity with cure, and the permeability and compressibility of the fiber preform and distribution media. The traditional trial-and-error methods for optimizing the mold and process design can be too time-consuming and economically prohibitive for a complex part. As a consequence, the numerical simulation of resin flow emerges as one of the most effective ways to optimize the VARTM mold-filling process. Successful computer simulations are able to improve the mold design in virtual space without the expensive and time-consuming trial-and-error approach (see Plate I between pages 240 and 241).

In VARTM mold-filling simulations, the fiber preforms are viewed as porous media, hence the liquid resin impregnating the dry fiber preform can be modeled using Darcy's law and the mass-balance equation. Change in viscosity due to temperature and resin cure can be predicted from the accompanying energy (resin temperature) and resin-cure transport equations. Application of vacuum to the mold with a flexible wall leads to compression of preform and distribution media, which in turn leads to a reduction in the permeability in the two layers arranged in series (Fig. 3.1). Impregnation of the compressed preform and distribution medium with the resin, as shown in Fig. 3.3, leads to gradual increase in the resin pressure inside the bag to the atmospheric pressure. Such an increase is accompanied by a relaxation of the two-layer 'spring', which in turn leads to an increase in the thickness of mold. Hence, the thickness, which is smallest in the dry mold, gradually increases as the mold gets filled up. Lack of constant thickness in the VARTM mold translates into a composites part with a variable thickness, and that can be a problem if the structural integrity of the VARTM part is very sensitive to its thickness. Hence mold-filling simulations can be a useful tool to predict the thickness of the VARTM-produced part and thus helps to reduce uncertainty associated with the part thickness.

Impregnation of resin through the fiber preform is a typical problem of free surface flow involving a moving boundary. How to describe and advance the moving boundary is the core issue in such problems. Generally, the moving boundary problems can be tackled using either the Lagrangian method (Zienkiewicz and Godbole, 1974; Li and Gauvin, 1991; Trochu and Gauvin, 1992; Schmidt et al., 1999) or the Eulerian method (Chiu and Lee, 2002; Young, 1994; Serrano-Perez and Vaidya, 2005; Simacek and Advani, 2007; Song and Youn, 2008). Unlike the Lagrangian method, which attaches the computational mesh to the fluid domain to capture the moving boundary, the Eulerian method employs the fixed mesh to describe the moving fluid. Since the Eulerian method circumvents the remeshing problem often associated with the Lagrangian method, it is highly efficient computationally, especially for large problems with complex geometry. As a result, the Eulerian method has been very attractive in practical engineering applications. So far, all the flow simulations of VARTM (Govignon et al., 2010; Grujicic et al., 2005; Serrano-Perez and Vaidya, 2005; Simacek and Advani, 2007; Song and Youn, 2008) are based on the Eulerian method because of its high computational efficiency.

Various front-tracking techniques based on the Eulerian method have been proposed in the past decades. The most common are the marker-and-cell (MAC) method (Harlow and Welch, 1965), the flow analysis network (FAN) method (Gutfinger et al., 1974; Tadmor et al., 1974), the pseudo-concentration method (Thompson, 1986), and the volume of fluid (VOF) method (Hirt and Nichols, 1981; Rider and Kothe, 1998). Due to robustness and efficiency, the VOF method has been widely used in the mold-filling simulations in the fields of die casting, polymer processing, and composites manufacturing.

Once the computational domain is determined by the front-tracking method, a proper numerical technique needs to be used to solve the governing equations of the VARTM process. Different numerical methods have been proposed to solve the equations. The most popular ones among them are the finite difference method (Trochu and Gauvin, 1992; Gauvin and Trochu, 1993), the boundary element method (Soukane and Trochu, 2006), and the finite element method (Grujicic et al., 2005; Serrano-Perez and Vaidya, 2005; Simacek and Advani, 2007; Song and Youn, 2008). Because of its ability to handle complex irregular geometries, the finite element method has become the most popular tool for discretization in mold-filling simulations for VARTM (Grujicic et al., 2005; Serrano-Perez and Vaidya, 2005; Trochu et al., 2006; Simacek and Advani, 2007; Song and Youn, 2008; Li et al., 2008; Govignon et al., 2010). A hybrid of the finite element method with the front-tracking technique FAN, often referred to as the control-volume method, has been widely used to model the resin flow in VARTM, which is commonly referred to as the finite element/control volume (FE/CV) method (Bruschke and Advani, 1989; Buchan and Chen, 2007; Mohan et al., 1999; Joshi et al., 2000; Trochu et al., 2006).

To model resin flow in VARTM, the transient resin impregnation through fiber preforms is treated as a continuous quasi-steady process with the moving boundaries and the changing preform and distribution-layer thicknesses, so the whole process is divided into quasi-steady time-steps and the flow front progresses step by step. The finite element is first used to solve for the pressure field with rate of change of mold thickness on the right-hand side of the discretized set of equations. The computed pressure field is then used to calculate flow rates at control volume faces and advance the flow front. The control volume method is then used to track the flow front. Then by balancing the sum of the preform stress and the resin stress with the outside atmospheric pressure, the current mold thickness at finite elements is determined. From the current and previous mold thicknesses, the rate of change of mold thickness at various finite elements is estimated. The pressure field for the new domain is now computed and the procedure repeated until the mold is full. This algorithm is often called explicit mold-filling simulation, because time integrations for the transient terms is evaluated numerically in an explicit manner. In order to ensure the stability of the numerical solution based on the explicit time integration, the time-step increment is determined such that only one control volume is filled at each time-step. In other words, the time-step size for each time-step is automatically determined in the simulation process based on the calculation of the mass flux and empty volume-fraction of a control volume that is filled most quickly among all the border control volumes at that time-step.

The mold-filling in VARTM under vacuum pressure becomes more difficult as the complexity or size of the part increases or when fiber performs exhibiting low permeability are used. Although the optimum process design, which can be achieved through numerical modeling, is crucial to achieve high quality parts (Hsiao *et al.*, 2004), process and material parameter uncertainties as well as real-time variabilities often cause deviations from the design targets. Therefore, robust fabrication necessitates real-time control strategies that ensure reliable mold-filling in VARTM in the face of practical variabilities.

Recently, significant research attention has been focused on the control of resin flow in VARTM. Walsh and Mohan (1999) proposed a control approach of using real-time flow sensing to determine the optimum time to activate a second resin inlet mid-fill as a means of compensating for low permeability in the VARTM process. Heider and Gillespie Jr (2004) developed VARTM work cells to evaluate control strategies and sensors, such as SMART weave, for providing feedback for an intelligent control system. They presented a continuously controlled vacuum actuator system and evaluated the influence of vacuum gradients on resin flow-front control. An automated sequential injection scheme for producing large-scale composite parts was also implemented and evaluated. Pitchumani *et al.* (Johnson and Pitchumani, 2007; Nielsen and Pitchumani, 2002) reported model-based control schemes for RTM and VARTM processes, including controls for both flow-rate controlled and pressure controlled injection systems. They used neural networks and real-time numerical process simulations to actively control the mold-filling stage. One of the control strategies used an online permeability estimation using fuzzy logic theory. The control schemes were shown to steer the flow and reproduce multiple desired fill patterns reliably in the face of permeability uncertainty. Bender *et al.* (2006) presented a flow-rate control system for VARTM which allowed for vacuum pressure to be applied on the injection bucket to generate a computer-controlled vacuum-pressure differential between the injection and vent gate. A fuzzy-logic controller was implemented and the system was optimized virtually using a VARTM flow simulation.

Although the studies discussed differ in approach used to achieve the control decisions, it is commonly recognized that a careful choice of the process control parameters in real time is needed for fabricating high-quality composite parts using VARTM. The control strategy should accomplish the goals such as driving the flow to a desired location, avoiding the formation of dry spots and weld lines, and decreasing the process time, while keeping the control effort and the necessary hardware expenditure at a reasonable level.

3.6 Pultruded composites

The pultrusion process is particularly suited for civil engineering applications because of the ability to generate near prismatic shapes. In the pultrusion process, different types of reinforcements (fiber tows, continuous filament mats, etc.) are guided from a creel into a resin impregnation unit for wetting the reinforcements (Fig. 3.5). Typically glass fibers are used due

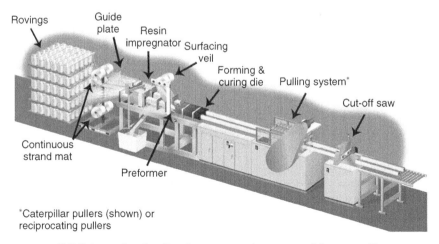

3.5 Schematic of pultrusion process (courtesy of Strongwell).

to cost considerations, but recently carbon fibers have seen increasing use. The wetted preforms are then drawn into heated die elements where curing occurs. The tension force used to pull the fibers is influenced by a variety of factors such as the change in the stresses as the resin is cured in the process. The tooling costs tend to be lower than those for metallic components but they are highly dependent on the part complexity. Most parts produced using pultrusion are straight, although it is possible to manufacture parts that have some curvature. A major advantage of the pultrusion process is that any length can be produced that can then be transported in one section. Longitudinal ribs and corrugations can also be manufactured using pultrusion with specially created dies.

Highlighting some of the civil engineering developments, significant research is seen in pultruded bridge decks replacing steel reinforced concrete. I-beams, channels, and angled pultrusion replacements are seen in many structures around factories, especially where corrosion is an issue. Several residential and commercial products are available using pultruded shapes for door and window framing. In addition, replacement of steel rebar with pultruded rebar is also receiving some attention for applications where durability near corrosive environments is of interest. There have been several projects using pultruded composites in bridge construction. One of the interesting studies is presented by Bank *et al.* (2006), where they presented a case study of a successful prefabricated double-layer pultruded fiber-reinforced polymer (FRP) grid for bridge deck construction in Wisconsin. The entire bridge slab was placed in 10 hours with a crane by four construction workers. The authors found that the cost of the project can be improved if connection approaches are developed which would bring it within the range of steel-reinforced decks.

Typically glass or carbon fibers are used in addition to polymer materials such as polyester or vinyl ester. If moisture issues are of concern, epoxies are less hygroscopic and offer superior mechanical properties (Maji *et al.*, 1997). Most of the parts produced tend to use polyester resins. Phenolic resins have been used where issues related to smoke from flames and toxicity is of importance. Toughness has been improved over some of the traditional resins by using two-component polyurethane pultrusions, producing two to five times increases in elongation and toughness over polyester (Miracle and Donaldson, 2001). Recent studies have also experimented using cements for pultrusion (Peled and Mobasher, 2006), where woven fabrics made from low modulus polypropylene (PP) and glass meshes were used to produce the pultruded cement composites. Fillers are also added to the resin in large quantities (up to 50% of the total resin by weight; Miracle and Donaldson, 2001) during the pultrusion process. These include materials such as calcium carbonate, glass beads, talc, alumina silicate, and alumina trihydrate. These are usually added for various reasons such as flame suppression or specific property improvement. However, some fillers can have detrimental effects favoring crack growth which results in reduced mechanical properties after impact (Paciornik et al., 2003).

Pultruded composites are thicker than their laminated counterparts in a per layer estimate. Since the fiber spools are pulled in one direction, the pultruded composites tend to be highly orthotropic. To alleviate the bias of properties in one direction, typically different forms of reinforcement are repeated through the thickness such as continuous filament mats or woven fabrics. The addition of these layers is necessary to provide material continuity and to provide additional reinforcement in areas experiencing multi-directional stress states such as near material discontinuities, holes, etc. Modeling of geometric parameters of the injection ports and their axial location within the injection chamber related to the final thickness can be used to improve the productivity of resin injection pultrusion (Jeswani et al., 2009; Rahatekar et al., 2005). Numerous parameters affect the resin pressure rise in the die inlet. The geometry of the tapered die inlet region can have a significant effect on the pressure rise in the pultrusion die and in turn the quality of the pultruded product. Finite element analysis can be used to predict the effect of different shaped die inlets to provide insight on the design of the die inlet (Sharma et al., 1998). Other design parameters explored using finite element analysis are the aspect ratio of the final composite, injection slot width and location from inlet of the injection chamber, and the location of the ports (Jeswani and Roux, 2006). Process modeling results in improved fiber wet-out that ultimately results in improved mechanical properties.

The production of pultruded composites may result in a number of manufacturing defects such as porosity, fiber waviness, matrix cracks, or areas of resin richness or starvation. Microcracks exposed to repeat environmental and mechanical loading may result in the cracks coalescing to form areas where delaminations may grow and cause failure. The anisotropic nature of pultruded composites also results in anisotropic fracture properties (El-Hajjar and Haj-Ali, 2005). For through-thickness cracks, bridging from the roving fibers is observed for a crack propagating in the roving direction, but a crack in the transverse direction relies on the secondary reinforcements for crack growth resistance. Pultruded composites offer important advantages when stiffness controlled design is used, which highlights the importance of achieving a high fiber volume-fraction ratio, low void content and careful control of manufacturing anomalies such as fiber waviness. This is highlighted in the use of pultrusion in thin walled composite columns (Bank et al., 1994). Creative connection and stiffening details to reinforce and increase the structural capacity of pultruded open-web profiles and beams can result in a 65% increase of the ultimate loading capacity in comparison to unstiffened profiles (Mosallam, 1995). Building systems using interlocking components have also been introduced to improve the performance and costs of load-bearing panels (Fig. 3.6). In adhesive bonding, special surface preparations are usually required and are highly dependent on the types of adhesives used. Research in this area has been limited, although some studies have attempted to quantify the stress transfer in bonded pultruded joints (Boyd *et al.*, 2008).

Pultruded composites exhibit a nonlinear response when subjected to multiaxial loadings resulting from the soft response of the polymer matrix and/or interaction of the secondary reinforcement with voids, microcracks and other defects (Haj-Ali and Kilic, 2002). At the structural level, this nonlinearity in the material behavior may be amplified due to geometric and material discontinuities. The practical use of pultruded composites has usually relied on conservative assumptions that usually do not account for the nonlinear response behavior. Micromechanical constitutive models for pultruded composites have successfully been integrated in commercial FE software to capture the material nonlinearity in a general purpose nonlinear analysis (Haj-Ali and El-Hajjar, 2003). The micromechanical models use the *in-situ* fiber and matrix properties to determine the effective nonlinear material response. Typically, simple coupon level testing is used for calibration of the modeling approach. This methodology can be combined with cohesive layer models for fracture prediction in cases where crack growth is of concern (Haj-Ali *et al.*, 2006). Recently, improvements in the modeling of viscoelastic response of pultruded composites can result in greater reliability of composite pultruded structures that can be exposed to seasonal temperature variations while still being exposed to various mechanical loadings (Muddasani *et al.*, 2010; Haj-Ali and Muliana, 2003). The studies have shown how exposure to higher temperatures and stresses accelerate

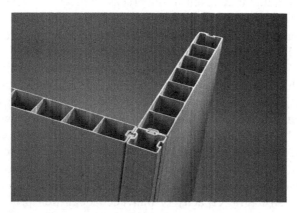

3.6 Interlocking joints for pultruded building panels (courtesy of Strongwell).

the nonlinear deformations of pultruded composites. Time-dependent scaling methods were used to create long-term creep responses from the available short-term creep data. The modeling indicates that the nonlinear responses of the composites are more affected in tension due to enlargement of voids in the matrix. A nonlinear viscoelastic constitutive model implemented in finite element analysis based on the convolution integral equation for orthotropic materials shows overall good predictions with the experimental data available.

3.7 Automated fiber placement

The automated fiber placement procedure offers great advantages especially for manufacturing large composite structures. The automated fiber placement (AFP) machine consists of a computer-controlled robotic arm depositing prepreg strips in order to form a composite layup (Fig. 3.7). The

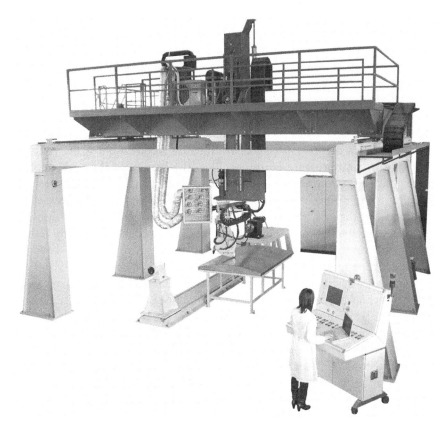

3.7 An advanced automated fiber placement machine (photo courtesy of MIKROSAM A.D. – Macedonia).

computer controller represents an important part of this process, with recent methods using nonlinear optimization techniques based on artificial neural networks to predict material quality as a function of process set points (Heider *et al.*, 2003). The robotic arm performs multiple synchronous functions at various speeds, which requires accurate machine programming to ensure close coordination of cutting, motion, and position control with highly dynamic reversal movements (Marsh, 2011). The prepreg strips (usually made from carbon fiber) or tows are aligned side by side over the mold to form the composite structure.

The AFP head can be designed to accommodate multiple tows to speed up the manufacturing process. The width of the prepreg strips can go down to levels of approximately 3 mm for more accurate contour capture. The strips are laid generally in orientations of 0°, +45°, −45°, and 90°; however other directions are also possible. Strips of approximately 12 mm can be used if the structure is not contoured and if the throughput is of concern. In some processes, a compaction roller is used for further consolidation and compaction of the prepreg plies. Prepreg, which refers to a carbon or glass fabric containing resin in a semi-cured state, offers important advantages in handling while containing the resin content of the final part. The AFP process, if combined with a rotating mandrel, can speed up the depositing process and resembles to some extent the filament winding process. The AFP technology is currently extensively used for aerospace composite structures, such as in the wing and fuselage of the Boeing 787 and the Airbus A350 commercial airliners. Once the part is laid up, it is then transferred into autoclaves or press curing operations for application of the prepreg temperature and pressure curing profiles. The benefit of the AFP process lies in the ability to produce optimized structures with varied stiffness and strength along different parts of the structure. It is also possible to use the technique in regions having curvature such as in the aerospace fuselage sections described above. The AFP process is also characterized by reduction of material scrap when compared to hand layup and also reduced manufacturing times and costs in some cases.

Manufacturing of composite structures can result in various types of defects such as porosity, resin migration, and fiber waviness (El-Hajjar and Petersen, 2011). Figure 3.8 shows an example of the waviness defects present in some carbon fiber composite structures. Such defects may be present in many manufacturing procedures with the addition of other unique defects peculiar to the AFP technology such as gaps, overlaps, or twisted tows. These gaps between the material deposited can have significant effects on notched and unnotched mechanical performance, especially at the lamina level (around 5%) with laminate level effects at up to 13% (Croft *et al.*, 2011). Several finite element analysis studies were also conducted on waviness defects and manufacturing defects unique to the AFP process.

Advanced processing techniques for composite materials

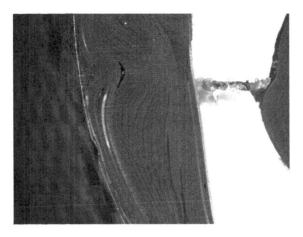

3.8 Fiber waviness defect in a composite structural component.

Sawicki and Minguet (1998) studied the effect of intraply gaps and overlaps on compression strength in composite laminates using experiments and finite element analysis (FEA) modeling. They found that the failure of the specimens was driven by the interaction of in-plane compression and interlaminar shear stresses in the wavy zero-degree plies. Thermosetting prepreg may also cause resin migration when used at room temperature conditions and using cooled prepregs usually takes care of this problem. Resin migration may affect prepreg buckling or fiber waviness in the final structure. Thus, it is crucially important to characterize and monitor manufacturing defects in large composite structure produced using the AFP process.

3.8 Future trends

Use of VARTM in civil engineering applications is likely to grow in the future due to the significant advantages offered by the technology vis-à-vis the traditional hand lay-up method. The power of VARTM mold-filling simulation for optimizing the layout of resin supply and air vent lines as well as for predicting the final part thickness has not been used in civil engineering applications till now. This may change in the near future. Moreover, the VARTM mold-filling simulation may itself become more accurate after incorporating the dual-length-scale effects observed in woven and stitched fiber mats when used with the RTM process (Tan and Pillai, 2010, 2012). In conjunction with the simulation, use of advanced control concepts (such as active control of resin supply lines) and feedback techniques (such as use of SMART weave) will perhaps be tried in civil engineering applications such as repair and retrofit of old bridges. The application of appropriate control strategy will overcome problems associated with variability in preform properties due to handling and will accomplish goals such as avoiding the formation of dry spots and weld lines, decreasing the process time, etc., while keeping the necessary hardware expenditure at a reasonable level.

It is likely that there will be continued expansion of the analysis methods for using pultrusion in infrastructure for bridge deck applications (Yu *et al.*, 2008). The prospects of using rapid prototyping in pultrusion manufacturing can improve the time taken from introducing a part from concept to fabrication. Rapid prototyping techniques, such as selective laser sintering, can be used to produce sample parts from resin powders and also construction of complex parts, which may be used to pultrude interlocking and fastenerless joint designs (Lesko *et al.*, 2008). The development of full-field strain measurement technologies offers a potentially practical approach for structural health monitoring using techniques such as thermoelastic stress analysis (Boyd *et al.*, 2008; El-Hajjar and Haj-Ali, 2003). The development of hybrid processes combining one or more of the methods discussed is something likely to be seen in the future, especially for larger structures containing 3D geometries or preforms. Effects of defects from the various manufacturing processes will continue to be an area of major concern, especially that there is little data on the interaction of defects and long-term durability. This puts more requirements on testing in the absence of reliable analytical methods that can predict these types of behaviors produced in manufacturing and how they affect the actual structural performance of the as-built structure. Speed of manufacturing will continue to be a major driver in all the composite processing discussed, with the AFP machines pushed to speeds higher than 2,000 inches per minute. This will need to be accomplished with more advanced control systems having high precision to enable high volume manufacturing of large and complex composite structures.

3.9 Sources of further information

- *ASM Handbook 21*, Composites. ASM International 2011.
- *Strongwell Design Manual*, Strongwell Corporation, 2011.
- *Fundamentals of Composites Manufacturing*, A. Brent Strong, 2nd edn, Society of Manufacturing Engineers (SME), 2008.

3.10 References

Bank, L. C., Mosallam, A. S. & McCoy, G. T. 1994. Design and performance of connections for pultruded frame structures. *Journal of Reinforced Plastics and Composites*, 13, 199–212.

Bank, L. C., Oliva, M. G., Russell, J. S., Jacobson, D. A., Conachen, M., Nelson, B. & McMonigal, D. 2006. Double-layer prefabricated FRP grids for rapid bridge deck construction: Case study. *Journal of Composites for Construction*, 10, 204–212.

Bender, D., Schuster, J. & Heider, D. 2006. Flow rate control during vacuum-assisted resin transfer molding (VARTM) processing. *Composites Science and Technology*, 66, 2265–2271.

Boyd, S. W., Dulieu-Barton, J. M., Thomsen, O. T. & Gherardi, A. 2008. Development of a finite element model for analysis of pultruded structures using thermoelastic data. *Composites Part A: Applied Science and Manufacturing*, 39, 1311–1321.

Bruschke, M. V. & Advani, S. G. 1989. Numerical simulation of the resin transfer mold filling process. *Proceedings Society of Plastics Engineers 47th Annual Technical Conference*, 1–4 May 1989, New York. Society of Plastics Engineers, Brookfield Center, CT, 1769–1773.

Buchan, P. A. & Chen, J. F. 2007. Blast resistance of FRP composites and polymer strengthened concrete and masonry structures – a state-of-the-art review. *Composites Part B: Engineering*, 38, 509–522.

Cheng, Z.-Q., Howson, W. P. & Williams, F. W. 1997. Modelling of weakly bonded laminated composite plates at large deflections. *International Journal of Solids and Structures*, 34, 3583–3599.

Chiu, H.-T. & Lee, L. J. 2002. Simulation of liquid composite molding based on control-volume finite element method. *Journal of Polymer Engineering*, 22, 155–175.

Croft, K., Lessard, L., Pasini, D., Hojjati, M., Chen, J. & Yousefpour, A. 2011. Experimental study of the effect of automated fiber placement induced defects on performance of composite laminates. *Composites Part A: Applied Science and Manufacturing*, 42, 484–491.

El-Hajjar, R. & Haj-Ali, R. 2003. A quantitative thermoelastic stress analysis method for pultruded composites. *Composites Science and Technology*, 63, 967–978.

El-Hajjar, R. & Haj-Ali, R. 2005. Mode-I fracture toughness testing of thick section FRP composites using the ESE (T) specimen. *Engineering Fracture Mechanics* 72, 631–643.

El-Hajjar, R. F. & Petersen, D. R. 2011. Gaussian function characterization of unnotched tension behavior in a carbon/epoxy composite containing localized fiber waviness. *Composite Structures*, 93, 2400–2408.

Garden, H. N., Quantrill, R. J., Hollaway, L. C., Thorne, A. M. & Parke, G. A. R. 1998. An experimental study of the anchorage length of carbon fibre composite plates used to strengthen reinforced concrete beams. *Construction and Building Materials*, 12, 203–219.

Gauvin, R. & Trochu, F. 1993. Comparison between numerical and experimental results for mold filling in resin transfer molding. *Plastics, Rubber and Composites Processing and Applications*, 19, 151–157.

Govignon, Q., Bickerton, S. & Kelly, P. A. 2010. Simulation of the reinforcement compaction and resin flow during the complete resin infusion process. *Composites Part A: Applied Science and Manufacturing*, 41, 45–57.

Grujicic, M., Chittajallu, K. M. & Walsh, S. 2005. Non-isothermal preform infiltration during the vacuum-assisted resin transfer molding (VARTM) process. *Applied Surface Science*, 245, 51–64.

Gutfinger, C., Broyer, E. & Tadmor, Z. 1974. Analysis of a cross head die with the flow analysis network (FAN) method. *Transport & Road Research Laboratory (Great Britain), TRRL Report*, 20, 339–343.

Haj-Ali, R. & El-Hajjar, R. 2003. Crack propagation analysis of mode-I fracture in pultruded composites using micromechanical constitutive models. *Mechanics of Materials*, 35, 885–902.

Haj-Ali, R. & Kilic, H. 2002. Nonlinear behavior of pultruded FRP composites. *Composites Part B: Engineering*, 33, 173–191.

Haj-Ali, R. M. & Muliana, A. H. 2003. A micromechanical constitutive framework for the nonlinear viscoelastic behavior of pultruded composite materials. *International Journal of Solids and Structures*, 40, 1037–1057.

Haj-Ali, R., El-Hajjar, R. & Muliana, A. 2006. Cohesive fracture modeling of crack growth in thick-section composites. *Engineering Fracture Mechanics*, 73, 2192–2209.

Harlow, F. H. & Welch, J. E. 1965. Numerical calculation of time-dependent viscous incompressible flow of fluid with free surface. *Physics of Fluids*, 8, 2182–2189.

Heider, D. & Gillespie Jr, J. W. 2004. Automated VARTM processing of large-scale composite structures. *Journal of Composite Materials*, 36, 11–17.

Heider, D., Piovoso, M. J. & Gillespie, J. W. 2003. A neural network model-based open-loop optimization for the automated thermoplastic composite tow-placement system. *Composites Part A: Applied Science and Manufacturing*, 34, 791–799.

Hirt, C. W. & Nichols, B. D. 1981. Volume of fluid (VOF) method for the dynamics of free boundaries. *Journal of Computational Physics*, 39, 201–225.

Hsiao, K.-T., Devillard, M. & Advani, S. G. 2004. Simulation based flow distribution network optimization for vacuum assisted resin transfer molding process. *Modeling and Simulation in Materials Science and Engineering*, 12, 175–190.

Jeswani, A. L. & Roux, J. A. 2006. Numerical modelling of design parameters for manufacturing polyester/glass composites by resin injection pultrusion. *Polymers & Polymer Composites*, 14, 651–669.

Jeswani, A. L., Roux, J. A. & Vaughan, J. G. 2009. Multiple injection ports and part thickness impact on wetout of high pull speed resin injection pultrusion. *Journal of Composite Materials*, 43, 1991–2009.

Johnson, R. J. & Pitchumani, R. 2007. Flow control using localized induction heating in a VARTM process. *Composites Science and Technology*, 67, 669–684.

Joshi, S. C., Lam, Y. C. & Liu, X. L. 2000. Mass conservation in numerical simulation of resin flow. *Composites Part A: Applied Science and Manufacturing*, 31, 1061–1068.

Kim, J.-Y., Choi, J.-H. & Joo, J.-W. 2008. A study on robust optimization of layered plates bonding process based on inverse analysis. *Journal of Materials Processing Technology*, 201, 261–266.

Lesko, J. J., Peairs, D. M., Zhou, A., Moffitt, R. D., Mutnuri, B. & Zhang, W. 2008. Rapid prototyping and tooling techniques for pultrusion development. *SAMPE Journal*, 44, 65–68.

Li, J., Zhang, C., Liang, R., Wang, B. & Walsh, S. 2008. Modeling and analysis of thickness gradient and variations in vacuum-assisted resin transfer molding process. *Polymer Composites*, 29, 473–482.

Li, S. & Gauvin, R. 1991. Numerical analysis of the resin flow in resin transfer molding. *Journal of Reinforced Plastics and Composites*, 10, 314–327.

Maji, A. K., Acree, R., Satpathi, D. & Donnelly, K. 1997. Evaluation of pultruded FRP composites for structural applications. *Journal of Materials in Civil Engineering*, 9, 154–158.

Marsh, G. 2011. Automating aerospace composites production with fibre placement. *Reinforced Plastics*, 55, 32–37.

Miracle, D. & Donaldson, S. 2001. *ASM Handbook Composites*. ASM International, Materials Park, OH.

Mohan, R. V., Ngo, N. D. & Tamma, K. K. 1999. On a pure finite-element-based methodology for resin transfer mold filling simulations. *Polymer Engineering and Science*, 39, 26–43.

Mosallam, A. S. 1995. Connection and reinforcement design details for pultruded fiber-reinforced plastic (PFRP) composite structures. *Journal of Reinforced Plastics and Composites*, 14, 752–784.

Muddasani, M., Sawant, S. & Muliana, A. 2010. Thermo-viscoelastic responses of multilayered polymer composites: experimental and numerical studies. *Composite Structures*, 92, 2641–2652.

Nielsen, D. & Pitchumani, R. 2002. Closed-loop flow control in resin transfer molding using real-time numerical process simulations. *Composites Science and Technology*, 62, 283–293.

Oehlers, D. J. 2001. Development of design rules for retrofitting by adhesive bonding or bolting either FRP or steel plates to RC beams or slabs in bridges and buildings. *Composites Part A: Applied Science and Manufacturing*, 32, 1345–1355.

Paciornik, S., Martinho, F. M., De Mauricio, M. H. P. & D'almeida, J. R. M. 2003. Analysis of the mechanical behavior and characterization of pultruded glass fiber-resin matrix composites. *Composites Science and Technology*, 63, 295–304.

Parnas, R. S. 2000. *Liquid Composite Molding*, Hanser Gardner Publications, Cincinnati, OH.

Peled, A. & Mobasher, B. 2006. Properties of fabric-cement composites made by pultrusion. *Materials and Structures*, 39, 787–797.

Rahatekar, S. S., Roux, J. A., Lackey, E. & Vaughan, J. G. 2005. Multiple injection port simulation for resin injection pultrusion. *Polymers & Polymer Composites*, 13, 559–570.

Rider, W. J. & Kothe, D. B. 1998. Reconstructing volume tracking. *Journal of Computational Physics*, 141, 112.

Sawicki, A. & Minguet, J. 1998. The effect of intraply overlaps and gaps upon the compression strength of composite laminates. Collection of Technical Papers – AIAA/ASME/ASCE/SHS/ASC Structures, Structural Dynamics & Materials Conference, pp. 744–754.

Schmidt, F. M., Lafleur, P., Berthet, F. & Devos, P. 1999. Numerical simulation of resin transfer molding using linear boundary element method. *Polymer Composites*, 20, 725–732.

Serrano-Perez, J. C. & Vaidya, U. K. 2005. Modeling and implementation of VARTM for civil engineering applications. *SAMPE Journal*, 41, 20–31.

Sharma, D., McCarty, T. A., Roux, J. A. & Vaughan, J. G. 1998. Pultrusion die pressure response to changes in die inlet geometry. *Polymer Composites*, 19, 180–192.

Simacek, P. & Advani, S. G. 2007. Modeling resin flow and fiber tow saturation induced by distribution media collapse in VARTM. *Composites Science and Technology*, 67, 2757–2769.

Song, Y. S. & Youn, J. R. 2008. Modeling of resin infusion in vacuum assisted resin transfer molding. *Polymer Composites*, 29, 390–395.

Soukane, S. & Trochu, F. 2006. Application of the level set method to the simulation of resin transfer molding. *Composites Science and Technology*, 66, 1067–1080.

Strong, A. B. 2008. *Fundamentals of Composites Manufacturing*, Society of Manufacturing Engineers, Dearborn, MI.

Tadmor, Z., Broyer, E. & Gutfinger, C. 1974. Flow analysis network (FAN) – a method for solving flow problems in polymer processing. *Polymer Engineering & Science*, 14, 660–665.

Tan, H. & Pillai, K. M. 2010. Fast liquid composite molding simulation of unsaturated flow in dual-scale fiber mats using the imbibition characteristics of a fabric-based unit cell. *Polymer Composites*, 31, 1790–1807.

Tan, H. & Pillai, K. M. 2012. Multiscale modeling of unsaturated flow in dual-scale fiber preforms of liquid composite molding III: reactive flows. *Composites Part A: Applied Science and Manufacturing*, 43, 29–44.

Thompson, E. 1986. Use of pseudo-concentrations to follow creeping viscous flows during transient analysis. *International Journal for Numerical Methods in Fluids*, 6, 749–761.

Trochu, F. & Gauvin, R. 1992. Limitations of a boundary-fitted finite difference method for the simulation of the resin transfer molding process. *Journal of Reinforced Plastics and Composites*, 11, 772–786.

Trochu, F., Ruiz, E., Achim, V. & Soukane, S. 2006. Advanced numerical simulation of liquid composite molding for process analysis and optimization. *Composites Part A: Applied Science and Manufacturing*, 37, 890–902.

Uddin, N., Vaidya, U., Shohel, M. & Serrano-Perez, J. C. 2004. Cost-effective bridge girder strengthening using vacuum-assisted resin transfer molding (VARTM). *Advanced Composites Materials*, 13, 214–235.

Walsh, S. M. & Mohan, R. V. 1999. Sensor-based control of flow fronts in vacuum-assisted RTM. *Plastics Engineering*, 55, 29–32.

Young, W.-B. 1994. Three-dimensional nonisothermal mold filling simulations in resin transfer molding. *Polymer Composites*, 15, 118–127.

Yu, W. R., Lee, J. S. & Baek, S. K. 2008. Designing textile lay-up sequence of pultruded composite bridge deck using three-dimensional finite element analysis. *Composites Science and Technology*, 68, 17–26.

Zienkiewicz, O. C. & Godbole, P. N. 1974. Flow of plastic and visco-plastic solids with special reference to extrusion and forming processes. *International Journal for Numerical Methods in Engineering*, 8, 3–16.

4
Vacuum assisted resin transfer molding (VARTM) for external strengthening of structures

N. UDDIN, S. CAUTHEN, L. RAMOS and U. K. VAIDYA, The University of Alabama at Birmingham, USA

DOI: 10.1533/9780857098955.1.77

Abstract: High-quality and expedient repair methods are necessary to address concrete deterioration that can occur in bridge structures. Most infrastructure-related applications of fiber-reinforced plastics (FRPs) use hand layup methods. Hand layup is tedious, labor-intensive and results are sensitive to personnel skill level. An alternative method of FRP application is vacuum assisted resin transfer molding (VARTM). VARTM uses single-sided molding technology to infuse resin over fabrics wrapping large structures, such as bridge girders and columns. There is no research currently available on the interface developed, when VARTM processing is adopted to wrap fibers such as carbon and/ or glass over concrete structures. This chapter investigates the shear and flexural strength gains of a beam by carbon fiber cast on concrete using the VARTM method. The carbon fiber composite was made using Sikadur HEX 103C and low viscosity epoxy resin Sikadur 300. Tests were conducted to determine and document the gains of FRP rehabilitated beams applied by the VARTM method compared to the hand layup method of application. This newly introduced technique to repair and retrofit a simple span I-565 prestressed concrete bridge girder in Huntsville, Alabama, was implemented in the field within two days without any traffic interruption, and the field demonstration of this newly introduced technique to civil infrastructure is presented to the end.

Key words: vacuum assisted resin transfer molding (VARTM), hand layup, flexural fiber-reinforced polymer (FRP), wrapped beams, bonding, repair.

4.1 Introduction

High-quality and expedient repair methods are necessary for enhancing the service life of bridge structures, and simultaneously making them less vulnerable to vandalism and fire. Reinforced concrete bridge girders are the superstructure elements between the abutments and can suffer from surface deterioration problems. Deterioration of concrete can occur as a result of structural cracks, corrosion of reinforcement, and freeze-thaw degradation. Cost-effective methods with potential for field implementation are required to address the issue of repair and strengthening of bridge structures. Numerous researchers have studied the behavior of FRP strengthened concrete beam and the Departments of Transportation (DOTs) across the USA have implemented FRP strengthening techniques by the hand layup process. For example, carbon FRP (CFRP) laminates have already been adopted in many practical cases (i.e., Highway Appia near Rome; Bridge A10062, St. Louis County, Missouri; Bridge A5657, south of Dixon, Missouri) of PC girders accidentally damaged to restore the original flexural strength of the member (see Nanni, 1997; Nanni *et al.*, 2001; Parretti *et al.*, 2003). Mayo *et al.* (1999) applied bonded FRP laminates to strengthen and lift load restriction from a simple span, reinforced concrete slab bridge, Bridge G270, in Missouri for Missouri DOT. Kachlekev *et al.* (2000) upgraded the capacity of the historic Horsetail Falls Bridge for Oregon DOT. Shahrooz and Serpil (2001) examined four 76-year old T-reinforced beams and retrofitted a 45-year-old, three-span reinforced concrete slab bridge for Ohio DOT. Besides these, there have been Seismic upgrading of Bridge Columns for I-57 in Illinois, I-580 in Reno, Nevada and I-80 in Salt Lake City, Utah.

When FRP is chosen for external strengthening, the most common bonding method is hand layup. Hand layup involves the application of toxic and flammable epoxy resin by hand. This method has several shortcomings including the shortage of skilled workers, improper wetting of fabrics and difficulty of maintaining quality control when applying fabric to long bridge girders or high columns. A weak bond can occur when resin is not properly applied between the contact surface of the structure and the FRP. In addition, this method is tedious, expensive, and labor-intensive. Investigation is still needed to truly understand the relationship between the FRP and the contact surface which could include concrete, steel, or other components. Research in both the laboratory and field testing is required to verify assumptions within the areas of bonding and failure.

This chapter will address two key areas associated with bonding of FRP utilizing the VARTM process. First, comparisons and contrasts between the hand layup versus VARTM will be presented. This newly introduced technique to repair and retrofit a simple span I-565 prestressed concrete bridge girder in Huntsville, Alabama, was implemented in the field within two days without any traffic interruption. Much research has been completed in this area and is described in detail in Uddin *et al.* (2004). Additional information on the effectiveness of VARTM in conjunction with its failure mechanisms will be discussed, including load, deflection, and strain for seven specially designed specimens.

4.2 The limitations of hand layup techniques

Prior to discussing VARTM, some disadvantages regarding the hand layup application method will be detailed. In the hand layup method, uneven application of the epoxy agent to the surface of the FRP can occur for many reasons including: inadequate working conditions, faulty equipment, and uneven surfaces. Inadequate working conditions are prevalent, since the FRP is often required in a cramped, confined area. Faulty equipment can be a problem and the surface on which the FRP is spread can be uneven or lumpy due to the underlying surface. The hand layup methods are shown in Figs 4.1 and 4.2.

Faulty equipment can be a factor in the hand layup method, including broken or imperfect consolidation rollers, sporadic spray from spray applicators, etc. Each can lead to either an imperfect 'air ridden' or inconsistent application of the epoxy. Uneven coating of the surface creates areas without epoxy for bonding or areas of deep epoxy which lead to lumps or poor bonding. Additional localized shear stress can develop within these areas causing failure of the bonding agent. Also, when utilizing a spray technique, volatile organic compounds (VOC) can be emitted into the atmosphere.

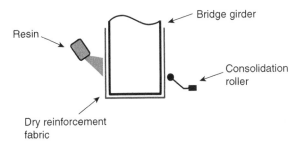

4.1 Hand layup (roller) method.

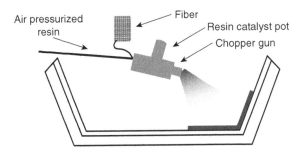

4.2 Hand layup (spray) method.

VARTM, on the other hand, provides for a uniform coverage of epoxy. Multiple layers can be bonded to a single surface in one application. Other advantages include: low process volatile emissions, high fiber-to-resin ratio, good quality, process repeatability, and ability to install while in use.

VARTM begins by preparing the surface, usually through sandblasting, to allow the epoxy resin to evenly distribute and bond. Layers of FRP are then placed on the surface. A layer of release film followed by a layer of distribution mesh is placed on top of the FRP. Inlet runner channel(s), connected to a premixed epoxy resin source, are located and placed for optimal epoxy infusion. Vacuum lines, connected to a vacuum pump, evacuate the epoxy. The vacuum system is engaged and the resin is allowed to flow until the fiber is completely saturated with epoxy. The epoxy is then allowed to cure, under vacuum, before the vacuum bag, release film, and distribution mesh are removed. The VARTM application process is presented in greater detail in Uddin *et al.* (2004). The VARTM method is illustrated in Fig. 4.3.

By infusing the fiber with epoxy resin through a vacuum process, the resin is forced to transfer through the medium from the inlet tube to the vacuum, completely covering and infusing the FRP. Infusion can be visually examined during the process through the vacuum plastic. VARTM application is consistent and verifiable.

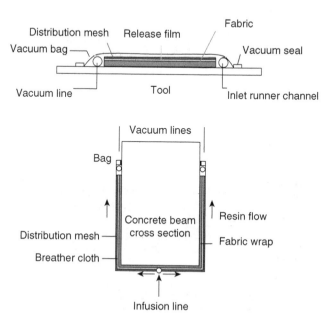

4.3 VARTM method of FRP application.

4.3 Comparing hand layup and vacuum assisted resin transfer molding (VARTM)

Seven samples were experimentally tested in The University of Alabama at Birmingham structural laboratory utilizing the 50-kip load machine. 4 ksi concrete RC beams were designed and built at Sherman Concrete in Birmingham, AL. There are two types of beams: flexural strengthened beams (Fig. 4.4), designed to fail in flexure, and shear strengthened beams (Fig. 4.5), designed to fail in shear.

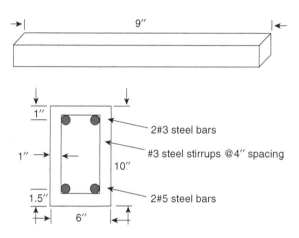

4.4 Flexural strengthened beam.

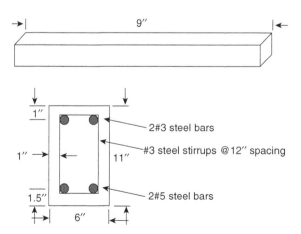

4.5 Shear strengthened beam.

One of each beam type, not wrapped in FRP material, acted as controls for the experiment. Two flexural type beams were wrapped, one using the VARTM method and one using the hand layup method. The flexural CFRP wrap is shown in Fig. 4.6.

Three shear-type beams were wrapped with uniaxial CFRP, two using the VARTM method and one using the hand layup method. The shear CFRP wrap is shown in Fig. 4.7.

Beam testing was performed as follows:

- Strain gages were glued using epoxy at various points along the side and underneath the beam.
- Beams were then placed in the 50-kip load-testing machine.
- Beams were supported and loaded where shown in Fig. 4.8.
- Load was applied to the beam while strain and deflection data were collected.

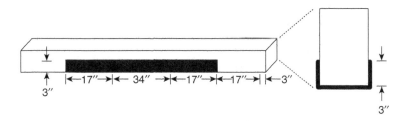

4.6 CFRP wrapping scheme for flexural beam.

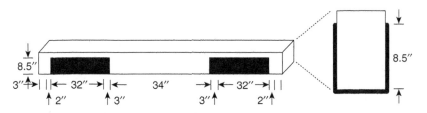

4.7 CFRP wrapping scheme for shear beam.

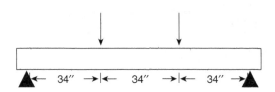

4.8 Loading points.

- Cracks were marked as they appeared and the load at that time denoted at the top of the respective crack. Photographs were taken during loading.
- Results were compared and correlated with respective photographs. Additionally, the failure mode of each beam was determined based on the collected data.

4.4 Analyzing load, strain, deflections, and failure modes

Tests were conducted in order to collect load versus strain and load versus deflection data. Additionally, maximum loads for the respective experimental setup were determined using the appropriate methodology. From this information, comparisons have been drawn concerning the overall strengthening for each possible failure mode. Additionally, conclusions were drawn concerning any changes in the failure mode observed for each set of specimens.

Prior to testing, maximum load conditions for both flexural and shear capacity were calculated for each of the two distinct control beams. Pre-calculating these limiting parameters allowed for better control of forces applied to the beams. It also provided information as to when certain structural phenomena, such as steel yielding and concrete crushing, should occur during the loading of the specimen.

The load capacity for both flexural and shear strengthened beams was calculated. Calculated capacities were used to set limits on testing. Calculations also helped to predict steel yielding and concrete crushing. Values are indicated in Table 4.1. Testing was performed. Data collected per unit time included: load, deflection, and strain. The collected load versus deflection for the control flexural and control shear beams are shown in Figs 4.9 and 4.10, respectively. These values are included in Table 4.1. Beam failure capacities were determined from these figures with additional evidence provided by the strain data.

Based on the theoretical values presented in Table 4.1, the control flexural beam is expected to fail in flexure. This assumption is confirmed in Fig. 4.9(b) by the marked failure crack occurring at mid-span. Also noted are additional hairline cracks occurring at the mid-span of the beam signifying impending failure of the concrete. From Fig. 4.9(c) and information presented in Wang and Salmon (1993), tensile concrete cracking can be identified as occurring when slightly greater than 4,100 lbs is applied. Table 4.1 shows the calculated control flexural beam theoretical load of 17,006 lb. When compared to the experimental value of 16,400 lb, this yields a difference of –3.6%. Through examination of Fig. 4.9(a), it can be said that the flexural reinforcement bar yielded at 16,000 lb at a total deflection of 0.4

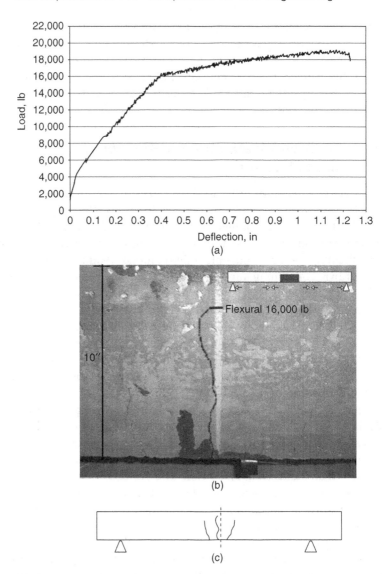

4.9 (a) Control flexural, load versus deflection. (b) Control flexural, failure at mid-span. (c) Control flexural, cracking.

inches. This is indicated by the sudden slope change denoting a stiffness loss. In addition, strain data indicate a major event occurring close to 1,630 pounds. At this point, the strain gage is stretched past its measurement limit. Additional load endured past 16,000 lb resulted in a much larger increase in deflection until failure at 16,400 lb and a deflection of slightly less than

VARTM for external strengthening of structures 85

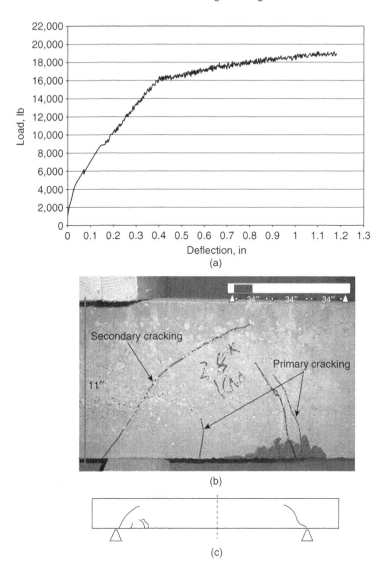

4.10 (a) Control shear, load versus deflection. (b) Control shear, shear failure at ends. (c) Control shear, cracking.

Table 4.1 Comparison of theoretical and experimental results (control)

	Theoretical (lb)[a]				
	Flexural w/o FRP	Flexural with FRP	Shear w/o FRP	Shear w/o FRP	Experimental (lb)
Control flexural	17,006	n/a	47,741	n/a	16,400[b]
Control shear	19,192	n/a	17,660	n/a	18,900[c]

[a] Analysis method from Wang and Salmon (1993).
[b] See Fig. 4.9(a).
[c] See Fig. 4.10(a).

1.2 inches. Based on Fig. 4.9(b), it can be asserted that the failure mode for the control flexural was flexural cracking at mid-span. Failure cracking is detailed in Fig. 4. 9(c).

Stirrups in the shear beam were placed at 12-inch centers intentionally to force a shear failure in the beam. As illustrated in Fig. 4.10(b), a shear failure, noted by the primary and secondary cracks, is prominent. By examining Fig. 4.10(a), it can be said that yielding of the reinforcement steel began at 16,000 lb. At this point the primary cracking, shown in Fig. 4.10(b) began to appear. After reinforcement yielding, the beam continued to load until it reached failure. As noted by the increased slope of the load versus the deflection, post yielding the beam began to deflect at almost twice the rate it did prior to yielding. According to Fig. 4.10(a), it can be asserted that failure of the beam occurred close to 18,900 lbs. This is supported by the secondary shear crack denoted in Fig. 4.10(b). Additional evidence exists in the slope change of the strain data. Based on the calculated theoretical values, the beam was predicted to fail in shear. Failure cracking is detailed in Fig. 4.10(c). Table 4.1 reveals the theoretical flexural failure is almost two times the load as a shear failure. The experimentally determined capacity was 7.0% greater than the theoretical. This could result from the concrete curing to a greater capacity than the cement utilized.

In the following section, the results determined from both the control flexural and control shear will be compared with the results determined for the beam wrapped externally with FRP. The loading capacities and failure modes of each will be discussed.

4.5 Flexural fiber-reinforced polymer (FRP) wrapped beams

The following two beams were wrapped in FRP. Figures 4.11(a)–(c) utilized the hand layup method. Figures 4.12(a)–(c) utilized the VARTM method. The theoretical capacities for each beam were computed using equations provided in ACI 440.2R-02 *Guide for the Design and Construction of Externally Bonded FRP Systems for Strengthening Concrete Structures*. The capacities for flexural with FRP and shear with FRP are shown in Table 4.2. Additionally, flexural without FRP and shear without FRP for each beam were calculated utilizing standard RC equations. These values are also shown in Table 4.2. Experimental values shown in the last column were determined from Fig. 4.11(a) and Fig. 4.12(a) for flexural hand layup and flexural VARTM, respectively. A discussion of these values follows.

As shown in Table 4.2, the shear capacity without FRP reinforcement is 66% greater than the reinforcement provided for flexural. Failure was

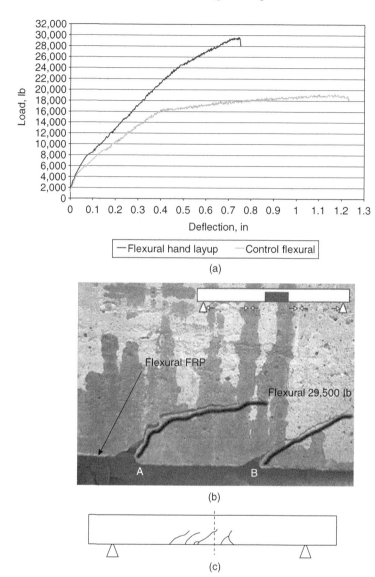

4.11 (a) Flexural hand layup and flexural control, load versus deflection. (b) Flexural hand layup, failure at mid-span side of beam. (c) Flexural hand layup, cracking.

expected in flexural near the mid-span. Strain gage data were collected at the mid-span point. From Fig. 4.11(a), it can be seen that failure of the member occurred at 29,500 lbs. This is reinforced by the strain data, which show a drop occurring at the same load. Utilizing Fig. 4.11(b), one can see

88 Developments in FRP composites for civil engineering

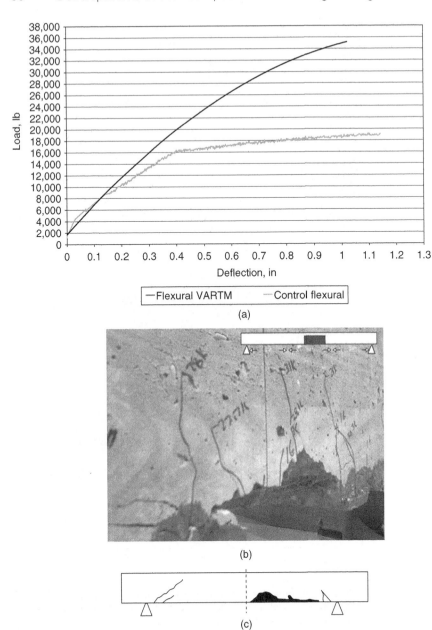

4.12 (a) Flexural VARTM and control flexural, load versus deflection. (b) Flexural VARTM, failure zoom front side. (c) Flexural VARTM, cracking.

Table 4.2 Comparison of theoretical and experimental results (flexural)

	Theoretical (lb)[a]				
	Flexural w/o FRP	Flexural with FRP	Shear w/o FRP	Shear with FRP	Experimental (lb)
Flexural hand layup	17,006	28,745	47,741	n/a	29,500[b]
Flexural VARTM	17,006	28,745	47,741	n/a	35,050[c]

[a] Analysis method from Wang and Salmon (1993).
[b] See Fig. 4.11(a).
[c] See Fig. 4.12(a).

a flexural failure occurring at the mid-span. At the mid-span, cracks appear before failure. Figure 4.11(c) portrays the failure cracking of the beam. The cracks propagated through the core of the beam. Measurement indicates that the cracks reach the bottom steel reinforcement. The failure mode appears to be steel yielding.

In order to understand the contribution of the FRP reinforcement applied by hand layup, load versus deflection data for the control flexural beam was superimposed on Fig. 4.11(a). Prior to the load reaching 4,000 lbs, both experience relatively the same deflection per pound. Then the FRP fiber begins carrying load. Between 4,000 lbs and 16,000 lbs, the slope of the two lines differs only slightly, with the deflection of the control flexural increasing only slightly greater than the flexural hand layup per pound of load. At 16,000 lbs, a considerable difference can be seen between the slopes of the lines. At failure, the flexural hand layup beam is carrying a load 79.9% greater than the control flexural. This value is only slightly less than the calculated theoretical capacity increase of 69.0%. Bonding of both the reinforcement bar to the concrete and the FRP to the concrete in reality will not form a perfect bond; therefore, loss is expected.

It was assumed that the application of the FRP utilizing VARTM would increase the failure strength of the beam. From Fig. 4.12(a), the failure load was determined to be 35,050 lbs. Unlike the control flexural at a deflection of 0.4 inches, the FRP reinforced beam did not experience a change in load over deflection or slope. Therefore, it can be said that at the yield of the steel reinforcement, the external FRP assumed the role of the steel, carrying the load applied to the beam. Data show no significant acceleration of strain in the FRP reinforcement at any point throughout the loading process. In comparison, the VARTM applied FRP allowed for a 113% increase in capacity over the control flexural. A theoretical increase based on current ACI equations yielded a 69.0% increase when utilizing FRP as compared to a standard reinforced beam. In turn, a 21.9% increase was determined between the theoretical failure and the experimental failure. This difference will be discussed further in Section 4.7.

Figure 4.12(b) provides a vivid image of the failure mode. At slightly greater than 31,000 lbs, concrete cracking began to occur on the bottom of the beam at mid-span. Concrete cracking continued to occur until an abrupt failure at 35,050 lbs. At this loading, the concrete was ripped from the internal steel reinforcement bars. The concrete remained attached to the VARTM applied FRP throughout and post failure. Therefore, it can be concluded that bonding between the concrete and FRP was stronger than the forces acting on the bottom of the mid-span of the specimen. The concrete displaced from the bottom of the beam measured 1.5 inches, visible in Fig. 4.12(b). From inspection, the bottom of the reinforcement bar could be seen in the hole produced by the displaced concrete. The mode of failure therefore was different from that experienced by both the control flexural and the flexural hand layup. The failure can best be described as mid-span debonding as explained in Section 4.4 and illustrated in Fig. 4.12(c).

4.6 Shear and flexural fiber-reinforced polymer (FRP) wrapped beams

Once again, maximum load conditions for both flexural and shear capacity were calculated for each of the beams. As before, ACI 440.2R-02 *Guide for the Design and Construction of Externally Bonded FRP Systems for Strengthening Concrete Structures* was utilized. The calculated values are shown in Table 4.3. Experimental values were determined from Figs 4.13(a) and Fig. 4.14(a).

Shear flexural beam testing proceeded in the same fashion as the flexural beam testing. Data were collected, as noted, in Section 4.4. FRP was applied to the shear flexural hand layup beam in a hand application method. From load testing, the capacity of the beam was determined from Fig. 4.13(a) to be 33,650 lbs. Due to the jagged nature of the collected data, a best-fit line has been shown. The steel reinforcement bar yielding began at 23,000 lbs as determined by Fig. 4.13(a) and by the changed strain versus load slope in the data. When compared with the theoretical value, the experimental load was 18% greater. When compared with a similar beam without FRP applied, the experimental load was 90.5% greater. Based on the theoretical data collected on the control shear beam, the theoretical to experimental increase was 79.0%. Such increase should be expected when FRP is externally bonded.

Table 4.3 Comparison of theoretical and experimental results (shear flexural)

	Theoretical (lb)[a]				
	Flexural w/o FRP	Flexural with FRP	Shear w/o FRP	Shear with FRP	Experimental (lb)
Shear flexural hand layup	19,192	32,752	17,660	28,381	33,650[b]
Shear VARTM	19,192	32,752	17,660	28,381	37,780[c]
Shear VARTM[d]	19,192	32,752	17,660	28,381	42,980

[a] Analysis method from Wang and Salmon (1993).
[b] See Fig. 4.13(a).
[c] See Fig. 4.14(a).
[d] Discussion omitted from this chapter due to faulty data collected during experiment.

The failure mode for the shear flexural hand layup beam was shear cracking at the ends. This is illustrated in Fig. 4.13(b). When comparing the failure of the shear flexural hand layup with that of a control shear, only a slight variation of the angle of the shear cracking can be seen. In all respects, both beams exhibit the same failure mode. Failure cracking for the shear flexural hand layup is detailed in Fig. 4.13(c). Based on information defined in Section 4.4, the cracking method can best be described as either a shear failure or mid-span debonding. Closer examination would be needed to identify positively the failure.

The shear flexural VARTM beam was tested following the previously outlined procedures. From Fig. 4.14(a), the failure load was determined to be 37,780 lbs. Figure 4.14(a) allows for identification of steel yielding occurring internally. This is shown by a slight change in slope of the load versus defection and load versus strain fitted curves. Figure 4.14(b) shows shear failure just right of the left support. Based on the values in Table 4.3, a shear failure mode was expected. The next possible failure mode, flexural with FRP, was 15% greater than that of a shear failure with FRP. The experimental capacity was 33.1% greater than the expected theoretical value. With the VARTM applied FRP, the capacity was increased 57.1%, when comparing the theoretical shear value with the experimental shear flexural VARTM1 value in Tables 4.1 and 4.3.

Figures 4.14(b) and (c) reveal the whole and localized failures, respectively. Figure 4.14(b) details the shear cracking just right of the left pin support. The cracking begins as flexural cracks running perpendicular to the bottom of the sample at 29,000 lbs and proceeds expanding until failure at 37,000 lbs. At 37,000 lbs, additional shear cracking forms a 45° angle with

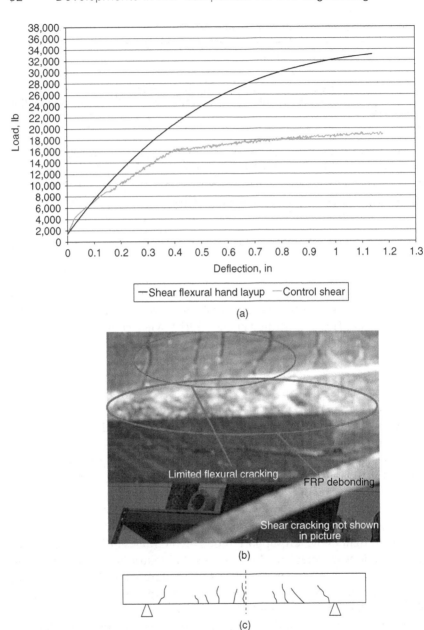

4.13 (a) Shear flexural hand layup and control shear, load versus deflection. (b) Shear flexural hand layup, failure. (c) Shear flexural hand layup, cracking.

the bottom face. When comparing this result with Fig. 4.10(b), it was noted that both beams failed in the same manner. Primary and secondary cracking can be seen in these figures. A localized concrete crushing failure occurred directly above the global failure. No localized failure occurred under the opposite support. Failure cracking is detailed in Fig. 4.14(d).

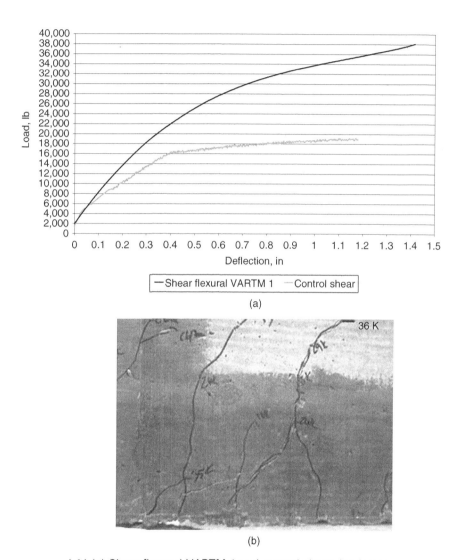

4.14 (a) Shear flexural VARTM 1 and control shear, load versus deflection. (b) Shear flexural VARTM 1, shear failure at end. (c) Shear VARTM 1, crushing at loading point. (d) Shear flexural VARTM, cracking.

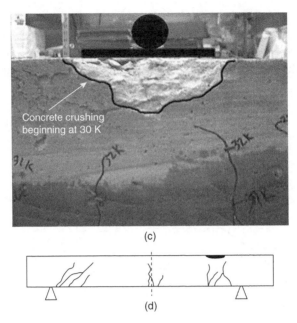

4.14 Continued

In the next section, a discussion will be provided on the benefits of utilizing VARTM application of FRP to concrete beams. Furthermore, information presented in Section 4.6 will be discussed in detail pertaining to the increased capacity that can be attributed to the VARTM method.

4.7 Comparing hand layup and vacuum assisted resin transfer molding (VARTM): results and discussion

Flexural and shear flexural reinforcement utilizing FRP has been proven to increase the capacity of a beam. In this section, application methods of the hand layup and the VARTM method will be explored. Data collected and explained above will be presented in this section to reinforce the claim that VARTM application of FRP provides for a higher capacity for a beam when used in both flexural and shear flexural applications.

Flexural reinforcement included an 85-inch single FRP layer applied evenly spaced from the midpoint of the beam. Strain data sensors were placed to record strain on the underside of the beam at mid-span. The location of the strain gage is the same as for the flexural VARTM beam. When comparing the trend lines in Fig. 4.15, two items should be noted: first, deflection at 29,500 lbs was greater for the hand layup (0.78 inches) than that of the VARTM (0.74 inches); and second, the VARTM member

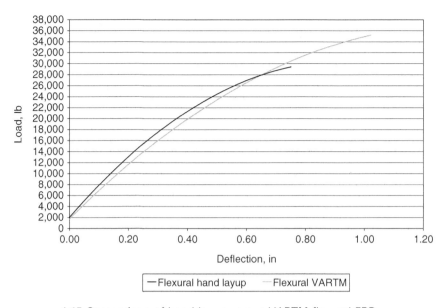

4.15 Comparison of hand layup versus VARTM flexural FRP.

Table 4.4 Flexural FRP and experimental comparison

	Theoretical (lb)		Experimental (lb)
	Flexural with FRP	Shear with FRP	
Flexural hand layup	28,745	n/a	29,500
Flexural VARTM	28,745	n/a	35,050

(35,050 lb) was able to withstand more load than the hand layup beam (29,500 lb), as shown in Table 4.4. Therefore, the VARTM resulted in an 18.8% increase in capacity over hand layup. Both hand layup and VARTM beams exceed theoretical capacity expectations by 2.6% and 21.9%, respectively. The greater increase with the VARTM method is expected due to more even coating of the fiber and surface with the epoxy resin.

For a strain at the mid-span of a flexural beam, distinct changes in the direction of strain within the specimen would not be expected as it would with a shear face gage. For a loading of 29,500 lbs for the hand layup beam, a strain of 5,000 lbs was recorded. At the same load for the VARTM beam, a strain of only 3,800 lbs was recorded. The VARTM beam experienced less stress on the bottom tensile fibers. Therefore, it can be concluded that the utilization of the VARTM method for external flexural reinforcement using FRP does provide a higher loading capacity of approximately 20% over the hand layup application method. Also, there is no notable change in the load versus strain curve between the two application methods. Next, it is important to compare the use of hand layup and VARTM for shear flexural external reinforcement as shown in Table 4.5.

Shear flexural FRP consisted of a 34-inch single external layer of FRP applied evenly spaced from the midpoint of the beam, and two 34-inch single external FRP layers applied 5 inches from the supports. Strain sensors were placed to record strain at both the underside of the beam at the mid-span and at the shear face located just right of the beam south

Table 4.5 Shear flexural FRP and experimental comparison

	Theoretical (lb)		Experimental (lb)
	Flexural with FRP	Shear with FRP	
Shear flexural hand layup	32,752	28,381	33,650
Shear VARTM	32,752	28,381	37,780
Shear VARTM[a]	32,752	28,381	42,980

[a] Discussion omitted from this chapter due to faulty data collected during experiment.

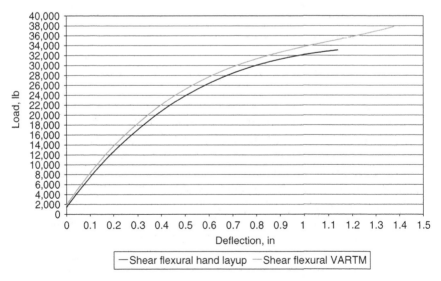

4.16 Comparison of hand layup versus VARTM shear flexural FRP.

end. The location of the gages is the same as for the hand layup beam. When comparing Fig. 4.16, two items should be noted. First, the deformation of the two beams was nearly the same up to failure of the hand layup beam. A much greater difference of deflection was noted for the flexural reinforced specimens. Second, the VARTM beam was able to take both an increased loading and increased deflection. From Table 4.5, the VARTM beam had an increased capacity of nearly 12.3% over the hand layup beam. Hand layup and VARTM flexural beams showed a 2.7% and 15.4% increase over theoretical values. Again, the VARTM beam provided a greater capacity than that expected from ACI 440.2R-02 calculations. This increase was slightly less than that of the flexural reinforced only VARTM member. As previously discussed, an increase in capacity was expected utilizing the VARTM method due to its ability to evenly saturate both the FRP material and the beam surface.

Mid-span strain and shear face strain data were collected. The mid-span strain is similar for both specimens until hand layup failure. For loads from 26,000 lbs to 79,000 lbs, the two plots are equal. This shows that the composite action between both FRP cases is similar. No additional stress is induced by either application method.

For shear face strain, the hand layup beam has one well-defined shift in strain direction, which most likely occurred at the point the FRP took over reinforcement in the shear plane. For the VARTM beam, three well-defined shifts occur: 20,000 lb, 26,000 lb, and 31,000 lb. At failure of the hand layup beam, a difference in strain of 10 exists. However, the direction trend of the strain at failure for both specimens is similar.

Using the VARTM method for external shear and flexural reinforcement provides a higher loading capacity of approximately 15% over the hand layup method. The mid-span strain for both hand layup and VARTM are nearly equal. For shear face strains, two additional shifts in strain direction occur on the hand layup beam. However, at failure, the shear face strain of both specimens appears to be concurrent.

For both analyses, the VARTM method provides a uniform interface between the fiber and the surface of the application leading to an increased capacity for the beam. Therefore, when utilizing VARTM instead of hand layup, an increased capacity can be expected. Modification of the current ACI 440.2R-02 should be made to predict better the overall loading capacity of a steel RC beam with VARTM FRP.

4.8 Case study: I-565 Highway bridge girder

The bridge selected for demonstration of the VARTM strengthening technology is Bridge I-565 Highway located on Route/Bin 52 in Madison County. It is a 27.13 m (89 ft) long solid pre-stressed concrete girder, cast *in situ* slab bridge built in 1991. The bridge has a load restriction of 71.2 kN (16 kips) for HS20 trucks. After visiting the site of some cracked and deteriorated girders, the authors and the Alabama Department of Transportation (ALDOT) selected this bridge girder for evaluation.

Figure 4.17 is the I-565 Highway bridge in Huntsville. At the present condition it was seen that the open spaces in between some girders were sealed. As a result, cracks developed very close to the support due to bending caused by the temperature variation shown in Fig. 4.18. (Uddin *et al.*, 2003). As shown in Fig. 4.19, some preliminary repair implemented by injecting polyurethane foam in the cracks is seen to be ineffective in containing the growth of stress and environment induced cracks. This technology only seals the crack. By sealing the cracks, no infusion of resin can take place inside the cracks and there is no actual improvement in the structural integrity to carry the imposed load coming from the deck. Crack propagation is still in the micro phase. In order to improve this situation, one need not only seal the cracks but also stop the crack propagation.

4.8.1 CFRP design calculation

The cross section of the 1,600.2 mm (63″) bulb-tee girder is shown in Fig. 4.20. Table 4.6 shows the sectional properties of the 1,600.2 mm (63″) bulb-tee girder. Table 4.7 shows the sectional properties of 1,600.2 mm (63″) bulb-tee girder and 203.2 mm (8″) cast *in-situ* slab. Here it was assumed that after developing the flexural cracks, the concrete section is fully cracked and concrete is no longer able to carry any tensile force. To retrofit such a section, it is assumed that the external reinforcement carries all of the nominal capacity of the section. First critical section was calculated as

4.17 I-565 Highway bridge in Huntsville.

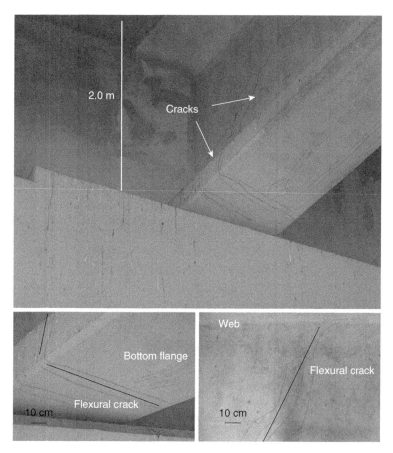

4.18 Initial cracks in analyzed girder, flexural cracks in flange, and shear crack in the web. Black lines illustrate crack direction.

4.19 Some preliminary repair implemented by injecting polyurethane foam in the cracks is seen to be ineffective at I-565 Highway bridge girder in Huntsville.

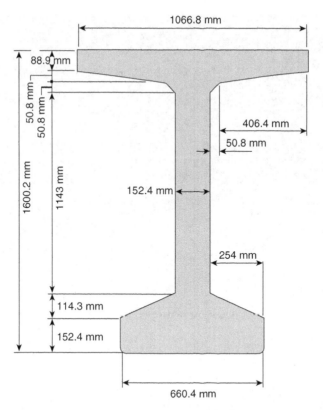

4.20 The cross section of bulb tee girder.

Table 4.6 Bulb-tee girder sectional properties

Girder type	Beam height mm (in.)	Web height mm (in.)	Section area mm² (in².)	Center of gravity mm (in.)	Moment of inertia mm⁴ (in⁴)	Weight N/mm (lb/ft)	Max. span* mm (ft.)
BT-63	1,600.2 (63)	1,143 (45)	459,999.08 (713)	815.848 (32.12)	1.6343×10^{11} (392,638)	10.85 (743)	3,302 (130)

*Based on simple span, HS-25 loading and f'_c = 48.3 MPa (7,000 psi).

h_c, centroid of the pre-stressing + 254 mm (10″) i.e., h_c + 254 mm (h_c + 10″) from the support. At this critical section, both the pre-stressing force and temperature variation was considered for calculating the required number of CFRP layers. All the calculations for the required CFRP are shown

Table 4.7 Sectional properties of 1600.2 mm (63″) bulb-tee girder and 203.2 mm (8″) cast *in-situ* slab

Element	A mm² (in²)	Y mm (in)	AY mm³ (in³)	AY² mm⁴ (in⁴)	I_0 mm⁴ (in⁴)	I_{gc} mm⁴ (in⁴)
Slab	348,644.4 (540.4)	1682.75 (66.25)	586,681,311.1 (35,801.49)	9.87238×10^{11} (2,371,848.72)	791,945,973.6 (1,902.66)	9.88×10^{11} (205,601.04)
Girder	459,999.1 (713)	815.848 (32.12)	375,289,329.4 (22,901.56)	3.06179×10^{11} (735,598.11)	1.63428×10^{11} (392,638)	1.63×10^{11} (547,025.9)
Total	808,643.4 (1,253.4)		961,970,640.5 (58,703.05)	1.29342×10^{11} (3,107,446.83)	1.6422×10^{11} (394,540.7)	1.15×10^{11} (752,626.94)

Table 4.8 Pre-cast AASHTO 1,600.2 mm (63″) bulb-tee girder properties

Pre-cast beam	Composite beam (with transformed slab)
A_c = 459,999.08 mm² (713 in²)	A_{cc} = 808,643.4485 mm² (1,253.4 in²)
y_t = 784.352 mm (30.88 in)	y_{tc} = 14,622.51633 mm (22.67 in)
y_b = 815.848 mm (32.12 in)	y_{bc} = 1,189.610381 mm (46.84 in)
h = 1,600.2 mm (63 in)	h_c = 1,765.3 mm (69.5 in)
I_g = 1.63428×10^{11} mm⁴ (392,638 in⁴)	I_{gc} = 3.13267×10^{11} mm⁴ (752,626.9 in⁴)
Z_t = 208,360,882 mm³ (12,714.96 in³)	Z_{tc} = 544,159,517.9 mm³ (33,206.65 in³)
Z_b = 200,317,062.1 mm³ (12,224.1 in³)	Z_{bc} = 263,335,787 mm³ (16,069.74 in³)
k_t = −435.4727451 mm (−17.14 in)	k_{tc} = −325.651296 mm (−12.82 in)
k_b = 452.959345 mm (17.83 in)	k_{bc} = 672.9288649 mm (26.49 in)
b_v = 1,066.8 mm (42 in)	h_f = 165.1 mm (6.5 in)
b_w = 152.4 mm (6 in)	b_e = 2,438.4 mm (96 in)
	$b = b_{tr}$ = 2,111.716345 mm (83.14 in)

in Tables 4.8–4.11. From Table 4.11, two layers of CFRP were required for flexural reinforcement. In order to avoid shear failure, shear reinforcement was calculated based on shear force developed at the critical section due to live load. According to the design guideline ACI 440, this gave two layers of CFRP for shear strengthening of that section, shown in Table 4.11. These shear reinforcements were placed along with the flexural reinforcement in 0/90/0/90 sequence.

4.8.2 VARTM in I-565 Highway bridge girder: processing detail

After performing strengthening calculations, two flexure layers and two shear layers of unidirectional carbon fiber and epoxy resin were selected for retrofitting the T-bulb girder. Retrofitting strategy and schematic of

Table 4.9 Bending moments and shear forces for a typical beam

Element	Magnitude	Unit	Location
Weight of pre-cast beam =	10.8 (0.74)	N/mm (klf)	
Weight of the slab =	9.5 (0.65)	N/mm (klf)	
Superimposed dead load =	3.5 (0.25)	N/mm (klf)	
HS20 truck P =	71,200 (16)	N (kip)	
Moment due to single loading lane =	1,808,869 (1,330.1)	N.m (kips-ft)	
M_{lane} =	1,808,869 (1,330.1)	N.m (kips-ft)	at mid-span
D.F =	1.45		
I =	0.23		
M_{L+I} =	1,622,619.0 (1,193.1)	N.m (kips-ft)	
First critical section	1,009.65 (3.313)	Mm (ft)	From support
V_{lane} =	274,978.7 (61.8)	N (kips)	
V_{L+I} =	246,665.5 (55.4)	N (kips)	

Table 4.10 Moment and shear

Loading	Moment at midspan N.m (kips-ft)	Shear forces at first critical section N (kips)	Moment at first critical shear section N.m (kips-ft)	Resisting Section
Precast beam	1,004,679.1 (738.7)	136,462.5 (30.7)	143,690.8 (105.7)	Precast
Cast on place slab	879,270.3 (646.5)	119,428.6 (26.8)	125,062.9 (92)	
$M_p =$	1,883,949.4 (1,385.25)	255,891.1 (57.5)	268,753.6 (197.6)	
Asphalt	338,180.9 (248.7)	45,934.1 (10.32)	48,367.2 (35.6)	Composite
Live load + impact	1,622,619.0 (1,193.1)	246,665.5 (55.4)	278,377.3 (204.9)	
$M_c =$	1,960,799.9 (1,441.8)	292,599.6 (65.7)	326,744.4 (240.3)	

Table 4.11 Number of CFRP layers calculation

Area/No. of CFRP layer	Nominal moment capacity	For temperature	At critical section	
			Temperature and pre-stressing effect	Shear force for HS 20 truck
Required area of CFRP A_f mm² (in²)	11,548.4 (17.9)	1,871 (2.9)	4,839 (7.5)	890.3 (1.38)
No. of layers of CFRP	3.4	0.47	1.42	1.72
Remarks (CFRP need)	4	1	2	2

process detail is shown in Figs 4.21–4.23. A crew of four students carried out processing on site for a period of two days on a 1.524 m × 4 m area. The first step in the processing was the surface preparation of the girder using a sandblaster and compressed air; after the surface cleaning, a coat of resin was applied in the border of the part to ensure an adequate seal between the concrete and the vacuum sealant tape as shown in Fig. 4.24.

After the layer of primer was cured, the vacuum sealant tape was placed into the surface of the concrete following the pattern of the part to be bagged. The layup sequence of [0/90/0/90] was adopted and the flexure plies were laid first in order to have the shear plies (laid up to 0.2 m above the neutral axis to ensure anchoring effects) help in containing the edge peel stresses that usually initiate de-bonding of the flexure plies. All plies were held in place against the concrete surface with the aid of 3M spray adhesive as shown in Fig. 4.25. Once all the plies were put in place, a porous bleeder release film made of tightly woven silicone coated polyester was placed on top of the layup so that the distribution mesh would not bond to the fabric (Fig. 4.26).

Then the distribution mesh, made from polyethylene terephthalate (PET), was placed with the infusion and vacuum lines on the top of the plies and on the upper part of the bottom flange. Following the distribution mesh and the inlet and outlet channels, the bag was put in place on top of the whole arrangement as illustrated in Fig. 4.27. De-bulking took place for 20 minutes under vacuum, and the bag was straightened and checked for leaks. The infusion strategy proposed was to place an infusion channel at the middle of the bottom flange, two infusion/vacuum channels at the web and bottom flange, and two final vacuum channels at the top of the part in

VARTM for external strengthening of structures 105

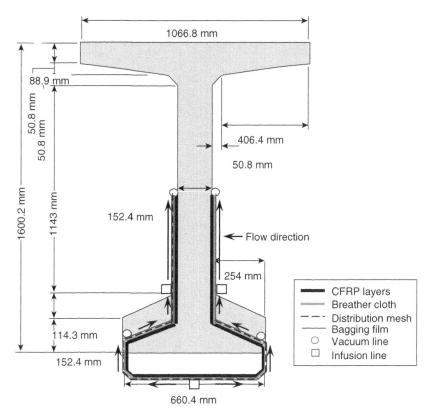

4.21 Schematic of bulb tee section with retrofitting strategy.

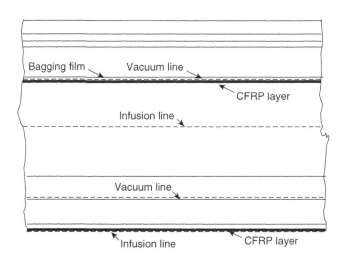

4.22 Longitudinal view of the 1,600.2 mm (63″) bulb tee girder, not to scale.

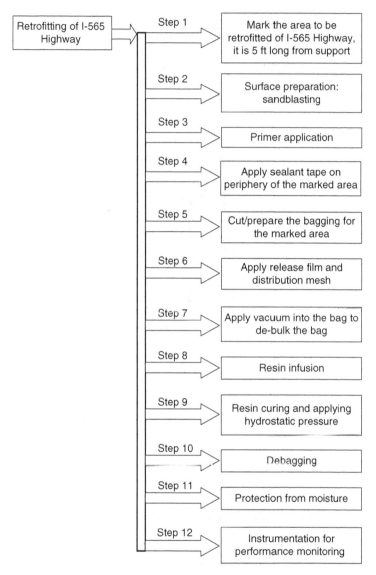

4.23 Schematic of retrofitting scheme of I-565 Highway at a glance.

the web as seen in Fig. 4.28. Finally, infusion was performed and monitored for 90 minutes; the flow front was very homogeneous and symmetrical to each side of the girder showing a linear pattern. The part was kept under vacuum for the following 12 hours to ensure proper curing under 635 mm Hg vacuum pressure. The on-site filling time for the girder was approximately 60 min. After debagging, the part had conformed readily to the shape of

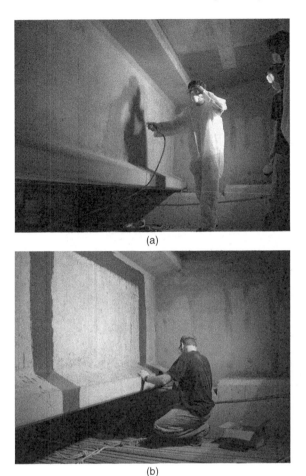

4.24 (a) Surface preparation and (b) pre-priming of vacuum bag sealant tape zone.

the girder; details of the web and beam appearance at debagging are presented in Fig. 4.29. A coat of latex-based paint was applied on the surface of the finished part to act as a seal against any environmental attack such as moisture, dust or particles (Fig. 4.30). A final overall view is shown in Fig. 4.31.

4.8.3 Cost evaluation for field implemenation of VARTM

A generalized analysis of the main factors influencing the economics of the process was performed (Serrano-Perez, 2003) and the total price for the initial stage of the project was estimated using three different processing

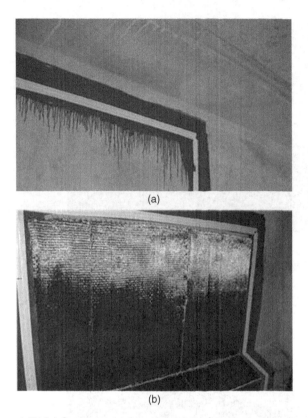

4.25 (a) Sealant tape layout and (b) fabric placement with 3M spray adhesive.

techniques: VARTM, hand layup and steel plate bonding. The total initial cost for the project is found to be lower for the case of steel plate bonding. However, maintenance costs and durability of the bonded steel plates in terms of corrosion and bond line degradation increase the overall cost of using steel. Another important issue to take into consideration is that when a long span is going to be repaired, steel plates need to be welded in place and thereby welding labor, machinery, and consumables increase cost. These combined facts make the use of composite materials more attractive, competitive, and permanent. Hand layup and VARTM are very similar in terms of costing, since the relatively higher tooling cost in VARTM is made equivalent in the hand layup process by the increased labor rates and longer processing times. The final product advantages of VARTM over hand layup (e.g., higher quality product, which might help reduce the number of layers needed and improved bond and durability) discussed in detail by Serrano-Perez (2003) make it an attractive and cost-effective processing route for the repair of large reinforced concrete structures.

VARTM for external strengthening of structures 109

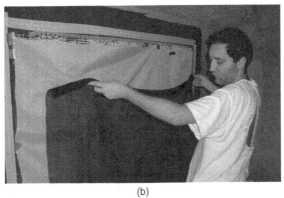

4.26 (a) Release film and (b) distribution mesh placement over placed fabric.

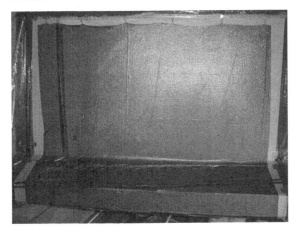

4.27 Bagging and de-bulking of layup.

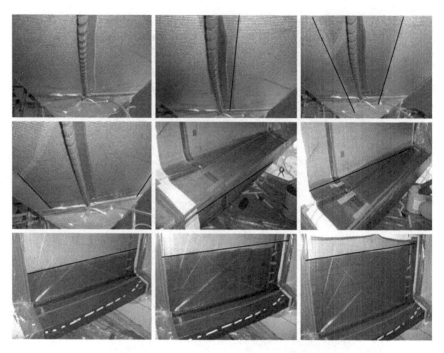

4.28 Infusion progression and flow front throughout flange and web. Straight line indicates flow front, dark areas indicate wet area, and light grey represents dry areas.

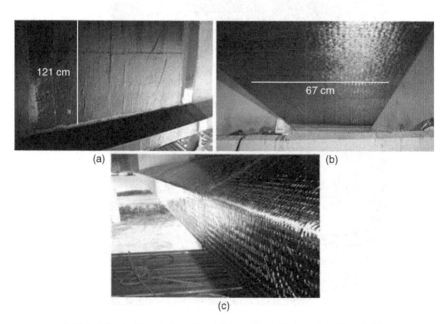

4.29 (a) Part after debagging: (b) bottom flange detail and (c) bottom flange side detail.

VARTM for external strengthening of structures   111

4.30 Latex-based coating application.

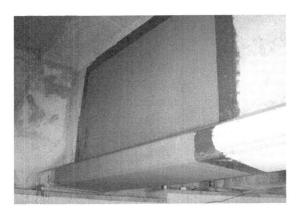

4.31 Final overall view of reinforced section.

4.9 Conclusion and future trends

The focus of this chapter has included an introduction to VARTM and a thorough investigation of several large-scale beam tests with external FRP. Wrapping schemes including both shear and shear flexural were detailed along with application procedures and details. Seven large-scale beam tests were conducted, six of which are discussed in Sections 4.4–4.7 with detailed conclusions for each beam drawn and recorded. Based on the experimental data collected, one can draw three conclusions pertaining to the use of externally bonded FRP. First, the use of externally bonded FRP increases the load capacity of a beam. Second, the use of externally bonded FRP decreases the deflection of the beam as related to the load applied. Third, the use of externally bonded FRP decreases the elongation prior to failure.

Bridges across the United States are experiencing increased loading by modern vehicles. Older bridges were not designed for increased loading. Many bridges in the United States are becoming structurally deficient. Externally bonded FRP is a viable method to rehabilitate these structures. In addition, the VARTM method increases capacity and safety over that of hand layup methods. This has been shown for both shear and shear flexural reinforcement as in Figs 4.11 and 4.13. In these figures, one can see that the use of VARTM in most cases doubles the loading capacity of the beam.

FRP is an advantageous method of repair, which is growing in popularity. VARTM maximizes the potential of FRP by allowing fewer layers of fabric for greater strength gain. VARTM increases the overall capacity of a beam for flexural or shear between 99% and 113% over that of a reinforced beam. VARTM increases the overall capacity of a beam strengthened in flexural or shear by 15–19% more than that of a hand layup FRP. Therefore, VARTM provides a higher degree of strengthening per wrap of fiber. Second, VARTM and hand layup differ only slightly at deflections when compared for similar loading at a defined point. Of the beams compared, one utilized the VARTM method and one utilized the hand layup method of application. Though only a small difference, the VARTM beam possessed an advantage. This deflection difference can be seen visually in Figs 4.15 and 4.16. Third, VARTM increases the overall deflection at failure. When VARTM is considered, it has been proven to provide higher failure strength. At failure, it is noted that a deflection of nearly five-tenths greater than that of a traditional hand wrapped beam is documented. This advantage can be seen in Figs 4.15 and 4.16. A greater deflection prior to failure gives warning of an impending failure, signaling that repairs need to be performed. Therefore, the practice of VARTM for adhering FRP offers a variety of safety benefits including greater strength and ductility of the whole structure (compared to hand layup).

VARTM provides a high bond ratio of the FRP to the concrete. First, with the substrate bonded securely to the concrete, the instances of type 2 failures or debonding-related failures are reduced. Debonding failures are generally in the category of a brittle failure, a mode that must be avoided. VARTM reduces the likelihood of a debonding failure by securely bonding, with little or no voids, the FRP to the concrete. As shown in the failure modes for all samples, none was a result of premature FRP debonding. The conclusion can then be drawn that failure did not occur due to debonding, but by a type 1 failure mode that is detailed for each beam specimen. Second, monetary considerations must be taken into account. Since VARTM provides a larger load yield per wrap over hand layup, it is more efficient than the traditional method.

Additional research is necessary to obtain conclusive evidence as to the effectiveness of VARTM in the long term. For instance, topics such as durability and resin flow should be addressed in detail. It has been shown that VARTM capacity exceeds that of hand layup. Furthermore, it has been proven that capacities predicted utilizing the traditional equations, presented in ACI 440.2R-02, are exceeded utilizing VARTM. It is suggested that factors be applied to adjust for this improved quality and strength. For these factors to be determined, additional large-scale rectangular beam testing will be required. Such testing will provide additional results enabling one to provide a value that statistics show is safe for public usage. Results and statistical assumptions must be approved by the ACI 440.2R-02 committee, before use.

4.10 Acknowledgment

The authors gratefully acknowledge funding and support provided by Alabama Department of Transportation ALDOT research project 930–549, and graduate students M. Shohel, J.C. Serrano-Perez, and Dr. Brian Pillay.

4.11 References

ACI 440.2R-02. *Guide for the Design and Construction of Externally Bonded FRP Systems for Strengthening Concrete Structures*. American Concrete Institute.

Kachlekev, D., Green, B.K., and Barnes, W. (2000), 'Behavior of Concrete Specimens Reinforced with Composite Materials – Laboratory Study', Report No. FHWA-OR-RD-00-10, Oregon Department of Transportation.

Mayo, R., Nanni, A., Gold, W., and Baker, M. (1999), 'Strengthening of Bridge G270 with Externally Bonded Carbon Fiber Reinforced Polymer Reinforcement', *SP-188, American Concrete Institute Proceedings of the Fourth International Symposium on FRP for Concrete Structures (FRPRCS4)*, Baltimore, MD, pp. 429–440.

Nanni, A. (1997), 'Carbon FRP strengthening: new technology becomes mainstream', *Concr. Int.*, 19(6): 19–23.

Nanni, A., Huang, P.C., and Tumialan, J.G. (2001), 'Strengthening of impact damaged bridge girder using FRP laminates', *Proc. 9th Int. Conf. on Structural Faults and Repair*, Engineering Technics, Edinburgh.

Parretti, R., Nanni, A., Cox, J., Jones, C., and Mayo, R. (2003), 'Flexural strengthening of impacted PC girder with FRP composites', *Field Applications of FRP Reinforcement: Case Studies, SP-215*, S. Rizkalla and A. Nanni (eds), American Concrete Institute, Detroit, 249–262.

Serrano-Perez, J.C. (2003), 'Implementation of vacuum assisted resin transfer molding in the repair of reinforced concrete structures', MS thesis, University of Alabama at Birmingham.

Shahrooz, B.M. and Serpil, B. (2001), 'Retrofitting of Existing Reinforced Concrete Bridges with Fiber Reinforced Polymer Composites', Ohio Department of Transportation and the US Department of Transportation, Federal Highway Administration, Report No. UC-CII 01/01.

Uddin, N., Vaidya, U.K., Shohel, M., and Serrano-Perez, J.C. (2003), 'ALDOT Research Project #930-549', Alabama Department of Transportation.

Uddin, N., Vaidya, U., Shohel, M., and Serrano-Perez, J.C. (2004), 'Cost-effective bridge girder strengthening using vacuum-assisted resin transfer molding (VARTM)'. *Advanced Composite Materials*, 13(3–4): 255–281.

Wang, C.-K. and Salmon, C.G. (1993), *Reinforced Concrete Design*, 6th edn. California: Addison-Wesley Longman Publishing Company.

5
Failure modes in structural applications of fiber-reinforced polymer (FRP) composites and their prevention

O. GUNES, Cankaya University, Turkey

DOI: 10.1533/9780857098955.1.115

Abstract: Fiber-reinforced polymer (FRP) composite materials have been increasingly used in civil engineering applications in the past two decades. Their wide ranging use, however, is still not realized due to a few fundamental issues including high material costs, relatively short history of applications and the gaps in the development of established standards. Design safety requires that all possible modes and mechanisms of failure are identified, characterized, and accounted for in the design procedures. This chapter provides a review of the failure types encountered in structural engineering applications of FRP and the preventive methods and strategies that have been developed to eliminate or delay such failures. As part of preventive measures, various non-destructive testing (NDT) and structural health monitoring (SHM) methods used for monitoring FRP applications are discussed with illustrative examples.

Key words: fiber-reinforced polymer (FRP), strengthening, retrofitting, failures, prevention, non-destructive testing (NDT) methods, structural health monitoring (SHM), embedded sensors, fiber optic, Bragg grating.

5.1 Introduction

Fiber-reinforced polymer (FRP) composite materials have been increasingly used in civil engineering applications in the past two decades. This relatively new class of materials, offers several mechanical and durability properties superior to the conventional construction materials, which make them the material of choice in many applications despite their relatively high cost. The structural engineering applications range from all composite structural systems and components to the use of FRP as internal or external reinforcement in conventional structures for improved durability and structural performance. Already accounting for the second largest share in FRP composites shipments in North America, the construction market is likely to take the lead as the use of FRPs in structural engineering applications becomes mainstream.

Failure, although an undesired phenomenon, is at the root of structural engineering design. In an effort to design against failures, engineers learn from failures encountered in controlled or in-service environments. Design safety requires that all possible modes and mechanisms of failure are identified, characterized, and accounted for in the design procedures included in standards. As promising as FRPs are for use in structural engineering applications, characterization of the respective failure modes and development of preventive measures is a prerequisite for the development of comprehensive standards that will enable their widespread use.

This chapter provides a review of the failure types encountered in structural engineering applications of FRP and the preventive methods and strategies that have been developed to eliminate or delay such failures. The scope is mostly limited to failure of FRP strengthened concrete and steel structures which represent the most common applications of FRP in structural engineering. As part of preventive measures, various non-destructive testing (NDT) and structural health monitoring (SHM) methods used for monitoring FRP applications are discussed with illustrative examples.

5.2 Failures in structural engineering applications of fiber-reinforced polymer (FRP) composites

FRP materials may exhibit quite sophisticated failure modes based on their highly diverse composition, structure, and geometry. Characterization of failure modes and development of associated failure criteria for various FRP materials is an active field of research around the world (Hinton *et al.*, 2004; Davila *et al.*, 2005; Knops, 2010). The composition and structure of FRP composites are very briefly reviewed below as a basis for the following discussions on their material and failure behavior.

The basic composition of FRP composites is formed by high strength and stiffness fibers embedded in a polymer matrix with distinct interfaces between them (Hull and Clyne, 1996; Peters, 1997; Barbero, 2010). In this form, both fibers and matrix retain their physical and chemical identities, yet they produce a combination of properties that cannot be achieved with either of the constituents alone. The fibers serve as the principal load-carrying members. The surrounding matrix keeps the fibers in a desired location and orientation, acts as a load transfer medium between them, and protects them from environmental effects. FRP composites can be produced in various types depending on the volume fraction, length, orientation, and type of fibers in the polymer matrix. A laminate is the most common form of composites for structural applications, fabricated by stacking a number of thin layers of unidirectional laminae. Maximum strength and stiffness properties are achieved in the fiber axis direction when all the fibers are unidirectional. This arrangement is highly anisotropic and is suited for applications where the laminate will be subjected to tension in the fiber direction only. To obtain more isotropic properties, alternate layers

Failure modes in structural applications of FRP composites 117

of fibers may vary between 0 and 90°, resulting in less directionality, but at the expense of decreased properties in the absolute fiber direction. Alternatively, woven fabrics can be used with various weave patterns based on the specific application. Biaxially woven fabrics provide good strength in the fiber directions and allow fast composites application through wet layup. However, they are generally less effective in strength compared to laminates due to their three-dimensional structure. Figure 5.1 shows various types of commercially available carbon FRP (CFRP) and glass FRP (GFRP) composite systems used in structural strengthening applications (Gunes, 2004).

Failure behavior of a unidirectional FRP laminate under loading in different directions provides an informative illustration of the complex failure modes in FRP composite materials in general. Figure 5.2(a) shows the stress–strain behavior of unidirectional carbon and glass FRP plates and sheets in comparison with that of reinforcing steel. As can be seen from the figure, a linearly elastic stress–strain behavior followed by an undesirably brittle failure is common to all presented types. Figure 5.2(b) conceptually illustrates the variation in the failure behavior of unidirectional FRP composites as the direction of the loading changes (Peters, 1997). The strength properties, shown in normalized form with respect to the tensile strength, rapidly deteriorate as the loading deviates from the fiber axis and reach very low values for angles more than 20°. The failure modes are fiber rupture and buckling in tension and compression, respectively, for loading

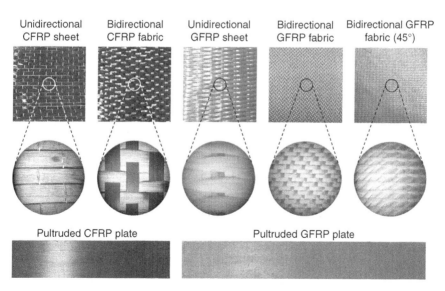

5.1 Various commercially available FRP systems used in structural strengthening.

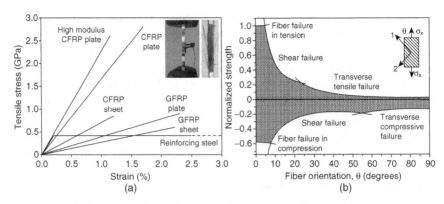

5.2 Tensile stress–strain diagrams of selected unidirectional FRP systems (a) and conceptual failure behavior at different angles to fiber axis (b).

close to the fiber axis, shear failure for intermediate angles, and transverse tensile and compressive failures for large deviations from the fiber axis. Analysis of deformation and failure for multi-directional laminates requires use of more involved analysis methods and associated failure criteria (Hull and Clyne, 1996; Peters, 1997; Ochoa and Reddy, 2010).

Despite several favorable characteristics of FRP composites, their brittle failure behavior described in Fig. 5.2 is a major concern for structural engineers. Ductility, defined – at different scales – as the ability to undergo inelastic deformation before failure, is a very important safeguard against failures in structural engineering. Ductility not only results in warning before ultimate failure, but also reduces the dynamic load demand through increased energy dissipation and damage. A measure of ductility is the ratio of inelastic and elastic deformation called the ductility ratio (μ). For structural and reinforcing steel, the ductility ratio is typically greater than one hundred ($\mu_s > 100$) and concrete has a much lower ductility ratio of around two ($\mu_c \approx 2$). At the structural level, a ductility ratio (calculated by the ratio of inelastic and elastic drift) $\mu_\Delta = 4$–6 is generally needed for satisfactory structural performance. Hence, a fundamental concern regarding use of FRPs in structural applications is their ductility ratio of $\mu \approx 1$, even more brittle than concrete, which at first sight does not promise a favorable contribution to the system ductility. A partially compensating property of FRP composites is their typically much higher ultimate strain ($\varepsilon_{uf} \approx 0.012$–$0.023$) compared to that of concrete ($\varepsilon_{uc} \approx 0.003$) and the yield strain of reinforcing steel ($\varepsilon_{ys} \approx 0.002$) (Fig. 5.2(a)). Since the ductility of reinforcing steel in a properly designed reinforced concrete (RC) member is never fully realized due to concrete failure in compression, the additional FRP reinforcement, if properly designed and installed, acts as additional reinforcement that contributes to the load capacity and/or ductility depending on the application (Gunes et al., 2013a,b). As proper design requires a thorough knowledge of the failure modes and mechanisms, comprehensive experimental and modeling research is needed for general and application-specific failure types of structural systems that include FRP composites as a basis for development of associated design codes.

5.2.1 Failures in compression members

Use of FRP composites for additional reinforcement and/or confinement in compression members has been a very effective and cost-efficient type of composites application in structural engineering. Its application is much easier and faster than the alternative conventional methods such as reinforced concrete or steel jacketing of columns. In confinement strengthening, FRP composite plate or fabrics are wrapped or fiber strands are wound in the shear reinforcement direction to enhance shear strength and confinement. The wrapped or wound FRP reinforcement confines the concrete to improve the concrete compressive strength as well as the ductility, resulting in improvement of performance in compression. The lateral confining pressure depends on the thickness and orientation of the FRP reinforcement and the corresponding failure stress. Figure 5.3 shows the stress–stain behavior and failure of six axially loaded concrete cylinders wrapped with one or two layers of GFRP sheets with three different fiber orientations (Au and Büyüköztürk, 2005). All FRP-wrapped cylinders displayed an improvement in both load capacity and ductility compared to the plain concrete cylinder. The degree of improvement and the failure

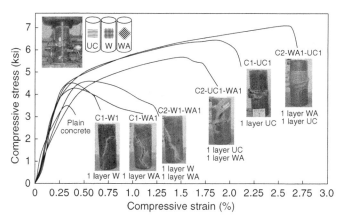

5.3 Failure behavior of GFRP-wrapped concrete cylinders under axial compression.

mode were affected by the number of layers, fiber orientation, and the stack sequence. Similar improvements were obtained in lateral load and deformation capacity of FRP-wrapped columns (Colomb et al., 2008).

The success of column strengthening using FRPs stems from the ability to form closed hoops with multiple layers of FRP reinforcement which provides effective confinement for concrete. The confinement is most effective for circular columns and reduces for rectangular columns, especially for high aspect ratios of the column cross-section (Maalej et al., 2003). Promising results obtained from numerous experimental studies have resulted in several FRP-confined concrete models for use in design practice (Bisby et al., 2005). Although there is little agreement between the developed models, their characteristic contribution to the structural performance is very similar and their use in design generally does not make a significant difference in terms of the structural performance evaluation of the retrofitted system (Gunes et al., 2013a,b).

5.2.2 Failures in tension members

In structural engineering applications, tension is most generally resisted by steel and there are several issues regarding the use of FRPs in conjunction with steel without concrete as a load transfer medium. In a strengthening application, the strengthening material is generally expected to have a similar or higher stiffness compared to the base material of the member being strengthened. In this respect, FRP composites have been a viable choice of materials for use in conjunction with wood, masonry, concrete, and even aluminum (in aviation and aerospace industries) as the typical elastic modulus of FRPs is higher than those of these materials (Triantafillou, 1998). Structural steel, however, has an elastic modulus higher than most commercially available FRP composites used in structural strengthening applications (Fig. 5.2(a)). Hence, as far as elastic deformations are concerned, use of FRPs for strengthening of steel structures does not make much sense from both mechanics and economy perspectives since more FRP reinforcement would be needed than, for instance, additional steel reinforcement to perform the strengthening. Nevertheless, FRP reinforcement for steel becomes effective in the inelastic deformation stage during which steel yields under constant stress while FRPs continue their linearly elastic deformation behavior until failure (Fig. 5.2(a)). Hence, FRP composites are more suitable for improving the ultimate load capacity of steel structures than improving their serviceability. Additional issues include high interfacial stresses in FRP-bonded steel members and potential durability concerns such as galvanic corrosion of the steel substrate in contact with CFRP which is a conductive material. Despite the concerns, FRP strengthening of steel structures has received significant research attention in recent years due to the potential of eliminating welding and bolting in steel members, and ease of installation (Büyüköztürk *et al.*, 2004; Zhao and Zhang, 2007). Continued research in this area focusing on the interface stresses and durability problems together with the recent advances in nano-fiber and nanotube composites (Coleman, *et al.*, 2006) are likely to boost FRP applications to steel structures.

Repair of fatigue-damaged steel members using FRP composites is a specific type of application that is both mechanically and economically well justified. Fatigue cracks can develop in tension members or in the tension regions of flexural members under repeated loading. Repair of a fatigue-damaged member aims at restoring or improving the fatigue resistance of the member and increasing its remaining service life. Commonly used repair techniques involve drilling a hole at the crack tip to eliminate stress concentrations and to control crack growth, or welding and/or bolting of a steel angle or plate over the cracked area. The problems associated with such repairs is that they may inflict further damage to the structure in terms of reducing its load carrying capacity due to drilling of holes, or may introduce further local stress concentrations which can promote further fatigue cracking. Due to the uncertainties in the reliability of repaired members, replacement is generally the preferred action unless replacement cost is very high. Conceptually, composite patches reinforced with high modulus fibers can be expected to restrain opening deformations of cracks occurring in steel components to which the patches are bonded. Thus, the bonded patch reduces the stress intensity factor at the crack tip through bridging the stresses between the cracked plate and composite patch. It reduces the stress field in the vicinity of the crack leading to retardation of crack growth and an improvement in fatigue life.

Figure 5.4 shows the results of an experimental study that involves tension fatigue testing of side-notched steel specimens patched with CFRP laminates having different lengths and widths (Gunes, 2004). As shown in the figure, CFRP patches have resulted in significant improvements in the fatigue life of the specimens which would have failed right away without the patch due to the high amplitude of cyclic loading. The improvement in fatigue life was dependent on the size of the CFRP patch, which lends itself to modeling of the delayed crack propagation and calculation of the remaining fatigue life after patching. Preliminary environmental exposure studies carried out on similar specimens concluded that environmental exposure had insignificant influence on the effectiveness of the repair and that the CFRP patch had prevented rather than accelerated the corrosion of the steel substrate. Continued studies using coupled environmental exposure and fatigue loading should provide more reliable information regarding the potential durability problems in FRP-patched steel members. The method in general was found to be a very effective, cost efficient and easy

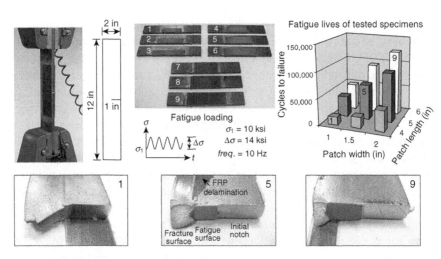

5.4 Tensile fatigue failures of FRP-patched 3/8 in (9.25 mm) thick side-notched steel specimens.

application of FRP composites to repair fatigue cracks in steel structures in order to prolong their service life considerably before replacing the damaged members.

5.2.3 Failures in flexural members

Studies on the use of FRP composites to improve the serviceability and ultimate capacity of flexural members were the earliest FRP applications in structural engineering due to the debonding and durability problems encountered with conventional methods such as strengthening using bonded steel plates. Early experimental studies have shown the high potential and characteristic contribution of FRP reinforcement to the flexural and shear capacity of flexural members and have identified the possible failure modes (Büyüköztürk and Hearing, 1998). Despite the proven potential of the method, FRP strengthening of flexural members raised some concerns regarding the effectiveness, safety, and reliability of the method. Figure 5.5(a) shows the flexural capacity increase in reinforced concrete beams upon strengthening with CFRP plates and sheets. Depending on the FRP reinforcement ratio, the flexural capacity of the member can be dramatically increased, but this increase in capacity is accompanied by a reduction in the deformation capacity, hence the ductility of the member which is an undesired behavior. The figure also indicates that wet layup applications of FRP sheets may not be as effective as intended, especially below a certain thickness, due to premature failures caused by possible stress concentrations. Identified failure modes of FRP strengthened RC and steel flexural members can be classified into three groups as flexural, shear, and

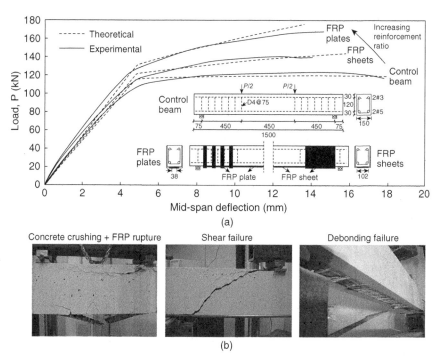

5.5 Load deflection behavior of FRP strengthened beams (a) and possible failure modes (b).

debonding failure modes as illustrated for RC members in Fig. 5.5(b) (Büyüköztürk *et al.*, 2004). The flexural and shear failure modes are very similar to those encountered in reinforced concrete flexural members and the analysis of the strengthened member can be performed using the classical ultimate strength design principles with modifications to account for the additional FRP reinforcement. The debonding failure modes, which cannot be characterized by ultimate strength analysis, were of particular concern due to their premature and brittle nature. Hence, if the strengthening design and application is not performed by properly considering all failure modes, the strengthening may not only be ineffective, but also it may harm the originally ductile member by causing its brittle failure. These concerns, which are not valid for FRP applications to columns, require more attention to the failure behavior of FRP-strengthened flexural members.

5.3 Strategies for failure prevention

Engineers learn from failures to prevent them from happening in real-life applications. Structural engineering applications of FRP composites have been researched and applied in pilot and field applications for more than two decades and have produced a fairly large database of failures. Significant progress has been made in characterization and modeling of failure types and development of associated design procedures, but there is still much room for improvement. For certain applications, such as FRP wrapping of columns for improvement of column shear capacity and ductility, the positive contribution of the FRP reinforcement to the application objectives is guaranteed (Fig. 5.3). The potential risk for this application may be that the strengthening may be less effective than intended, or in the worst case scenario, may be ineffective. For other applications, such as FRP strengthening of beams, insufficient consideration to debonding problems may lead to a performance lower than that of the member before strengthening. Figure 5.6 shows an illustrative example of the possible consequences of FRP strengthening of the beam shown in Fig. 5.5, depending on the shear capacity of the beam and the anchorage conditions of the FRP reinforcement. As can be seen from the figure, FRP strengthening can turn a perfectly ductile beam to a brittle one if the shear capacity and anchorage conditions cannot accommodate the increase in flexural capacity. Figure 5.6 shows the evolution of beam behavior with increasing shear capacity, through internal steel or external FRP shear reinforcement and improving anchorage condition. As important as the wide range of ductility behavior is the position of the ACI 440 provision for limiting FRP strain (ACI 440, 2002) shown with a dashed line on the figure. This provision provides the limiting FRP strain that should not be exceeded in design calculations, even if the FRP

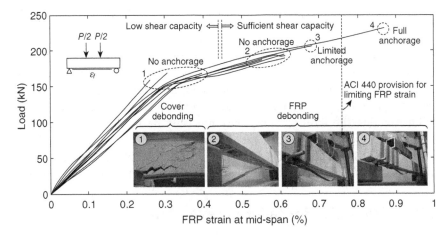

5.6 Failure behavior of FRP-strengthened beams depending on beam shear capacity and bond anchorage conditions.

material can accommodate a higher strain. As can be seen in Fig. 5.6, most failure strains of FRP reinforcement are below the strain limit, which indicates that a design based only on the strain limit could be unconservative by a large margin. Hence, specification of FRP materials in structural engineering applications should note the possible failure modes associated with the specific application and the safety risks involved at the current state of knowledge.

Before examples of methodologies developed or applied for failure prevention in FRP applications in structural engineering are presented, the framework strategies are described below to set the stage.

5.3.1 Safe-life versus fail-safe approaches

Among various failure prevention approaches, it is of interest to note two distinct generic approaches to dealing with damage in structures, primarily those under dynamic loading, in an effort to prevent failures: the safe-life and fail-safe approaches (Suresh, 1998). In the safe-life approach, the typical service load spectra experienced by the structure under service conditions is determined. Based on this information, the components are analyzed or tested in the laboratory under load conditions that are similar to service spectra, and a useful service life is estimated for the component. The estimated service life, modified by a safety factor, is called the safe life for the component. At the end of the safe operational life, the component is automatically retired from service, even if it has considerable residual service life. The safe-life approach depends on achieving a specified life without the development of damage leading to failure (fatigue crack, delamination, debonding etc.), so the emphasis is on the prevention of damage initiation.

The fail-safe concept, on the other hand, is based on the argument that even if an individual member of a large structure fails, there should be sufficient structural integrity in the remaining parts to enable the structure to operate safely until the damage is detected and repaired. Components that have multiple load paths are generally fail-safe because of structural redundancy. In addition, the structure may contain additional safeguards to prevent undesirable levels of damage occurrence. This approach mandates periodic inspection along with the requirement that the damage detection techniques be capable of identifying flaws to enable prompt repairs or replacements.

Selection between safe-life and fail-safe approaches in maintenance of structures, particularly those under dynamic loading, is a problem of economy versus safety. In civil engineering, the fail-safe approach is generally preferred for economic reasons, although there are exceptions. In Canada, for instance, safe-life techniques have been used for bridges at the expense of higher maintenance costs based on the principle that 'it is better to replace a number of bridge components a few years too soon rather than have one bridge replaced too late' (Sweeney, 1990).

The safe-life approach, despite the higher costs involved, is possible for FRP applications in structures under dynamic loading such as bridges. However, for other FRP applications performed to resist design loadings, such as retrofitting a building or even the columns of a bridge against earthquakes, it is difficult, if not impossible, to determine a safe life based on the design event. While it is possible to define a safe life based on durability or cyclic debonding tests or to treat FRP applications as an interim solution during a predefined safe life until a more comprehensive solution is planned and executed, this is usually not even an option due to lack of necessary funds. Hence, the failure prevention practices for FRP applications must be closer to fail-safe approaches that require periodic inspections as part of the failure prevention strategy.

5.3.2 Redundant design against failure

A possible failure prevention strategy for FRP applications in structural engineering is to conservatively provide added redundancy in the design especially for known brittle failure modes even if the design procedure does not call for it. For the debonding failures presented in Fig. 5.6, a debonding model was developed that performed better than the ACI 440 provision for the limiting strain (Gunes et al., 2009). This study concluded that until a generally applicable model is included in the codes, the FRP flexural reinforcement ends should be anchored along a length equal to the effective depth of the beam as an added precaution against brittle debonding failures. A conservative approach in shear resistance of the beam is also important in preventing brittle failures. If the beam requires shear strengthening, most of the configurations shown in Fig. 5.7 also provide additional bond anchorage for the flexural reinforcement and reduce the probability of debonding failures. For beams with sufficient shear resistance, mechanical anchors at the ends of the bonded FRP flexural reinforcement can also help to prevent brittle debonding failures (Gunes, 2004). An alternative strategy is to use thermal curing to gradually reduce stresses at plate ends to avoid stress concentrations. Such an example is shown in Fig. 5.8, where a full-scale prestressed girder was strengthened by bonding prestressed FRP plates using the gradient method that produced a force gradient in the FRP plate over a length of 0.5 m at the plate ends to gradually reduce the force in the plate to zero (Czaderski and Motavalli, 2007). This type of measure is also possible for FRP

Failure modes in structural applications of FRP composites 127

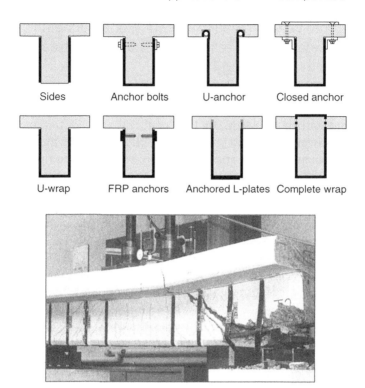

5.7 FRP shear strengthening configurations that mostly also provide bond anchorage for the flexural FRP reinforcement.

5.8 Full-scale prestressed girder strengthened through bonding prestressed FRP plates using the gradient method to eliminate stress concentrations at the plate ends (Photos taken by the author at EMPA, Switzerland).

applications to steel structures since mechanical anchorage by drilling holes would defeat the purpose of using bonded reinforcement. For such measures, inspection and monitoring practices become more important unless a safe-life approach is taken.

5.3.3 Local versus global monitoring approaches: non-destructive testing (NDT) and structural health monitoring (SHM)

Verifying and monitoring the effectiveness of FRP composites applications in structural engineering is an indispensible means of failure prevention at the current state of knowledge on FRP materials and their interaction with the conventional construction materials. Verification of the effectiveness of FRP strengthening/retrofit applications may be possible in the short term through monitoring the behavior of the structure before and after the application. Figure 5.9 shows such a verification for a bridge on I-65 in Louisville, Kentucky, USA (Harik and Peiris, 2012). Cracks that developed close to the supports of some of the girders in the elevated spans of the bridge were repaired using bonded CFRP sheets. Monitoring of both relative horizontal and vertical movements between the girders and the supporting piers was carried out through LVDTs mounted on the girders. The figure shows the girders as well as the measured relative horizontal movement before and after the retrofit, which clearly verifies the effectiveness of the repair. It should be noted that this verification was a targeted implementation for a specific application at a known location on the structure

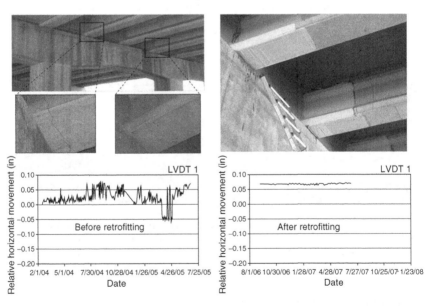

5.9 Short-term monitoring of relative horizontal movement between a girder and the supporting pier of a bridge before and after retrofitting with bonded CFRP sheets for verification of effectiveness (Courtesy of Prof. Issam Harik, University of Kentucky, USA).

with known deformations to monitor. Once it is determined that the application works, verifying the long-term effectiveness must be performed either by continued monitoring or through periodic inspections depending on the resources. Inspection of known locations on the bridge can be performed using various NDT techniques developed and adapted for civil engineering applications (Büyüköztürk, 1998). Several such techniques were also adapted for inspection and evaluation of FRP applications in civil structures (Dong and Ansari, 2011).

Global monitoring approaches within the area of SHM measure the deformation and vibrations at strategic locations on the structure and attempt to relate any changes in the dynamic characteristics of the structure with damage occurrence. Advanced applications in this area employ system identification routines for localization and sizing of damage (Gunes and Gunes, 2012, 2013). Both NDT and SHM approaches are discussed in greater detail in the next section.

5.4 Non-destructive testing (NDT) and structural health monitoring (SHM) for inspection and monitoring

NDT and SHM are generally associated with local and global inspection and monitoring of structures with certain areas of overlap in between. Another distinction can be that the former is generally focused on physical properties of materials and detection of anomalies, while the latter is generally focused on deformation behavior of structures and components for damage detection. Both approaches and the specific methods included therein have their strengths and limitations with respect to applicability, sensitivity, and accuracy depending on the application. The following subsections discuss and provide illustrative examples of SHM and NDT methods that are mostly applied to bonded FRP-strengthened structures.

5.4.1 Stress wave methods

The stress wave or acoustic methods include the traditional sounding techniques such as hammer tapping or chain dragging and more advanced methods such as impact-echo (IE), ultrasonic testing (UT), and acoustic emission (AE). In principle, the stress wave methods are based on elastic wave propagation in solids which takes place in forms of compression (P) waves and shear (S) waves in the solid, and surface waves or Rayleigh (R) waves along the surface. Inhomogeneity in the material causes scattering of sound waves which can be recorded and interpreted to extract information about the condition and elastic properties of materials (Blitz and

Simpson, 1996; Kundu, 2007). Limitations of these methods include low directivity and resolution at low frequencies and the requirement of intimate contact between the test equipment and the object under test.

Impact-echo involves transmission of a transient pulse into concrete by a mechanical impact, and analysis of the reflected waves recorded at the concrete surface in the frequency domain. Due to its low frequency range, this method is more suitable for rapid preliminary survey of areas for locating the anomalies (Tawhed and Gassman, 2002). Images of these anomalies can then be captured using more comprehensive ultrasonic testing methods. Impact-echo is sometimes used in quality control of bond and detection of bond defects in structural engineering applications of FRP composites.

Acoustic emission is a passive condition monitoring technique which allows continuous testing of a structure rather than at regular intervals (Blitz and Simpson, 1996). Acoustic emission refers to the pulses due to the change in the elastic strain energy, which occurs locally in the material and at material interfaces as a result of deformation, debonding, and fracture. Part of this energy propagates through the material which can be detected by highly sensitive transducers placed on the surface of the structure. AE is generally used for detection and monitoring purposes rather than providing an imaging capability (Grosse and Ohtsu, 2010). This technique was used in monitoring damage build-up and failure of FRP-wrapped compression members in the laboratory (Mirmiran et al., 1999), quality control of full-scale hybrid FRP-concrete bridge spans in the manufacturing plant prior to construction (Ramirez et al., 2009), and field monitoring of an FRP strengthened non-prismatic reinforced concrete beam (Carpinteri et al., 2007). While promising results were reported by these studies, use of this technique for long-term monitoring of FRP applications is hindered by the requirement for continuous field monitoring and the difficulty of differentiating between the stress waves from damage and those due to other factors such as vehicular traffic. Use of this method is likely to increase in the future in parallel with the developments in hybrid and all composite structural members as well as effective signal processing algorithms.

Ultrasonics refers to the study and application of ultrasound which is sound of a pitch too high to be detected by the human ear, i.e. of frequencies greater than about 18 kHz (Blitz and Simpson, 1996; Kundu, 2007). The technique involves transmission of ultrasound waves into the material using a transducer in contact with the surface of the object. The scattered signals are then recorded and interpreted. The data obtained from ultrasonic experiments can be used to reconstruct an image of the inclusions and inhomogeneity in the material using tomographic imaging algorithms. Applications of this technique include thickness determination, measurement of elastic modulus, and detection and imaging of cracks, voids, and delaminations.

Ultrasonic pulse-echo involves introduction of a stress pulse into the material at an accessible surface by a transmitter. The pulse propagates into the material and is reflected by cracks, voids, delaminations, or material interfaces. The reflected waves, or echoes, are recorded at the surface and the receiver output is either displayed on an oscilloscope or stored for further processing. There are several methods of examining a test specimen using the pulse-echo technique (Cartz, 1995). The A-scan or A-scope method is a 1D view of the defects in the material. Figure 5.10 shows results obtained from ultrasonic A-scans of a partially debonded FRP-patched steel specimen. The response from the debonded region lacks the reflections from substrate material interfaces which is a clear indication of debonding at the FRP–steel interface (Gunes, 2004). The B-scan or B-scope method involves a series of parallel A-scans and produces a 2D view of the defects in the material. The C-scan or C-scope method involves a series of parallel A-scans performed over a surface. For high frequency ultrasound imaging applications which can be used for NDT of steel or FRP composites, display of B- or C-scans can provide significant information about the interior defects due to the high directivity of the waves (Kundu, 2012). For concrete, however, the presence of coarse aggregate, often exceeding 10 mm in diameter, requires that ultrasonic testing be conducted at relatively low frequencies in order to avoid excessive attenuation caused by scattering (Blitz and Simpson, 1996; Büyüköztürk, 1998). Thus, the ultrasonic beam has virtually no directional characteristics, which makes it difficult to infer the size of the defects.

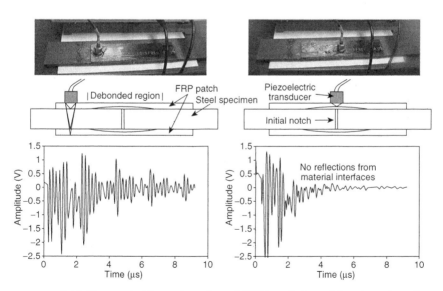

5.10 Ultrasonic (A-scan) testing of FRP-bonded steel for detection of debonding.

5.4.2 Acoustic-laser technique

The acoustic-laser technique is a combined acoustic and laser optical technique which remedies an important limitation of acoustic techniques by enabling non-contact, even remote NDT of bonded FRP applications in structural engineering. The technique is based on the principle that local damage such as debonding and voids in the FRP–concrete or FRP–steel interface vibrate differently than intact regions upon an acoustic excitation (Fig. 5.11(a)) (Büyüköztürk et al., 2012). This difference in the vibration response can be related to the size of damage and the mechanical properties of the FRP. Vibration anomalies remotely measured at the target surface using a laser vibrometer can be used to detect, map, and characterize defects. The technique is effective for bonded thin layered systems, since debonding or voids larger than a certain size respond to the acoustically induced surface waves and display a vibration response depending on the size of debonding. Use of laser vibrometry for sensing the vibration response provides a remote, accurate, and fast method that does not alter the target characteristics. Figure 5.11(b) shows the experimental setup for detecting debonding in GFRP-wrapped concrete cylinders and Fig. 5.11(c) shows the vibration velocities measured at intact and debonded regions as a function of acoustic excitation frequency. As can be seen from the figure, the vibration signatures over the void have higher velocity amplitudes than those of the intact region, the difference being related to the size of the void. Hence, the method has high potential for conducting rapid inspection of bonded FRP applications in structural engineering.

5.4.3 Infrared thermography and digital shearography

Infrared thermography (IRT) and digital shearography (DISH), also known as speckle pattern shearing interferometry (SPSI), are both optical techniques that enable non-contact, full-field, real-time, and rapid non-destructive testing; but these methods are fundamentally different in their damage detection principles. Thermography measures a material's heat transfer response to thermal effects while shearography measures a material's mechanical response to stress (Hung et al., 2009). Both methods are applicable to metals, non-metals, and composite materials for detection of damage and flaws and can be used in a complementary fashion to improve detection capability and accuracy.

Infrared thermography is based on the principle that subsurface anomalies in a material result in localized differences in surface temperature caused by different rates of heat transfer at the defect zones. Thermography senses the emission of thermal radiation from the material surface and produces a visual image from this thermal signal which can be related to

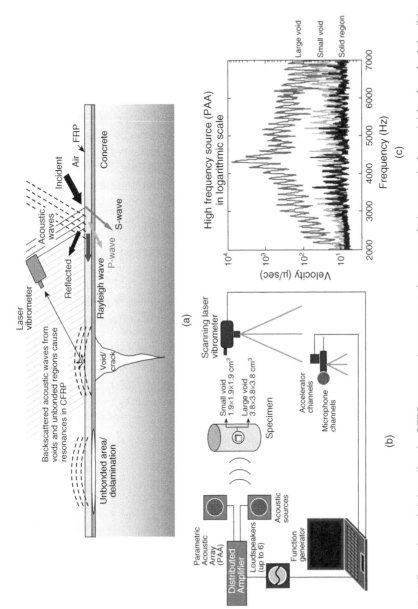

5.11 Detection of debonding in GFRP-wrapped concrete using acoustic-laser technique: (a) basic principle, (b) test setup and the specimen, (c) vibration signatures of solid and debonded regions (Courtesy of Prof. Oral Büyüköztürk, Massachusetts Institute of Technology, USA).

the size of an internal defect. Most infrared thermography applications use a thermographic camera in conjunction with an infrared-sensitive detector which images the heat radiation contrasts. Thermographic imaging may involve active or passive sources such as a flash tube or solar radiation. Active thermography can be further divided into four groups based on the excitation techniques: transient pulse thermography, step heating (long pulse thermography), periodic heating (lock-in) thermography, and thermal mechanical vibration thermography (vibrothermography) (Hung et al., 2009). Use of active thermography improves the applicability and accuracy of the technique, enabling quantitative information regarding subsurface defects. Figure 5.12 shows active thermographic imaging of FRP-bonded concrete for detection of debonding. The obtained images show the potential of the method in detection and sizing of the defects behind single or multiple layers of FRP reinforcement (Cantini et al., 2012).

Shearography is an interferometric imaging technique that directly measures the selected first derivatives of specific surface displacement components (components of surface strains) using coherent laser illumination and a charge-coupled device (CCD) camera for recording (Hung et al., 2009; Lai et al., 2009). The technique has been used for detection of delaminations, residual stresses, vibration modes, and leakage detection and has gained industrial acceptance as a practical and reliable NDT method. Non-destructive testing using digital shearography involves recording of two states of an object, before and after the application of certain stresses using thermal, acoustic, or pressure loading.

Both IRT and DISH were successfully applied to detection of debonding in FRP-bonded concrete in several studies with up to 90% accuracy in determining the sizes of artificial defects (Hung et al., 2009; Lai et al., 2009; Taillade et al., 2011; Cantini et al., 2012). Complementary use of these two

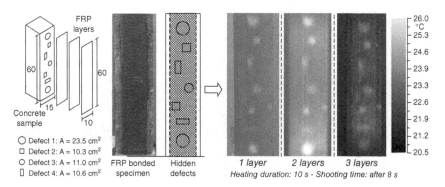

5.12 Infrared thermographic imaging of CFRP-bonded concrete specimen for detection of delamination (Courtesy of Dr. Lorenzo Cantini, Politecnico di Milano, Italy).

methods in field applications has the potential to provide effective and cost-efficient non-destructive evaluation of bonded FRP applications in structural engineering.

5.4.4 Microwave NDT

The principle of microwave NDT is to generate and transmit electromagnetic (EM) short pulses or time harmonic waves through a transmitter antenna towards a target medium and record the scattered signals at the receiver antenna. When the transmitted EM waves encounter an object or another medium with different EM properties, some portion of the transmitted energy is reflected from the boundary and the rest is transferred into the new medium undergoing some refraction depending on the material properties of the new medium and the angle of incidence. Thus, the scattered signals recorded at the receiver contain some information about the target's EM properties, which can be extracted by processing and interpreting the recorded signals (Gunes and Büyüköztürk, 2012). In the microwave NDT method, the ability to image buried inclusions in concrete such as rebars and delaminations requires understanding of concrete as a dielectric material (Rhim and Büyüköztürk, 1998) and application of advanced imaging techniques. The resolution of the image improves with the larger bandwidth of the incident wave (shorter pulse) at the expense of reduced penetration depth due to increased wave attenuation inside the dielectric material. Hence, there is a trade-off between the image resolution and the penetration depth that limits the resolution of the image that can be obtained within a certain depth inside the dielectric material. Conductive materials such as steel rebars do not allow penetration of microwaves, hence, it is not possible to see inside a steel or densely reinforced concrete structure.

Applications of the microwave NDT method to FRP-bonded concrete have revealed mixed results. Feng *et al.* (2002) concluded that debonding at the FRP–concrete interface is difficult to detect using plane waves and used dielectric lenses for focusing EM waves on a point in the bonded surface which led to detection of debonded areas. Yu and Büyüköztürk (2008) performed step-frequency radar measurements using plane wave (far field) excitation for detection of debonding in GFRP-bonded concrete cylinders and obtained mildly successful results. The fundamental problem with debonding detection in FRP-bonded concrete is the small thickness of the debonding. Figure 5.13 shows finite difference-time domain simulation of plane microwave scattering by a GFRP-bonded reinforced concrete target with and without a void (Büyüköztürk *et al.*, 2003). The simulation results clearly indicate the reflection from the void. However, both the measurement conditions and the void size can be quite different in the field.

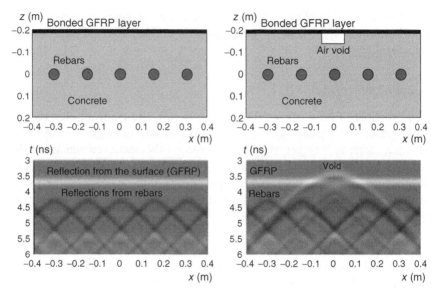

5.13 Simulated microwave scattering response of GFRP-bonded reinforced concrete with and without a subsurface air void to incident plane wave excitation.

Proper detection (range resolution) of debonding requires that the incident wavelength be half the thickness of debonding, which is typically in the order of a millimeter. A simple calculation shows that the frequency of the incident wave must be about an order of magnitude higher than those used in the above-cited studies for proper detection. An additional limitation for microwave NDT of FRP is that conductive composites such as CFRP do not allow sufficient penetration of microwaves for assessment of bond quality. The method has the advantage of being non-contact and rapid, but further research and applications are needed for successful evaluation of FRP applications in structural engineering.

5.4.5 Embedded sensors for SHM

Structural health monitoring of civil engineering structures using embedded sensors for measuring deformation, pressure, or temperature and detecting damage has been a popular interdisciplinary research field in the last two decades. The embedded sensors include, but are not limited to, passive types of sensors such as fiber optic sensors with many variations such as the fiber optic Bragg grating sensors, pre-embedded concrete bar (PECB) sensors, corrosion sensors and active types of sensors such as the piezoelectric wafer active sensors (Lau, 2003; Giurgiutiu, 2008).

The main component of the optical fiber sensor is a small diameter glass fiber that guides light by confining it within regions having different optical indices of refraction. The sensor is basically formed by the optical fiber, a light source, sensing element, and a detector. When subjected to external perturbations such as strain, pressure, or temperature, the sensing element modulates some parameter of the optical system such as the intensity, wavelength, polarization, or phase which changes the characteristics of the optical signal received at the detector. These changes can then be related to the parameter being measured. The sensors can be embedded within the structural material or bonded to the member surface for real-time damage assessment. As the sensor can serve as both a sensing element and a medium for signal transmission, the electronic instrumentation can be located away from the sensor allowing remote monitoring of structures in localized, multiplexed, or distributed arrangements (Méndez and Csipkes, 2012; Lau, 2003).

The optical fiber Bragg grating (FBG) type sensor is one of the most promising and popular type of sensor that is well suited for structural health monitoring of composite materials and their structural applications. The operating principle of the FBG sensor is illustrated in Fig. 5.14(a) (Miller and Méndez, 2011). Any form of broad-spectrum light passing through the grating has a portion of its energy transmitted through and the rest reflected back as a narrow light signal centered at a certain Bragg wavelength. Any change in the characteristics of light due to external perturbations results in a Bragg wavelength shift, which can be related to the parameter being measured. Small size and environmental durability of these sensors have led to their use in many SHM applications, although alkaline attack to the fiber core in embedded applications and damaging of fibers during field installation is still a serious concern to be addressed (Lau, 2003).

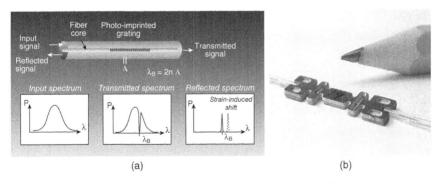

5.14 Principle of operation and transmission and reflection spectra of a fiber Bragg grating (a) and a commercial fiber Bragg grating strain sensor (b) (Courtesy of Dr. Alexis Méndez, Micron Optics, Inc., Atlanta, GA, USA).

138   Developments in FRP composites for civil engineering

Numerous successful laboratory and field applications of fiber optic sensors have been reported in the literature proving their high potential for long-term structural health monitoring of structures (Tennyson et al., 2000, 2001; Labossiere et al., 2000; Lau et al., 2001; Zhao and Ansari, 2002; Lau, 2003; Watkins et al., 2007; Mehrani et al., 2009; Jiang et al., 2010). Figure 5.15 shows one of the longest running SHM applications in structural engineering (Meier, 2012; Meier et al., 2012). The Stork Bridge in Winterthur, Switzerland, is a cable stay bridge constructed in 1996, which has two stay cables made of CFRP. The cable type used consists of 241 wires, each with a diameter of 5 mm. Although the cable type was fatigue tested in the laboratory for 10 million load cycles, several times greater than that expected during the service life of the bridge, under a load three times greater than the permissible load of the bridge, monitoring the long-term performance of this new material is still of interest to bridge engineers. Figure 5.15(a) shows the bridge, the CFRP cable section and the data acquisition board located in the box girder for SHM. The location and arrangement of FBG sensors are shown in Fig. 5.15(b) and the strain measurement data obtained from both FBG and resistive strain gages (RSG) for a period of 14 years is shown in Fig. 5.15(c). The fluctuation of the strains was observed to be synchronous with the temperature changes and the FBG sensors were observed to be more stable than RSG sensors.

Use of embedded sensor technologies with emphasis on fiber optic sensors for SHM is likely to be one of the fastest growing application areas

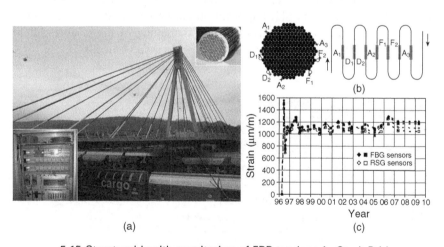

5.15 Structural health monitoring of FRP tendons in Stork Bridge, Switzerland, using embedded sensors: (a) the bridge, CFRP tendons and the data acquisition board inside the box girder; (b) FBG sensor locations and the meander structure; (c) long-term strain measurements over 14 years (Courtesy of Prof. Urs Meier, EMPA, Switzerland).

in civil engineering due to its high impact potential in dealing with infrastructure sustainability problems in a cost-efficient manner.

5.4.6 Combined use of different methods

Combination of the capabilities of more than one NDT and/or SHM method in a complementary fashion in order to increase the accuracy and reliability of characterization or damage assessment is a commonly voiced approach in the NDT/SHM research community (Maierhofer *et al.*, 2010). A RILEM technical committee (TC 207-INR) was recently devoted to this goal. A quick review of the NDT and SHM methods presented in the preceding sections immediately reveals some possible combinations of methods that can be used in a complementary fashion. Infrared thermography and digital shearography methods are an obvious combination that are very similar in their test procedures but fundamentally different in the way they characterize damage. Infrared thermography measures the heat transfer response of materials to thermal excitation while digital shearography measures the mechanical response of materials to stresses. Coherent superposition of information obtained from both methods is likely to produce better characterization of materials and detection of inherent damage.

Another combination of methods that provides complementary information is ultrasonic imaging and microwave (radar) methods. Ultrasonic testing involves propagation of stress waves in materials, governed by the material's mechanical properties, whereas the microwave method involves propagation of electromagnetic waves in materials, governed by the material's electromagnetic (dielectric) properties. Stress waves travel easily in dense and moist materials and are reflected by voids in the material. EM waves travel more easily in dry materials, not significantly affected by voids, but are completely reflected by metals. Ultrasonic testing requires contact with the material whereas microwave antennas enable remote testing. Hence, the test procedures and information content provided by these two materials are truly complementary and their combination leads to better evaluation of materials.

Sometimes, combination of methods in the same category facilitates more efficient evaluation. Impact-echo and ultrasonic testing are both stress wave methods that require measurements in contact with the material. Impact-echo provides faster but less accurate evaluation of large areas. Hence, a preliminary survey of a large area using the impact-echo technique can map areas of low wave velocity which can then be tested using the more accurate ultrasonic testing technique. Replacing impact-echo with the acoustic-laser technique can further speed up the evaluation. Potential damage areas mapped by the acoustic-laser technique can be verified using local ultrasonic evaluation.

Combination of SHM and NDT methods can also lead to effective and cost-efficient evaluation of materials and structures. In condition evaluation or damage detection of structures, vibration-based SHM methods that use system identification techniques can detect and locate potential areas of damage which can then be effectively evaluated using local NDT methods. More research in this area is needed for identification of optimum combinations of methods and strategies for fast and accurate evaluation of structures.

5.5 Future trends

FRP applications in structural engineering have come a long way since their inception about three decades ago. There is a growing consensus on the tremendous potential of these materials in addressing the challenges associated with the aging infrastructure all around the world. It is only a matter of time before the civil engineering community becomes comfortably familiar with the behavior and specific failure modes of these materials and their applications in structural engineering, and have at their disposal comprehensive standards that provide proper procedures for safe and reliable design practices. Compared to the likely scene fifty years from now, the current state of knowledge and practice of FRP applications in structural engineering can be considered as only in its infancy. Even now, exciting applications that include innovative hybrid or all FRP solutions to pressing infrastructure challenges continue to spur, instilling optimism about what is to come in the near future.

A key feature of FRP composites that makes them so promising as engineering materials is the opportunity to tailor the material properties through the control of fiber and matrix combinations and the selection of processing techniques. As fiber diameters are reduced, their stiffness and strength, and hence those of the composite material, are increased due to the highly aligned microstructure of the fibers and reduction of the flaws to very small sizes. The glass and carbon fibers commonly used in structural engineering applications have diameters in the order of 10 microns and produce composite elastic modulus in the range of 20–200 GPa (between those of concrete and steel) and tensile strength in the range of 400–2800 MPa (approximately 1–7 times the yield strength of reinforcing steel) (Gunes, 2004). Recent progress in the development of nanofibers and especially nanotubes with diameters in the order of 100 nm and 1–100 nm, respectively, has achieved such drastic improvements in mechanical properties that significant increases in the elastic and strength properties of FRP composites are likely in the near future (Coleman et al., 2006).

Recent progress in the development of FRP composite materials and embedded sensor technologies are probably the most exciting developments for the future of civil engineering. Marriage of these two technologies paves the way for the development of smart composites and constitutes a foundation for biomimetics in structural design. An intelligent composite structure wired with a network of embedded sensors and actuators that monitors and responds to its own health no longer seems a distant possibility. This popular multi-disciplinary research area is likely to grow rapidly in the near future, leading to more efficient, smart, and sustainable structures.

NDT methods with emphasis on advanced, non-contact, and fast techniques are and will be an indispensable tool for condition evaluation and damage assessment of existing structures. There is much room for further research and development in an effort to improve the capabilities of current NDT techniques and their combined use with complementary methods and strategies for better material characterization and damage assessment.

5.6 Conclusion

Civil engineering is in the midst of a transition era in which FRP composite materials gradually gain acceptance as a new class of materials, superior in many ways to conventional construction materials. Realizing the potential of these materials in rehabilitation and upgrading of existing structures and building more durable and sustainable new structures depends on the characterization of their behavior and failure modes and the development of comprehensive codes and guidelines covering all aspects of their specification and design. Failure behavior of FRP applications to compression, tension, and flexural members are reviewed in this chapter together with various preventive measures. Developments in NDT and SHM methods for effective materials characterization, damage assessment and health monitoring of FRP applications are an essential component of their acceptance and common use by the civil engineering community.

5.7 Acknowledgment

The author gratefully acknowledges the mentorship provided by Prof. Oral Büyüköztürk, who was the principal investigator in many projects presented herein, and the collaboration and support by the former and current members of the Infrastructure Science and Technology research group at MIT. Appreciation is extended to Prof. Urs Meier, Prof. Issam Harik, Dr. Lorenzo Cantini and Dr. Alexis Méndez for their kind contributions with illustrative images from their research. Special thanks go to Prof. Nasim Uddin, the editor, and Ms. Lucy Beg, the publisher's coordinator, for their incredible patience and support.

5.8 Sources of further information

Additional information regarding the contents of this chapter can be obtained from the publications and websites listed below.

5.8.1 Books

Büyüköztürk O, Taşdemir M A, Güneş O and Akkaya Y (eds) (2012), *Nondestructive Testing of Materials and Structures*, RILEM Bookseries Vol. 6, Dordrecht, Springer.

Karbhari V M and Lee L S (eds) (2011), *Service Life Estimation and Extension of Civil Engineering Structures*, Cambridge, Woodhead Publishing.

Barbero E J (2010), *Introduction to Composite Materials Design*, 2nd ed, Boca Raton, FL, CRC Press.

Grosse C U and Ohtsu M (eds) (2010), *Acoustic Emission Testing*, Berlin, Springer.

Knops M (2010), *Analysis of Failure in Fiber Polymer Laminates: The Theory of Alfred Puck*, New York, Springer.

Ochoa O O and Reddy J N (2010), *Finite Element Analysis of Composite Laminates*, Dordrecht, Kluwer Academic Publishers.

Giurgiutiu V (2008), *Structural Health Monitoring with Piezoelectric Wafer Active Sensors*, Burlington, MA, Elsevier Academic Press.

Kundu T (ed.) (2007), *Advanced Ultrasonic Methods for Material and Structure Inspection*, London, ISTA Publishing.

Hinton M, Soden P D and Kaddour A-S (2004), *Failure Criteria in Fibre-Reinforced-Polymer Composites*, London, Elsevier.

Oehlers D and Seracino R. (2004), *Design of FRP and Steel Plated RC Structures: Retrofitting Beams and Slabs for Strength, Stiffness and Ductility*, Oxford, Elsevier.

Teng J G, Chen J F, Smith S T and Lam L (2002), *FRP strengthened RC structures*, Hoboken, NJ, John Wiley and Sons.

Hollaway L C and Head P R (2001), *Advanced Polymer Composites and Polymers in the Civil Infrastructure*, Oxford, Elsevier.

Peters S T (1997), *Handbook of Composites*, 2nd edn, London, Chapman and Hall.

5.8.2 Papers

Dong Y and Ansari F (2011), 'Non-destructive testing and evaluation (NDT/NDE) of civil structures rehabilitated using fiber reinforced polymer (FRP) composites', *Service Life Estimation and Extension of Civil Engineering Structures*, Karbhari V M and Lee L S (eds), Cambridge, Woodhead Publishing, pp. 193–222.

Maierhofer C, Kohl C and Wöstmann J (2010), 'Combining the results of various non-destructive evaluation techniques for reinforced concrete: data fusion', *Non-destructive Evaluation of Reinforced Concrete Structures, Volume 2: Non-destructive Testing Methods*, Maierhofer C, Reinhardt H-W and Dobmann G (eds), Cambridge, Woodhead Publishing, pp. 95–107.

Hollaway L C (2010), 'A review of the present and future utilisation of FRP composites in the civil infrastructure with reference to their important in-service properties', *Construction and Building Materials*, 24(12), 2419–2445.

Pendhari S S, Kant T and Desai Y M (2008), 'Application of polymer composites in civil construction: a general review', *Composite Structures*, 84(2), 114–124.

Zhao X-L and Zhang L (2007), 'State-of-the-art review on FRP strengthened steel structures', *Engineering Structures*, 29(8), 1808–1823.

Coleman J N, Khan U, Blau W J and Gun'ko Y K (2006), 'Small but strong: a review of the mechanical properties of carbon nanotube-polymer composites', *Carbon*, 44(9), 1624–1652.

Büyüköztürk O, Gunes O and Karaca E (2004), 'Progress review on understanding debonding problems in reinforced concrete and steel members strengthened using FRP composites', *Construction and Building Materials*, 18(1), 9–19.

Lau K T (2003), 'Fibre-optic sensors and smart composites for concrete applications', *Magazine of Concrete Research*, 55(1), 19–34.

Tennyson R C, Coroy T, Duck G, Manuelpillai G, Mulvihill P, Cooper D J F, Smith P W E, Mufti A A and Jalali S J (2000), 'Fibre optic sensors in civil engineering structures', *Canadian J of Civil Engineering*, 27(5), 880–889.

5.8.3 Theses

Zadeh A A (2009), Practicality and capability of NDT in evaluating long-term durability of FRP repairs on concrete bridges, M.Sc. Thesis, Department of Civil Engineering, University of British Columbia, Vancouver, Canada.

Rao M R P D (2007), Review of nondestructive evaluation techniques for FRP composite structural components, M.Sc. Thesis, Department of Civil and Environmental Engineering, West Virginia University, Morgantown, WV.

Gunes O (2004), A fracture based approach to understanding debonding in FRP bonded structural members, PhD thesis, Massachusetts Institute of Technology, Cambridge, MA.

5.8.4 Websites

http://www.mdacomposites.org/mda/
http://www.quakewrap.com/
http://www.ndt.net

5.9 References

ACI 440 (2002), Guide for the design and construction of externally bonded FRP systems for strengthening concrete structures, ACI 440.2R-02, American Concrete Institute, Farmington Hills, MI.

Au C and Büyüköztürk O (2005), 'Effect of fiber orientation and ply mix on fiber reinforced polymer-confined concrete', *J Composites for Construction*, 9(5), 397–407.

Barbero E J (2010), *Introduction to Composite Materials Design*, 2nd edn, Boca Raton, FL, CRC Press.

Bisby L A, Dent A J S and Green M F (2005), 'Comparison of confinement models for fiber-reinforced polymer-wrapped concrete', *ACI Structural J*, 102(1), 62–72.

Blitz J and Simpson G (1996), *Ultrasonic Methods of Non-destructive Testing*, London, Chapman & Hall.

Büyüköztürk O (1998), 'Imaging of concrete structures', *NDT & E International*, 31(4), 233–243.

Büyüköztürk O and Hearing B (1998), 'Failure behavior of precracked concrete beams retrofitted with FRP', *J Composites for Construction*, 2(3), 138–144.

Büyüköztürk O, Park J and Au C (2003), 'Non-destructive evaluation of FRP-confined concrete using microwaves', *Proceedings of the International Symposium on Non-Destructive Testing in Civil Engineering (NDT-CE 2003)*, Berlin, Germany, September 13, 2003.

Büyüköztürk O, Gunes O and Karaca E (2004), 'Progress review on understanding debonding problems in reinforced concrete and steel members strengthened using FRP composites', *Construction and Building Materials*, 18(1), 9–19.

Büyüköztürk O, Haupt R, Tuakta C and Chen J (2012), 'Remote detection of debonding in FRP-strengthened concrete structures using acoustic-laser technique', *Nondestructive Testing of Materials and Structures*, Büyüköztürk O, Taşdemir A, Güneş O and Akkaya Y (eds), RILEM Bookseries Vol. 6, Dordrecht, Springer.

Cantini L, Cucchi M, Fava G and Poggi C (2012), 'Damage and defect detection through infrared thermography of fiber composites applications for strengthening of structural elements', *Nondestructive Testing of Materials and Structures*, Büyüköztürk O, Taşdemir A, Güneş O and Akkaya Y (Eds.), RILEM Bookseries Vol. 6, Dordrecht, Springer.

Carpinteri A, Lacidogna G and Paggi M (2007), 'Acoustic emission monitoring and numerical modeling of FRP delamination in RC beams with non-rectangular cross-section', *Materials and Structures*, 40(6), 553–566.

Cartz L (1995), *Nondestructive Testing: Radiography, Ultrasonics, Liquid Penetrant, Magnetic Particle, Eddy Current*, Materials Park, OH, ASM International.

Coleman J N, Khan U, Blau W J and Gun'ko Y K (2006), 'Small but strong: A review of the mechanical properties of carbon nanotube-polymer composites', *Carbon*, 44(9), 1624–1652.

Colomb F, Tobbi H, Ferrier E and Hamelin P (2008), 'Seismic retrofit of reinforced concrete short columns by CFRP materials', *Composite Structures*, 82, 475–487.

Czaderski C and Motavalli M (2007), '40-Year-old full-scale concrete bridge girder strengthened with prestressed CFRP plates anchored using gradient method', *Composites Part B – Engineering*, 38(7–8), 878–886.

Davila C G, Camanho P P and Rose C A (2005), 'Failure criteria for FRP laminates', *J Composite Materials*, 39(4), 323–345.

Dong Y and Ansari F (2011), 'Non-destructive testing and evaluation (NDT/NDE) of civil structures rehabilitated using fiber reinforced polymer (FRP) composites', *Service Life Estimation and Extension of Civil Engineering Structures*, Karbhari V M and Lee L S (eds), Cambridge, Woodhead Publishing, pp. 193–222.

Feng M Q, De Flaviis F and Kim Y J (2002), 'Use of microwaves for damage detection of fiber reinforced polymer-wrapped concrete structures', *J of Engineering Mechanics – ASCE*, 128(2), 172–183.

Giurgiutiu V (2008), *Structural Health Monitoring with Piezoelectric Wafer Active Sensors*, Burlington, MA, Elsevier Academic Press.

Grosse C U and Ohtsu M (eds) (2010), *Acoustic Emission Testing*, Berlin, Springer.

Gunes B and Gunes O (2012), 'Structural health monitoring and damage assessment. Part II: Application of the damage locating vector (DLV) method to the ASCE benchmark structure experimental data', *Int J Phys Sci*, 7(9), 1509–1515.

Gunes B and Gunes O (2013), 'Structural health monitoring and damage assessment. Part I: A critical review of approaches and methods', *Int J Phys Sci*, in press.

Gunes O (2004), A fracture based approach to understanding debonding in FRP bonded structural members, PhD thesis, Massachusetts Institute of Technology, Cambridge, MA.

Gunes O and Büyüköztürk O (2012), 'Simulation-based microwave imaging of plain and reinforced concrete for nondestructive evaluation', *Int J Phys Sci*, 7(3), 383–393.

Gunes O, Büyüköztürk O and Karaca E (2009), 'A fracture-based model for FRP debonding in strengthened beams', *Engineering Fracture Mechanics*, 76(12), 1897–1909.

Gunes O, Lau D, Tuakta C and Büyüköztürk O (2013a), 'Ductility of FRP-concrete systems: Investigations at different length scales', *Construction and Building Materials*, http://dx.doi.org/10.1016/j.comboildmat.2012.10.017.

Gunes O, Tumer R, Gunes B and Faraji S (2013b), 'Performance-based seismic retrofit design for RC frames using FRP composite materials', *Structural Engineering and Mechanics*.

Harik I and Peiris A (2012), 'Case studies of structural health monitoring of bridges', *Nondestructive Testing of Materials and Structures*, Büyüköztürk O, Taşdemir A, Güneş O and Akkaya Y (eds), RILEM Bookseries Vol. 6, Dordrecht, Springer.

Hinton M, Soden P D and Kaddour A-S (2004), *Failure Criteria in Fibre-Reinforced-Polymer Composites: The World-Wide Failure Exercise*, London, Elsevier.

Hull D and Clyne T W (1996), *An Introduction to Composite Materials*, Cambridge, Cambridge University Press.

Hung Y Y, Chen Y S, Ng S P, Liu L, Huang Y H, Luk B L, Ip R W L, Wu C M L and Chung P S (2009), 'Review and comparison of shearography and active thermography for nondestructive evaluation', *Materials Science & Engineering R-Reports*, 64(5–6), 73–112.

Jiang G, Dawood M, Peters K and Rizkalla S (2010), 'Global and local fiber optic sensors for health monitoring of civil engineering infrastructure retrofit with FRP materials', *Structural Health Monitoring*, 9(4), 309–322.

Knops M (2010), *Analysis of Failure in Fiber Polymer Laminates: The Theory of Alfred Puck*, New York, Springer.

Kundu T (ed.) (2007), *Advanced Ultrasonic Methods for Material and Structure Inspection*, London, ISTA Publishing.

Kundu T (2012), 'Guided waves for nondestructive testing – experiment and analysis,' *Proceedings of the International Symposium on Nondestructive Testing of Materials and Structures*, NDTMS-2011, Büyüköztürk O, Taşdemir A, Güneş O and Akkaya Y (eds), 15–18 May 2011, Istanbul Technical University, Istanbul, Turkey, RILEM Bookseries Vol. 6, Springer, Dordrecht, The Netherlands.

Labossiere P, Neale K W, Rochette P, Demers M, Lamothe P, Lapierre P and Desgagné G (2000), 'Fibre reinforced polymer strengthening of the Sainte-Emelie-de-l'Energie bridge: design, instrumentation, and field testing', *Canadian J of Civil Engineering*, 27(5), 916–927.

Lai W L, Kou S C, Poon C S, Tsang W F and Ng S P (2009), 'Characterization of flaws embedded in externally bonded CFRP on concrete beams by infrared thermography and shearography', *J of Nondestructive Evaluation*, 28(1), 27–35.

Lau K T (2003), 'Fibre-optic sensors and smart composites for concrete applications', *Magazine of Concrete Research*, 55(1), 19–34.

Lau K T, Yuan L B, Zhou L M, Wu J S and Woo C H (2001), 'Strain monitoring in FRP laminates and concrete beams using FBG sensors', *Composite Structures*, 51(1), 9–20.

Maalej M, Tanwongsval P and Paramasivam P (2003), 'Modelling of rectangular RC columns strengthened with FRP', *Cement & Concrete Composites*, 25, 263–276.

Maierhofer C, Kohl C and Wöstmann J (2010), 'Combining the results of various non-destructive evaluation techniques for reinforced concrete:data fusion', *Non-Destructive Evaluation of Reinforced Concrete Structures, Volume 2: Non-Destructive Testing Methods*, Maierhofer C, Reinhardt H-W and Dobmann G (eds), Cambridge, Woodhead Publishing, pp. 95–107.

Mehrani E, Ayoub A and Ayoub A (2009), 'Evaluation of fiber optic sensors for remote health monitoring of bridge structures', *Materials and Structures*, 42(2), 183–199.

Meier U (2012), 'Carbon fiber reinforced polymer cables: Why? Why Not? What If?', *Arabian J of Science and Engineering*, 37, 399–411.

Meier U, Brönnimann R, Anderegg P and Terrasi G P (2012), '20 years of experience with structural health monitoring of objects with CFRP components', *Nondestructive Testing of Materials and Structures*, Büyüköztürk O, Taşdemir A, Güneş O and Akkaya Y (eds), RILEM Bookseries Vol. 6, Dordrecht, Springer.

Méndez A and Csipkes A (2012), 'Overview of fiber optic sensors for NDT applications', *Nondestructive Testing of Materials and Structures*, Büyüköztürk O, Taşdemir A, Güneş O and Akkaya Y (eds), RILEM Bookseries Vol. 6, Dordrecht, Springer.

Miller J W and Méndez A (2011), 'Fiber Bragg grating sensors: market overview and new perspectives', in *Fiber Bragg Grating Sensors: Recent Advancements, Industrial Applications and Market Exploitation*, A. Cusano, A. Cutolo and J. Alberts (eds), London, Bentham Press.

Mirmiran A, Shahawy M and El Echary H (1999), 'Acoustic emission monitoring of hybrid FRP-concrete columns', *J of Engineering Mechanics – ASCE*, 125(8), 899–905.

Ochoa O O and Reddy J N (2010), *Finite Element Analysis of Composite Laminates*, Dordrecht, Kluwer Academic Publishers.

Peters S T (1997), *Handbook of Composites*, 2nd edn, London, Chapman and Hall.

Ramirez G, Ziehl P H and Fowler T J (2009), 'Nondestructive evaluation of full-scale FRP bridge beams prior to construction', *Research in Nondestructive Evaluation*, 20(1), 32–50.

Rhim H C and Büyüköztürk O (1998), 'Electromagnetic properties of concrete at microwave frequency range', *ACI Materials J*, 95(3), 262–271.

Suresh S (1998), *Fatigue of Materials*, 2nd edn, Cambridge, Cambridge University Press.

Sweeney R A P (1990), 'Update on Fatigue Issues at Canadian National Railways', Remaining Fatigue Life of Steel Structures, IABSE Workshop, Lausanne.

Taillade F, Quiertant M, Benzarti K and Aubagnac C (2011), 'Shearography and pulsed stimulated infrared thermography applied to a nondestructive evaluation of FRP strengthening systems bonded on concrete structures', *Construction and Building Materials*, 25(2), 568–574.

Tawhed W F and Gassman S L (2002), 'Damage assessment of concrete bridge decks using impact-echo method', *ACI Materials J*, 99(3), 273–281.

Tennyson R C, Coroy T, Duck G, Manuelpillai G, Mulvihill P, Cooper D J F, Smith P W E, Mufti A A and Jalali S J (2000), 'Fibre optic sensors in civil engineering structures', *Canadian J of Civil Engineering*, 27(5), 880–889.

Tennyson R C, Mufti A A, Rizkalla S, Tadros G and Benmokrane B (2001), 'Structural health monitoring of innovative bridges in Canada with fiber optic sensors', *Smart Materials & Structures*, 10(3), 560–573.

Triantafillou T C (1998), 'Composites: A new possibility for the shear strengthening of concrete, masonry and wood', *Composites Science and Technology*, 58(8), 1285–1295.

Watkins S E, Fonda J W and Nanni A (2007), 'Assessment of an instrumented reinforced-concrete bridge with fiber-reinforced-polymer strengthening', *Optical Engineering*, 46(5), 051010.

Yu T-Y and Büyüköztürk O (2008), 'A far-field airborne radar NDT technique for detecting debonding in GFRP-retrofitted concrete structures', *NDT&E Int*, 41(1), 10–24.

Zhao X-L and Zhang L (2007), 'State-of-the-art review on FRP strengthened steel structures', *Engineering Structures*, 29(8), 1808–1823.

Zhao Y and Ansari F (2002), 'Embedded fiber optic sensor for characterization of interface strains in FRP composite', *Sensors and Actuators A – Physical*, 100(2–3), 247–251.

6
Assessing the durability of the interface between fiber-reinforced polymer (FRP) composites and concrete in the rehabilitation of reinforced concrete structures

J. WANG, The University of Alabama, USA

DOI: 10.1533/9780857098955.1.148

Abstract: Strengthening reinforced concrete (RC) members using fiber reinforced polymer (FRP) composites through external bonding has emerged as a viable technique to retrofit/repair deteriorated infrastructure. The interface between the FRP and concrete plays a critical role in this technique. This chapter discusses the analytical and experimental methods used to examine the integrity and long-term durability of this interface. Interface stress models, including the commonly adopted two-parameter elastic foundation model and a novel three-parameter elastic foundation model (3PEF) are first presented, which can be used as general tools to analyze and evaluate the design of the FRP strengthening system. Then two interface fracture models – linear elastic fracture mechanics and cohesive zone model – are established to analyze the potential and full debonding process of the FRP–concrete interface. Under the synergistic effects of the service loads and environments species, the FRP–concrete interface experiences deterioration, which may reduce its long-term durability. A novel experimental method, environment-assisted subcritical debonding testing, is then introduced to evaluate this deteriorating process. The existing small cracks along the FRP–concrete interface can grow slowly even if the mechanical load is lower than the critical value. This slow-crack growth process is known as environment-assisted subcritical cracking. A series of subcritical cracking tests are conducted using a wedge-driven test setup to gain the ability to accurately predict the long-term durability of the FRP–concrete interface.

Key words: interface stress, linear elastic fracture mechanics, cohesive zone models, durability, environment-assisted subcritical debonding.

6.1 Introduction

Fiber-reinforced polymers (FRPs) have emerged as important structural materials in the last three decades. Among their many applications in civil infrastructure, retrofit/rehabilitate reinforced concrete (RC) structures is most popular due to many advantages of FRPs such as high corrosion resistance, high strength-to-weight ratio, and ease of handling. In this application, FRP plates/fabrics/strips are either externally bonded (EB) or near-surface mounted (NSM) on RC structures. To improve the strengthening efficiency, prestress can be applied to FRPs. This technique has evolved into one of the primary techniques to address the deterioration of the civil infrastructure system caused by severe environmental exposure, natural extreme events, excessive use, and intentional attacks. Stresses along the FRP–concrete interface are critical to the success of this technique because the high interface stress concentration can lead to debonding along the FPR–concrete interface, which has been shown to be one of the most common failure modes of the FRP-strengthened RC structures. Extensive studies have been conducted, and various models have been proposed to estimate the stresses along the FRP–concrete interface. A comprehensive review of these studies was given by Smith and Teng (2001) and a recent study by Wang and Zhang (2010). The long-term durability as well as damage and failure mechanisms of the FRP-to-concrete interface in aggressive environments is still an unresolved issue (Grace and Grace, 2005; Porter and Harries, 2007) and has become a major barrier to their wide acceptance in civil infrastructure. This is corroborated by the panelists of a Workshop on Research in FRP Composites in Concrete Construction (Porter and Harries, 2007). In this workshop, the durability studies into FRP strengthening techniques were identified as research priorities.

6.2 Interface stress analysis of the fiber-reinforced polymer (FRP)-to-concrete interface

6.2.1 Fundamental equations

Consider a concrete beam (adherend 1) reinforced by an FRP plate (adherend 2) through a thin adhesive layer (Fig. 6.1). Both the adherends and adhesive are linear elastic and orthotropic materials to account for the most general situation. The adherends are modeled as two beams with thickness h_1 and h_2, respectively, and are connected by an interface of thin adhesive layer with thickness of h_0. A simply supported beam is considered (Fig. 6.1)

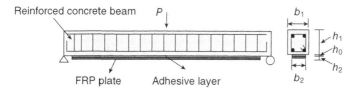

6.1 Simply supported FRP-strengthened RC beam under concentrated load.

in this paper for simplicity. Other boundary and loading conditions can also be solved by this model with very little modification.

Consider a typical infinitesimal isolated body of the bi-layered beam system (Fig. 6.2). The deformation of the FRP plate and concrete beam can be written as:

$$U_i(x, z_i) = u_i(x) + z_i \phi_i(x), W_i(x, z_i) = w_i(x_i), \quad [6.1]$$

where $u_i(x)$, $w_i(x)$, and $\phi_i(x)$ ($i = 1, 2$) are the axial, transverse displacements, and rotation of the neutral axis of beam i, respectively; $U_i(x,z_i)$ and $W_i(x,z_i)$ ($i = 1, 2$) are the axial and transverse displacements of beam i, respectively; subscript $i = 1, 2$, represent the beam 1 (concrete beam) and 2 (FRP plate) in Fig. 6.2, respectively; x and z_i are the local coordinates of beam i with x-axis along the neutral axes of the beam i.

By making use of the constitutive equations of individual layers, we can relate beam forces and displacements of beams as:

$$C_1 \frac{du_1(x)}{dx} = N_1(x), C_2 \frac{du_2(x)}{dx} = N_2(x), \quad [6.2a]$$

$$\frac{dw_1(x)}{dx} + \phi_1(x) = \frac{Q_1(x)}{B_1}, \frac{dw_2(x)}{dx} + \phi_2(x) = \frac{Q_2(x)}{B_2}, \quad [6.2b]$$

$$D_1 \frac{d\phi_1(x)}{dx} = M_1(x), D_2 \frac{d\phi_2(x)}{dx} = M_2(x), \quad [6.2c]$$

where $N_1(x)$ and $N_2(x)$, $Q_1(x)$ and $Q_2(x)$, and $M_1(x)$ and $M_2(x)$ are the internal axial forces transverse shear forces, and bending moments in beam

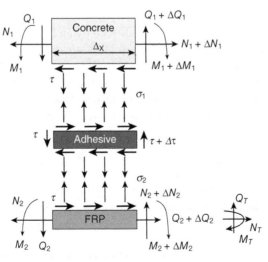

6.2 Free body diagram of the FRP strengthened RC beam.

1 and beam 2, respectively; C_i, B_i, and D_i (i = 1, 2) are the axial, shear, and bending stiffness, respectively, and they are expressed as $C_i = E_i b_i h_i$, $B_i = \frac{5}{6} G_i b_i h_i$, $D_i = E_i \frac{b_i h_i^3}{12}$, where E_i and G_i (i = 1, 2) are the longitudinal Young's modulus and shear modulus of beam i, respectively; b_i is the width of beam i.

Assuming that the shear stress is constant through the thickness of the adhesive layer, we can establish the following equilibrium equations by using the free body diagram shown in Fig. 6.2

$$\frac{dN_1(x)}{dx} = b_2 \tau(x), \quad \frac{dN_2(x)}{dx} = -b_2 \tau(x), \qquad [6.3a]$$

$$\frac{dQ_1(x)}{dx} = b_2 \sigma_1(x), \quad \frac{dQ_2(x)}{dx} = -b_2 \sigma_2(x), \qquad [6.3b]$$

$$\frac{dM_1(x)}{dx} = Q_1(x) - \frac{h_1}{2} b_2 \tau(x), \quad \frac{dM_2(x)}{dx} = Q_2(x) - \frac{h_2}{2} b_2 \tau(x), \qquad [6.3c]$$

where $\sigma_1(x)$, $\sigma_2(x)$ are the normal stresses along the AC interface and the normal stress along the PA interface, respectively; $\tau(x)$ is the shear stresses in the adhesive. Note that the overall equilibrium condition requires (Fig. 6.2)

$$N_1(x) + N_2(x) = N_T, \qquad [6.4a]$$

$$Q_1(x) + Q_2(x) + Q_a(x) = Q_T, \qquad [6.4b]$$

$$M_1(x) + M_2(x) + N_1(x) \frac{h_1 + h_2 + h_0}{2} = M_T, \qquad [6.4c]$$

where N_T, Q_T, and M_T are the corresponding resulting forces with respect to the neutral axis of the FRP plate; $Q_a(x)$ is the shear force of the adhesive layer, which is given by $\tau(x) b_2 h_0$.

6.2.2 Two-parameter elastic foundation model

The classical adhesively bonded joint model of Goland and Reissner (1944) (G-R model) was widely used to obtain the analytical solution of the stresses in the FRP-to-concrete interface (Smith and Teng, 2001; Wang, 2003). In this model, the z-dependency of the strain in the adhesive layer is assumed a constant. The through-thickness strain ε_z is an average in plane strain. Therefore,

$$\varepsilon_z(x) = \frac{w_1(x) - w_2(x)}{h_0}, \qquad [6.5a]$$

$$\gamma_{xz}(x) = \frac{1}{h_0}\left(u_1(x) - \frac{h_1}{2}\phi_1(x) - u_2(x) - \frac{h_2}{2}\phi_2(x)\right). \quad [6.5b]$$

Strain–stress relations are then given by:

$$\varepsilon_z(x) = \frac{\sigma(x)}{E_a}, \quad [6.6a]$$

$$\gamma_{xz}(x) = \frac{\tau(x)}{G_a}, \quad [6.6b]$$

where $\varepsilon_z(x)$ and $\gamma_{xz}(x)$ are the peel and shear strains, respectively; $\sigma(x)$ and $\tau(x)$ are the corresponding stress components. E_a and G_a are Young's modulus and shear modulus of the adhesive, respectively.

It can be seen that the adhesive layer is modeled as a layer of continuously distributed shear and vertical springs in Eqs [6.5] and [6.6]. No interactions are assumed between the shear and vertical springs. The force equilibrium conditions of the adhesive layer are also ignored. In this way, the adhesive layer was essentially modeled as a two-parameter elastic foundation (2PEF). Simple closed-form expressions of interface stresses and beam forces can be obtained by this 2PEF model as (Wang, 2003):

$$\frac{1}{h_0^2}\frac{d^6 N_1}{dx^6} + \frac{a_4}{h_0}\frac{d^4 N_1}{dx^4} + a_2\frac{d^2 N_1}{dx^2} + a_0 N_1 + a_M M_T + a_N N_T = 0, \quad [6.7]$$

where

$$a_4 = -\left(G_a\left(\frac{\xi h_1}{2} + \eta\right) + E_a\left(\frac{1}{B_1} + \frac{1}{B_2}\right)\right),$$

$$a_2 = G_a\left(G_a\left(\frac{1}{B_1} + \frac{1}{B_2}\right)\left(\eta + \frac{h_1}{2}\xi\right) + h_0\left(\frac{1}{D_1} + \frac{1}{D_2}\right)\right),$$

$$a_0 = -E_a G_a\left(\left(\frac{1}{D_1} + \frac{1}{D_2}\right)\eta + \frac{\xi(h_1 + h_2)}{2D_2}\right), \quad a_M = E_a G_a\left(\left(\frac{1}{D_1} + \frac{1}{D_2}\right)\frac{h_2}{2D_2} + \frac{\xi}{D_2}\right),$$

$$a_N = E_a G_a\left(\frac{1}{D_1} + \frac{1}{D_2}\right)\frac{1}{C_2}.$$

The interface stresses predicted by the 2PEF model reach good agreement with those obtained through continuum analysis such as finite element analysis (FEA) (Teng et al., 2002) except a small zone at the vicinity of the edge of the adhesive layer. In this small zone, continuum analysis (Rabinovitch and Frostig, 2000; Teng et al., 2002) reveals that the normal stresses along the concrete–adhesive (CA) interface and the FRP plate–adhesive (PA) interface are significantly different. The normal stress along the CA interface is tensile while the one along the PA interface is compressive. This feature is very important because it reveals where debonding will occur. Bearing in mind that compressive stress does not contribute to the debonding of the interface, tensile normal stress along the CA interface suggests that debonding should occur along the CA interface, instead of the PA interface. This has been confirmed by numerous experimental studies. The 2PEF model cannot capture this important feature. Another difficulty of the 2PEF model is that it cannot satisfy all the boundary conditions.

6.2.3 Three-parameter elastic foundation model

To overcome the drawbacks in the 2PEF model, a three-parameter elastic foundation model was later proposed by Wang and Zhang (2010). As shown in Fig. 6.3, the adhesive layer can be viewed as two linear normal spring layers with stiffness of $K = 2E_a/h_0$ interconnected by a shear layer with constants of G_a. By using this model, the strain–stress relations of the adhesive layer can be written as:

$$\sigma_1(x) = \frac{2E_a}{h_0}(w_1(x) - w_a(x)),\ \sigma_2(x) = \frac{2E_a}{h_0}(w_a(x) - w_2(x)), \quad [6.8a]$$

$$\tau(x) = \frac{G_a}{h_0}\left(u_1(x) - \frac{h_1}{2}\phi_1(x) - u_2(x) - \frac{h_2}{2}\phi_2(x)\right) + G_a\frac{dw_a x}{dx}, \quad [6.8b]$$

where E_a and G_a are the Young's modulus and shear modulus of the adhesive, respectively.

Ignoring the axial and bending moment of the adhesive layer, the equilibrium condition of the adhesive layer requires (Fig. 6.2):

$$\frac{d\tau(x)}{dx} = \frac{\sigma_2(x) - \sigma_1(x)}{h_0} \quad [6.9]$$

Equation [6.9] describes the interaction between the normal and shear stresses within the adhesive layer. Noting that $\tau(x)$ changes drastically at the vicinity of the FRP plate end, Eq. [6.9] suggests that $\sigma_1(x)$ and $\sigma_2(x)$ are

6.3 Three-parameter elastic foundation model of adhesive layer model.

significantly different at the vicinity of the end of the FRP plate. In the two-parameter elastic foundation model, $\sigma_1(x)$ is assumed equal to $\sigma_2(x)$ even though the left-hand side of Eq. [6.9] is not zero. Therefore, the force equilibrium condition of the adhesive layer is not satisfied in the two-parameter elastic foundation model. The governing equation of the 3PEF model for the FRP-strengthened RC beam in terms of N_1 has been obtained as (Wang and Zhang 2010):

$$F_{11}\frac{d^8N_1(x)}{dx^8} + F_{12}\frac{d^6N_1(x)}{dx^6} + F_{13}\frac{d^4N_1(x)}{dx^4} + F_{14}\frac{d^2N_1(x)}{dx^2} + F_{15}N_1(x) \\ + F_{16}N_T + F_{17}M_T = 0,$$

[6.10]

where coefficient F_{1i} are given by Wang and Zhang (2010). Equation [6.10] is two orders higher than Eq. [6.7] and allows for implementing all the boundary conditions.

6.2.4 Comparisons and verifications

As comparisons and verifications, a RC beam strengthened by a thin FRP plate under three-point bending (Fig. 6.1) studied by Smith and Teng (2001) is examined. The simply supported RC beam with a span of 3000 mm is subjected to a mid-span load of $P = 150$ kN. The distance from the support to the end of the FRP plate is 300 mm. The material properties are given as: adhereds, $E_1 = 30,000$ MPa (concrete), $E_2 = 100,000$ MPa (FRP); adhesive, $E_a = 2000$ MPa, $v_a = 0.35$. The geometries of the beam are given by: $h_1 = 300$ mm, $h_2 = 4$ mm, $h_0 = 1$ mm, $b_1 = b_2 = 200$ mm.

Numerical solutions by FEA are carried out as baseline for comparison with the commercial finite element package ANSYS. Shear and normal stress obtained by the present method and FEA are presented in Fig 6.4. Figure 6.4(a) shows the shear stress distribution near the end of the FRP plate. FEA provides three shear stress distributions, i.e., the shear stress distribution along the AC interface, PA interface, and centerplane of the adhesive layer, as shown in Fig 6.4(a). These three shear stress distributions are almost identical except a very small region at the edge of the adhesive layer due to the stress singularity at the corner edge. This suggests that it is reasonable to assume the shear stress is uniform through the thickness of the adhesive layer. Shear stress along the midplane reaches its maximum at a small distance to the edge and reduces to zero at the edge of the adhesive layer as anticipated. This feature is captured successfully by the 3PEF model, as demonstrated by the solid line in Fig. 6.4(a). The results of the 2PEF model (Wang, 2003; Smith and Teng, 2001) are very close to the 3PEF model except that they cannot satisfy the zero shear stress boundary condition at the edge of the adhesive layer (Fig. 6.4(a)).

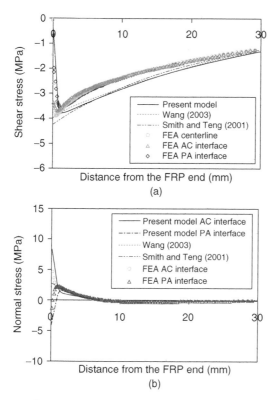

6.4 Comparison of interface stresses obtained by different methods: (a) shear stress; (b) normal stress.

Figure 6.4(b) compares the normal stress distributions obtained by the different methods. As illustrated by FEA predictions, the normal stress distribution along the AC interface is different from that along the PA interface. The normal stress along the AC interface is tensile while the one along the PA interface is compressive at the end of the FPR plate. This feature is important because it explains why debonding usually occurs along the AC interface (with a thin layer of concrete), not along the PA interface. As demonstrated by Fig. 6.4(b), the 3PEF model captures this feature very well, while the 2PEF model only predicts one value for both interfaces. Nevertheless, all the solutions converge to one value if the distance from the edge is big enough.

6.3 Fracture analysis of the fiber-reinforced polymer (FRP)-to-concrete interface

Usually, the FRP–concrete interface is not perfect. Some small cracks can exist within this interface zone due to manufacturing quality or induced by intermediate cracks in the concrete substrate. In this case, the integrity of the FRP–concrete interface should be assessed with interface fracture mechanics.

6.3.1 Linear elastic fracture mechanics

Linear elastic fracture mechanics (LEFM) is often used in the literature to study the FRP–concrete interface debonding. A typical LEFM model for the FRP–concrete interface was developed by Au and Büyüköztürk (2006) (referred as to 'AB' in the following). Their proposed tri-layer interface fracture energy model is essentially a direct application of the classical interface fracture models in bi-layered beams (Suo and Hutchinson, 1990) to three-layered beams. Therefore, the shear forces were ignored in their expression of energy release rate (ERR), which could lead to underestimation of the ERRs if shear forces exist. Moreover, mode mixity of the interface debond is not available for the tri-layer model.

Wang and Qiao (2004, 2005) improved the classical interface solution of Suo and Hutchinson (1990) through considering the transverse shear effect. To this end, Wang and Qiao (2004, 2005) introduced two crack tip deformation models, the semi-rigid joint model and flexibiel joint model (Qiao and Wang, 2005). By using this improved LEFM solution, the ERR for an interface crack shown in Fig. 6.5 can be given by:

$$G = \frac{1}{2}(C_N N^2 + C_Q Q^2 + C_M M^2 + C_{MN} MN + C_{NQ} NQ + C_{MQ} MQ), \qquad [6.11]$$

where C_N, C_Q, C_M, C_{MN}, C_{NQ}, and C_{MQ} are constants depending on the geometries and materials of the structure (Qiao and Wang, 2005). The detailed expressions of these coefficients and three loading parameters M, N, and Q can be found in Qiao and Wang (2005).

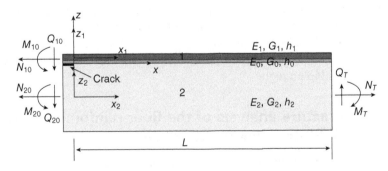

6.5 Crack tip element of FRP-strengthened concrete beam.

Equation [6.11] clearly shows that the transverse shear force is accounted for by Q. If all the terms with Q are ignored, Eq. [6.11] becomes identical to AB's existing solution (Au and Büyüköztürk, 2006).

Then the individual stress intensity factors can be given by:

$$K_I = \frac{P}{\sqrt{2}}\left(\sqrt{C_N}N\cos(\omega) + \sqrt{C_M}M\sin(\omega+\gamma_1) + \sqrt{C_Q}Q\sin(\omega+\gamma_2)\right), \quad [6.12a]$$

$$K_{II} = \frac{P}{\sqrt{2}}\left(\sqrt{C_N}N\sin(\omega) - \sqrt{C_M}M\cos(\omega+\gamma_1) - \sqrt{C_Q}Q\cos(\omega+\gamma_2)\right). \quad [6.12b]$$

The phase angle ψ is given by:

$$\psi = \tan^{-1}\left(\frac{\sqrt{C_N}N\sin(\omega) - \sqrt{C_M}M\cos(\omega+\gamma_1) - \sqrt{C_Q}Q\cos(\omega+\gamma_2)}{\sqrt{C_N}N\cos(\omega) + \sqrt{C_M}M\sin(\omega+\gamma_1) + \sqrt{C_Q}Q\sin(\omega+\gamma_2)}\right), \quad [6.13]$$

where

$$\sin(\gamma_1) = \frac{C_{MN}}{2\sqrt{C_M C_N}}, \quad \sin(\gamma_2) = \frac{C_{NQ}}{2\sqrt{C_N C_Q}}. \quad [6.14]$$

It should be emphasized that the beam model does not have enough information to determine the ω in Eq. [6.12]. Therefore, an extra continuum analysis such as FEA is needed to determine the angle ω.

To verify and demonstrate the enhanced accuracy of the new joint solutions, typical delaminated DCB specimens are analyzed using AB's model (Au and Büyüköztürk, 2006), the present model considering the shear force, and the FEA method. The FEA results are used as the baseline for comparison. Commercially available finite element software, ANSYS 10.0, is used to perform the FEA.

Analyses of the FRP-strengthened concrete DCB specimens (Fig. 6.5) are conducted. The Young's modulus and Poisson's ratios of the FRP plate, adhesive layer, and concrete specimens are $E_1 = 131$ GPa, $v_1 = 0.28$, $E_0 = 1.2$ GPa, $v_0 = 0.44$, $E_2 = 25$ GPa, $v_1 = 0.18$, respectively. The thicknesses of the FRP plate, adhesive layer, and concrete specimens are 1.4 mm, 1.5 mm, and 50.4 mm, respectively. The crack lies along the interface between the concrete substrate and the adhesive layer as shown in Fig. 6.5.

Results for the FRP-strengthened concrete DCB specimens with different crack lengths are presented in Fig. 6.6. As expected, the two new joint models give better estimations of the ERRs than the existing solution. The ERR solutions based on the semi-rigid joint model agree with the FEA results better than that of AB's method because shear deformation can be incorporated in this method. Because the crack length is much larger than the thickness of the FRP plate and adhesive layer, the contribution of shear force is not very significant in this case. However, due to the existence of the soft adhesive layer, both the semi-rigid joint model and AB's method

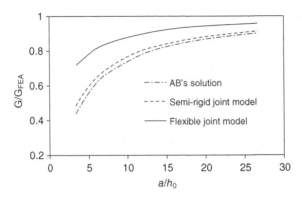

6.6 EERs versus crack lengths for a DCB specimen.

cannot accurately capture the local rotations at the crack tip. This difficulty can be overcome in the flexible joint model by introducing the interface compliances. Therefore, the flexible joint model predicts a much better solution than those of the semi-rigid joint model and AB's method.

Since one mode mixity solution is provided by AB's existing solution, to demonstrate the effect of shear force in phase angle, a phase angle was obtained by ignoring the terms with shear force in Eq. [6.13]. This phase angle is referred to as 'solution without shear' in the figures. Figure 6.6 compares the phase angles obtained by all four methods. It can be seen that all but the solutions without shear force agree well with the FEA results. This suggests that the DCB specimen is not a purely mode I specimen if the crack is along the interface between the concrete and adhesive, which seems to contradict the common belief that DCB is a purely mode I specimen. In this figure, we can find that the phase angle increases with the crack length according to the new analytical solutions and FEA, while the solution without shear cannot capture this feature. Because the transverse shear force is neglected in the solution without shear, the phase angle does not change with the length of crack, as shown in Fig. 6.7.

6.3.2 Nonlinear fracture mechanics

Recent studies show a trend whereby nonlinear fracture mechanics have gained more popularity and been adopted by more and more researchers. Here, nonlinear fracture mechanics refers to using a nonlinear traction-separation law, rather than the linear one as assumed in the LEFM, to describe the stress-deformation behavior of the FRP–concrete interface. The application of a nonlinear traction-separation law is supported by much experimental evidence obtained in the last decade (Chajes *et al.*, 1995; Dai *et al.*, 2005). The shear traction-separation law of the FRP–concrete

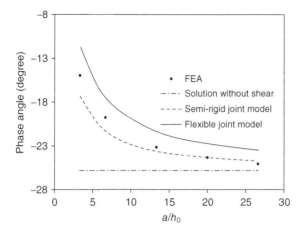

6.7 Phase angles versus crack lengths for a DCB specimen.

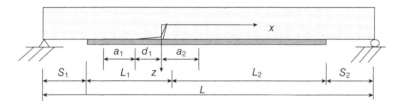

6.8 Flexural-shear crack induced debonding of an FRP-strengthened RC beam.

interface is generally referred to as the bond stress-slip law in the literature. Generally, this nonlinear relationship consists of two stages: an initially elastic stage in which the interfacial stress increases with the slip until it reaches a maximum value, and a softening stage in which interfacial stress decreases with the slip. This nonlinear relationship can be measured directly using a J-integral method as recently suggested by Wang (2007a). It should be pointed out that using a nonlinear bond stress-slip law in the analytical model, the debonding process is essentially approached through a cohesive zone model (CZM).

Consider the FRP–concrete interface layer in Fig. 6.8, which can be modeled as a large fracture processing zone with a nonlinear bond-slip law (Wang, 2006a). Various nonlinear bond-slip laws have been proposed (Chajes *et al.*, 1995; Yuan *et al.*, 2004; Dai *et al.*, 2005; Wang, 2007a,b). Among them, the bilinear law is the most popular for its simplicity and good agreement with experiment observations (Yuan *et al.*, 2004; Wang, 2006a,b), and therefore, is also adopted in this study. As shown in Fig. 6.9(a), the bilinear bond stress-slip law consists of a linearly elastic branch for bond slip less

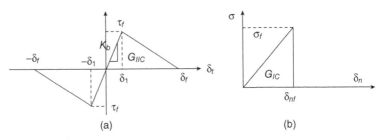

6.9 Traction-separation law used in this study: (a) shear traction-separation law, (b) normal traction-separation law.

than a particular value δ_1, and a linearly decreasing branch until complete delamination occurs. This law can be expressed by the following equations:

$$\tau = \begin{cases} 0 & \delta_t < -\delta_f \\ -\dfrac{\delta_f + \delta_t}{\delta_f - \delta_1}\tau_f & -\delta_f \leq \delta_t < -\delta_1 \\ \dfrac{\delta_t}{\delta_1}\tau_f & -\delta_1 \leq \delta_t < \delta_1 \\ \dfrac{\delta_f - \delta_t}{\delta_f - \delta_1}\tau_f & \delta_1 \leq \delta_t < \delta_f \\ 0 & \delta_f \leq \delta_t \end{cases} \qquad [6.15]$$

From the point of view of CZM, such a nonlinear relationship given by Eq. [6.15] is a material property of the FRP–concrete interface. τ_f and δ_f are the shear strength and the separation slip of the interface, respectively; $K_b = \tau_f/\delta_1$ is the initial elastic stiffness of the FRP–concrete interface.

Very few studies have been conducted on characterizing the mode I traction-separation law of the FRP–concrete interface (Dai et al., 2005). Existing modeling studies (Niu et al., 2006; Pan and Leung, 2007) used a triangular model to approximate the open traction-separation law of the FRP–concrete interface. This model is also adopted here to simplify formulation (Fig. 6.9(b)). In Fig. 6.9(b), σ_f and δ_{nf} are the maximum normal stress and open displacement of the FRP–concrete interface, respectively. The bond-slip law for normal stress simply reads:

$$\sigma_a = \begin{cases} 0 & \delta_n < -\delta_{nf} \\ \dfrac{\delta}{\delta_{nf}}\sigma_f & -\delta_{nf} \leq \delta_n < \delta_{nf} \\ 0 & \delta_{nf} \leq \delta_n \end{cases} \qquad [6.16]$$

It should be pointed out that it is an open question as to how the shear and open behaviors of the FRP–concrete interface couple, as very little experimental study has been carried out. For this reason, a mode-independent cohesive law is adopted in this study, which assumes the shear and opening traction-separation laws of the FRP–concrete interface are unrelated. The fracture energies of mode I and mode II of the interface, G_I and G_{II}, are given by the area below the traction-separation curves in Fig. 6.9(a) and (b):

$$G_I^f = \int_0^{\delta_{nf}} \sigma(\delta_n) d\delta_n, \quad G_{II}^f = \int_0^{\delta_f} \tau(\delta_t) d\delta_t, \qquad [6.17]$$

and the total fracture energy G_T of the interface reads:

$$G_T = G_I + G_{II}. \qquad [6.18]$$

The mode mixity of the debonding can be described by the phase angle ψ, which is defined by:

$$\tan \psi = \sqrt{G_{II}/G_I}. \qquad [6.19]$$

A simple linear debonding criterion (Hutchinson and Suo, 1992) is used in this study,

$$\frac{G_I}{G_{Ic}} + \frac{G_{II}}{G_{IIc}} = 1, \qquad [6.20]$$

where G_{Ic} and G_{IIc} are the mode I and II fracture toughness of the interface, respectively, given by the area under the total traction-separation laws shown in Fig. 6.9. Full debonding occurs as soon as the fracture energies of mode I and II satisfy Eq. [6.20].

By using nonlinear fracture mechanics, closed form solutions can be obtained for the whole debonding process of the FRP-strengthened RC beam induced by an intermediate crack shown in Fig. 6.8 (Wang and Zhang, 2008). Figure 6.10 shows the interface stress solutions for the debonding process of the structure shown in Fig. 6.8 subjected to a concentrated load P at the midspan. Different debonding stages can be easily identified from Fig. 6.10. Line 1 of Fig. 6.10(a) and (b) presents the interfacial shear and normal stress distributions along the FRP–concrete interface when $P = 1.0$ kN, respectively. In this case, both the left and right interfaces are in elastic stage. It can be observed that the normal stress is negative (compressive) along the right interface, while positive (tensile) along the left interface. The compressive normal stress does not contribute to the interface debonding. Therefore, the left interface is under mixed-mode loading; while the right interface is under pure mode II loading. Considering that the debonding is most difficult to occur under mode II loading, the flexural-shear crack induced debonding can only occur along the left interface for the case studied here.

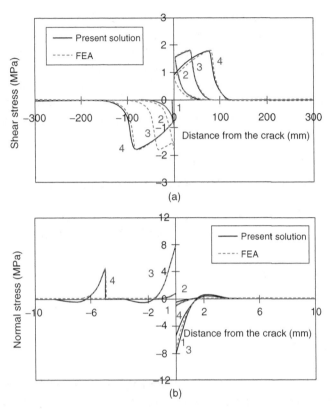

6.10 Interface stress distributions at different debonding stages: (a) interfacial shear stress, (b) interfacial normal stress.

If P is increased to a certain range, the maximum of the shear stress is higher than the shear strength. In this case, the left interface enters the elastic-softening stage (Wang, 2006b) while the right interface is still in the elastic stage, as demonstrated by line 2 in Fig. 6.10 when $P = 1.95$ kN. If the applied load is $P = 9.0$ kN, both the left and right interfaces enter elastic-softening stage as shown by the line 3 in Fig. 6.10. With the increase of load P, the energy release rate of the left interface increases too. Once Eq. [6.20] is satisfied, full debonding initiates and grows along the left interface, as demonstrated by line 4 in Fig. 6.10. In this case, the applied load $P = 24.4$ kN. A fully debonded zone of 5 mm is formed along the left interface.

Compared with the single-parameter fracture approach of LEFM, which ignores the microscopic details and discloses little about what happens within the damage zone, the CZM takes the behavior of the fracture processing zone into consideration and provides a way to examine the 'inner problem' of understanding, characterizing, and modeling the failure processes that actually lead to energy dissipation. Furthermore, the CZM unifies the crack initiation and growth into one model and can be easily formulated and implemented in numerical simulation, such as the 'interface element' method in finite element code.

6.4 Durability of the fiber-reinforced polymer (FRP)–concrete interface

6.4.1 Critical failure-based durability studies of the FRP–concrete interface

A number of studies have been dedicated to the durability of the FRP-to-concrete interface (Green et al., 2000; Au and Büyüköztürk, 2006; Wan et al., 2006; Ouyang and Wan, 2008). In these studies, test specimens were first conditioned in typical civil infrastructure environments such as various aqueous solutions, freeze-thaw cycling, wet-dry and temperature cycling, cyclic, and sustained loads. Synergetic effects of different environmental conditions were also considered in the existing studies. After environmental conditioning, specimens were loaded to failure to measure the deteriorated mechanical properties. Various specimens have been used in existing studies to evaluate the strengths of the interface before and after environmental conditions, including lap-joint specimens (Green et al., 2000) and beam-type fracture specimens (Au and Büyüköztürk, 2006; Wan et al., 2006). In the case of lap-joint specimens, the FRP sheet is loaded until full separation between the FRP and concrete occurs. The maximum load is recorded as the indicator of the strength of the interface. In the case of fracture specimens, the energy release rate G at catastrophic debond is measured and referred to as fracture toughness or critical energy release rate G_c of the interface. In all these studies, only the loads at the time of catastrophic failure are measured. Therefore, all these studies are essentially critical failure based.

Grace and Grace (2005) examined the effect on the FRP–concrete interface of various factors, and identified that moisture could do the most damage to the FRP–concrete interface. This observation was corroborated by recent studies of two groups (Au and Büyüköztürk, 2006; Ouyang and Wan, 2008). In these studies, the strength of the FRP–concrete interface degraded by moisture is assessed by the fracture toughness of interface debond. Two important conclusions can be drawn from these studies. First, substantial loss of the fracture toughness of the interface can be induced by moisture. Second, the debonding locus shifts from within the concrete cover in its dry state to along the adhesive–concrete interface in its wet state. These two conclusions are also found valid for structures exposed to other environmental conditions. Reay and Pantelides (2006) measured the bond strength of the CFRP–concrete interface exposed to indoor and real outdoor environments up to three years using a pull-off adhesion test. They also found that the debond locus shifts from within concrete after 6 months' environmental exposure to along the adhesive–concrete interface after 12 months' exposure. Studies on adhesive joints also suggested that debonding tends to move to the adhesive–adherend interface in harsh environmental conditions (Kinloch, 1979). Besides environmental exposures, time-dependent behaviors such as creep and creep fracture of FPR or adhesive layer has also been examined to evaluate the long-term durability of FRP-strengthened concrete structures (Wu and Diab, 2007; Meshgin *et al.*, 2009).

6.4.2 Environment-assisted subcritical debonding of adhesive joints

An inherent problem of all critical failure-based studies is that only the loads at the time of catastrophic failure are measured. Debond, however, is a gradual process where slow growth of cracks occurs at the interface. The most distinct feature of these slow cracks is that they grow at a very slow rate with an energy release rate G, and only a fraction of the critical energy release rate G_c if reactive environmental species exist. This slow crack growth is a long-term process of synergistic action of environments and mechanical loads. The catastrophic interface debond (critical crack) is only the ending point of this process. For any structure which requires long-term stability, a resistance to this slow crack growth would be needed. To understand the degradation mechanism of the interface and gain the ability to accurately predict the long-term durability ultimately require quantifying and appropriate analysis of the slow debond growth process.

The slow crack growth in adhesive joints in aggressive environments is referred to as environment-assisted subcritical cracking (debonding) (Wiederhorn, 1968), leading to critical cracking at catastrophic failure. Sometimes it is also called static fatigue or stress corrosion cracking in the literature (Krausz, 1978). Following the classic work of Wiederhorn (1968), the environment-assisted subcritical debond growth can be treated as a synergistic interaction between strained adhesion bonds and environmental species. A schematic illustration of debond growth rate (da/dt) versus driving energy release rate at the crack tip (G) curve is shown in Fig. 6.11. This curve consists of three debond growth regions and a threshold G_{th}. If the driving energy release rate G is less than G_{th}, subcritical debonding will not occur. In Region I, the debond growth is so slow that the environmental species have enough time to transport to the crack tip to enable the environmental attack mechanism to occur readily. As a result, the debond growth rate is dependent on both the reaction rate and the mechanical load G. In Region II, the debond growth is faster so that the debond growth rate

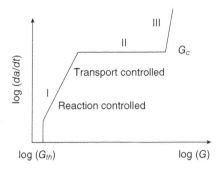

6.11 Schematic of typical environment-assisted debond.

is controlled by the availability of the environmental species. As a result, the debond growth rate is almost independent of the mechanical load G. In Region III, the debond growth is much faster than the transportation rate of environmental species so that environmental species cannot reach the crack tip. Consequently, the debond growth rate is only dependent on G. Measurements in this region do not provide any information about the interaction with environmental species at the crack tip. Clearly, Region III describes the critical debond growth in adhesive joint and the corresponding G is the critical energy release rate G_c.

Existing studies on the strength and durability of the FRP–concrete interface in aggressive environments only focus on Region III. Regions I and II are totally ignored. Since no environmental species can reach the crack tip in Region III, existing studies have to adopt a two-step approach. In the first step, test specimens are conditioned in designed accelerated environments so that environmental species can reach the interface through diffusion and capillary action. In the second step, the residual strengths of the conditioned specimens are measured at catastrophic failure (Region III). This approach suffers a few obvious drawbacks:

- The results of critical debong testing can be misleading to be used to evaluate the long-term durability of the interface. This is because catastrophic failure-based testing can lead to a different failure mode from that of real applications. As demonstrated in many studies (Diab and Wu, 2007; Singh *et al.*, 2008), interface debond may shift from adhesive failure at slow growth rate under service loads to cohesive failure at high growth rate under catastrophic failure.
- Interaction between the environmental species and mechanical loads is lost. As a result, besides the ultimate strength, little information about the degradation mechanism of the interface under environmental species attack can be obtained from the testing.
- The environmental conditioning process usually takes a fairly long time.

The subcritical debond testing focuses on Regions I and II rather than Region III. Compared with the critical failure-based method, the subcritical debonding test may provide a better way to characterize the long-term durability of the FRP–concrete interface because it closely simulates the failure occurring in real-life applications or service-life of the interface. The change of debond locus that occurs in catastrophic failure testing can be avoided. The interaction with environmental species is allowed in subcritical debond testing due to the slow debond growth rate. This makes it possible to measure important parameters of the reaction kinetics at the crack tip and to deepen our understanding of the degradation mechanism of the interface. Shorter time is needed in the subcritical debonding testing because the time-consuming process of conditioning specimens is unnecessary. The ambiguity associated with bond strength due to competitive effects of concrete curing, long-term concrete strength gaining, epoxy creep, and epoxy curing can be avoided because the debond locus is within the epoxy–concrete interphase zone and long-term environmental conditioning is not needed. Subcritical debond testing also provides a far more useful indicator of interface quality than the ultimate bond strength or fracture toughness because the durability of the interface is generally more important than ultimate bond strength in harsh and changing environmental conditions. New knowledge to improve the durability of the interface can be obtained.

6.4.3 Environment-assisted subcritical debond growth of the FRP–concrete interface

A series of wedge driving tests were conducted in various environments to characterize the subcritical crack growth along the epoxy–concrete interface. The wedge driving test specimens were prepared with the epoxy layers constrained between the CFRP plates and the concrete substrates. The test specimen size and structure are shown schematically in Fig. 6.12.

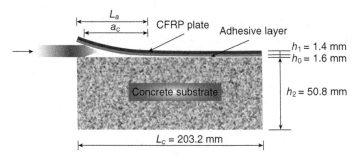

6.12 Wedge driving test specimen.

The specimens are inserted into a steel frame so that a steel wedge can be driven into the interface between the adhesive layer and concrete substrate. L_a is the total crack length from the edge of the specimen to the crack tip. a_c is the effective crack length, which is the distance from the contact point between the wedge and the FRP–epoxy layer to the crack tip. The width of the specimen is 2 in., which is the same as the width of the commercially available CFRP plate. The steel wedges were made according to the ASTM standard D3762-98.

The specimen was mounted onto a MTS testing machine using a homemade steel fixture. The steel wedge was then driven into the pre-crack between the epoxy and the concrete of the specimen at an intermediate speed (0.02 in./sec.) by the MTS machine. Once the thickest portion of the wedge reached the epoxy–concrete interface, the wedge was held there for one hour to reduce the instant effect of the high residue stress induced by the quick loading process. Then the wedge was driven by the MTS machine into the crack at a very low speed (0.00002 in./sec.). A high resolution digital camera was used to capture the crack length. The length of the crack can be identified from these images. A transparent glass vessel was used to enclose the whole specimen and the load fixture. After filling this vessel with water or other aggressive solutions, the whole specimen was submerged in these fluids. In this way, we were able to apply both the mechanical forces and the environment species to the specimen. Ignoring the deformation of the concrete substrate, the driving energy release rate at the crack tip provided by the wedge can be calculated as:

$$G = \frac{9D_c \Delta^2}{2ba_c^4} \quad [6.21]$$

where D_c is the bending stiffness of the FRP–epoxy composite layer; a_c is the effective crack length shown in Fig. 6.12; and Δ is the thickness of the wedge.

Subcritical debond testing of the FRP–concrete interface is conducted in the ambient condition, the tap water, the alkaline solution, and the deicing salt solution. Testing results for specimens exposed to ambient environment are shown in Figs 6.13 and 6.14. Figure 6.13 shows the relationship between the crack growth rate and the driven energy release rate in ambient environment. The crack growth rate versus energy release rate shown in Fig. 6.13 clearly exhibits two distinct regions (Regions I and II).

The strengths of the concrete substrate for Groups 1, 2, and 3 in Fig. 6.13 are 2824 psi, 4238 psi, and 4880 psi, respectively. No significant difference can be identified from the crack growth rate versus driven energy curves of these three groups. The major differences may be the value of the critical energy release rates, G_c. The G_c of Group 1 with lowest strength concrete substrates is lower than those of the other two groups.

168   Developments in FRP composites for civil engineering

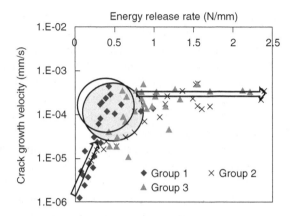

6.13 Subcritical debonding of the epoxy–concrete interface in ambient environment.

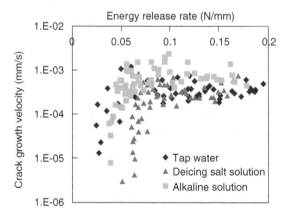

6.14 Comparisons of subcritical debonding of the epoxy–concrete interface in various aqueous environments.

In existing critical-debonding-based studies, the interface strength of the FRP–concrete bond is dependent on the strength of the concrete substrate because the interface debonding occurs within a thin layer of the concrete substrate. In subcritical debonding testing, the debond is more likely growing along the epoxy–concrete interface, as shown below. As a result, the effect of the strength of the concrete is not as significant as it is in the case of critical debonding.

Results of subcritical debonding testing in aqueous conditions (tap water, deicing salt solution, alkaline solution) are presented in Fig. 6.14. Similar to the case in ambient temperature, interface debonding growth is observed, although the driven energy release rate is much lower than the critical value, and there are two distinct regions on the crack growth rate versus driven energy release rate curves, as shown in Fig. 6.14. Compared with the case in ambient environment, we find that the energy release rate needed to drive the subcritical debonding in aqueous conditions is much lower, even though the crack growth rate is much faster than in the ambient environment. Clearly, water and chemical ions make the subcritical debonding along the epoxy–concrete interface much easier due to the interaction between the water molecules and the strained adhesion bonds between the epoxy and the concrete. Water molecules are much more abundant at the crack tip in water than in ambient environment. As a result, more epoxy chains strained by the wedge can be displaced by the water molecules, leading to much fast subcritical debonding in water, while in the critical failure-based studies, interaction between the environment species and the adhesion bonds between the epoxy and the concrete is not possible due to the fast crack growth rate.

Some difference can be observed in the reaction-controlled Region I for the three different aqueous environments. It can be seen that mechanical energy needed to drive debond growth in this region is lowest in tap water and highest in deicing salt solutions. However, this does not necessarily suggest that water can deteriorate the epoxy–concrete interface more than the other two solutions. One possible reason causing this difference could be the different levels of curing of epoxy. In this study, the epoxy was cured for about two weeks for testing in tap water, two months for testing in alkaline solution, and three months for testing in deicing salt. Due to different curing times, more adhesion bonds may develop along the epoxy–concrete interface in those specimens tested in deicing salt than those tested in tap water and alkaline solution. It is also noticed that although the concrete substrates have different strengths, their effects on the subcritical debonding in aqueous condition are very insignificant, as shown in Figs 6.13 and 6.14. This is because the subcritical debonding locus is mainly along the epoxy–concrete interface. As a result, the strength of the concrete substrate is not as important as it is in the critical debonding.

The debonded surfaces of the specimens (epoxy side) are shown in Fig. 6.15. In this figure, the regions within the dashed rectangles are the initial crack surfaces induced by the fast driving. Most of this region is covered by concrete for all four test conditions, suggesting that debonding is cohesive failure within the concrete for fast-driven cracking. This is in agreement with the existing critical debonding-based studies. In the subcritical debonding region of the specimen tested in ambient environment, there is much less concrete attached to the epoxy surface, indicating a change of failure mode from the cohesive failure in concrete to mainly adhesive failure along the epoxy–concrete interface. This change of failure mode can be seen more clearly in specimens tested in aqueous conditions as shown in Fig. 6.15.

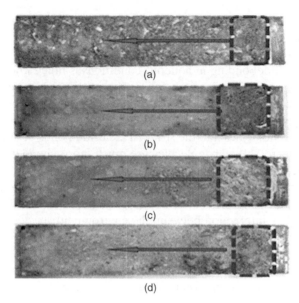

6.15 Fracture surfaces of wedge-driven tests: (a) in ambient condition; (b) in tap water; (c) in deicing salt solution; (d) in alkaline solution.

There is very little concrete attached to the epoxy layer of specimens tested in aqueous conditions, suggesting that debonding occurs along the epoxy–concrete interface in these specimens.

There are two reasons for the change of failure mode from the cohesive failure within the concrete in fast crack growth to adhesive failure in slow (subcritical) crack growth. First, the interaction of water molecules with the chemical bonds (most likely hydrogen bonds) between the epoxy and the concrete is allowed in subcritical debonding, significantly deteriorating the epoxy–concrete interface. Second, the strength of the chemical bonds (hydrogen bonds) between the epoxy and the concrete increases with the loading rate, as shown in Namkanisorn *et al.* (2001). For these two reasons, it is possible that the epoxy–concrete interface is weaker than that of the adjacent concrete layer in environment-assisted subcritical debonding. As a result, the failure mode is mainly adhesive. This can also explain why more concrete is attached to the epoxy in the specimens tested in ambient environment than those tested in aqueous conditions. In the former case, much less environment species are available at the crack tip than the latter cases. As a result, the epoxy–concrete interface is stronger in the former case than those in the latter cases, leading to more concrete attached to the surface of the epoxy in the specimens tested in ambient conditions. In the existing critical debonding-based studies, the specimens are loaded much faster and there is no interaction between environment species and epoxy–concrete interface bond. As a result, the epoxy–concrete interface can be stronger than the adjacent concrete layer. Consequently, most debonding modes observed in the existing studies are cohesive failure in concrete.

The change of failure mode is also observed in the existing durability testing on the FRP–concrete interface (Au and Büyüköztürk, 2006; Ouyang and Wan, 2008). In these studies, if the specimens were conditioned in water for sufficiently long, adhesive failure mode was observed even though the testing was based on critical debonding. This is because the environment species are available at the crack tip due to the long duration of conditioning, making the epoxy–concrete interface weaker than the adjacent concrete. Change of failure mode from cohesive failure at high loading rate to adhesive failure at low loading rate in epoxy–metal interface was observed by Rakestraw et al. (1995). The difference in debonding mode of the FRP–concrete interface in critical and environment-assisted subcritical debonding may imply that the existing critical debonding-based approach is unsuitable for predicting the long-term durability of the FRP–concrete interface.

6.5 References and further reading

Au, C. and Büyüköztürk, O., 2006. Peel and shear fracture characterization of debonding in FRP plated concrete affected by moisture. *Journal of Composites for Construction*, 10(1), 35–47.

Avramidis, I.E. and Morfidis, K., 2006. Bending of beams on three-parameter elastic foundation. *International Journal of Solids and Structures*, 43(2), 357–375.

Chajes, M.J., Januszka, T.F., Mertz, D.R., Thomson Jr., T.A. and Finch Jr., W.W., 1995. Shear strengthening of reinforced concrete beams using externally applied composite fabrics. *ACI Structural Journal*, 92(3), 295–303.

Dai, J., Ueda, T. and Sato, Y., 2005. Development of the nonlinear bond stress–slip model of fiber reinforced plastics sheet–concrete interfaces with a simple method. *Journal of Composites for Construction*, 9, 52–62.

Diab, H. and Wu, Z.S., 2007. Nonlinear constitutive model for time-dependent behavior of FRP–concrete interface. *Composites Science and Technology*, 67, 2323–2333.

Goland, M. and Reissner, E., 1944. The stresses in cemented joints. *Journal of Applied Mechanics*, 66, A17–A27.

Grace, N.F. and Grace M., 2005. Effect of repeated loading and long term humidity exposure on flexural response of CFRP strengthened concrete beams. *Proceedings of the International Symposium on Bonded Behavior of FRP in Structures (BBFS 2005)*, Chen, J.F. and Teng, J.G. (eds), 539–546.

Green, M.F., Bisby, L.A., Beaudoin, Y. and Labossiere, P.J., 2000. Effect of freeze-thaw cycles on the bond durability between fibre reinforced polymer plate reinforcementand concrete. *Canadian Journal of Civil Engineering*, 27(5), 949–959.

Hutchinson, J.W. and Suo, Z., 1992. Mixed mode cracking in layered materials. *Advances in Applied Mechanics*, 29, 63–199.

Kinloch, A.J., 1979. Interfacial fracture mechanical aspects of adhesive bonded joints – a review. *Journal of Adhesion*, 10, 193–219.

Krausz, A.S., 1978. The deformation and fracture kinetics of stress corrosion cracking. *International Journal of Fracture*, 14, 5–15.

Meshgin, P., Choi, K.K. and Reda Taha, M.M., 2009. Experimental and analytical investigations of creep of epoxy adhesive at the concrete–FRP interface. *International Journal of Adhesion & Adhesive*, 29, 56–66.

Namkanisorn, A., Ghatak, A., Chaudhury, M.K. and Berry, D.H., 2001. A kinetic approach to study the hydrolytic stability of polymer–metal adhesion. *Journal of Adhesion Science and Technology*, 15(14), 1725–1745.

Niu, H., Karbhari, V. and Wu, Z., 2006. Diagonal macro-crack induced debonding mechanisms in FRP rehabilitated concrete. *Composites Part B: Engineering*, 37, 627–641.

Ouyang, Z. and Wan, B., 2008. Experimental and numerical study of moisture effects on the bond fracture energy of FRP/concrete joints. *Journal of Reinforced Plastics and Composites*, 27, 205–223.

Pan, J. and Leung, C.K.Y., 2007. Debonding along the FRP–concrete interface under combined pulling/peeling effects. *Engineering Fracture Mechanics*, 74, 132–150.

Porter, M.L. and Harries, K.A., 2007. Future directions for research in FRP composites in concrete construction. *Journal of Composites for Construction*, 11, 253–257.

Qiao, P. and Wang, J., 2005. Novel joint deformation models and their application to delamination fracture analysis. *Composites Science and Technology*, 65, 1826–1839.

Rabinovitch, O. and Frostig, Y., 2000. Closed-form high-order analysis of RC beams strengthened with FRP strips. *Journal of Composites for Construction*, 4, 65–74.

Rakestraw, M.D., Taylor, M.W., Dillard, D.A. and Chang, T., 1995. Time dependent crack growth and loading rate effects on interfacial and cohesive fracture of adhesive joints. *Journal of Adhesion*, 55(1–2), 123–149.

Reay, J.T. and Pantelides, C.P., 2006. Long-term durability of state street bridge on interstate 80. *Journal of Bridging Engineering*, 11, 205–216.

Singh, H.K., Wan, K.T., Dillard, J.G., Reboa, P., Smith, J., Chappell, E. and Sharan, A., 2008. Subcritical delamination in epoxy bonds to silicon and glass adherends: effect of temperature and preconditioning. *Journal of Adhesion*, 84, 619–637.

Smith, S.T. and Teng, J.G., 2001. Interfacial stresses in plated beams. *Engineering Structures*, 23, 857–871.

Suo, Z. and Hutchinson, J.W., 1990. Interface crack between two elastic layers. *International Journal of Fracture*, 43, 1–18.

Taljsten, B., 1997. Strengthening of beams by plate bonding. *ASCE J Mater Civil Engng*, 9(4), 206–212.

Teng, J.G., Zhang, J.W. and Smith, S.T., 2002. Interfacial stresses in RC beams bonded with a soffit plate: a finite element study. *Constuction and Building Materials*, 16(1), 1–14.

Wan, B., Petrou, M.F. and Harris, K.A., 2006. The effect of the presence of water on the durability of bond between CFRP and concrete. *Journal of Reinforced Plastics and Composites*, 25, 875–890.

Wang, J., 2003. Mechanics and fracture of hybrid material interface bond. Ph.D. thesis, Department of Civil Engineering, The University of Akron, OH.

Wang, J., 2006a. Nonlinear bond-slip analysis of delamination failure of FRP reinforced concrete beam – Part I: Closed-form solution. *International Journal of Solids and Structures*, 43(21), 6649–6664.

Wang, J., 2006b. Cohesive zone model of intermediate crack-induced debonding of FRP-plated concrete beam. *International Journal of Solids and Structures*, 43(21), 6630–6648.

Wang, J., 2007a. Cohesive-bridging zone model of FRP–concrete interface debonding. *Engineering Fracture Mechanics*, 74, 2643–2658.

Wang, J., 2007b. Cohesive zone model of FRP–concrete interface debonding under mixed-mode loading. *International Journal of Solids and Structures*, 44, 6551–6568.

Wang, J. and Qiao, P., 2004. Interface crack between two shear deformable elastic layers. *Journal of the Mechanics and Physics of Solids*, 52(4), 891–905.

Wang, J. and Qiao, P., 2005. Beam-type fracture specimen analysis with crack tip deformation. *International Journal of Fracture*, 132(3), 223–248.

Wang, J. and Zhang, C., 2008. Nonlinear fracture mechanics of flexural-shear crack induced debonding of FRP strengthened concrete beams. *International Journal of Solids and Structures* 45, 2916–2936.

Wang, J. and Zhang, C., 2010. A three-parameter elastic foundation model for interface stresses in curved beams externally strengthened by a thin FRP plate. *International Journal of Solids and Structures*, 47, 998–1006.

Wiederhorn, S.M., 1968. Moisture assisted crack growth in ceramics. *International Journal of Fracture*, 4, 171–177.

Wu, Z. and Diab, H., 2007. Constitutive model for time-dependent behavior of FRP–concrete interface. *Journal of Composites for Construction*, 11, 477–486.

Yuan, H., Teng, J.G., Seracino, R., Wu, Z.S. and Yao, J., 2004. Full-range behavior of FRP-to-concrete bonded joints. *Engineering Structures*, 26, 553–565.

Part II
Particular types and applications

Part II

7
Advanced fiber-reinforced polymer (FRP) composites for civil engineering applications

S. MOY, University of Southampton, UK

DOI: 10.1533/9780857098955.2.177

Abstract: This chapter deals with the uses of advanced composite materials in the construction industry. After considering the advantages of using composites and methods of fabrication, it outlines the surprisingly wide range of applications of composites. Examples are given from around the world of components and complete buildings and bridges, railway and other infrastructure, geotechnical applications and pipes for the water sector. Finally a number of more unusual or future possibilities are presented.

Key words: composites in construction, advantages and fabrication, buildings and bridges, infrastructure, geotechnics, water sector.

7.1 Introduction

The construction industry, certainly in the UK, has a rather conservative image. It is felt to be reluctant to innovate in forms of contract, costing, and construction techniques. The UK government was so concerned about the industry that it commissioned two major enquiries into it. The first, chaired by Sir Michael Latham (Latham, 1994), addressed procurement and contractual arrangements. The second, chaired by Sir Peter Egan (Egan, 1998), dealt amongst other things with innovation in the whole construction process. It recommended the use of factory-built components and sub-assemblies delivered to site when needed. In some ways this was the driver in the UK for the development and use of fiber-reinforced composite (FRP) products in construction, an aspect of the construction industry which is becoming more and more important. It would not have been possible without technology transfer from the aerospace and shipbuilding industries, which has been ongoing for the last twenty years. The author of this chapter is based in the UK but his experiences of the use of FRP products in construction are repeated in the US, Canada, the rest of Europe, Asia, and Australasia.

This chapter will describe the uses of advanced FRP composites in construction applications, the fabrication techniques which have driven those uses, and the advice available to someone thinking of using these materials. Examples will be given of typical products and how they are used. The emphasis will be on practicalities rather than theory.

7.2 The use of fiber-reinforced polymer (FRP) materials in construction

7.2.1 General

As has been discussed in earlier chapters, FRP materials are generally two component composites. The first component is the reinforcing fibers which almost exclusively in construction will be carbon, aramid, or glass fibers. In some situations, two or more fiber types can be used, hence the use of 'generally' in the first sentence. The second component is a resin, an organic chemical which in the right circumstances will polymerize and solidify into long-chained molecules. The fibers and resin are intimately mixed together before the resin cures. The fibers give the composite strength and stiffness, the resin binds the fibers together and provides protection to the fibers. In this chapter many of the uses of composites involve mass production and the need to drive down cost. Thus glass fibers, the cheapest of the common fibers, will often be used. The resin will be a thermoset, a polymer in which the curing process cannot be reversed. Polyester, vinylester, and epoxy resins are all used in construction products. Polyester resins are cheapest but shrink during curing, vinylesters are more dimensionally stable, while epoxy resins produce the best quality FRP but are the most expensive of the three resin types. A problem with the use of polyester and vinylester resins is the styrene vapor given off during curing, which has been identified as a health risk. Another class of resin is the phenols whose mechanical properties are not as good as the others, but they do offer very good fire performance.

7.2.2 The benefits of using FRP composites

It is worth repeating the benefits from using FRP composite materials. There are four main advantages; high strength, the ability to customize the properties of the FRP, low density, and excellent long-term durability.

FRP composites have high strength; for example, typical glass fibers have a tensile strength of 2400–3500 N/mm^2. The effective strength of the composite will be much lower (depending on the ratio of reinforcing fibers to resin), but it would still be significantly greater than that of most steels.

The properties of FRP can be tailored as desired by changing the quantity and direction of the fibers. Fabricators have great flexibility in their choice of materials. They can choose the most appropriate reinforcing fibers, both type (glass, carbon, etc.) and also arrangement (chopped strand mat, individual fibers, woven or stitched fabrics of various geometries). Whilst many products can be obtained 'off the shelf', there is also the opportunity to have 'tailor-made' products in novel situations or for specific (large volume) applications. With careful design, FRP materials can be used very efficiently.

The density of FRP is about one-fifth that of steel and less than two-thirds that of concrete. The density of glass fiber-reinforced polymer composite (GFRP) is typically about 1800 kg/m^3, while the density of steel is about 7850 kg/m^3. This means that FRP materials require less falsework (scaffolding) and heavy lifting equipment than conventional materials and can be handled by a smaller workforce.

Extensive testing has shown that FRP is extremely durable when used appropriately with low long-term degradation and very good fatigue resistance. FRP composites are resistant to water, salt, and other chemicals, and are unaffected by oil and other heavy hydrocarbons; as a result they require little maintenance in comparison to conventional materials.

No material is perfect and there can be downsides to the use of FRP. Material costs on a weight for weight basis are higher than those for conventional materials, but this should be seen as an incentive to use the FRP in innovative, efficient, and cost-effective ways. FRP materials have low coefficients of thermal expansion (some CFRP has a negative coefficient). This is only a problem when the FRP is used in conjunction with a conventional material and careful design should eliminate any potential problems. The resin properties are temperature dependent; lower temperatures cause the strength to increase but make the resin more brittle. The converse is true as temperature is increased, but at a critical temperature (called the glass transition temperature and in reality a temperature range of which the glass transition temperature is the median) there is a rapid decrease in strength. Problems can be avoided by the choice of a resin whose glass transition temperature is significantly higher than the anticipated maximum operating temperature.

A frequent criticism of FRP composites is that they perform poorly in fire. However, this has been shown not to be the case (Cutter *et al.*, 2009). In a fire the exposed resin surface chars and produces a tar-like layer which protects the underlying fibers, delaying the onset of failure. Of course, it is also possible to use conventional fire protection on composite material.

It is worth mentioning two further points concerning FRP composites. Stiffness (measured as elastic modulus) can be an important property. Carbon fiber-reinforced polymer composites (CFRP) can have high stiffness; an elastic modulus of up to 300 kN/mm^2 compared to 200 kN/mm^2 for steel. However, GFRP has lower stiffness, typically in the range 72–87 kN/mm^2 and it is frequently stiffness rather than strength which drives the design of GFRP. Finally, in many applications surface finish and appearance is important; with composites the use of appropriate fabrication techniques and materials, for example automated systems with high quality moulds and the use of gel coats, will achieve the finishes required.

7.2.3 Fabrication techniques

These have been covered in earlier chapters but it is worth looking at them again in the context of construction applications. The basic methods such as wet-layup are labour intensive and appropriate for one-off products. For mass production, the automated processes are much better. FRP panels are fabricated to a high quality and finish using vacuum assisted resin transfer molding (VARTM) or vacuum infusion. Structural shapes (I-beams, channels, angles, etc.) are produced in large quantities using the pultrusion process; it is ideal for products produced in long straight lengths with a constant cross section and constant reinforcing fiber architecture. Pipes are produced in many different sizes using the fiber winding process, which is a development of pultrusion in which the fibers are wound to a predetermined helical shape within the resin.

A further benefit of the automated processes is that they produce the best quality FRP. Composites are inherently variable because of air trapped within the matrix; however, the automated processes squeeze out almost all the air (typically less than 2% air is left in the matrix compared to over 5% in wet layup composites) so that the resulting composite is much less variable. Also it is possible to incorporate more reinforcement into the same volume of composite (reinforcement volume fraction is about 0.6 in vacuum infused composite compared to about 0.4 in wet layup material). Strength and stiffness are proportional to the volume fraction of reinforcement.

7.2.4 Conclusion

The discussion above has emphasized the flexibility in the use of FRP composites and the scope to use them in novel or unconventional ways. The designer is able to benefit from the potential to customize the composite properties to suit a particular application.

Even a cursory search of the web will reveal the very wide range of composites applications in construction. In order to give an overview, this chapter will now present a variety of applications for FRP composites in construction. Illustrations will use photographs from particular companies but it should be emphasized that there is considerable choice in manufacturers and fabricators and a potential user should investigate the possibilities before making any choice. Applications are presented in sections to give an impression of what is available. Where appropriate, fabrication methods and materials are also discussed.

7.3 Practical applications in buildings

7.3.1 Building interiors and exteriors

Advanced composite materials have been used in buildings for many years. An early example was the cladding of Mondial House in London (Fig. 7.1) which was completed in 1974. The white cladding was still in excellent condition when the building was demolished in 1996. Cladding systems are widely available and come in a variety of textures and colors.

Advanced composite materials are available for almost every aspect of building interiors ranging from floors to doors. Since FRP composites are electrically non-conducting and have hard smooth surfaces, they are particularly applicable to situations where high levels of hygiene are required or where magnetic or electrical machinery is being operated. Figure 7.2

7.1 White GRP cladding of Mondial House, London.

7.2 FRP wall panels used in a clean room (courtesy of Plastruct Canada Inc.).

shows a clean room finished almost entirely with FRP composite. In this case, the walls have a smooth finish for cleaning but textured finishes are also available.

Typical construction of wall and ceiling panels has GRP faces as little as 1 mm thick with a polystyrene insulating core, whose thickness will depend on the level of insulation required.

There are various FRP grating type flooring systems on the market. They are in competition with metal gratings but score heavily in aggressive environments where their excellent durability is important. Figure 7.3 shows a grating system which is used in water treatment and offshore applications. This system uses a phenolic resin in the GRP which gives particularly effective fire resistance.

Floor grating systems are frequently used in industrial buildings for mezzanine or intermediate floors. Figure 7.4 shows a typical application in which the handrails are also GFRP.

7.3 Duragrid composite grating system (courtesy of Pipex Structural Composites).

7.4 GRP mezzanine flooring and handrail system (courtesy of Redman Composites).

Advanced FRP composites for civil engineering applications   183

Doors, door frames, and window frames are now being fabricated in GFRP in factories as far apart as the US and India. These range from internal and external doors for house construction through to heavy-duty industrial doors. Figures 7.5 and 7.6 show typical examples, but many different finishes are available. An Indian manufacturer points out that GFRP doors are termite proof, a significant factor in many parts of the world.

Pultruded composite sections (often similar in shape to standard steel sections) are being used for structures such as the floor support structure shown in Figure 7.7. In some ways this is a strange application of FRP because steel section shapes have been developed to use the material properties of steel which are different from those of FRP. However, FRP requires much lighter lifting gear and is a non-conductor, factors which may be important in a particular project.

Components fabricated from GRP composites are used as architectural features on building exteriors as shown in Figures 7.8 and 7.9, while outbuildings made from composite material are increasingly common (Fig. 7.10).

The potential for the use of advanced composites in construction has been demonstrated by the Eyecatcher Building in Basel, Switzerland. The

7.5 GFRP door for house construction (courtesy of Mitras Composites (UK) Ltd).

7.6 GFRP chemically resistant doors (courtesy of Chempruf Door Company Ltd).

7.7 Composite floor support structure (courtesy of Redman Composites).

Advanced FRP composites for civil engineering applications 185

7.8 GRP pillars and porch (courtesy of IJF Developments Ltd).

7.9 GRP dome on a mosque (courtesy of David Kendall).

load bearing structure of this striking five-storey building is entirely fabricated from composite material. The pultruded GRP profiles used were produced using E-glass fibers and polyester resin by Fiberline Composites in Denmark. Figure 7.11 shows the building which was originally produced for the Swissbau99 exhibition. An important feature is that the load-bearing structure forms part of the façade because the good thermal properties of the GRP do not produce any cold or warm bridges.

7.10 GRP gatehouse (courtesy of Fibaform Products Ltd).

7.11 The Eyecatcher Building (courtesy of Fiberline Composites).

7.3.2 Bridges

There are ongoing problems in the US and elsewhere due to salt-induced degradation of concrete bridge decks. Chloride ions migrate through the concrete and onto the steel reinforcing bar. This causes the steel to corrode and the concrete breaks up because the volume of the corrosion products is greater than that of the original steel. Since advanced composites are very durable and largely unaffected by salt, they offer an alternative to concrete decks.

There has been considerable research into composite bridge decks using pultruded GFRP sections. The original shape considered was the plank developed by Maunsell Structural Plastics in the UK. This section, shown in Fig. 7.12, had rectangular cells which could be connected together in-line or at right angles using adhesive and a connector strip. The planks were used in the innovative all composite Aberfeldy footbridge in Scotland. Unfortunately, Maunsell Structural Plastics no longer exists, but their planks (now called Composolite) are still fabricated by Strongwell in the US.

Although suitable for pedestrian loading, the rectangular cells have drawbacks with vehicle loading. Various pultruded sections have been developed to improve vehicle load-bearing characteristics. Figures 7.13 and 7.14 show, respectively, the Asset and DuraSpan profiles which span transverse to the main girders and can act compositely with the main girders. The Asset profile was first used in the all composite West Mill replacement bridge in Oxfordshire, UK. During its opening ceremony, its

7.12 Strongwell Composolite (formerly Maunsell) planks connected at right angles (courtesy of Strongwell).

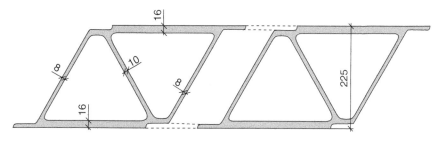

7.13 Asset bridge deck profile (courtesy of Fiberline Composites).

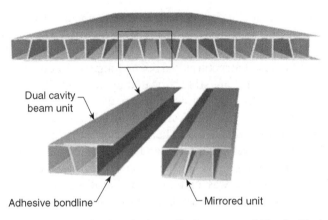

7.14 DuraSpan bridge deck profile (courtesy of Martin Marietta Materials).

7.15 FRP handrail system (courtesy of Pipex Structural Composites).

integrity was demonstrated by means of a Sherman tank. The four main girders consisted of hybrid GRP/CFRP box beams to which the Asset profile deck was adhesively bonded. The DuraSpan profile has been used in a number of deck replacement projects in the US and abroad, the thirtieth application being the Siuslaw Bridge in Florence, Oregon. Both systems come with a recommended wearing surface which has to bond to the FRP as well as provide traction for vehicles. Other profiles are also available.

Handrail and balustrade systems in advanced composites are readily available and competitive in cost with conventional materials. Figures 7.15 and 7.16 show typical details.

Advanced FRP composites for civil engineering applications 189

7.16 FRP balustrade system (courtesy of Pipex Structural Composites).

7.17 GFRP bridge enclosure to soffit of approach bridge, Second Severn Crossing (Maunsell Structural Plastics).

Concerns over steel girders in aggressive environments (coastal or industrial) have led to the development of bridge enclosure systems. These structures are attached to the underside of the bridge, protecting the girders and allowing access for inspection or maintenance. GFRP has proved the ideal material for the skin of the enclosure because of its low weight. Figure 7.17 shows the enclosure on one of the approach bridges to the Second Severn Crossing, Avon, UK.

Aramid (trade name Twaron or Kevlar) fibers have high tensile strength and have been formed into cables, one trade name being Parafil. Research at Cambridge University, UK (Burgoyne, 1993) has developed specialized end anchors for the FRP cables. As a result, these cables have been used on cable-stayed bridges, for example the Aberfeldy Bridge, Perth and Kinross, Scotland, shown in Fig. 7.18. This footbridge, already mentioned above, is particularly interesting. It is set in a golf course which straddles the River Tay and was designed by Maunsell Structural Plastics and constructed by a group of Dundee University students under the supervision of Professor Bill Harvey (Harvey, 1993). It was amongst the first all advanced composite bridges to be constructed anywhere in the world and, during almost twenty years in service, has required minimal maintenance.

7.3.3 Infrastructure

The uses of advanced composites in infrastructure applications are diverse and surprisingly large. This section will present a selection which will give an impression of that diversity.

7.18 General view of Aberfeldy footbridge.

Advanced FRP composites for civil engineering applications   191

The construction of roads in or close to built-up areas has resulted in the need for noise barriers which mitigate the noise pollution caused by traffic. Traditional barriers have been constructed of timber and concrete. However, FRP barriers are also available and come in two versions. Reflective barriers have solid surfaces and deflect incident sound back towards the road. Absorbent barriers have perforated GRP faces with a fiberglass wool infill. Sound passing through the perforations is absorbed by the infill, while some sound is reflected by the solid parts of the face. Figures 7.19 and 7.20 show both types of barrier, one manufactured in the UK, the other in China.

Pylons for supporting power and other utility cables often have to be installed in inhospitable terrain. The low weight and excellent durability of FRP composites helps to minimize the resulting installation and

7.19 GRP reflective noise barrier (courtesy of NCN-UK).

7.20 GRP absorbent noise barrier (courtesy of Jiangsu Shuangying Acoustics Equipment Co. Ltd).

maintenance difficulties. Figure 7.21 shows typical details. Signs for displaying information on motorways are also being fabricated in FRP composites as shown in Fig. 7.22. In Dubai there are a surprising number of GRP palm trees which conceal mobile phone masts.

FRP composites are widely used in the water treatment and chemical industries. Storage tanks up to 160,000 litres capacity are readily available.

7.21 FRP utility poles (a) installation in difficult terrain, (b) complex crosshead arrangement (courtesy of Creative Pultrusions Inc).

7.22 Motorway sign gantry (courtesy of NGCC).

Special lining resins are used to provide exceptional chemical resistance. Figure 7.23 shows a typical tank. Another possibility is the GRP sectional tank which is made up of panels connected together, possibly with an internal supporting structure, as shown in Fig. 7.24. FRP tank covers for storage tanks are a cost-effective alternative to conventional covers. The covers can be freestanding or supported by an FRP truss system allowing spans up to 30 m. Figure 7.25 shows a freestanding cover on an underground tank.

A common sight in built-up areas is the excavations made by utility companies as they repair or update their services. The resulting trenches can cause access problems for residents and delivery companies. A simple but very useful solution has been the GRP temporary access covers shown in Figs 7.26 and 7.27. These can be 1 m square and about 25 mm thick, and are suitable for light traffic or heavy duty capable of supporting lorry traffic. A useful feature is that they can be colored so that they are easily recognizable by their owner.

This section has presented a selection of advanced composite artefacts that are now being widely used in the construction industry. Most are competing with conventional materials such as steel or concrete and are gaining a market position because of their durability and low weight.

7.23 GRP storage tank (courtesy of Forbes Technologies).

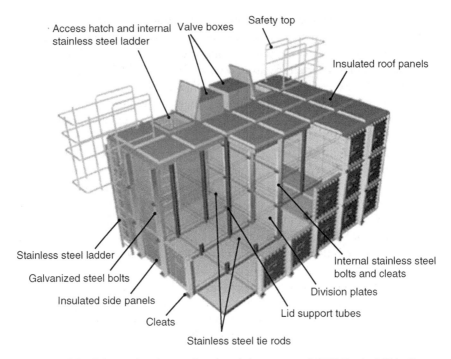

7.24 Schematic of a sectional tank (courtesy of GRP Tanks UK Ltd).

7.25 GRP tank cover (courtesy of Fiberglass Fabricator Inc).

7.26 Lightweight trench covers.

7.27 Heavy-duty trench covers (courtesy of Redman Composites).

7.3.4 Railway infrastructure

Railways have particular infrastructure needs which differ from mainstream construction. GRP composites have been used in a variety of applications. Figure 7.28 shows a station platform system made of GRP and with a non-slip wearing surface. Other applications have included trays and frames for carrying cables and electrical components (Fig. 7.29), water collection pits for drainage systems (Fig. 7.30) and ballast retaining supports.

7.28 Station platform system (courtesy of Pipex Structural Composites).

7.29 GRP cable tray (courtesy of Marshall Tufflex Ltd).

Advanced FRP composites for civil engineering applications

7.30 Water catchpit for rail track drainage applications (courtesy of Marton Geotechnical Services).

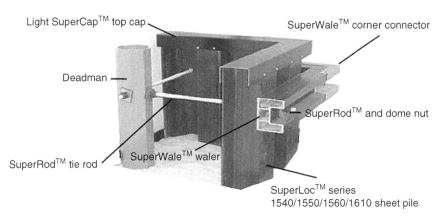

7.31 Components of GRP sheet piling system (courtesy of Redman Composites).

7.3.5 Geotechnical applications

FRP systems are being used in geotechnical applications in increasing numbers. The long-term durability of FRP can be very attractive in aggressive environments which would corrode steel and its low weight is an advantage in areas such as steep cliffs where access is difficult. This section will look at typical current applications.

Sheet piling has traditionally been constructed using steel sheets. GRP is now providing a viable alternative and Fig. 7.31 shows the components of a typical system.

A common requirement in construction is for slope or rockface stabilization, whether for relatively small embankments or large cliff faces. A typical approach is the use of soil nails or rockbolts with a structural grid attached to the nail heads. Holes are drilled into the face to be stabilized, the nails or bolts are inserted to the required length and are then grouted in place. Finally, if necessary, a net is attached to the nail or bolt heads. GRP systems can have significant advantages because traditional steel nails and bolts need corrosion protection and are heavy. Figure 7.32 shows typical GRP soil nails and rockbolts, and Figs 7.33–7.35 show their use in embankment, cliff face (with difficult access), and retaining wall stabilization.

An interesting development reported by Ortigao (1996) is the use of geobars as temporary soil nails in tunneling applications. A GRP geobar is a tube which may have valves at intervals along its length, which may be as long as 30 m. The geobars are inserted into drilled holes at the tunnel drive face and grouted in place by pumping resin through the tube and valves; they stabilize the drive face but are sacrificed as driving proceeds. The

7.32 GRP soil nails and rock bolts (courtesy of Minova Weldgrip).

7.33 GRP soil nail and netting, A638 near Wakefield, Yorkshire, UK (courtesy of Minova Weldgrip Ltd).

Advanced FRP composites for civil engineering applications 199

7.34 Installation of GRP soil nails and rock bolts, Dawlish cliff face stabilization, Devon, UK (courtesy of Minova Weldgrip Ltd).

7.35 Epoxy soil nails for retaining wall stabilization, Nant Ffrancon wall stabilization, A5 Trunk Road, North Wales (courtesy of Minova Weldgrip Ltd).

advantage of the hollow tubes is that they offer less resistance to the tunnel boring machine than solid steel or GRP rock bolts.

7.3.6 Pipes

Traditionally, water pipes have been made from cast iron (actually very early pipes were timber). FRP pipes offer considerable advantages; they are fabricated by the pultrusion/fiber winding process shown in Fig. 7.36 which results in an accurate cross-section shape and a very smooth internal finish with low friction. Hydraulically, FRP pipes are more efficient because they suffer little head loss due to internal friction so that for a given application smaller diameter GRP pipes are needed. Hence, it is frequently possible to replace older pipes by pushing the FRP replacement through the original. New pipes can range from 50 mm to 4 m in diameter and are used in the water, chemical, petroleum, and gas industries up to internal pressures of 175 bar. Figure 7.37 shows a large diameter water pipe.

A frequent problem with underground clay drainage pipes is that they are damaged and blocked by debris. Contractors can usually clear the blockage, but the pipe needs to be repaired. An ingenious system uses a pitch-based GFRP liner which is threaded through the pipe like a flat hose. The liner is then inflated using water or air pressure until it fits tightly to the inside of the pipe. At this stage the resin catalyst activates the curing process resulting in a solid lining within a few hours. This reduces the effective pipe cross section by about 6% but this is compensated for by the increased hydraulic efficiency. Figure 7.38 shows the benefit of this type of repair. Other localized pipe repair systems use GFRP tape wrapped around the pipe and cured *in situ*; some of these systems can even be used underwater.

7.3.7 Conclusion

The intention behind this section has been to give an impression of the wealth of applications for advanced composite materials within the

7.36 Production of GRP pipe (courtesy of Subor Pipe Production Inc).

7.37 Large diameter GRP water pipe (courtesy of Subor Pipe Production Inc).

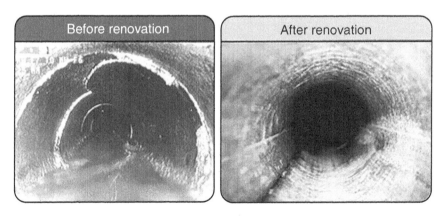

7.38 Pitch-based GRP pipe lining (courtesy of Drainline Southern Ltd).

construction industry. The examples presented have come from all over the world, thanks to the Internet! Fabricators are always on the lookout for new applications and would welcome enquiries from potential customers.

7.4 Future trends

It is always risky to predict what will happen in the future; however, it is possible to give some indications. At a basic level there is room for considerable expansion within existing markets. The durability of composite materials makes them very attractive in comparison to conventional materials which require regular maintenance. Also their low weight means that the installation of composite components is less labour intensive, another factor which can influence cost.

However, designers are becoming more confident in their use of composite materials and some exciting structures have been created as shown in the Brisbane river walkway (Fig. 7.39) and the GRP classroom in Fig. 7.40. Figure 7.41 shows a modular construction system for houses being developed by Startlink and based on just nine pultruded FRP profiles that bolt and snap-fit together enabling rapid assembly. It is predicted that this concept could provide more economic, thermally efficient, and sustainable housing than conventional materials.

It is considered feasible to fabricate much larger structures from FRP composites, and spans of over 200 m have been suggested. In fact, the Millenium Dome, now renamed the O_2 Arena, in London uses GFRP for the dome covering. The low weight of the material would mean that less supporting structure would be required allowing considerable freedom and flexibility in the internal layout. Potential applications could include schools, offices, retail, industrial, or exhibition buildings. Perhaps this is the way forward.

7.39 Brisbane river walkway constructed using GRP, Brisbane, Australia.

7.40 GRP classroom, complete and under construction (courtesy of White Young Green).

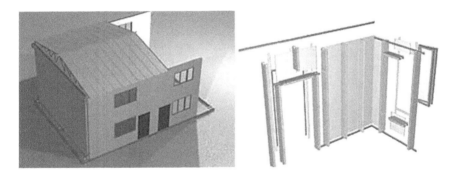

7.41 Startlink modular construction system.

7.5 Sources of further information

Most of the information presented in this chapter has been found by searching the web. That is as good a starting point as any. Companies give contact details and in many cases quite detailed information including specifications. The author of this chapter is based in the UK where very useful sources of information are the Network Group for Composites in Construction (NGCC) (www.ngcc.org.uk) and the National Composites Network (NCN) (www.ncn-uk.co.uk). Both organizations promote the use of advanced composite materials, organize technical and training events, and have helplines which can be accessed via their websites.

7.6 References

Burgoyne, C.J. (1993), Developments in the use of unbonded parallel-lay ropes for prestressing concrete structures. *Proceedings FIP Symposium '93, Modern Prestressing Techniques and their Applications*, Volume 2, 727–734, Kyoto, Japan, October 1993.

Cutter, P.A., Moy, S.S.J., Shenoi, R.A. (2009), Predictive Methods for the Fire Resistance of Single Skin and Sandwich Composite Materials. *Advanced Composites in Construction, ACIC2009, Proceedings of the Fourth International Conference*, University of Edinburgh, Scotland, September.

Egan, J. (1998), *Rethinking Construction: Report of the Construction Task Force*. Department of Trade and Industry, HMSO, London.

Harvey, W.J. (1993), A Reinforced Plastic Footbridge, Aberfeldy, UK. *Structural Engineering International*, 3 (4), 229–232.

Latham, M. (1994), *Constructing the Team – Final Report of the Government/Industry Review of Procurement and Contractual Arrangements in the UK Construction Industry*. Department of the Environment, HMSO, London.

Ortigao, J.A.R. (1996), FRP applications in geotechnical engineering. *ASCE 4th Materials Conference*, Washington, DC, November.

8
Hybrid fiber-reinforced polymer (FRP) composites for structural applications

D. LAU, City University of Hong Kong, P. R. China

DOI: 10.1533/9780857098955.2.205

Abstract: Fiber-reinforced polymer (FRP) has been a practical alternative construction material for replacing steel in the construction industry for several decades. However, some mechanical weaknesses of FRP are still unresolved, which limit the extensive use of this material in civil infrastructure. In order to mitigate the disadvantage of using FRP, the concept of hybridization is delivered here. The advantages of hybrid structural systems include the cost effectiveness and the ability to optimize the cross section based on material properties of each constituent material. In this chapter, two major applications of hybrid FRP composites are discussed: (1) the internal reinforcement in reinforced concrete (RC) structures, and (2) the cables in long-span cable-stayed bridges. In order to improve the flexural ductility of FRP-reinforced concrete (FRPRC) beam, the additional steel longitudinal reinforcement is proposed such that the hybrid FRPRC beams contain both FRP and steel reinforcement. In order to improve the vibrational problem in pure FRP cables used in bridge construction, an innovative hybrid FRP cable which can inherently incorporate a smart damper is proposed. The objective of this chapter is to deliver an up-to-date review of hybrid FRP composite structures, including both the industrial practice and the research in academia. The advantages of using hybrid FRP composites for construction will also be described with experimental support. It is hoped that the reader will appreciate the concept of hybridization, which leads to the efficient utilization of all constituent materials in a bonded system.

Key words: ductility, fiber-reinforced polymer (FRP), hybrid, reinforced concrete (RC).

8.1 Introduction

Fiber-reinforced polymer (FRP) has become a practical alternative construction material in various structural aspects. It can be used externally to improve the flexural, shear, and axial capacities of beams, slabs, columns, and shear walls made by reinforced concrete (RC) (Grace et al., 1996; Triantafillou, 1998; Deniaud and Cheng, 2003; Büyüköztürk et al., 2004; Bruno et al., 2007; Greco et al., 2007). Also, it can be used as internal reinforcement, replacing conventional steel bars in RC structures due to its advantages, such as corrosion resistance, non-conductivity, high strength, and light weight. Research related to experimental studies and *in-situ* applications of FRP bars in RC structures can be found in various publications (Hollaway, 1978; Aiello and Ombres, 2000; Peece *et al.*, 2000). A good review of practical applications of FRP bars can be found in Rizkalla and Nanni (2003). In the past two decades, FRP bars have been widely used in some countries, such as the United States of America, Canada, Germany, Switzerland, and Japan, in bridge deck and road construction owing to the seasonal use of de-icing salts which cause traditional steel reinforcement to corrode. Meanwhile, some concrete structures require non-metallic material as the constituent materials, such as the magnetic resonance imaging (MRI) rooms in hospitals or research laboratories, as well as the roads and bridge decks near electronic toll plazas. In all these special circumstances, FRP bars are good substitutions for the conventional steel in RC structures. Besides using FRP bars as internal reinforcement, these bars can also be made as cables due to their superior tensile strength compared to the conventional steel cables. Carbon FRP (CFRP), which has the best mechanical and chemical behavior among different kinds of common FRP materials, was initially proposed for use in long-span cable-stayed bridges. Several studies have already demonstrated its high static and dynamic performance (Meier, 1987). However, the high and continuously increasing cost of CFRP limits their applications in new structures, especially in large-scale constructions, such as long-span bridges. Additionally, the sensitivity of CFRP cables to wind load is difficult to control due to their extremely light weight and high strength. Considering the limitations of CFRP, the feasibility of using various FRP materials as stay cables has been investigated (Wu and Wang, 2008), with a conclusion that a new type of cable which combines both CFRP and basalt FRP (BFRP) is suitable to be used in bridge construction due to its chemical stability, high stiffness, and low cost.

Although FRP possesses many superior material properties, such as high specific stiffness, high specific strength, the high corrosion resistance, and durability, the high cost and brittle nature of FRP prevent it from being commonly used in the industry. In order to overcome these obstacles and to make the best use of the material, combinations of FRP and conventional materials have recently been investigated by a number of researchers (Alnahhal *et al.*, 2006). An improvement of the structural performances in buildings and bridges can be obtained by utilizing a combination of FRP and steel (Newhook, 2000; Aiello and Ombres, 2002) or, alternatively, by combining various types of FRP materials (Wang and Wu, 2011a). The advantages of hybrid structural systems include the cost effectiveness and the ability to optimize the cross section based on material properties of each constituent material. In other words, the purpose of hybridization is to create a new material that picks out the advantages of each constituent, while the weaknesses can be improved. In general, hybridization is a positive effect of any property obtained through the rule of mixture (Kretsis, 1987). For instance, if the ultimate tensile strength of an FRP member composed of more than two types of fiber is higher than that of a fiber with the lowest strength, it is regarded as a positive hybrid effect obtained by hybridization. Nanni *et al.* (1994a, 1994b) tested numerous bars of braided aramid fiber around a steel core in an epoxy matrix and obtained a bilinear stress–strain behavior. However, such hybrid bars have limited flexibility to the steel distribution within the cross section when they are used in RC structures. Bakis *et al.* (1996) suggested that the high modulus material must be dispersed over the entire cross-sectional area in order to maximize the ductile behavior. Adopting two different types of bars allows a more uniform distribution of stiffer material in a cross section, and this approach has recently been verified through a comprehensive experimental program (Lau and Pam, 2010).

In this chapter, two major applications of hybrid FRP composites are discussed: (1) the internal reinforcement in RC structures, and (2) the cables in long-span cable-stayed bridges. Although the hybrid FRP systems found in these two scenarios are very different, they both achieve the same goal, i.e. to improve the global structural behavior by optimizing the cost effectiveness. The objective of this chapter is to deliver an up-to-date review of hybrid FRP composite structures, including both the industrial practice and the research in academia. The advantages of using hybrid FRP composites for construction will also be described with experimental support.

8.2 Hybrid fiber-reinforced polymer (FRP) reinforced concrete beams: internal reinforcement

Research into FRP reinforcement bars (rebars) in RC structures has been conducted continuously since the late 1990s and is still ongoing, especially in the area of improving the structural performance of FRP reinforced concrete structures, especially their ductility and stiffness. Conventionally, RC structures are designed by the ultimate strength approach, in which the ductility is a great concern in the design process. Because of the inherent brittleness of FRP rebars, Mufti *et al.* (1996) suggested that the design of RC sections using FRP rebars should be based on the concept of deformability, rather than the usual concept of ductility adopted in steel-reinforced sections. Yet, the need for an FRP-reinforced concrete (FRPRC) section with ductile characteristics remains.

In general, the composite system constituted by FRP rebars and RC, which is called FRPRC, possesses less ductility. In order to improve the ductility of such composite system, a certain amount of ductile material is proposed to be added in the FRPRC system such that the brittleness of FRP rebars can be compensated. On the research front, FRPRC flexural members have been investigated over the last decade or so. However, the ductility issue of FRPRC beams is still a problem to be solved. Even though it is mentioned in ACI440.1R-06 (2006) that the flexural members should possess a higher reserved strength in order to compensate the lack of ductility, the ductility of beam members cannot be neglected because it is closely related to the safety, especially when there is a catastrophic load (e.g., earthquake load) leading to a structural failure. Ductility is important and cannot be overlooked in any situations because it gives ample observable warnings before failure so that the loss of human life can be greatly reduced.

With the limited ductility of FRPRC beams, practising engineers are reluctant to adopt FRP rebars in the construction industry. Although some researchers have carried out experimental testing on FRPRC beams recently (Newhook, 2000; Aiello and Ombres, 2002), those experimental data were limited to a certain design range and it is still unclear in what situation the ductility improvement can be effectively achieved by the addition of steel rebars. Recently, Lau and Pam (2010) have conducted an extensive experiment which enables us to understand the flexural behavior of FRPRC beams in an approximate construction scale. Based on their valuable experimental results, several guidelines on the ductility improvement of FRPRC beams have become available. In what follows, some discussions and comparisons on the structural behavior of various flexural concrete beams, including normal reinforced concrete beams (SRC), pure FRPRC beams, and hybrid FRPRC beams, will be described based on the experiments conducted recently (Lau and Pam, 2010). These beams were designed such that both under-reinforced and over-reinforced design scenarios were covered.

8.2.1 Analysis of FRPRC beams

Traditionally, the balanced reinforcement ratio is a very important parameter for designing RC beams. For an SRC beam, the balanced steel reinforcement ratio (ρ_{bs}) is a condition for which the beam is designed to fail by crushing of concrete in compression and yielding of steel in tension simultaneously. For a pure FRPRC beam, the balanced FRP reinforcement ratio (ρ_{bf}) refers to the condition that the beam is designed to fail by crushing of concrete and rupture of FRP rebars simultaneously. Combining these two cases, the balanced reinforcement ratio for a hybrid FRPRC beam should refer to a failure condition in which crushing of concrete, yielding of steel and rupture of FRP happen simultaneously. However, in practice, it is almost impossible for this situation to happen. The steel reinforcement will have yielded long before the rupture of the FRP reinforcement. Hence, the balanced condition for a hybrid FRPRC beam is proposed in a way such that concrete crushing in compression and rupture of FRP reinforcement occur at the same time, while its steel counterpart has already yielded. Hence, the balanced reinforcement ratio provided in ACI440.1R-06 (2006) can be used for both the pure and hybrid FRPRC beams as shown in Eq. [8.1].

$$\rho_{bf} = 0.85\beta_1 \frac{f'_c}{f_{fu}} \frac{E_f \varepsilon'_c}{E_f \varepsilon'_c + f_{fu}} \quad [8.1]$$

where β_1 = the ratio between the depth of equivalent rectangular concrete stress block and the neutral axis depth, $\varepsilon'_c = 0.003$ = extreme fiber concrete compressive strain in conjunction with f'_c. For both the pure and hybrid FRPRC beams, the equivalent rectangular stress block of concrete recommended in ACI318M-02 (2002) should be used. It is reminded that the reinforcement ratio (ρ) for all the FRPRC beams is proposed to be named the effective reinforcement ratio, and with the definition shown in Eq. [8.2]:

$$\rho = \frac{A_s m + A_f}{bd} = \rho_s m + \rho_f \quad [8.2]$$

where A_s = area of steel reinforcement, A_f = area of FRP reinforcement, $m = f_y/f_{fu}$, ρ_s = steel reinforcement ratio, and ρ_f = FRP reinforcement ratio. Hence, the balanced reinforcement ratio for a hybrid FRPRC beam can be calculated from Eqs [8.1] and [8.2] with any combination of steel and FRP reinforcement content under the condition that the steel has yielded. The beam specimens which are going to be discussed in this chapter were all designed based on the balanced reinforcement ratio defined above. Based on the respective actual tensile strength(s) of the rebars, the 28th day concrete strength and the effective reinforcement ratio, the theoretical moment capacity (M_n) of hybrid FRPRC beams can be evaluated by the section analysis. It has to be noted that the above discussion does not include the partial safety factors for material strengths.

Based on the laboratory testing on FRPRC beam specimens having a length scale comparable to structural components found in buildings and bridges, three important aspects related to the structural behavior will be discussed, namely: (1) flexural strength and ductility improvement; (2) minimum flexural FRP reinforcement content; and (3) effectiveness of 135° hooks compared to 90° hooks in stirrups. It has to be noted that all these beams were designed to fail in flexure around the midspan so as to evaluate the contribution of FRP rebars to the flexural capacity of FRPRC beams, and the shear failure was prevented by providing excessive shear reinforcement (double the amount required) at the critical locations.

8.2.2 Flexural strength and ductility improvement of FRPRC beams

During the experimental test, the measured maximum moment (M_{exp}) or the actual moment at midspan can be calculated from the corresponding measured maximum load (P_{exp}) that was obtained by Eq. [8.3]:

$$M_{exp} = \frac{P_{exp}L}{4} \qquad [8.3]$$

Based on the experimental result, it is noticed that most of M_{exp} were still over-estimated when compared with the respective theoretical counterpart (M_n). The over-estimation or reserved strength ranges from 0 to 33%, and this will even be greater if the partial safety factors are taken into account. Therefore, the provisions of ACI440.1R-06 (2006) to determine the flexural strength of FRPRC members are sufficient.

Even though FRP is a brittle material, once it is embedded in concrete, the beam can possess a certain amount of yielding when it is over-reinforced (failed upon concrete crushing). The general load–displacement curves of FRPRC beams with the same moment capacity under a three-point bending situation are shown in Fig. 8.1. As shown in Fig. 8.1, the pure FRPRC beam which only contains FRP bars as reinforcement was the most brittle compared to the others. But still, it behaved slightly ductile because the beam was designed as an over-reinforced section. Hence, contrary to the common design practice in SRC beams, the over-reinforced design is preferred to the under-reinforced design in FRPRC beams. The structural performance of the hybrid FRPRC specimen has been improved in terms of flexural ductility and final flexural strength. When considering the two hybrid FRPRC beams, it is obvious that their stiffness was dictated by the value

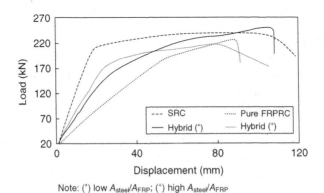

Note: (*) low A_{steel}/A_{FRP}; (+) high A_{steel}/A_{FRP}

8.1 Typical load–displacement curves for pure and hybrid FRPRC over-reinforced beam.

of the ratio of steel and FRP reinforcement content (A_{steel}/A_{FRP}). A higher A_{steel}/A_{FRP} value results in a higher stiffness value. This observation is important and useful for determining a suitable ratio of A_{steel}/A_{FRP} for a hybrid FRPRC beam, so that its stiffness can meet the required performance demand during the serviceability limit state. It is also evident that the ductility of FRPRC beams can be improved by adding steel rebars. By comparing the load–displacement curves of the pure and hybrid FRPRC beams, it is obvious that those hybrid ones have a greater ductility. Although it is possible that there is a sudden drop of load capacity in hybrid FRPRC beams caused by the rupture of FRP rebars, a certain amount of residual ductility can still be maintained as the steel rebars are still far below their breaking point. The experimental results confirm the effectiveness of steel reinforcement improving significantly both the stiffness and ductility of hybrid FRPRC beams when compared to those pure FRPRC beams.

The flexural ductility of FRPRC beam can be measured in terms of two recommended parameters, namely the displacement ductility factor (μ) and the ultimate displacement ratio (Δ_u/L) (Lau and Pam, 2010). The displacement ductility factor is a dimensionless number (μ) to study the ductility improvement of the FRPRC beams and is defined as the ratio of midspan displacement at ultimate stage (Δ_u) and at yield stage (Δ_y), where the former is obtained at $0.8P_{exp}$ after reaching the peak, while the latter is obtained by linearly interpolating the displacement at $0.75P_n$ to the level of P_n. It is obvious that the balanced-reinforced (or under-reinforced) pure FRPRC member has a value of μ very close to 1, which means that the beam has no ductility in general. The addition of steel rebars can effectively increase the ductility of pure FRPRC beams, and the increase is considerably larger in the over-reinforced member than the balanced-reinforced one in which the ductility improvement can be more than 100%. Although there is a ductility improvement in terms of μ in the balanced- or under-reinforced member by adding the steel rebars, such improvement is solely due to the decrease in the yield displacement (Δ_y), rather than the increase of the ultimate displacement (Δ_u). In fact, the over-reinforced hybrid members have high structural efficiency compared to the balanced- and under-reinforced counterparts, since all materials (concrete, FRP, and steel) reach their strength capacities at failure.

It is rather disadvantageous to use the displacement ductility factor because the yield deflection (Δ_y) is difficult to assess theoretically. In addition, in over-reinforced FRPRC beams, as shown in Fig. 8.1, the stiffness may reduce during the elastic stage and this will make the obtained yield displacement rather inaccurate. For the balanced-reinforced FRPRC section, the yield deflection is almost similar to the ultimate deflection. In order to avoid these problems, it is suggested to use the displacement ratio (Δ/L) to measure the beam deformability.

By definition, the effective reinforcement ratio in the over-reinforced hybrid beam is higher than its balanced reinforcement ratio. It should be mentioned that the addition of steel rebars improves the ultimate displacement ratio of the over-reinforced hybrid FRPRC beams under a situation which fulfills the following two conditions. Firstly, the FRP reinforcement content should be higher than the minimum amount in the ACI code, which means that the FRP rebars have some reserved strain when the beam is about to reach the maximum flexural capacity. Secondly, the A_{steel}/A_{FRP} ratio should not be too high such that the FRP rebars can still play an important role to resist the loading even after the yielding of steel rebars. In the presence of these two conditions, the hybrid FRPRC beam can achieve a larger ultimate deflection when compared with the pure FRPRC beam having the same moment capacity.

In conclusion, there is a ductility improvement when steel rebars are added to pure FRPRC beams. The ductility improvement is higher in over-reinforced FRPRC beams than in the under-reinforced counterparts. There are three recommendations for the design of hybrid FRPRC beams:

(1) the amount of FRP rebars should be larger than the minimum FRP reinforcement content recommended by ACI 440.1R-06 (2006);
(2) the amount of FRP reinforcement should be larger than that of steel reinforcement; and
(3) the effective reinforcement ratio should be larger than the balanced FRP reinforcement ratio (ρ_{bf}).

8.2.3 Minimum flexural FRP reinforcement content

Based on the results of the experiments conducted previously, it is found that there is room for the reduction of the minimum flexural FRP reinforcement content in FRPRC beams (Lau and Pam, 2010). The concept of having a minimum content of flexural reinforcement, whether steel or FRP, is to ensure that the beam has a moment capacity greater than its cracking moment. It has been reported that the moment capacity of a pure FRPRC beam with only 75% of the minimum FRP content recommended by ACI440.1R-06 (2006) is still nearly double that of the SRC beam with the minimum reinforcement content as recommended by ACI318M-02 (2002) and about quadruple that of the plain concrete (PC) beam counterpart. Typical load–displacement curves for the RC beams with different degrees of the minimum reinforcement content, together with the plain concrete beam counterpart, are shown in Fig. 8.2. Figure 8.3 shows the failure snapshots of the beams having 75% of the minimum FRP content and 100% of the minimum FRP content recommended by ACI440.1R-06 (2006). Figure 8.4 shows the deflection profiles of these two beams. It is noticed the shape of the deflection profile in these two types of beam resembles a parabola

Hybrid FRP composites for structural applications

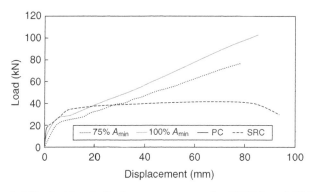

8.2 Typical load–displacement curves for FRPRC and SRC with minimum reinforcement content based on ACI code.

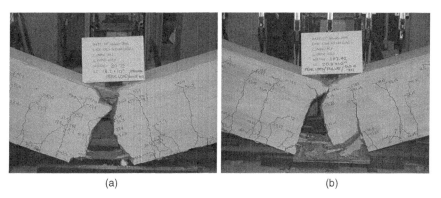

8.3 (a) The failure mode of FRPRC beam with 75% A_{min} which is very brittle; (b) the failure mode of FRPRC beam with 100% A_{min} which is very similar to that of 75% A_{min}.

and they behave very similarly to each other. More importantly, this information further validates that the reduction in the minimal FRP content is allowable which leads to a more economical design of both pure and hybrid FRPRC beams.

8.2.4 Effectiveness of 135° hook in stirrup

The 135° hook in the stirrups is very effective in improving the ductility of an over-reinforced FRPRC beam. Typical load–displacement curves of an over-reinforced FRPRC beam with different hook angles are shown in Fig. 8.5 (Lau and Pam, 2010). From Fig. 8.5, it is observed that the maximum loads of the two specimens are approximately the same. However, it is obvious that the over-reinforced FRPRC beam with 135° hook in the stirrups has much better deformability than that with 90° hook in the stirrups

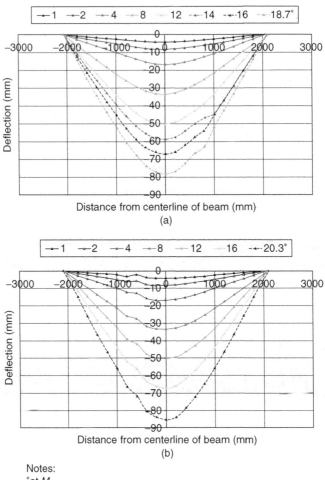

Notes:
*at M_{exp}
All values in the legend are $\Delta/L \times 1000$.

8.4 (a) The deflection profile of FRPRC beam with 75% A_{min}; (b) the deflection profile of FRPRC beam with 100% A_{min}.

due to better confinement as a result of its 135° hook stirrups. This phenomenon is corroborated in Table 8.1, in which the ultimate displacement ductility factor and displacement ratio of the over-reinforced FRPRC beam with 135° hook in the stirrups are respectively 77% and 35% higher than that with 90° hook in the stirrups. By comparing the ultimate displacement ductility factor of the under-reinforced FRPRC beam and that of the over-reinforced FRPRC beam with 135° hook in the stirrups, it is found that there is a significant improvement in the ductility (an increase of μ by 184%) by using both the over-reinforced design approach and the stirrups

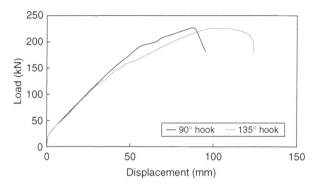

8.5 Typical load–displacement curves for over-reinforced FRPRC beam with different degrees of hook in stirrups.

Table 8.1 Properties and test results for the study of minimum reinforcement content

Unit	Min. reinf. area (mm²)		Mom. capacity (kNm)					
	Required	Actual	M_{exp}	M_n	M_{exp}/M_n		Δ_u (mm)	Δ_u/L
75% A_{min}	468	339	80.4	64.3	1.25		78.36	18.7×10^{-3} (*)
100% A_{min}	470	452	107.3	85.2	1.26		85.32	20.3×10^{-3} (*)
SRC	138	226	44.0	36.4	1.21		91.09	21.7×10^{-3} (#)
PC	–	–	19.6	22.2	0.88		0.88	0.21×10^{-3} (+)

Notes on failure modes: (*) FRP rupture; (#) Steel rupture; (+) Concrete breakage.

of 135° hooks (Lau and Pam, 2010). Hence, it is recommended that the stirrups of 135° should be applied whenever the over-reinforced design approach is adopted in the FRPRC beams.

All the above findings prove the effectiveness of 135° hook stirrups in improving the deformability and ductility of FRPRC beams. However, such improvement can only happen in over-reinforced FRPRC beams. In over-reinforced beams, concrete crushing will happen prior to failure. Stirrups with 135° hooks can improve the confinement of concrete in the compression region and hence the ductility improvement can be achieved.

8.2.5 Design philosophy of hybrid FRPRC beams

According to ACI440.1R-06 (2006), the flexural design method of a pure FRPRC member is similar to that of a conventional SRC member by adopting the strength design approach with a different strength reduction factor. However, due to the brittle nature of FRP bars, the design philosophies of FRPRC and SRC members are very different. SRC beam sections are normally designed for under-reinforcement to ensure yielding of steel prior to crushing of the concrete. The yielding of steel causes a ductile failure with prior ample warnings. Tension failure in an FRPRC member due to FRP reinforcement rupture is sudden and brittle with hardly any warning. The non-ductile behavior of FRPRC reinforcement makes it suitable for an FRPRC member to have compression failure by concrete crushing, which exhibits some warning prior to failure. This requires the FRPRC member to be designed for over-reinforcement. Tension failure in an FRPRC member is acceptable only if the member possesses higher reserved strength. The balanced FRP reinforcement ratio ρ_{bf} is an important factor in the flexural strength design of FRPRC members. If the FRP reinforcement ratio ρ_f is less than ρ_{bf}, failure mode by FRP rupture will govern or the section is under-reinforced.

Let us consider two pure FRPRC beams with the same cross section but different FRP reinforcement content. Also, let us assume the constituent material properties (concrete and FRP strengths) at the design stage are originally identical, so that ρ_{bf} of these two beams should also be the same. At the design stage, the tensile strength and elastic modulus of FRP reinforcement recommended by the manufacturer are adopted. However, on the testing day, the actual compressive strength of the concrete was different from the value adopted at design. In addition, the actual tensile strength and elastic modulus of the GFRP are different from those recommended by the manufacturer. As a result, there should be a discrepancy between the ρ_{bf} evaluated during the design stage and on the testing date. According to ACI440.1R-06 (2006), section failure is uncertain if $\rho_{bf} \leq \rho_f \leq 1.4\rho_{bf}$. The uncertainty is due to variation in the actual strengths of the concrete and FRP. Table 8.2 summarizes all the parameters that lead to ρ_{bf} at the design stage and on the testing day for these two beams. It is noticed that the increase of ρ_{bf} from the design to testing stage for the

Table 8.2 Comparison of ρ_{bf} at design and testing stage

Unit	Stage	f_{fu} (MPa)	E_f (GPa)	f'_c (MPa)	ε_{cu}	ρ_{bf} (%)	ρ_{bf} testing ρ_{bf} design
Under-reinforced	Design	670	40	32.0	0.003	0.52	–
Over-reinforced							
Under-reinforced	Testing	593	40	36.6	0.003	0.75	1.44
Over-reinforced	Testing	582	38	41.3		0.84	1.62

under-reinforced and over-reinforced beams is about 44% and 62%, respectively, both of which are greater than 40% as recommended by ACI440.1R-06 (2006). Hence, the limit of $1.4\rho_{bf}$ for concrete crushing failure should be increased.

It is recommended that the over-reinforced beam design should be adopted, as a more ductile manner in the beam and sufficient warning prior to beam failure can be obtained. Conventional steel reinforcement in addition to FRP reinforcement could enhance the ductility of pure FRPRC specimens. It is recommended that concrete crushing failure should happen before FRP rupture for pure FRPRC members, while for hybrid FRPRC members, steel yielding should happen first, followed by concrete crushing and lastly by FRP rupture. Also, the amount of steel and FRP reinforcement in a hybrid FPRRC member should be combined such that a certain amount of ductility can be maintained. To prevent excessive elongation that causes rupture of the FRP reinforcement, the amount of GFRP reinforcement should be larger than that of the steel reinforcement and should also be greater than the minimum FRP reinforcement content recommended by ACI440.1R-06 (2006). It is also important to note that the effective reinforcement ratio should be larger than the balanced FRP reinforcement ratio (ρ_{bf}).

In an FRPRC beam, FRP reinforcement is responsible for taking up the strength, while the role of steel reinforcement is mainly for ductility improvement. Hence, it is proposed that the member should be firstly designed as a pure FRPRC member in accordance to ACI440.1R-06 (2006). Subsequently, the member section is checked for the degree of over- or under-reinforcement. If the section is well over-reinforced such that $\rho_f \geq 1.4\rho_{bf}$, it is not necessary to add steel reinforcement. However, if the section is under-reinforced or fairly close to balance-reinforced, it is necessary to add steel reinforcement in order to improve ductility of the member. The most important objective is to make sure that either pure or hybrid FRPRC beams have sufficient warning before failure.

This existing experimental result has demonstrated the superior performance of the hybrid FRPRC beams in comparison with classical SRC, in terms of their ultimate behaviors (Lau and Pam, 2010). The hybrid FRPRC beams possess a higher ultimate strength capacity under the same reinforcement content, especially for the CFRP rebars. Also, the hybrid FRPRC beams can have sufficient ductility when compared to the classical SRC beams. With an appropriate design, hybrid FRPRC beams can combine the advantages of both classical SRC beams (large ductility) and pure FRPRC beams (high ultimate strength capacity). By using the hybrid reinforcement approach, it is expected that a larger allowable strength of the FRP rebars can be used in the design process as the brittleness problem can now be mitigated.

8.3 Hybrid fiber-reinforced polymer (FRP) composites in bridge construction

FRP is always regarded as one of the most suitable materials for building long-span bridges due to the outstanding mechanical properties in longitudinal direction compared with conventional steel material. Figure 8.6 shows the application of FRP in rebars and tendons which can be found in some existing cable-stayed bridges. When FRP is used in stay cables for long-span cable-stayed bridges, it can exhibit essential advantages that address the weaknesses of conventional steel cable (Caetano de Sa, 2007). The major disadvantages of conventional steel cables in a super long-span cable-stayed bridge lie in its pronounced sag effect, which will lower the material utilization and the overall stiffness of the bridge, and the durability deficiency induced by corrosion, which will greatly limit the initial advantages of a cable-stayed bridge with super long span. CFRP cables were initially most investigated to replace steel cables (Meier, 1987; Cheng and Lau, 2006; Kao *et al.*, 2006). Although the superior static and dynamic performance of long-span cable-stayed bridges with CFRP cables was

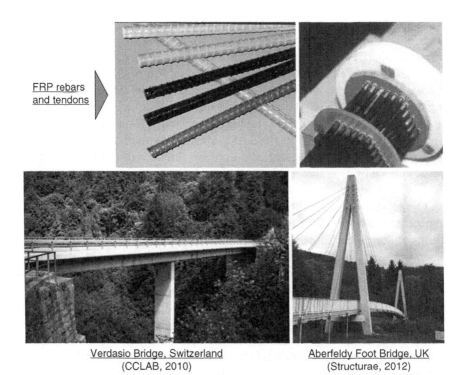

Verdasio Bridge, Switzerland (CCLAB, 2010)

Aberfeldy Foot Bridge, UK (Structurae, 2012)

8.6 FRP rebars and tendons used in some existing bridges.

verified by both theoretical and numerical analyses, the consistently high cost of CFRP cables restricted their practical application in the construction industry. Moreover, CFRP cables are also sensitive to wind effects due to their extremely high strength-to-weight ratio (Wang and Wu, 2009). Therefore, the concept of hybrid FRP cables, consisting of basalt FRP (BFRP) and a small proportion of CFRP, has been developed. It should be mentioned that basalt continuous fibers are an environmentally friendly and nonhazardous material, which is produced from basalt rock by using a single-component raw material and then drawing and winding fibers from the melt. BFRP composites display not only a higher strength and modulus, but also a similar cost and a greater chemical stability compared to E-glass FRP composites (Wu *et al.*, 2009). This hybrid cable possesses high static and dynamic performances (Wu, 2004; Wang and Wu, 2010; Wu *et al.*, 2010), as well as a superior aerodynamic stability and relatively low cost compared to CFRP cables. To further explore the advantages of hybrid FRP cables, the potential ability of vibration control will be described with emphasis on the designable characteristics of hybrid FRP cables.

The design principle of hybrid FRP cable is innovative and can incorporate a smart damper within the hybrid cable. In general, a typical cable consists of parallel steel wires or twisted steel strands as shown in Fig. 8.7(a) and (b). The usual FRP cables such as CFRP, BFRP, and hybrid B/CFRP can also be composed of paralleled wires or twisted strands in a similar manner, where individual wires or stands are made of CFRP, BFRP, or B/CFRP instead of steel. When constructed in this way, the damping properties of FRP cables will be similar to those of steel cables from a structural perspective. Although the material damping of FRP cable is probably higher than that of steel due to the viscoelasticity of the matrix (Wei *et al.*, 2001; Berthelot and Sefrani, 2007; Berthelot *et al.*, 2008), it is still insufficient

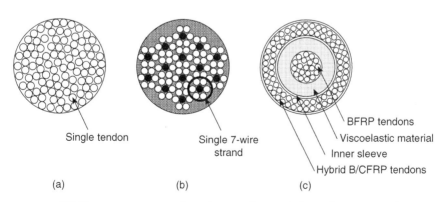

8.7 The arrangement of tendons and strands in various types of cables.

to improve the overall structural damping because of the small amplitude of vibrational strain in the matrix. In a hybrid B/CFRP cable, besides the conventional structure of the cable, a special arrangement of tendons or strands can be designed with an emphasis on enhancing the internal damping, as shown in Fig. 8.7(c). A gap is left between the inner and outer cable where a viscoelastic material can be inserted. The objective of this design is to generate an interaction between the inner FRP tendons and the outer hybrid tendons when the entire cable is excited to vibrate because of the different dynamic characteristic of these two portions of cables. This interaction can act on the inserted viscoelastic material, which will dissipate vibration energy intelligently with respect to the occurrence and the amplitude of vibration.

The inserted viscoelastic material is distributed according to its minimum influence on the cable mechanical behavior and the maximum effect on dissipation of vibration energy. Because the inserted viscoelastic material can be regarded as an additional weight applied along the cable due to its much lower elastic modulus compared with the cable material, it will lead to a negative effect on mechanical behavior such as the sag. Hence, an alternative approach should be introduced such that the inserted material can be discontinuously distributed. In this situation, the amount of inserted viscoelastic material will be greatly reduced and the corresponding influence to the static and dynamic behavior of the cable will be minimized. Furthermore, the energy dissipated by the discontinuous distribution can also be equivalent to that dissipated by the continuous distribution. For the detailed theoretical analysis of the modal damper and the corresponding parametric and example studies, interested readers can refer to a recent publication by Wang and Wu (2011b).

Due to the general deficiency in damping of stay cables in long-span cable-stayed bridges, a smart damper has been developed to improve the internal damping of a hybrid B/CFRP cable. Hybrid FRP cables not only exhibit integrated advantages in static and dynamic behavior for long-span cable-stayed bridges, but they can also provide superior vibration control ability through a proper design of the sectional structure as compared with conventional steel cable and CFRP cable. The principle of smart damper is to generate an interaction between the inner FRP cable and the outer hybrid cable when the entire cable is excited to vibration because of their different dynamic characteristics. This interaction will dissipate vibration energy. The discontinuous distribution of the viscoelastic material along the longitudinal direction of the cable can not only possess an equivalent energy consumption achieved by the continuous distribution method, but can also benefit the static behavior of cable, especially for long stay cable. The effectiveness of using smart damper designed hybrid FRP cable for mitigating large magnitude of in-plane vibration has been demonstrated by an existing bridge structure. It is believed that further experiments on evaluating the practical damping ratio of the cable materials and smart dampers of small-scale stay cables should be conducted before applying this kind of cable commonly in the construction industry.

8.4 Future trends

The use of FRPRC members in the construction industry is increasing in popularity. However, the related design guidelines and provisions still need to be improved for reliable implementation. More research, particularly involving experimental studies, should be performed to improve the understanding of the fundamental behavior of FRPRC members and address various problems (e.g., ductility, tensile strength, compression strength, etc.) before the widespread use of FRP bars in structural applications. Prior to exploiting the use of FRP in construction applications, it is important for future research to include the following:

- Further research on the over-reinforced pure FRPRC members is necessary to revise the degree of over-reinforcement, as it has been proven in this study that ρ_f should be larger than 1.4 times the balance FRP reinforcement ratio (ρ_{bf}).
- Further research on the hybrid FRPRC members is necessary for the aspects of area ratio of FRP to steel and reserved strength. In addition, arrangement of steel and FRP reinforcement needs to be investigated. For example, if the longitudinal steel reinforcement is installed at the outermost layer, the advantage of the FRP bars of being non-corrosive is not fully utilized.
- Research on FRPRC members containing high-strength concrete is necessary to investigate their flexural strength, ductility, and the balance between these two aspects.
- As stated in ACI440.1R-06, the behavior of FRPRC members with tension and compression FRP reinforcement is one of the future research areas. Beam members in a frame structure are required to resist hogging moment when there exist lateral loads, such as wind and earthquake load. Therefore, FRP reinforcement is necessary to resist compression alternately with tension. Thus, research on FRPRC members subjected to cyclic loading is needed.
- In addition to the minimum FRP flexural reinforcement content that has been investigated in this study, research should also be conducted on the minimum FRP reinforcement content for temperature and shrinkage in order to understand the whole picture of the minimum FRP reinforcement content towards both short- and long-term behavior of FRPRC members.

- As the fire resistance of FRP is low, it is necessary to study the minimum concrete cover in order to satisfy the requirement of fire resistance. The effects of adding a protective layer around FRP bars towards the bonding between FRP and concrete should be carefully studied in order to avoid excessive slip.
- Bond strength between concrete and FRP rebars is weak compared to that of steel and concrete. Sand-coated FRP rebars are usually adopted in order to minimize bond slip. Bond strength is very important in FRPRC members to make sure the FRP rebar and concrete on the same level have a similar strain. Therefore, comprehensive research should be carried out on this aspect.
- The different coefficients of thermal expansion between two different materials affect their interfacial bond. In the combination of steel and concrete, a perfect bond is usually assumed because their coefficients of thermal expansion are fairly similar. The thermal expansion of steel is about $13 \times 10^{-6}/°C$, while that of concrete $10 \times 10^{-6}/°C$. However, FRP has a coefficient of thermal expansion much higher in the transverse direction than in the longitudinal direction, and the value is also much higher than that of concrete. The transverse coefficient of thermal expansion of GFRP bars tested by Masmoudi et al. (2005) was equal to $33 \times 10^{-6}/°C$. In order to avoid debonding failure, further research on thermal expansion of FRPRC members should be carried out.
- The difference in Young's moduli between steel and FRP results in uneven load distribution among the rebars. The stiffer material (steel) takes more load due to strain compatibility. Hence, the addition of steel rebars in an FRPRC member decreases the proportion of load carried by the FRP rebars. This is not desirable as it defeats the purpose of using FRP to replace steel as longitudinal reinforcement. Research should be carried out in order to find out the optimum amount of steel that can be added, while FRP is still regarded as the main reinforcement in hybrid FRPRC members.

8.5 Sources of further information

For those interested readers who want to obtain further information about the most up-to-date information on hybrid FRP composites, the following journals are highly recommended:

- *Journal of Composites for Constructions, ASCE*
- *Composites Part B: Engineering*
- *Composites Structures*
- *Engineering Structures*

In fact, the references shown below are all good and current publications which are at the frontier of this research field. The author would like to recommend the readers to go through the references below for more in-depth information related to the application of hybrid FRP materials in the construction industry.

8.6 References

ACI318M-02 (2002). *Building Code Requirements for Structural Concrete*. Michigan, USA, American Concrete Institute (ACI), Committee 318.

ACI440.1R-06 (2006). *Guide for the Design and Construction of Concrete Reinforced with FRP Bars*. Michigan, USA, American Concrete Institute (ACI), Committee 440.

Aiello, M. A. and Ombres, L. (2000). 'Load-deflection analysis of FRP-reinforced concrete flexural members.' *Journal of Composites for Construction, ASCE* **4**(4): 164–171.

Aiello, M. A. and Ombres, L. (2002). 'Structural performances of concrete beams with hybrid (fiber reinforced polymer-steel) reinforcements.' *Journal of Composites for Construction, ASCE* **6**(2): 133–140.

Alnahhal, W., Chiewanichakorn, M., Aref, A. and Alampalli, S. (2006). 'Temporal thermal behavior and damage simulations of FRP deck.' *Journal of Bridge Engineering, ASCE* **11**(4): 452–465.

Bakis, C. E., Nanni, A. and Terosky, J. A. (1996). *Smart, pseudo-ductile reinforcing rods for concrete: manufacture and test*. Proceeding of First International Conference on Composite in Infrastructure, University of Arizona, Tucson, Arizona.

Berthelot, J.-M. and Sefrani, Y. (2007). 'Longitudinal and transverse damping of unidirectional fibre composites.' *Composites Structures* **79**(3): 423–431.

Berthelot, J.-M., Assarar, M., Sefrani, Y. and El Mahi, A. (2008). 'Damping analysis of composite materials and structures.' *Composites Structures* **85**(3): 189–204.

Bruno, D., Carpino, R. and Greco, F. (2007). 'Modeling of mixed mode debonding in FRP reinforced beams.' *Composites Science and Technology* **67**: 1459–1474.

Büyüköztürk, O., Gunes, O. and Karaca, E. (2004). 'Progress in understanding debonding problems in reinforced concrete and steel members strengthened using FRP composites.' *Construction Building Material* **18**: 9–19.

Caetano de Sa, E. (2007). *Cable Vibrations in Cable-Stayed Bridges*. IASBSE-AIPC-IVBH, Zurich.

CCLAB (2010). http://cclab.epfl.ch/files/content/sites/cclab/files/shared/images/projects/verdasio1.jpg

Cheng, S. and Lau, D. T. (2006). Impact of using CFRP cables on the dynamic behaviour of cable-stayed bridges. IABSE Symposium Report, pp. 19–26.

Deniaud, C. and Cheng, J. J. R. (2003). 'Reinforced concrete T-beams strengthened in shear with fiber-reinforced polymer sheets.' *Journal of Composites for Construction, ASCE* **7**(4): 302–310.

Grace, N. F., Sayed, G. A., Soliman, A. K. and Saleh, K. R. (1996). 'Strengthening reinforced concrete beams under fiber-reinforced polymer (FRP) laminates.' *ACI Structural Journal* **96**(5): 865–875.

Greco, F., Nevone Blasi, P. and Lonetti, P. (2007). 'An analytical investigation of debonding problems in beams strengthened using composite plates.' *Engineering Fracture Mechanics* **74**(3): 346–372.

Hollaway, L. (1978). *Glass Reinforced Plastics in Construction: Engineering Aspects*, John Wiley & Sons, New York.

Kao, C., Kou, C. and Xie, X. (2006). 'Static instability analysis of long-span cable-stayed bridges with carbon fiber composite cable under wind load.' *Tamkang Journal of Science and Engineering* **9**(2): 89–95.

Kretsis, G. (1987). 'A review of the tensile, compressive, flexural and shear properties of hybrid fibre-reinforced plastics.' *Composites* **18**(1): 13–23.

Lau, D. and Pam, H. J. (2010). 'Experimental study of hybrid FRP reinforced concrete beams.' *Engineering Structures* **32**(12): 3857–3865.

Masmoudi, R., Zaidi, A. and Gérard, P. (2005). 'Transverse thermal expansion of FRP bars embedded in concrete.' *Journal of Composites for Constructions, ASCE* **9**(5): 377–387.

Meier, U. (1987). 'Proposal for a carbon fiber reinforced composite bridge across the Strait of Gibraltar at its narrowest site.' *Proceedings of the Institution of Mechanical Engineers* **201**(B2): 73–78.

Mufti, A. A., Newhook, J. P. and Tadros, G. (1996). *Deformability versus ductility in concrete beams with FRP reinforcement*. International Conference on Advanced Composite Materials in Bridges and Structures, Montreal, Canada, Canadian Society of Civil Engineering.

Nanni, A., Henneke, M. J. and Okamoto, T. (1994a). 'Behavior of concrete beams with hybrid reinforcement.' *Construction and Building Materials* **8**(2): 89–95.

Nanni, A., Henneke, M. J. and Okamoto, T. (1994b). 'Tensile properties of hybrid rods for concrete reinforcement.' *Construction and Building Materials* **8**(1): 27–34.

Newhook, J. P. (2000). *Design of under-reinforced concrete T-sections with GFRP reinforcement*. The 3rd International Conference on Advanced Composite Materials in Bridges and Structures, Montreal, Canadian Society for Civil Engineering.

Peece, M., Manfredi, G. and Cosenza, E. (2000). 'Experimental response and code models of GFRP RC beams in bending.' *Journal of Composites for Construction, ASCE* **4**(4): 182–190.

Rizkalla, S. H. and Nanni, A. (2003). *Field Applications of FRP Reinforcement: Case Studies*. American Concrete Institute (ACI) Special Publication SP-215.

Structurae (2012). http://files2.structurae.de/files/photos/2094/1087_aberfeldy_s0002215.jpg

Triantafillou, T. C. (1998). 'Shear strengthening of reinforced concrete beams using epoxy-bonded FRP composites.' *ACI Structural Journal* **95**(2): 107–115.

Wang, X. and Wu, Z. (2009). 'Dynamic behavior of thousand-meter scale cable-stayed bridge with hybrid FRP cables.' *Journal of Applied Mechanics, ASCE* **12**: 935–943.

Wang, X. and Wu, Z. (2010). 'Evaluation of FRP and hybrid FRP cables for super long-span cable-stayed bridges.' *Composites Structures* **92**(10): 2582–2590.

Wang, X. and Wu, Z. (2011a). 'Integrated high-performance thousand-meter scale cable-stayed bridge with hybrid FRP cables.' *Composites Part B: Engineering* **41**(2): 166–175.

Wang, X. and Wu, Z. (2011b). 'Modal damping evaluation of hybrid FRP cable with smart dampers for long-span cable-stayed bridges.' *Composites Structures* **93**(4): 1231–1238.

Wei, Y.-T., Gui, L.-J. and Yang, T.-Q. (2001). 'Prediction of the 3-D effective damping matrix and energy dissipation of viscoelastic fiber composites.' *Composites Structures* **54**(1): 49–55.

Wu, Z. (2004). *Structural strengthening and integrity with hybrid FRP composites (Keynote paper)*. Proceedings of the Second International Conference on FRP Composites in Civil Engineering (CICE-2), Adelaide, Australia.

Wu, Z. and Wang, X. (2008). *Investigation on a 1000-m scale cable-stayed bridge with fiber composite cables*. The Fourth International Conference on FRP Composites in Civil Engineering (CICE-4). Zurich, Switzerland.

Wu, Z., Wang, X. and Wu, G. (2009). *Basalt FRP composite as a reinforcement in infrastructure (Keynote paper)*. The Seventeenth Annual International Conference on Composites/Nano Engineering (ICCE-17), Hawaii, USA.

Wu, Z., Wang, X., Iwashita, K., Sasaki, T. and Hamaguchi, Y. (2010). 'Tensile fatigue behavior of FRP and hybrid FRP sheets.' *Composites Part B: Engineering* **41**(5): 396–402.

9
Design of hybrid fiber-reinforced polymer (FRP)/autoclave aerated concrete (AAC) panels for structural applications

N. UDDIN, M. A. MOUSA, U. VAIDYA and F. H. FOUAD,
The University of Alabama at Birmingham, USA

DOI: 10.1533/9780857098955.2.226

Abstract: This chapter discusses design for fiber-reinforced polymer (FRP)/autoclaved aerated concrete (AAC) sandwich panels for structural applications. The chapter first presents the finite element analysis (FE) of FRP/AAC panels. The FE results are compared with the experimental results showing acceptable agreement. Next, analytical models are presented to predict the deflection and strength of the panels. Finally, design graphs have been developed to help in designing the floor and wall panels made from FRP/AAC panels. Also, those panels have been compared to the commercially used reinforced AAC panels demonstrating that FRP/AAC panels offer a relatively cost-effective solution for longer life cycle.

Key words: fiber reinforced polymer (FRP), finite element analysis (FEA), autoclaved aerated concrete (AAC), design graphs, sandwich panel.

9.1 Introduction

Building materials and labor to construct the structural design are the largest cost components in house constructions, so the need for lower cost and time-efficient technology becomes urgent. This can be achieved by using panelized construction. Panelized systems are pre-manufactured components or sub-elements (e.g., FRP/AAC panels) that are brought to the site and assembled into the finished house. Panelized construction can bring the benefits of mass production into the highly customized residential market through the pre-production of components and systems. There are many advantages to panelizing structures, including cost reductions, possible through mass production, ease of assembly, and a lower skill set required for field construction.

AAC is an ultra-lightweight concrete with a distinct structure. The raw materials used in production are simply cement, lime, aluminum paste/powder, and water, plus sand or fly ash as a silica source. The dry bulk density of the material ranges from 25 to 50 pcf (0.4 to 0.8 g/cc) about one fifth the weight of normal weight concrete and a compressive strength range from 300 to 1000 psi (2–7 MPa) (Shi and Fouad, 2005). Entrained air bubbles are the main reason behind the result of a chemical reaction between the cement hydration products and the aluminum paste/powder in which hydrogen gas is liberated. Hydrogen gas causes the fresh material to rise in the molds and expand to about twice the original volume, thus creating cellular structure. Due to its cellular structure, porosity, and reduced weight, the material is highly fire resistant and very durable compared with conventional construction material, and has unique thermal and sound insulation properties. In addition, AAC is a proven building material that will become much more widely used in the US for both residential and commercial construction. It is also important that the building material be cost effective, energy-efficient, and available throughout the world. It is brittle in nature and has much lower flexural and compressive strengths than normal weight concrete. It requires significantly less energy for heating and cooling. Few studies have been conducted to understand the structural behavior of plain and reinforced AAC structural floor and wall panels with internal reinforced rebars (Snow, 1999; Dembowski, 2001). The main conclusions of these studies are that the failure of floor panels occurred suddenly due to the sudden pull-out of steel bars while that for wall panels was cracking of the concrete cover at the top and/or the bottom of the panels. Further more, no signs of steel buckling were observed.

FRP composites take the concept of combing materials to create a new system having some of the advantages of each constituent. Since FRP composites are characterized by high tensile strength in the direction of fibers and high strength-to-weight ratio, they have been used in the aircraft and automotive industries. Recently, fiber-reinforced composites are also being used to repair and/or strengthen reinforced concrete bridges and other structures. They offer high corrosion and flexural resistance, and therefore should perform better than other construction materials in terms of weathering behavior. Accordingly, since AAC is ultra lightweight in nature and FRP is so stiff with high specific strength, the two could be used together to form hybrid structural panels.

9.2 Performance issues with fiber-reinforced polymer (FRP)/autoclave aerated concrete (AAC) panels

AAC is currently used in the form of steel reinforced panels using rebars as an internal reinforcement. These rebars are expensive and subject to corrosion in the long run. Further, these rebars do not play any role in the shear strength of the panels. Both shear and flexural strength can be enhanced by wrapping the plain AAC with FRP laminates as discussed in this chapter. Khotpal (2004) has investigated the compressive strength of plain AAC wrapped by FRP. The objectives were to evaluate the load-carrying capacity of the confined AAC cube and to observe the mode of failure of FRP/AAC panels. The results showed that the FRP wrap has increased significantly the compressive strength of FRP/AAC panels by about 80% over the plain AAC. Uddin and Fouad (2007) investigated the behavior of FRP/AAC panels using small size specimens under four point load test. The experimental results for that research showed a significant influence of FRP on the flexural strength and stiffness of the hybrid panels. Further, Mousa and Uddin (2009) developed theoretical formulas to calculate the shear and flexural strengths of FRP/AAC panel and the results were in good agreement with experimental ones. A cost analysis for reinforced AAC and FRP/AAC panels previously conducted by the authors showed that a thinner FRP/AAC structural panel (about half the size) can be as cost-effective as original reinforced thicker AAC panel (Mousa and Uddin, 2009).

The design concept of a sandwich panel in general is that the bending moments are resisted by an internal couple composed of forces in the facings, while the shearing forces are carried by the core and wraps if available. In the case of FRP/AAC panel, the panel is composed from core, like AAC, which is capable of carrying the shear stresses, and skins, like FRP laminates, which are capable of carrying the normal stresses. In other words, the sandwich structure is similar to an I-section in which the core has the same function of web in carrying shear stresses and the flange has the same function of skin which carries the normal stresses (Zenkert, 1995). The core-to-skin adhesive rigidly joins the sandwich components and allows them to act as one unit with high torsional and bending rigidity. In the case of FRP/AAC interface, the mechanism of adhesion is the mechanical interlocking of the resin in the irregularities of the concrete surface (Karbhari and Zhao, 2000). Although, both AAC and FRP are brittle materials, they have shown synergetic results in combination with each other in terms of shear and flexural strengths as demonstrated in this paper. Due to the higher strength resulting from this combination, the strength is not the criterion governing the design of the panel (Mousa, 2007), but the deflection is the one that controls the design of the proposed hybrid panels.

The objectives of the research summarized in this chapter were to analyze the behavior of FRP/AAC panels under out-of-plane loading and also to validate the numerical results obtained by ANSYS for the FRP/AAC panels. Analytical models for predicting the defection and strength for FRP/AAC panels are also developed in this study. Based on the FE results, design guides will be then developed for both floor and walls panels made of FRP/AAC sandwich panels. Structural comparison will also be performed between the proposed hybrid composite panels and the currently used reinforced AAC panels. It is anticipated that the experimental and the FE results presented here would be a step towards the long-term goal of providing a practical method to predict deflections, stress, and ultimate load, with the intent of developing tools for the design of FRP/AAC panels for building construction. The chapter also includes comparisons between the proposed hybrid panels and the reinforced AAC panels currently used in the housing market.

9.3 Materials, processing, and methods of investigation

The properties of AAC being used in this research are summarized in Table 9.1. Both carbon and glass FRP have been used as a skin in this research, in which both carbon and glass FRP laminates were used in the design of the proposed FRP/AAC panels. SIKA Carbon Fiber laminates (Sika Corporation, 2002) and TYFO Glass Fiber (www.fyfeco.com) have been used as skin materials. SIKA WRAP HEX 103C unidirectional carbon fibers, SIKAWRAP HEX 113C bidirectional carbon fibers, and SIKADUR HEX 300 resin were used. The mechanical properties of resin as well as laminates, as provided by the manufacturer, are listed in Table 9.2. For glass fiber laminates, TYFO SHE-51A Composite using Tyfo S Epoxy was used. The mechanical properties of the epoxy and the laminate, as provided by the company, are listed in Table 9.3.

Vacuum assisted resin transfer molding (VARTM) processing was used in this study to produce all FRP/AAC panels, as shown in Fig. 9.1, to reduce the processing time due to the construction and surface preparation efforts. As an alternative to labor-intensive hand layup, VARTM is an attractive process because it saves processing time (especially when several FRP layers are being applied). VARTM is a process for molding fiber-reinforced composite structures in which a sheet of flexible transparent material such as nylon or Mylar plastic is placed over the preform and sealed. A vacuum is applied between the sheet and the preform to remove the entrapped air. On proper application, VARTM ensures the complete wet-out of fiber and

Table 9.1 Mechanical properties of plain autoclaved aerated concrete (AAC)

Property	Value
Density	40 pcf (640 kg/m^3)
Compressive strength	456 psi (3.2 MPa)
Modulus of elasticity	256,000 psi (1800 MPa)
Shear strength	17 psi (0.12 MPa)
Poisson's ratio	0.25

Table 9.2 Mechanical properties of SIKA carbon fiber composite*

Property	SIKA HEX 300 (Resin)	Unidirectional laminate	Bidirectional laminate
Tensile strength	10,500 psi (72.4 MPa)	123,200 psi (849 MPa)	66,000 psi (456 MPa)
90° tensile strength	–	3,500 psi (24 MPa)	66,000 psi (456 MPa)
E_x	459,000 psi (3,170 MPa)	10,239,800 psi (70,552 MPa)	6×10^6 psi (41,400 MPa)
E_y	459,000 psi (3,170 MPa)	705,500 psi (4,861 MPa)	6×10^6 psi (41,400 MPa)
E_z	459,000 psi (3,170 MPa)	459,000 psi (3,170 MPa)	459,000 psi (3,170 MPa)
G_{xy}	–	362,500 psi (2,498 MPa)	249,400 psi (1720 MPa)
G_{yz}	–	176,900 psi (1220 MPa)	176,900 psi (1220 MPa)
G_{xz}	–	340,605 psi (2,394 MPa)	176,900 psi (1220 MPa)
PR_{xy}	–	0.35	0.3
PR_{yz}	–	0.45	0.35
PR_{xz}	–	0.35	0.35
Tensile elongation	4.8 %	1.12 %	1.2 %
Ply thickness	–	0.04 in (1.016 mm)	0.01 in (0.25 mm)

*E_x and E_y are from the manufacturers, the tensile modulus (E_z) has been considered as the matrix modulus while the Poisson's ratios and shear moduli are from the literature.

is not as tedious as the hand layup technique. Therefore, it improves the bond characteristics between FRP facesheets and AAC substrate leading to enhancing the whole stiffness of the CFRP/AAC panel. VARTM is typically a three-step process, including the layup of a fiber preform, impregnation of the preform with resin, and cure of the impregnated preform. In addition to improving strength and ductility, the reinforcement of the AAC panels with FRP composite facesheets is also expected to enhance durability performance leading to reduced maintenance costs of structures.

Two types of panels have been developed. The first one is a panel reinforced by uniaxial FRP for flexural reinforcement (UFFS) and the second type was three panels reinforced by biaxial FRP for flexural reinforcement (BFFS). After processing, all panels measured 47.24 in. × 6.88 in. × 3.94 in. (1200 mm × 175 mm × 100 mm). The UFFS panel was constructed from one block measuring 47.24 in. × 6.88 in. × 3.94 in. (1200 mm × 175 mm × 100 mm), and was reinforced by top and bottom unidirectional carbon fiber lamina

Table 9.3 Mechanical properties of TYFO glass fiber composite*

Property	Tyfo S Epoxy	Cured laminate (test values)
Tensile strength	10,500 psi (72.4 MPa)	83,400 psi (575 MPa)
90° tensile strength	–	3,750 psi (25.8 MPa)
E_x	461,000 psi (3,180 MPa)	3.79×10^6 psi (26,100 MPa)
E_y	461,000 psi (3,180 MPa)	461,100 psi (3,180 MPa)
E_z	461,000 psi (3,180 MPa)	461,100 psi (3,180 MPa)
G_{xy}	–	290,000 psi (2,000 MPa)
G_{yz}	–	176,900 psi (1,220 MPa)
G_{xz}	–	290,000 psi (2,000 MPa)
PR_{xy}	–	0.256
PR_{yz}	–	0.086
PR_{xz}	–	0.256
Ply thickness	–	0.05 in (1.3 mm)
Tensile elongation	5.0%	1.76%

* E_x and Poisson's ratios are from the manufacturers; both E_y and E_z have been considered as the matrix modulus, while the shear moduli are from the literature.

9.1 VARTM processed FRP-AAC specimens.

(Sika Corporation, 2002) (i.e., the fiber orientation is 0°) for flexural reinforcement and then wrapped by unidirectional carbon lamina (the fiber orientation is 90°) for shear reinforcement. Since there was an extended investigation focused on the strengthening of AAC by unidirectional FRP laminate using small-scale panel (Uddin and Fouad, 2007), only one UFFS panel has been used in this investigation.

BFFS panels have the same dimensions as UFFS panel. Three panels were wrapped with one layer of biaxial CFRP (Sika Corporation, 2002), which behaves as shear reinforcement as well as flexural reinforcement. This reinforcement for the panel would be exactly the same as the sandwich panel with complete shear and flexural reinforcement. The reason for fabricating three panels having the same properties for each was to investigate the effect of processing variation due to VARTM. Table 9.4 shows the dimensions and reinforcement type for each panel.

The panels were tested under four point loading test according to the ASTM C393 'Standard Test Method for Flexural Properties of Sandwich Constructions' (ASTM C393, 2000). The load was applied at a uniform rate of 0.025 in./min (0.635 mm/min). The 60 Kip (13.5 kN) Tinius Olsen machine was used to conduct all of the tests. An electronic dial gage was placed at the mid-span of the section to record the mid-span deflection and the strain gages were hooked up at the mid-span as well as the shear span to record the strains.

Geometric nonlinear FEA for the panels has been conducted using the nonlinear finite element program ANSYS. The facesheets (FRP laminates) were modeled as shell elements with orthotropic material properties, while the core (AAC) was modeled as a solid element with isotropic material properties. SHELL 99, a linear layered structural shell, was used for defining the skins. The element has six degrees of freedom at each node; translations in the nodal x, y, and z directions; and rotations about the nodal x, y, and z-axes. It is also defined by eight nodes (the mid plane and corner nodes), average or corner layer thicknesses, layer material direction angles, and orthotropic material properties. The element allows up to 250 layers. SOLID 186 was used for modeling the core. SOLID 186 is a higher order three-dimensional solid element that exhibits quadratic

Table 9.4 Dimensions and reinforcement type of the tested panels

Panel no.	Dimension, in. (mm)	Reinforcement type
UFFS	47.24 × 6.88 × 3.94 (1200 × 175 × 100)	Unidirectional
BFFS1	47.24 × 6.88 × 3.94 (1200 × 175 × 100)	Bidirectional
BFFS2	47.24 × 6.88 × 3.94 (1200 × 175 × 100)	Bidirectional
BFFS3	47.24 × 6.88 × 3.94 (1200 × 175 × 100)	Bidirectional

displacement behavior. This element is defined by 20 nodes having three degrees of freedom per node; translations in the nodal x, y, and z directions. The element supports plasticity, creep, stress stiffening, large deflection, and large strain capabilities. Geometric nonlinearity was pursued for modeling of sandwich panels to compare the numerical and the experimental results. This was a unique procedure compared with most of the previous studies on the modeling of sandwich structure (Haibin *et al.*, 2007).

9.4 Comparing different panel designs

9.4.1 Experimental results: UFFS panel

The uniaxial specimen underwent complete failure due to shear at a load of 3.5 Kips (15.54 kN), with a maximum deflection of 0.47 in. (11.97 mm) at the mid-span. As observed in the experiment, once the AAC cracked in the maximum shear zone, the FRP failed immediately; i.e., the general mode of failure was cracking of AAC in the shear zone followed by FRP shearing out. Wrinkling of FRP in compression was observed at a load of 1.8 Kips (8.00 kN). The behavior of the UFFS panel under static four-point bending is shown in Figs 9.2 and 9.3. At load 3.5 Kips (15.54 kN), the AAC cracked at the maximum shear followed immediately by FRP shearing out. Figure 9.4 shows the load deflection curve obtained from the test. As can be seen from Fig. 9.4, the deflection has increased significantly after wrinkling of FRP in compression from 0.15 to 0.23 in. (3.73–5.73 mm), which means that the wrinkling has a significant effect on deflection, but the panel has sustained carrying load until failure at load 15.54 kN.

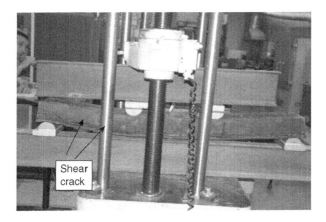

9.2 Failure of UFFS specimen in shear span.

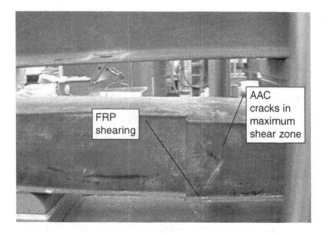

9.3 Failure mode of UFFS specimen.

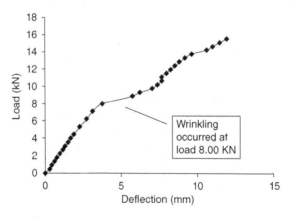

9.4 Experimental load versus deflection curve for UFFS panel.

9.4.2 Experimental results: BFFS panels

The BFFS panels showed a variation in ultimate load-carrying capacity (from 3.0 Kips to 3.65 Kips, 13.5 kN to 16.2 kN), although BFFS2 and BFFS3 show a similar load–deflection response. The discrepancy may be due to variations in the absorption of resin during the panels' fabrication and also due to the pores distribution of AAC. All biaxially reinforced panels failed in the flexural span, as shown in Fig. 9.5. In the experiment, there was a significant increase in deflection, while there was no sign of cracks in AAC or FRP, but once the failure load was close, both FRP and AAC failed at the same time. It was not a progressive failure, but an instant failure, in which the failure initiated in the FRP skin followed immediately by AAC crushing. No debonding has occurred between CFRP laminates

Design of hybrid FRP/AAC panels for structural applications   235

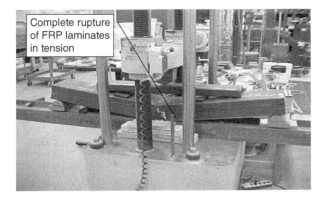

9.5 Failure of BFFS specimens.

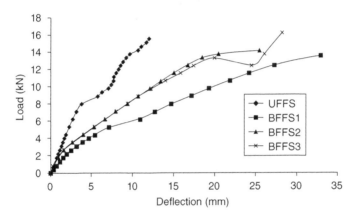

9.6 Comparison of the load versus deflection graphs for UFFS and BFFS panels.

and AAC substrate. For BFFS3, wrinkling of FRP in compression was observed at about 2.81 Kips (12.5 kN) until the panel underwent complete failure at load 3.64 Kips (16.2 kN). Figure 9.6 represents the load versus deflection curves of UFFS and BFFS panels. As shown in Fig. 9.6, the BFFS panels recorded maximum deflections ranging from 1.00 to 1.11 in. (25.5–33 mm) which are considerably more than the UFFS panel because of the high modulus of elasticity of unidirectional carbon fiber rather than the woven FRP (bidirectional FRP).

9.4.3 Finite element results: UFFS panel

The top and bottom layers of CFRP laminate were modeled using the same real constant that contains two layers. The first one, 0° layer, represents the longitudinal lamina, while the second, 90° layer, represents the transverse lamina (wrapped lamina) used for shear strengthening. The side layers were modeled using the same real constant, which contains one 0° layer with respect to its local axes. Static loading was carried out for all panels. For the UFFS panel, a total of 2,400 elements were used in the modeling, including 1,344 solid elements and 1,056 shell elements. The loading and boundary conditions are shown in Plate II (between pages 240 and 241). A load of 3.5 Kips (15.54 kN) was applied on twelve nodes. The boundary conditions provided were to prevent the translation in the Y-direction only to represent the real situation of the tested panels. As shown Plate III (between pages 240 and 241), the maximum deflection from this analysis was 0.38 in. (9.77 mm). The numerical load deflection curve developed from the analysis attached with that obtained from the experiments is shown in Fig. 9.7. Table 9.5 summarizes the results developed from FEA compared with those developed from the experiment. As shown in Fig. 9.7, there is a clear convergence between the two curves (experimental and FE). The difference in the deflection at the experimental failure load was 18.4%. However, the difference at the sub-step prior to failure load was 1.5%. The most probable

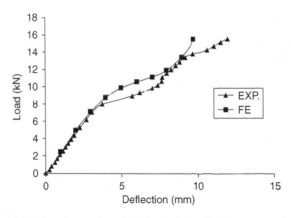

9.7 Experimental and FE load deflection curves.

Table 9.5 Deflection of UFFS panel FE analysis and experiment

Panel no.	Loading condition	Failure load, Kips (kN)	Maximum deflection, in. (mm) (Experimental)	Maximum deflection, in. (mm) FEA	Difference %
UFFS	Simply supported	3.5 (15.54)	0.47 (11.97)	0.38 (9.77)	18.38

reason for the difference at the failure is because, prior to the failure load, a movement occurred at the support points that caused a slip in support. The movement also includes the dislocation at the loading points. The finite element model cannot capture those movements.

9.4.4 Finite element results: BFFS panels

The FEA for the BFFS panels has been conducted with the same procedure as for the UFFS panel. The loading and boundary conditions are shown in Plate IV (between pages 240 and 241), while the deflection counters are shown in Plate V (between pages 240 and 241). In addition, Fig. 9.8 shows the comparison between the numerical and experimental load deflection curves. As shown in Fig. 9.8, the main reason for the difference between the numerical and experimental deflection is the movement of the panels at the supports prior to failure load. However, by observing the point prior to failure load, it can be noted that the difference is only about 7%.

9.5 Analytical modeling of fiber-reinforced polymer (FRP)/autoclave aerated concrete (AAC) panels

9.5.1 Deflection

The total central deflection of a sandwich panel under out-of-plan loading is composed of bending deflection and shear deflection. The general formula for the deflection of a sandwich panel under out-of-plan loading is given by:

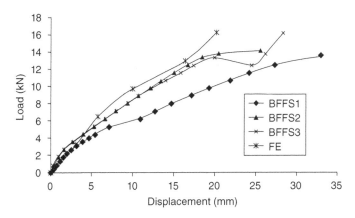

9.8 Comparison between numerical and experimental load deflection curves for BFFS panels.

$$\Delta = \frac{k_b P L^3}{D} + \frac{k_s P L}{U} \qquad [9.1]$$

The first term on the right-hand side of Eq. [9.1] is the deflection due to bending, and the second term is the deflection due to shear. k_b and k_s are the bending and shear deflection coefficients, respectively. The values of both k_b and k_s depend on the loading and boundary conditions. For the FRP/AAC panels tested in this study (two-point load, one-third span with simply supported boundary conditions), k_b and k_s are 1/56 and 1/8, respectively. For the sake of brevity, the derivation of k_b and k_s is not included in this chapter and the basis for deriving them can be found elsewhere (e.g., Allen, 1969). D and U are flexural and shear rigidities of the sandwich panel, respectively. According to ASTM C-393, D and U can be determined as follows:

$$D = \frac{E_{face}(d^3 - c^3)b}{12} \qquad [9.2]$$

$$U = \frac{G(d+c)^2 b}{4c} \qquad [9.3]$$

It is generally recognized that the ordinary theory of bending and resulting deflection can be applied to a homogeneous panel (e.g., a panel that is made from one material). For a sandwich panel, the behavior is different and this can be demonstrated by considering two extreme cases. First, when the core is rigid in shear, the sandwich panel is subjected to the same argument as those applied to a homogeneous panel (except for the difference in the flexural rigidity) and the deflections are expected to be small. Second, when the core is weak in shear, the faces act as two independent plates and the resulting deflections are expected to be much greater than in the first case. It was demonstrated by Allen (1969) that the parameter λ represents the transition from one extreme to the other, varying from $(-t/c)$ when the core is weak to $(+1)$ when the core is rigid in shear. Thus, an empirical formula was developed by Allen (1969) to define the nominal thickness of a sandwich panel which varies from t ($G = 0$) to d ($G = \infty$):

$$d_{nom} = 1 + \left(1 + \frac{c}{d}\right) + \frac{c^2}{d}\lambda \qquad [9.4]$$

Therefore, d_{nom} should be used instead of d in Eqs [9.2] and [9.3] when determining the flexural and shear rigidities. The parameter λ is determined experimentally based on the deflection. Since unidirectional FRP was used for designing the FRP/AAC panels in the next section, λ was determined for UFFS panel and was found to be 0.8. Since AAC is a relatively rigid core compared with other cores (e.g., foam cores), the obtained λ had a higher value and this recommends using AAC as a core material. Detailed calculations for determining λ are presented in Appendix A. Thus, Eq. [9.4] can be rewritten as follows:

$$d_{nom} = 1 + \left(1 + \frac{c}{d}\right) + 0.8\frac{c^2}{d} \quad [9.5]$$

9.5.2 Strength

Flexural and shear strengths for FRP/AAC structural panels were previously determined by the authors in an extensive experimental and analytical study (Mousa and Uddin, 2009). According to that study, the nominal flexural strength for FRP/AAC section is determined using the following equation:

$$M_n = b \cdot t \cdot \varepsilon_{face} \cdot E_{face} \cdot d \quad [9.6]$$

The total shear strength of the panel is composed of the contribution of the AAC core plus that of the FRP wraps. The total shear strength of FRP/AAC panel can be determined as follows (Mousa and Uddin, 2009):

$$V_n = V_{AAC} + V_f \quad [9.7]$$

The shear strength contribution from AAC is determined according to an ACI special publication (Shi and Fouad, 2005) and is given by:

$$V_{AAC} = 0.8\sqrt{fc_{AAC}'}bd \quad [9.8]$$

while the contribution from FRP wraps is calculated according to ACI-440.2R-02 (American Concrete Institute, 2002b) using the formula of U-wrapped scheme and is given by:

$$V_f = \frac{A_{fv} f_{fe} (\sin\alpha + \cos\alpha) d_f}{S_f} \quad [9.9]$$

It was demonstrated by Mousa and Uddin (2009) that FRP wraps increase the shear strength by about 300% more than the plain AAC or the unwrapped panels.

9.6 Design graphs for fiber-reinforced polymer (FRP)/ autoclave aerated concrete (AAC) panels

The design graphs provided in this section have been developed using ANSYS. These graphs could be used as a guide for preliminary determination of the AAC core thickness required for a particular project. AAC panels in these graphs are strengthened by one unidirectional ply top and bottom of CFRP or GFRP and the cores were designed to carry the shear stresses. The graphs provide a relation between span versus live load for floor panels and wall height versus wind load for wall panels.

9.6.1 Design limits

The sandwich panels must be designed to satisfy both stiffness and strength criteria. Stiffness means to satisfy both deflection and slenderness ratio, while the strength criterion means both skin strength and core shear strength criteria. However, the deflection is the main criterion governing the design of sandwich panels (Mousa, 2007). According to the Building Code for Structural Concrete (ACI-318 and ACI-318 R-02) (American Concrete Institute, 2002a), for roof or floor construction supporting or attached to nonstructural elements not likely to be damaged by large deflections, the allowable total deflection is $L/240$ for total load and $L/360$ for live load only. In addition to these limits, the slenderness ratio must be taken into consideration when the walls are designed. Table 9.6 lists the limits for the different design criteria. The limit for the wall slenderness ratio listed in Table 9.6 was obtained from Building Code Requirements for Masonry Structures (MSJC, 2002). All deflections used in the design were the immediate deflections only. Since there is no guidance to calculate the long-term deflection for FRP/AAC sandwich panels, long-term deflection components were ignored in the calculation. It should be noted that the calculated deflection was significantly below the allowable limit.

9.6.2 Floor panels

As known, floor panels are designed for dead and live loads only. Dead loads include the self weight of the entire structure in addition to the superimposed dead loads. Generally, a value of 35 psf (1.7 kN/m²) has been taken for the superimposed dead load for floor panels, which includes flooring, ceiling, and partitions (www.aercon.com). Live loads ranged up

Table 9.6 Design limits

Criterion	Design limit
Deflection	For total load: L/240
	For live load only : L/360
	For wind load only: L/360
CFRP strength	104,000 psi (717 MPa)
GFRP strength	66,720 psi (460 MPa)
AAC shear strength	17 psi (0.117 MPa)
Slenderness ratio (for walls only)	h/r < 99

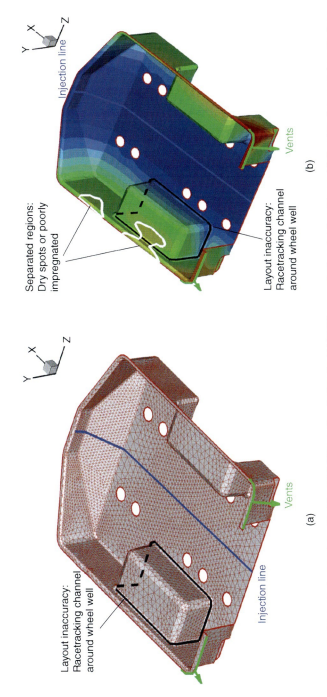

Plate I A case study on the use of mold-filling simulation in VARTM to optimize the location of injection lines and vents for a complex part. (a) Mesh details. (b) A plot of flow-front progress with time showing possible formation of dry spots for a particular choice of injection line and vent locations (courtesy of Prof. Advani and Dr. Simacek of the University of Delaware).

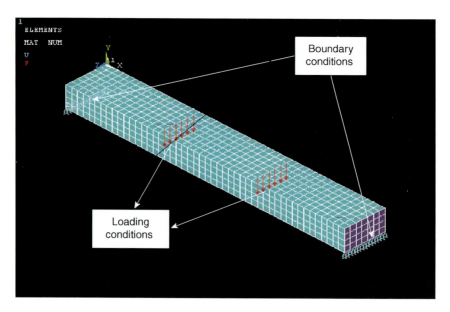

Plate II Panel model with loading and boundary conditions for UFFS panel.

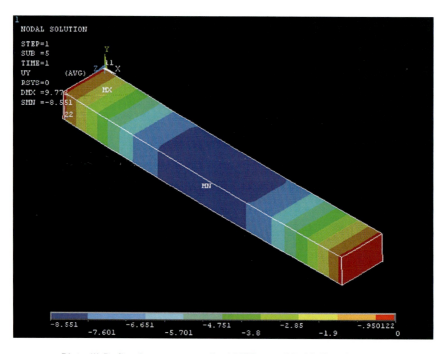

Plate III Deflection contours for UFFS panel in Y-direction indicate the maximum deflection between the loaded nodes.

© Woodhead Publishing Limited, 2013

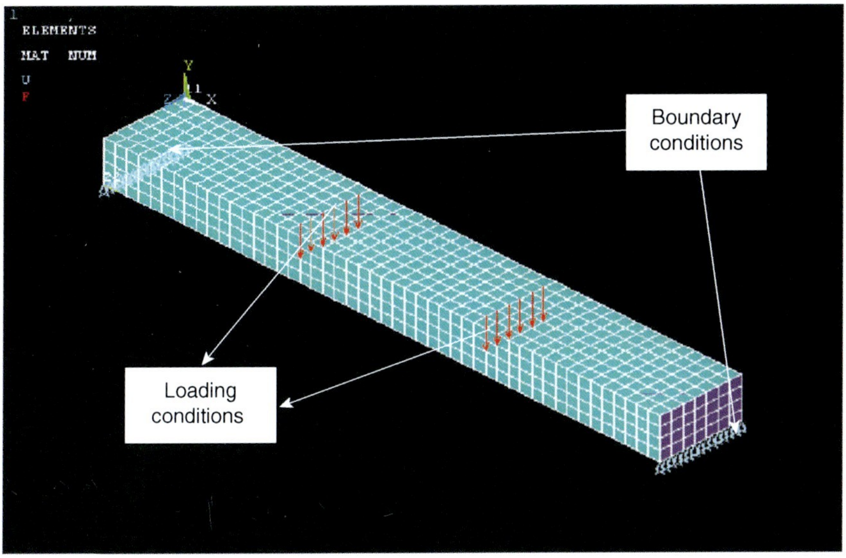

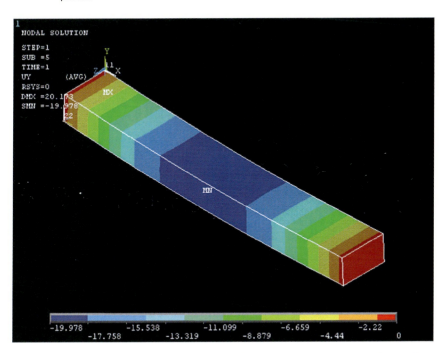

Plate IV Panel model with loading and boundary conditions for BFFS panel.

Plate V Deflection contours for BFFS panel in Y-direction indicate the maximum deflection between the loaded nodes.

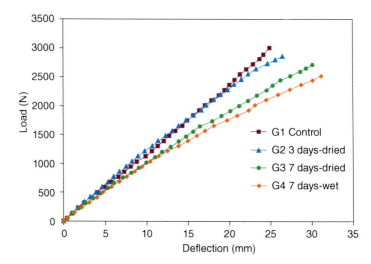

Plate VI Average load–deflection curves for all groups.

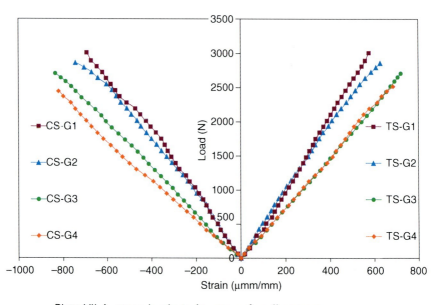

Plate VII Average load–strain curves for all groups.

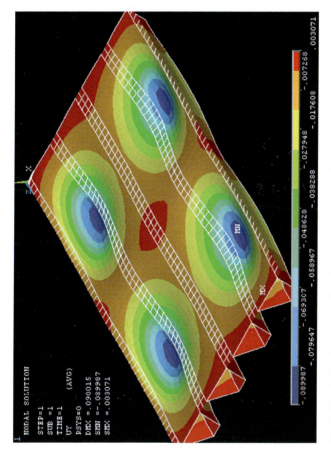

Plate VIII Deflection (in inches) for case 4 of a single-lane deck model.

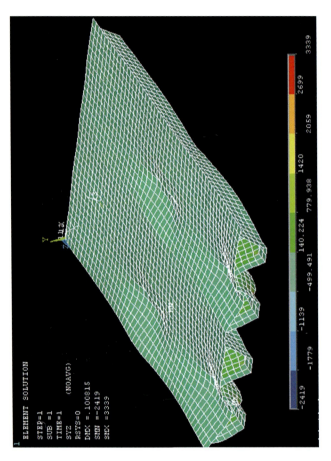

Plate IX Shear stress for case 4 of a single-lane deck model.

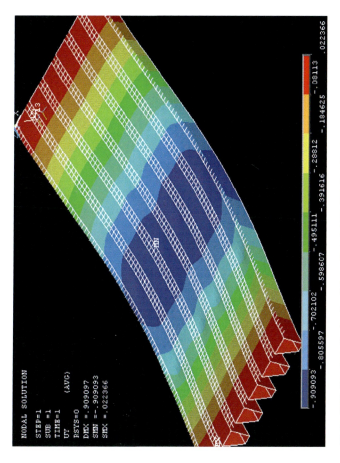

Plate X Deflection (in inches) for case 4 of a double-lane deck model.

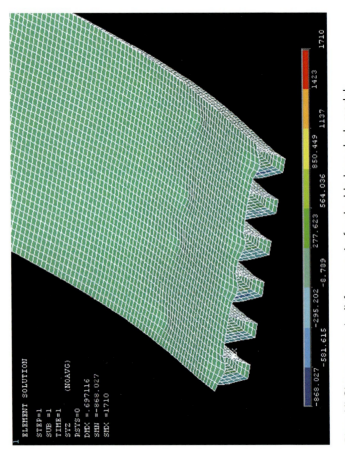

Plate XI Shear stress (psi) for case 4 of a double-lane deck model.

to 100 psf (4.8 kN/m^2) which represent the live load required for commercial buildings. The boundary conditions were adapted; the two sides of the panel that connected to the support (i.e., wall) were restrained for Y translations only to model the simple support situation. The reason for restraining in the Y direction only and not all directions is that the core has been molded using SOLID 186 element which has 20 nodes and each node has three degrees of freedom (three translations) only. Thus, restraining all translations means fixed support, not simple, and therefore the deflection obtained would be much lower than the actual. Moreover, the restraining for all translations is difficult to achieve in residential buildings, especially the connection between the floor and wall. The minimum bearing length is 2.5 in. (25.4 mm) for panelized construction (www.aercon.com). However, it is recommended to use a bearing length not less than 4 in. (101.6 mm) from both sides of the panel. The graphs shown in Figs 9.9 and 9.10 may be used as a preliminary design for both CFRP/AAC

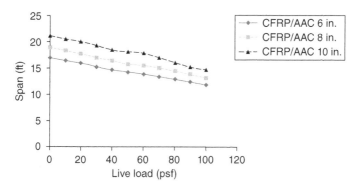

9.9 Design graph for CFRP/AAC floor panels.

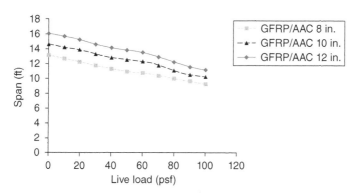

9.10 Design graph for GFRP/AAC floor panels.

and GFRP/AAC floor panels, respectively. Since the core also works as an insulation material, its thickness is restricted by the manufacturer, especially for AAC, so the minimum thickness of AAC to be used as core material is 4 in. (101.6 mm). The panel dimensions have been chosen based on the available sizes used in the housing market. The span ranges up to 20 ft (6100 mm) and depth ranges up to 12 in. (304.8 mm), while the width mainly depends on the manufacturer, usually 2 ft (610 mm). The graph provides a relation between span versus live load. Although the deflection is the main criterion controlling the design, all stresses in the skins were checked and found less than the allowable limits as listed in Table 9.6.

9.6.3 Wall panels

The wall panels are designed mainly for axial load in conjunction with transverse load due to wind or earthquake loads. The boundary and loading conditions for the wall panels have been adapted. A value of 900 plf (4 kN/m) has been chosen to represent the axial load that equals the loads carried by an external load-bearing wall carrying a two-storey building and supporting a floor slab of 12 ft × 12 ft (3.66 m × 3.66 m) subjected to live load of 100 psf (4.8 kN/m^2). The boundary conditions were fixed at the base of the wall that connected to the foundation and simple at the end that connected the floor. The slenderness ratio and deflection were the two main criteria that controlled the design of wall panels. Figure 9.11 provides a relation between wall height and wind load and may be used for determination of the AAC thickness for both CFRP/AAC and GFRP/AAC wall panels. The step part in Fig. 9.11 represents the limitation of slenderness ratio for different panels' thicknesses.

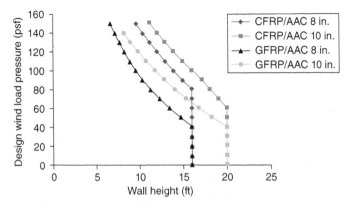

9.11 Design graph for FRP/AAC wall panels.

9.6.4 Comparison graphs

The proposed panels were compared with the reinforced AAC panels being used in the housing market. The comparison has focused on the structural standpoint. Figure 9.12 compares reinforced AAC floor panels and CFRP/AAC for the same AAC thickness. As shown in Fig. 9.12, the CFRP/AAC can support a span greater than that of reinforced AAC by about 15%. Figure 9.13 shows a comparison graph between reinforced AAC and both CFRP/AAC and GFRP/AAC wall panels having the same AAC thickness. As seen in Fig. 9.13, the difference in the wind load capacity is clear between the reinforced AAC and the hybrid panels. The reason is that the reinforced AAC is designed based on the tensile stresses carried by the allowable flexural tensile strength of AAC, but in case of

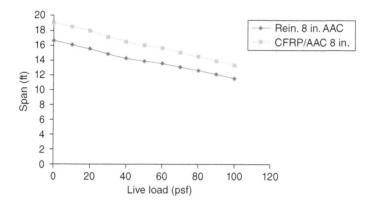

9.12 Comparison design graph between reinforced AAC and CFRP/AAC floor panels.

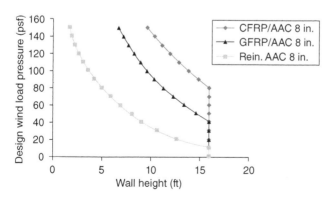

9.13 Comparison design graph between reinforced AAC and FRP/AAC wall panels.

FRP/AAC panels, the flexural compressive and tensile stresses are carried by the top and bottom FRP skins and shear stresses are resisted by the AAC core. As noted in Fig. 9.13, for AAC thickness of 8 in., the GFRP and CFRP laminates have increased the capacity of AAC walls by about 400% and 800% over the reinforced AAC, respectively.

9.7 Conclusion

The reasonable agreement between the FE and experimental results illustrates that the FE model which accurately predicts the performance of the panel is likely to minimize the need for full-scale testing of panels for code acceptance. In addition, a modeling tool would also allow the behavior of the panels to be investigated under different load combinations. In this chapter, analytical modeling for predicting the deflection and strength of FRP/AAC structural panels was developed. Further, design graphs have been developed to be used as a guide in designing the FRP/AAC panels for both floors and walls, and the comparative study showed how the proposed panels are economical compared to reinforced AAC panel currently used in the housing market. Based on the design for floor panels, the deflection and shear stresses are the main criteria that govern the design, while for wall panels, the wall slenderness ratio is the most important design criterion.

9.8 Acknowledgment

The authors gratefully acknowledge funding and support provided by National Science Foundation research project NSF-CMMI-825938.

9.9 References

Allen, H. G. (1969) *Analysis and Design of Structural Sandwich Panels*, Pergamon Press, London.

American Concrete Institute (2002a) *Building Code Requirements for Structural Concrete (ACI 318-02) and Commentary (ACI 318 R-02)*.

American Concrete Institute (2002b) *ACI 440.2R-02. Guide for Design and Construction of Externally Bonded FRP Systems for Strengthening Concrete Structures*, American Concrete Institute.

ASTM C393 (2000) *Annual Book of ASTM standards C393-00, Standard test method for flexural properties of sandwich construction*, ASTM, West Conshohocken, PA.

Dembowski, J. (2001) A Study of the material properties and structural behavior of plain and reinforced AAC components. MS Thesis, CEE Department, University of Alabama at Birmingham.

Haibin, N., Janowski, G. M., Vaidya, U. K. and Husman, G. (2007) 'Thermoplastic sandwich structure design and manufacturing for the body panel of mass transit vehicle', *Journal of Composites Structures*, 80(1), 82–91.

Karbhari, V. and Zhao, L. (2000) 'Use of composites for 21st century civil infrastructure', *Computer Methods in Applied Mechanics and Engineering*, 185, 433–454.

Khotpal, A. (2004) Structural characterization of hybrid fiber reinforced polymer (FRP)-autoclaved aerated concrete (AAC) panels. MS Thesis, CEE Department, University of Alabama at Birmingham.

Masonry Standards Joint Committee (MSJC) (2002) *Building Code Requirements for Masonry Structures (ACI 530-02/ASCE 5-02/TMS 402-02)*.

Mousa, M. (2007) Optimization of structural panels for cost-effective panelized construction. MS Thesis, CEE Department, University of Alabama at Birmingham.

Mousa, M. and Uddin, N. (2009) 'Experimental and analytical study of carbon fiber-reinforced polymer (FRP)/autoclaved aerated concrete (AAC) sandwich panels', *Journal of Engineering Structures*, 31(10), 2337–2344.

Shi, C. and Fouad, H. F. (2005) *Autoclaved Aerated Concrete – Properties and Structural Design*, Special Publication 226, American Concrete Institute.

Sika Corporation (2002) *Construction Products Catalog*, Sikadur, available at: http://www.sikausa.com.

Snow, C. (1999) A comprehensive study of the material properties and structural behavior of AAC products. MS Thesis, CEE Department, University of Alabama at Birmingham.

Uddin, N. and Fouad, H. (2007) 'Structural behavior of FRP reinforced polymer-autoclaved aerated concrete panels', *ACI Structural Journal*, 104(6), 722–730.

Zenkert, D. (1995) *An Introduction to Sandwich Construction*, Engineering Materials Advisory Service Ltd, West Midlands, United Kingdom.

9.10 Appendix A: λ calculations for fiber-reinforced polymer (FRP)/autoclave aerated concrete (AAC) using unidirectional fiber-reinforced polymer (FRP) facesheets (UFFS)

The experimental defection = 0.47 in. (11.97 mm)
$L = 48$ in. (1200 mm), $b = 7$ in. (175 mm), $c = 3.94$ in. (100 mm), $P = 3.5$ Kips (15,540 N)
$k_b = 1/56$
$k_s = 1/8$

Theoretical deflection:

To obtain λ, we have to assume $\lambda = 1$ and then get the theoretical deflection according to the following equation:

$$\Delta = \frac{k_b P L^3}{D} + \frac{k_s P L}{U}$$

$$D = \frac{E_{face}(d^3 - c^3)b}{12} = \frac{70552(102.03^3 - 100^3)175}{12} = 4.056 \times 10^{10} \text{ N} \cdot \text{mm}^2$$

$$U = \frac{G(d+c)^2 b}{4c} = \frac{720(102.03+100)^2 \, 175}{4 \times 100} = 1.03 \times 10^7 \text{ N}$$

$$\Delta = \frac{15540 \times 1200^3}{56 \times 4.056 \times 10^{10}} + \frac{15540 \times 1200}{1.03 \times 10^7} = 7.63 \text{ mm}$$

After several tries, it was found that the nominal thickness that can provide the experimental deflection is 3.17 in. (80.4 mm). Accordingly, λ can be determined as:

$$80.4 = 1 + \left(1 + \frac{100}{102.03}\right) + \frac{100^2}{102.03} \lambda$$

This results in $\lambda = 0.8$.

9.11 Appendix B: symbols

b	= width of FRP/AAC panel
t	= thickness of FRP facesheet
L	= panel length
c	= core thickness
d	= thickness of sandwich panel
D	= panel flexural rigidity
U	= panel shear rigidity
P	= total load applied to the panel
ε_{face}	= ultimate strain in FRP facesheet
E_{face}	= modulus of elasticity of FRP facesheets
G	= shear modulus of AAC core
M_n	= nominal flexural strength of the FRP/AAC panel
V_n	= total shear strength of the FRP/AAC panel
V_{AAC}	= shear strength of AAC core
V_f	= shear strength carried by FRP wraps

10
Impact behavior of hybrid fiber-reinforced polymer (FRP)/autoclave aerated concrete (AAC) panels for structural applications

N. UDDIN, M. A. MOUSA and F. H. FOUAD,
The University of Alabama at Birmingham, USA

DOI: 10.1533/9780857098955.2.247

Abstract: The low velocity impact response of plain autoclaved aerated concrete (AAC) and FRP/AAC sandwich panels has been investigated. The structural sandwich panels composed of a FRP/AAC combination have shown excellent characteristics in terms of high strength and high stiffness-to-weight ratios. In addition to having adequate flexural and shear properties, the behavior of FRP/AAC sandwich panels needs to be investigated when subjected to impact loading. During service, the structural members in the building structures are subjected to impact loading that varies from object-caused impact, blast due to explosions, to high velocity impact of debris during tornados, hurricanes, or storms. Low velocity impact (LVI) testing serves as a means to quantify the allowable impact energy that the structure is able to withstand and to assess the typical failure modes encountered during this type of loading. The objectives of this chapter are: to study the response of plain AAC and CFRP/AAC sandwich structures to low velocity impact and to assess the damage performance of the panels; to study the effect of FRP laminates on the impact response of CFRP/AAC panels; to study the effect of the processing method (hand layup versus VARTM) and panel stiffness on the impact response of the hybrid panels. Impact testing was conducted using an Instron drop-tower testing machine. Experimental results showed a significant influence of CFRPs laminates on the energy absorbed and peak load of the CFRP/AAC panels. Further, a theoretical analysis was conducted to predict the energy absorbed by the CFRP/AAC sandwich panel using the energy balance model, and the results found were in good accordance with the experimental ones.

Key words: fiber-reinforced polymer, autoclaved aerated concrete, low velocity impact, sandwich structures.

10.1 Introduction

The concept of the sandwich structures can be compared to that of I-sections, in which the facing skins of a sandwich panel can be compared to the flanges, as they carry the bending stresses and provide bending rigidity. The core corresponds to the web, as it resists the shear loads and stabilizes the faces against bulking or wrinkling and provides shear rigidity (Zenkert, 1995). It also increases the stiffness of the structure by holding the facing skins apart. However, the difference is that the core of the sandwich is made from different material than the faces and it provides a continuous support for the faces producing a uniformly stiffened panel rather than being concerted in a narrow web. The core-to-skin adhesive rigidly joins the sandwich components and allows them to act as one unit with high torsional and bending rigidity.

In the research reviewed in this chapter, the CFRP and AAC have been used as a skin and core respectively. In case of the FRP/AAC interface, the mechanism of adhesion is the mechanical interlocking of the resin in the irregularities of the concrete surface (Karbhari and Zhao, 2000). Although, both AAC and CFRP are brittle materials, they have shown excellent results in combination with each other in terms of increasing the shear and flexural strengths and also enhance the ductility of sandwich panels (Mousa, 2007). In addition, since this combination is lightweight in nature, it has the potential to be used for speedy panelized construction purposes to reduce labor-intensive construction.

This chapter is concerned with the application of CFRP/AAC sandwich panels for emergency shelter houses and related issues. In order for such materials to gain acceptance in such applications, their impact response has to be investigated. These sandwich panels may be either used for load bearing or cladding purposes. Under these circumstances, impacts from small objects, such as hand tools at low velocities, or blast fragments at high velocities, will involve local indentation and damage without global deformation of the sandwich panels. The damage tolerance of sandwich structures to localized loading is of concern and should be evaluated in each application prior to its implementation. In addition, the research presents an empirical method using the energy balance model to predict the energy absorbed by the sandwich panel.

The main objectives of the research reviewed in this chapter were:

- to study the response of plain AAC panels and CFRP/AAC sandwich panels to low velocity impact (LVI) loading
- to assess the damage performance and mode of failure of panels under LVI
- to study the effect of CFRP laminates on the impact response of AAC panels
- to study the effect of the processing method (hand layup versus VARTM) on the performance of the CFRP/AAC panels
- to study the effect of core thickness or panel stiffness on the impact response of the hybrid panels.

10.2 Low velocity impact (LVI) and sandwich structures

Low velocity impact (LVI) response, also known as quasi-static impact response, is a boundary controlled impact response in which the flexural and shear waves have a sufficient time to come to and be reflected by the boundary many times. The LVI usually results in less damage than other types of impact responses such as high velocity impact that is dominated by dilatational waves and transversely reflected several times (Olsson, 2000).

Typical failure modes of sandwich structures have been documented and analyzed by Abrate (1998). These studies confirmed the importance of the core-facing interface in explaining the damage progression of a sandwich plate subjected to impact. Abrate found that sandwich structures subjected to localized impact behave radically different from composite monolithic laminates and that the indentation of sandwich plates is dominated by deformation of the core. He also summarized and evaluated the different theories and empirical models that predict the indentation of composite materials.

Fatt and Park (2001) investigated the damage initiation of sandwich panels under LVI loading. They found that the initial impact damage depended upon the panel support conditions, projectile nose geometry, and material properties of the skin and core. Hazizan and Cantwell (2002) used the energy balance model to predict the energy absorbed for foam-based sandwich panels, while Ambur and Cruz (1995) used first-order shear deformation theory to model the low velocity impact response of composite panels. Sun and Wu (1991) used shear-deformable plate finite elements to model the composite skins in an impact loaded sandwich structure.

Sets of LVI impact tests were conducted by Cantwell *et al.* (1994) on a number of foam core and balsa core sandwich structures; the conclusion of that research was that skin shearing is the primary energy absorber under localized impact conditions. Rhodes (1975) conducted impact tests on a range of sandwich systems and the results showed that enhancing the crush strength of the core material can serve to increase the impact resistance of the sandwich structure. Mines *et al.* (1998) conducted low velocity impact tests on square panels based on polymer composite sandwich structures. The results showed that much of the incident energy of the projectile is absorbed in crushing of the core material within a localized region immediate to the point of impact.

Horrigan *et al.* (2000) conducted experimental and theoretical investigations on Nomex honeycomb sandwich structure with glass fiber-reinforced epoxy skins. The results showed that a soft, compliant projectile results in shallow crushing of the core, whereas hard bodies create deeper damage that conforms to the shape of the projectile. Li and Jones (2007) and Li and Venkata (2008a) investigated the low velocity impact behavior of both cement-based syntactic foam and sandwich panels made of cement-based syntactic foam as a core and E-glass fiber as facesheet. The results compared with the behavior of the panels made with pure cement paste core. Scanning electron microscopy (SEM) was also used to examine the energy dissipation mechanisms in the micro-length scale. The results showed that the cement-based syntactic foam has a higher capacity for dissipating impact energy with an insignificant reduction in strength as compared to the control cement paste core. When compared to a polymer-based foam core having similar compositions, it was found that the cement-based foam has a comparable energy dissipation capacity. Li and Venkata (2008b) investigated the impact characterization of sandwich structures with an integrated orthogrid stiffened syntactic foam core. Their results showed that the integrated core enhances impact energy transfer, energy absorption, and positive composite action, and insures quasi-static response to impact.

10.3 Materials and processing

The FRP/AAC panel discussed in this chapter is composed of CFRP laminates as a facesheet (skin) and AAC as a core. Fiber-reinforced composites offer high corrosion and flexural resistance. Accordingly, since AAC is an ultra lightweight material in nature and CFRP is stiff with a high specific strength, the two could be used together to form strong hybrid structural panels. Several studies have been conducted at the University of Alabama at Birmingham (UAB) to investigate the behavior of CFRP/AAC structural panels under axial and out-of-plane loading. Khotpal (2004) has investigated the compressive strength of plain AAC wrapped by CFRP. The objectives were to evaluate the load-carrying capacity of the confined AAC cube and to observe the mode of failure of CFRP/AAC panels. The results showed that the CFRP wraps increased significantly the compressive strength of CFRP/AAC panels by about 80% over the plain AAC. Uddin and Fouad (2007) investigated the behavior of CFRP/AAC panels using small size specimens under a four-point load test. The experimental results for that research showed a significant influence of the FRP on the flexural strength and stiffness of the hybrid panels. Mousa (2007) also used finite element modeling to analyze and design CFRP/AAC structural panels to be used as floor and wall panels. Mousa and Uddin (2009) developed theoretical formulas to predict the shear and flexural strengths of CFRP/AAC panels and the results found were in a good agreement with experimental ones. Further, a comparative study between the hybrid CFRP/AAC panel and the currently used reinforced AAC panels was conducted by Mousa (2007). The comparative study showed how the proposed panels are economical compared to reinforced AAC panels currently used in the housing market. Due to the higher strength resulting from this combination, the strength is not the criterion governing the design of the panel, but the deflection is the one that controls the design of the proposed hybrid panels (Mousa, 2007).

As mentioned previously, the CFRP/AAC panel is made from CFRP laminates as facesheets bonded to an AAC core using thermoset epoxy polymers producing a stiff panel. In general, autoclaved aerated concrete (AAC) is an ultra lightweight concrete with a distinct cellular structure. It is approximately one-fifth the weight of normal concrete with a dry bulk density ranging from 400–800 kg/m^3 (25–50 pcf) and a compressive strength ranging from 2 to 7 MPa (300–1000 psi) (Shi and Fouad, 2005). The low density and porous structure gives AAC excellent thermal and sound insulation properties which make it an excellent choice to be used as a core material for building applications. Due to the cellular structure and reduced weight, the material is highly fire resistant and very durable when compared with conventional construction material, and has unique thermal insulation properties.

AAC is currently used in the form of steel-reinforced panels using pre-treated rebars as an internal reinforcement. These rebars will be subjected to corrosion in the long term and are also expensive compared with those used for normal reinforced concrete. Furthermore, these rebars do not play any role in the shear strength of the panels. Therefore, the panels are required to be thick in order to overcome the shear and lower flexural strength problems. Mousa (2007) has demonstrated that the shear strength of CFRP/AAC can be much improved by wrapping the plain AAC with CFRP laminates. Therefore, the overall cost of reinforced AAC panels can be decreased by using the FRP laminates as external reinforcement (compared to CFRP/AAC sandwich panels) instead of the internal steel rebars in conjunction with low cost processing techniques which will be explained in this chapter. Table 10.1 lists the mechanical properties of AAC being used

Table 10.1 Mechanical properties of plain autoclaved aerated concrete (AAC)

Property	Value
Density	40 pcf (640 kg/m^3)
Compressive strength	456 psi (3.2 MPa)
Modulus of elasticity	256,000 psi (1,800 MPa)
Shear strength	17 psi (0.12 MPa)
Poisson's ratio	0.25

in the current research. In the present research, SIKA WRAP HEX 103C unidirectional carbon fibers, and SIKADUR HEX 300 resin were used. The mechanical properties of resin as well as laminates, as provided by the manufacturer (Sika Corporation, 2002), are listed in Table 10.2.

In this study, three groups of panels have been prepared and tested under low velocity impact. The first one is plain AAC specimens which are considered as the control panels. The second one is CFRP/AAC panels processed using the hand layup technique; the panels were sandwiched by top and bottom unidirectional carbon fiber lamina (i.e., the fiber orientation is 0°) for flexural reinforcement and then wrapped by another unidirectional carbon lamina (the fiber orientation is 90°, Fig. 10.1) for shear reinforcement. The third one is CFRP/AAC panels having the same characteristics as the second group, but processed using the vacuum assisted resin transfer molding (VARTM) technique. As an alternative to the labor-intensive hand layup process, VARTM is an attractive process because it saves processing time, especially when several CFRP layers are being applied. VARTM is a

Table 10.2 Mechanical properties of SIKA carbon fiber composite

Property	SIKA HEX 300	Unidirectional Laminate
Tensile strength	10,500 psi (72.4 MPa)	123,200 psi (849 MPa)
90° tensile strength	–	3,500 psi (24 MPa)
Modulus of elasticity, E_x	459,000 psi (3,170 MPa)	10,239,800 psi (70,552 MPa)
Modulus of elasticity, E_y	459,000 psi (3,170 MPa)	705,500 psi (4,861 MPa)
Shear modulus, G_{xy}	–	362,500 psi (2,498 MPa)
Tensile elongation	4.8%	1.12%
Ply thickness	–	0.04 in. (1.016 mm)

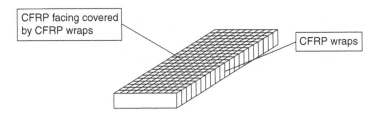

10.1 Schematic diagram for CFRP/AAC sandwich panel.

process for molding fiber-reinforced composite structures in which a sheet of flexible transparent material such as nylon or Mylar plastic is placed over the preform and then sealed to prevent any air from getting inside the preform (Perez, 2003). A vacuum is applied between the sheet and the preform to remove the entrapped air. VARTM ensures the complete wet-out of fiber, ensures that the fiber is fully impregnated with resin, and is not as tedious as the hand layup technique. VARTM is typically a three-step process consisting of the layup of a fiber preform, impregnation of the preform with resin, and cure of the impregnated preform. The complete procedure for processing the FRP/AAC panel using the VARTM technique is not included in this chapter for the sake of brevity and is described elsewhere (Uddin and Fouad, 2007). To avoid the excessive resin absorption by AAC due to its pores surface, the AAC surface is painted with block filler. The block filler consists of water, calcium carbonate, vinyl acrylic latex, amorphous silica, titanium dioxide, ethylene gyclone, and crystalline silica. The purpose of the block filler is to fill the surface pores present on AAC panel surfaces and to minimize the excessive resin absorption by AAC panels. It has a density of 1,461 kg/m^3. It is generally used for filling the pores of masonry or block-walls. It must be applied to clean, dry surfaces completely free from dirt, dust, chalk, rust, grease, and wax. It can be applied using a premium quality nylon or polyester brush, or spray equipment. The drying time of block filler is 2–3 hours. A waiting time of 4–6 hours is required before applying the FRP layer.

Table 10.3 shows the types of specimens used in this study with brief descriptions for each one. All specimens tested in this study were 609.8 mm (24.0 in.) long and 203.3 mm (8.0 in.) wide. In the specimen notation, the first letter indicates the type of manufacturing process used for specimen preparation and the second letter indicates the thickness of the specimen in inches. For example, in P-1 specimen, 'P' represents a plain AAC specimen, while '1' represents the thickness of the specimen, 25.4 mm (1.0 in.). Similarly, 'H' represents a hand layup processed specimen and 'V' represents a VARTM processed specimen. The dimensional accuracy of all specimens was close to ± 2.5 mm (0.1 in.). The AAC specimens were oven dried at 70°C (158 °F) to reach the moisture content specified by ASTM C 1386 (2007) which is 5–15% by weight.

10.4 Analyzing sandwich structures using the energy balance model (EBM)

The impact response of the sandwich structures can be modeled using an energy balance model. The model that was previously used by Hazizan and Cantwell (2002) was proven to be adequate in predicting the behavior of sandwich constructions that incorporate foam materials. The energy

Table 10.3 Details of test specimens

Specimen ID	Length, mm (in.)	Width, mm (in.)	Depth, mm (in.)	Core material	Face sheet	Preparation process
P-1	609.8 (24)	203.2 (8)	25.4 (1)	AAC	None	–
P-2	609.8 (24)	203.2 (8)	50.8 (2)	AAC	None	–
P-3	609.8 (24)	203.2 (8)	76.2 (3)	AAC	None	–
H-1	609.8 (24)	203.2 (8)	25.4 (1)	AAC	Carbon fiber Sikawrap Hex-103C	Hand layup
H-2	609.8 (24)	203.2 (8)	50.8 (2)	AAC	Carbon fiber Sikawrap Hex-103C	Hand layup
H-3	609.8 (24)	203.2 (8)	76.2 (3)	AAC	Carbon fiber Sikawrap Hex-103C	Hand layup
V-1	609.8 (24)	203.2 (8)	25.4 (1)	AAC	Carbon fiber Sikawrap Hex-103C	VARTM
V-2	609.8 (24)	203.2 (8)	50.8 (2)	AAC	Carbon fiber Sikawrap Hex-103C	VARTM
V-3	609.8 (24)	203.2 (8)	76.2 (3)	AAC	Carbon fiber Sikawrap Hex-103C	VARTM

absorption in sandwich structures subjected to impact loading is dominated by the independent contribution of three major factors; the energy absorbed by:

1. the bending of the face sheets,
2. the energy absorbed by shear in the core,
3. the energy absorption due to contact effects.

Accordingly, the kinetic energy due to a target impact can be expressed as:

$$E_{impact} = \frac{1}{2}mv^2 = E_{bs} + E_c \qquad [10.1]$$

where the subscripts b, s and c refer to energy dissipation in bending, shear and contact effects, respectively. The force–displacement relationship, $P - \delta$, for a sandwich beam subjected to centric flexural concentrated load, related to the concentrated impact loading, is given by (Allen, 1969):

$$\delta = \frac{PL^3}{48D} + \frac{PL}{4AG} \quad [10.2]$$

The first term in the above equation accounts for deflection due to bending of the sandwich panel while the second term accounts for deflection due to shear. From the maximum deflection of the sandwich panel and the maximum load P_{max}, the energy absorbed in bending and shear can be expressed by:

$$E_{bs} = \frac{1}{2} \cdot P_{max} \cdot \delta_{max} = \frac{P_{max}^2}{2}\left[\frac{L^3}{48D} + \frac{L}{4AG}\right] \quad [10.3]$$

The contribution of contact effects to the energy absorbed by the sandwich panel can be calculated by using Meyer's indentation law (Yang and Sun, 1982) which states that the relationship between the applied load and the indentation depth can be expressed by:

$$P = C\alpha^n \quad [10.4]$$

where C and n are the parameters obtained from Meyer's indentation law while α is the indentation due to load P. The expression for the contact effects is obtained by integrating Eq. [10.4] between 0 and the maximum indentation value (α_{max}) as follows:

$$E_c = \int_0^{\alpha_{max}} P\,d\alpha = \frac{C(P_{max}/C)^{(n+1)/n}}{n+1} \quad [10.5]$$

Meyer's law is an empirical expression representing the relationship between indentation load and the resultant indentation depth. C and n are the parameters that adapt this relation. The procedure to determine C and n is not included in this chapter for the sake of brevity and is explained elsewhere (Kedar, 2006). From the study, C and n were found to have average values of 5,800 N/mm (32,000 b/in.) and 1.0, respectively.

Therefore, and from Eqs [10.3] and [10.5], the energy balance model for the sandwich panel is:

$$E_{impact} = \frac{P_{max}^2}{2}\left[\frac{L^3}{48D} + \frac{L}{4AG}\right] + \frac{C(P_{max}/C)^{(n+1)/n}}{n+1} \quad [10.6]$$

The above equation represents the total impact energy absorbed by a sandwich panel.

10.5 Low velocity impact (LVI) testing

In LVI testing, a known weight is dropped on the specimen from a desired height. From the known weight and height, the impact energy can be calculated which means that the energy of impact can be varied by changing the height from which the impactor is falling. In this study, the testing was conducted using a Dynatup 8250 impact-testing machine with a hemispherical shape assembly of 19.05 mm (0.75 in.) diameter and with a load cell of capacity of 1,590 kg (3,500 lbs). A flat striker (tup) with an impact area of 76 mm × 102 mm (3.0 in. × 4.0 in.) was attached to the hemispherical impactor to duplicate the impact conditions similar to a falling object, such as a hand tool. A schematic diagram for the impact machine is illustrated in Fig. 10.2. Figure 10.3 shows the jig that was used to hold the specimen under the impact test machine. It consists of two holding steel plates that were half an inch thick. It has four bolts at each corner that pass through one plate to the other. The double nuts and washers were used inside and outside the holding plates to stabilize and to fix the position of the specimen in the frame. These plates are made of mild steel.

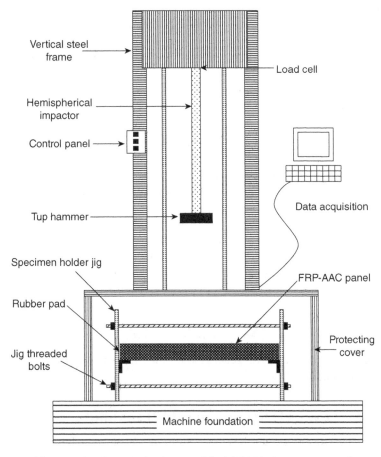

10.2 Schematic diagram for Instron Model 8250 drop tower testing machine.

10.3 Jig or specimen holder.

A fixed drop height was used for all of the specimens to investigate and compare their response to the same applied impact energy. The total mass of the impactor used was 25 kg (55.11 lb) dropping from a height 400 mm (15.75 in.) resulting in an impact energy of 100 J. The impact energy is a result of the product of the impactor mass and height dropped. It should be pointed out that the 100 J impact energy was chosen based on previous calculations to predict the maximum energy absorbed by the tested specimens. The testing machine used DynaTup software to record the data produced during impact loading. Load versus time, energy absorbed by the specimen versus time and impact velocity were measured directly by the software and acquired by integrated data acquisition system and measured directly by the software. For a forced object traveling in a straight line, the software uses standard equations of motion to express velocity, load, and energy absorption during the impact event. These equations are functions of many variables including hammer mass, acceleration due to gravity, velocity of the hammer, and kinetic energy of the hammer, etc.

For the analysis presented, initiation energy is defined as the energy at maximum load and the total impact energy is the impact energy when the projectile separates from the target or when the impact force is zero, while propagation energy is defined as the difference between total energy and initiation energy. These definitions have been used previously by many authors (Agarwal *et al.*, 2006; Li and Jones, 2007; Li and Venkata, 2008a, 2008b). The initiation energy is basically a measurement of the capacity for the target to transfer energy elastically and higher initiation energy means a higher load-carrying capacity. On the other hand, the propagation energy represents the energy absorbed by the target for creating and propagating gross damage of the panel.

10.6 Results of impact testing

10.6.1 Plain AAC specimens

The objectives for testing the plain AAC specimens were to evaluate their damage initiation and crack propagation and to compare their response with CFRP/AAC specimens. All plain AAC specimens failed in brittle manner. The reasons for the brittle mode of failure were attributed to the low tensile strength of the AAC and the porous structure of the AAC.

According to the information extracted from the energy versus time data and the load versus time graphs, the failure of plain AAC panels can be described by the following main events. The load versus time curve shows a significant amount of noise in the initial region of the curve (i.e., the onset of internal damage). Then, after initial damage, the maximum (peak) load is recorded (at which the AAC fails at the mid-span). The time it takes to reach maximum load is defined as time to max load. From that point on, the propagation phase begins and fracture through the AAC propagates in a brittle manner (shear failure of the AAC). Once the shear failure occurs in the AAC, the load drops sequentially to zero and the impact event ends.

Table 10.4 lists the relevant impact data as obtained by DynaTup software for all plain AAC specimens while load versus time and energy absorbed by the specimen versus time graphs are shown in Figs 10.4–10.6 for each specimen. As seen in Table 10.4 and Fig. 10.4, the P-1 specimen recorded the lowest maximum load due to its lower stiffness, while the P-3 specimen has shown the highest maximum load due to the higher stiffness compared to the other plain AAC specimens. It should be noted that the second peak in the load curves indicates the impact reaction of the striker (tup) following the failure of the specimens. For the P-1 specimen, as seen from Table 10.4 and Figs 10.4–10.6, the energy absorbed during the propagation phase was less than that absorbed during the initiation phase, while the energy

Table 10.4 LVI data for plain AAC specimens

Impact feature	Specimen		
	P-1	P-2	P-3
Energy of impact	100	100	100
Max. load, kN (Kips)	7.18 (1.62)	10.32 (2.32)	15.12 (3.41)
Time to max. load (ms)	0.35	0.37	0.4
Energy to max. load (J)	4.5	5.82	26
Total energy (J)	8.2	25.4	93.5
Impact velocity m/s	2.9	2.88	2.48
Prop. energy J	3.7	19.58	86.3

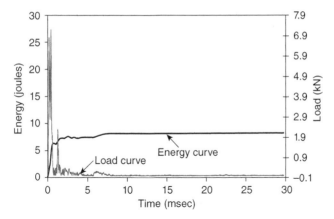

10.4 Energy versus time and load versus time curve for LVI testing for P-1 specimen.

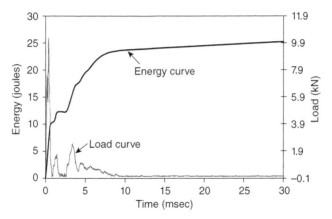

10.5 Energy versus time and load versus time curve for LVI testing for P-2 specimen.

behavior was the opposite for both P-2 and P-3 specimens because of the lower stiffness of P-1 compared to the other specimens.

Due to the lower stiffness of P-1 and P-2 specimens compared to P-3, P-1 and P-2 underwent complete failure under the LVI loading, in which they did not show any resistance to the impact loading and were completely destroyed by the end of the test. The complete damage of P-1 and P-2 specimens is the main reason for less energy absorbed (8.2 J for P-1 and 25.4 J for P-2) than the impact energy (100 J), unlike P-3, which absorbed most of the impact energy because of its higher stiffness. Figure 10.7 shows the failure modes of the specimens. It was noted that all the specimens failed at the mid-span where the impact occurred, followed immediately by shear failure at the support locations where the maximum shear stresses are located.

260 Developments in FRP composites for civil engineering

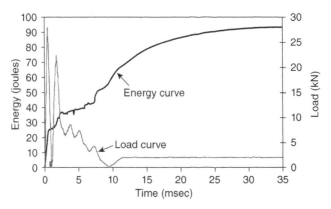

10.6 Energy versus time and load versus time curve for LVI testing for P-3 specimen.

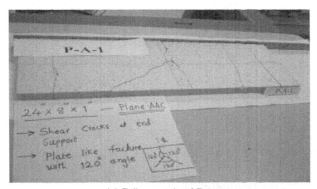

(a) Failure mode of P-1.

(b) Failure mode of P-2. (c) Failure mode of P-3.

10.7 Failure mode for plain AAC specimens; cracks occurred at mid-span (flexural failure) followed by cracks at near to support locations (shear failure).

© Woodhead Publishing Limited, 2013

10.6.2 CFRP/AAC panels processed using hand layup technique

As shown in the load versus time and energy absorbed by the specimen versus time graphs for each specimen (Figs 10.8–10.10), failure of the hand layup CFRP/AAC sandwich panels can be summarized as follows: again, the load versus time curve shows a significant amount of noise in the initial region of the curve. Then, after initiation of damage, the maximum load is recorded (failure of the CFRP facesheet in the compression side). After that point, the propagation phase starts by a sudden drop in the load (core crushing at mid-span) until the bottom facesheet (in the tension side) fails. Once there is failure in this facesheet, the load drops to zero and the impact event ends.

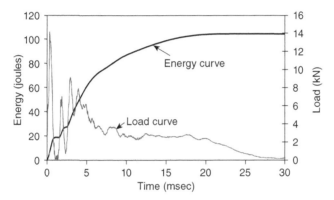

10.8 Energy versus time and load versus time curve for LVI testing for H-1 specimen.

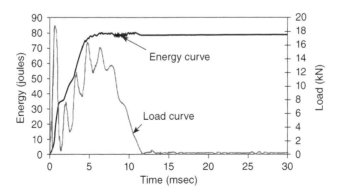

10.9 Energy versus time and load versus time curve for LVI testing for H-2 specimen.

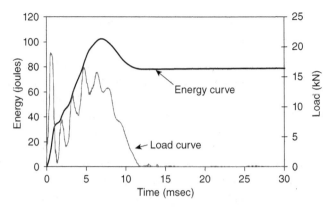

10.10 Energy versus time and load versus time curve for LVI testing for H-3 specimen.

Table 10.5 LVI data for hand layup processed FRP/AAC specimens

Impact feature	Specimen		
	H-1	H-2	H-3
Energy of impact	100	100	100
Max. load, kN (Kips)	14.2 (3.2)	18.9 (4.26)	19.1 (4.3)
Time to max. load (ms)	0.35	0.59	0.6
Energy to max. load (J)	6.8	14.8	14.8
Total energy (J)	100	80	80
Impact velocity m/s	2.81	2.81	2.9
Prop. energy J	93.2	65.2	65.2

The CFRP/AAC specimens recorded higher peak load than the plain AAC specimens because of the significant contribution of the CFRP laminates in both shear and flexural strengths of the panel. The time taken to reach the peak load was greater than the plain AAC specimens. As seen from Figs 10.8–10.10 and Table 10.5, the energy absorbed during the propagation phase is higher than that absorbed during the initiation phase for all specimens. This suggests that the CFRP/AAC panels are not brittle and show obvious ductility that can be attributed to the conical propagation of damage observed in the AAC core and the delamination from the facesheets. Due to the high stiffness of both H-2 and H-3, they did not absorb the total applied impact energy and they end up with 80 J total absorbed energy. As a result of its high contact stiffness, it was observed that H-3 absorbed 100 J at time 7.5 ms and dropped to 80 and stayed at 80 J until the impact ended. Impact data for this group are summarized in Table 10.5.

As shown in Fig. 10.11, all of the specimens in this group have failed in flexural at the mid-span where the maximum moment occurs. No shear cracks at support locations were observed and the reason for this is the higher shear strength of the CFRP/AAC panels with FRP wraps which can increase the shear strength of the entire panel by 300% (Mousa and Uddin, 2009). The top facesheet failed by rupture and split into two parts of equal sizes at the middle of the panel where the impact loading was located. The

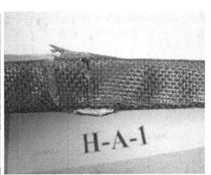

(a) Failure mode of H-1. Note the core completely crushed at mid-span.

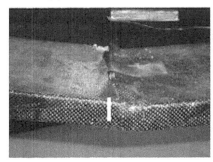

(b) Failure mode of H-2.

(c) Failure mode of H-3.

10.11 Failure mode for hand layup processed FRP/AAC specimens; failure initiated by rupture of the top facesheet, then localized crushing of the core, and finally debonding of the bottom facesheet.

core crushed along the mid-span due to the contact effect. The bottom facesheet showed no evidence of damage except in the case of H-1; however, the interface between the bottom facesheet and core failed. This can be attributed to the tensile failure along the adhesive line (debonding) between the facesheet and core substrate. This type of failure is a common mode for sandwich panels, especially when the panels are processed using the hand layup method which has more possibility of insufficient resin absorption than VARTM. In summary, the failure was initiated with the rupture of the top facesheet, followed by the localized crushing of the core, and finally debonding of the bottom facesheet. The failure was excessive in the case of H-1 due to its lower bending stiffness.

As shown by the energy versus time graphs for this group, the total energy absorbed by CFRP/AAC panel decreases as the core thickness, or panel stiffness, increases. This is mainly because when the core thickness increases, both bending and shear stiffness of the panel increase too. Thus, the ability of the panel to exhibit significant flexure and shear, and then absorb more energy, is reduced. On the other hand, when the core thickness decreases, the ability of the panel to bend becomes greater and the energy absorbed is expected to be higher. However, it should be mentioned that as the panel's stiffness increases, less damage due to impact loading is obtained.

10.6.3 CFRP/AAC panels processed using VARTM technique

The response of the VARTM processed panels was similar to that of hand layup panels in terms of peak load and failure events, although V-3 has shown a higher peak load over H-3. This is mainly due to the fact that the impact loading is localized, not spread over the panel. Thus, the processing method does not affect too much the peak load. Table 10.6 lists

Table 10.6 LVI data for VARTM processed FRP/AAC specimens

Impact feature	Specimen		
	V-1	V-2	V-3
Energy of impact	100	100	100
Max. load, kN (Kips)	13.65 (3.07)	17 (3.83)	23.4 (5.27)
Time to max. load (ms)	0.33	0.33	0.46
Energy to max. load (J)	7.7	8.5	16.7
Total energy (J)	100	96	67
Impact velocity m/s	2.98	2.9	2.88
Prop. energy J	92.3	87.5	50.3

the impact data for this group and Figs 10.12–10.14 show the load versus time and energy absorbed by the specimen versus time graphs for each specimen.

It was noted that the specimens did not fail or yield at the mid-span, except the V-1 specimen. The V-1 specimen failed due to FRP rupture at the tension face followed by crushing of the AAC core at the mid-span. No failure has been observed for the top facesheet. The failure of V-1 can be attributed to its lower bending stiffness compared to the other panels. The V-2 and V-3 specimens stayed intact after the impact. This observation recommends VARTM as a processing technique, as it improves the bond characteristics between CFRP facesheets and AAC substrate over any other hand layup techniques. Therefore, VARTM enhances the whole

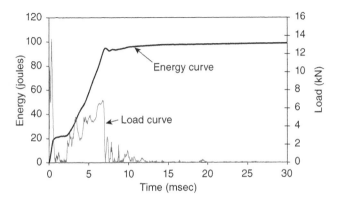

10.12 Energy versus time and load versus time curve for LVI testing for V-1 specimen.

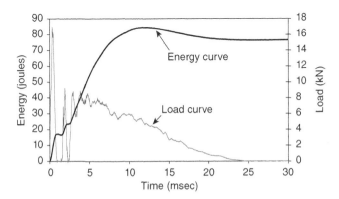

10.13 Energy versus time and load versus time curve for LVI testing for V-2 specimen.

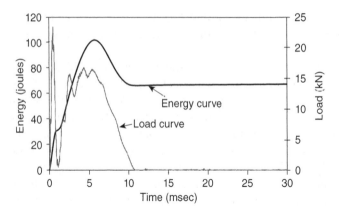

10.14 Energy versus time and load versus time curve for LVI testing for V-3 specimen.

stiffness of the CFRP/AAC panel. Further, VARTM ensures the complete wet-out of fibers and also provides uniform distribution of resin on the AAC substrate. This is the main reason for not having some debonding problems when using VARTM. The specimens after the LVI test are shown in Fig. 10.15. It should be mentioned that the total energy absorbed decreased as the core thickness increased due to the increase in the panel stiffness, which is the same as in the hand layup group.

10.7 Analysis using the energy balance model (EBM)

The energy absorbed by the CFRP/AAC sandwich panels was predicted using the energy balance model. From the properties of the skin and core materials and using the maximum load value obtained from LVI testing, the energy absorbed by CFRP/AAC sandwich panel can be calculated using Eq. [10.6]. All data and parameters used to predict the energy absorbed by CFRP/AAC sandwich panel tested in this study are listed in Table 10.7.

The following is a sample calculation for the predicted energy absorbed for H-1 panel under LVI:

$E_{impact} = 100$ J, as recorded by the LVI testing machine.

$$E_{impact(Calculated)} = \frac{P_{max}^2}{2}\left[\frac{L^3}{48D} + \frac{L}{4AG}\right] + \frac{C(P_{max}/C)^{(n+1)/n}}{n+1}$$

$$D = \frac{E_f \cdot t \cdot b(h+t)^2}{2} = \frac{70552 * 1.016 * 203.2(25.4+1.016)^2}{2} = 5.08 \times 10^9 \text{ N/mm}$$

$$AG = b(h+t)G = 203.2 \times (25.4+1.016) \times 706 = 3.79 \times 10^6 \text{ N/mm}$$

Impact behavior of hybrid FRP/AAC panels

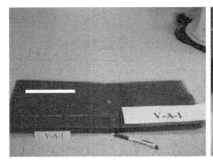
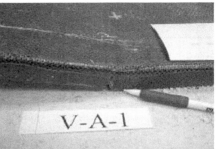

(a) Failure mode of V-1.

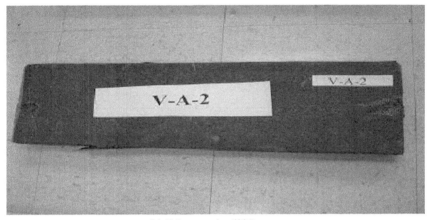

(b) Failure mode of V-2.

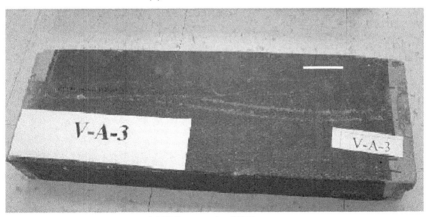

(c) Failure mode of V-3.

10.15 Failure mode for VARTM processed FRP/AAC specimens: (a) V-1 specimen failed by FRP rupture at the tension side followed by crushing of the AAC core at the mid-span. (b) and (c) V-2 and V-3 specimens stayed intact after impact loading.

Table 10.7 Data and parameters used to calculate the energy absorbed using the energy balance model

Specimen	Max. load (N)	E_f (N/mm²)	G (N/mm²)	C (N/mm)	n
H-1	14,200	70,552	706	5,800	1
H-2	18,900	70,552	706	5,800	1
H-3	19,100	70,552	706	5,800	1
V-1	13,650	70,552	706	5,800	1
V-2	17,000	70,552	706	5,800	1
V-3	23,400	70,552	706	5,800	1

Table 10.8 Comparison between the experimental and predicted absorbed energy for FRP/AAC panels

Specimen	E_{impact} Exp. (J)	E_{impact} per EBM (J)	Diff. %
H-1	100	116.61	14.2
H-2	80	78.84	1.5
H-3	80	56.61	41
V-1	100	108	7.4
V-2	76	67	13.4
V-3	67	54	24

$$E_{impact(Theoretical)} = \frac{14200^2}{2}\left[\frac{609.6^3}{48\times 5.0\times 10^9} + \frac{609.6}{4\times 3.79\times 10^6}\right] + \frac{5800(14200/5800)^{(1+1)/1}}{1+1} = 116.61 \text{ J}$$

Table 10.8 lists all the predicted energies absorbed for all CFRP/AAC sandwich panels tested in this study. As seen from Table 10.8, there is a good convergence between the experimental values and the predicted ones using the energy balance model for all specimens except H-3. This difference can be attributed to energy losses due to delamination, especially when using hand layup as in the case of the H-3 panel. During the hand layup method, the possibility of insufficient resin absorption is higher than using the VARTM method. Thus, this recommends the use of VARTM method over hand layup when the hybrid panels are subjected to impact loading. In addition, it can also be attributed to the empirical nature of Meyer's contact law, noise, and other factors that are not considered in the simplified model but contribute to energy absorption. According to the energy balance model, the energy absorbed due to bending was the major one for all CFRP/AAC panels tested in this study.

10.8 Conclusion

Low velocity impact testing provides meaningful information about the energy absorbing characteristics of a sandwich structure. It illustrates how the sandwich structures will respond to the drop or impact of an object falling from a certain height. This loading scenario is of significant importance since during service, the structural members in the building structures are subjected to impact loading that varies from object-caused impact, blast due to explosions, to high velocity impact of debris during tornados, hurricanes, or storms.

The failure mode of plain AAC is flexural followed by shear cracks at the support locations due to the poor flexural and shear strengths of the AAC when subjected to impact loading.

In the case of CFRP/AAC panels, the panels showed apparent ductility because the energy absorbed during the propagation phase is significantly higher than the energy absorbed during the initiation phase. Typical failure mode of CFRP/AAC sandwich panels wrapped by CFRP wrap is flexural, in which the failure is initiated with the cracking of the top facesheet, and followed by the localized crushing of the core, and finally debonding and breakage of the bottom facesheet. This failure mode is successful in applications that require energy absorption, since the damage is spread around a larger area on the backside of the sandwich.

The total energy absorbed by CFRP/AAC panels decreases as the core thickness increases, in which the ability of the panel to exhibit significant flexure and shear, and then absorb more energy, is reduced. VARTM processing improves the bond characteristics between FRP facesheets and AAC substrate, especially with stiff panels. However, it does not have much effect on the peak load, since the impact loading is localized. The reasonable agreement between the predicted and experimental impact results illustrates that the energy balance model could be used as an adequate tool to predict the energy absorbed by the CFRP/AAC sandwich panel. In addition, a modeling tool would also allow the maximum impact load to be calculated for a given impact energy.

10.9 Acknowledgment

The authors gratefully acknowledge funding and support provided by National Science Foundation research project NSF (NSF-CMMI-825938).

10.10 References

Abrate, S. (1998) *Impact on Composite Structures*. Cambridge: Cambridge University Press.

Agarwal, B.D., Broutman, L.J. and Chandrashekhara, K. (2006) *Analysis and Performance of Fiber Composites*, 3rd edn, Chichester: John Wiley & Sons.

Allen, H.G. (1969) *Analysis and Design of Structures Sandwich Panels*, Oxford: Pergamon Press.

Ambur, D.R. and Cruz, J. (1995) 'Low speed impact response characteristics of composite sandwich panels', *Proc. 36th AIAA/ASME/ASCE/AHS/ASC Structural, Structural Dynamics Material Conference*, 2681–2689.

ASTM C 1386 (2007) 'Standard specification for precast autoclaved aerated concrete (AAC) wall construction units', *Annual Book of ASTM Standards C 393-00*, West Conshohocken, PA.

Cantwell, W.J., Dirat, C. and Kausch, H.H. (1994) 'Comparative study of the mechanical properties of sandwich materials for nautical construction', *SAMPE Journal*, 30, 45–51.

Fatt, M.S.H. and Park, K.S. (2001) 'Dynamic models for low-velocity impact damage of composite sandwich panels', *Composite Structural*, 52, 335–351.

Hazizan, MdA. and Cantwell, W.J. (2002) 'The low velocity impact response of foam based sandwich structures', *Composite Part B: Eng*, 33, 193–204.

Horrigan, D.P.W., Aitken, R.R. and Moltschaniwskyj, G. (2000) 'Modelling of crushing due to impact in honeycomb sandwiches', *J. Sandwich Struct. Mater.*, 2, 131–151.

Karbhari, V. and Zhao, L. (2000) 'Use of composites for 21st century civil infrastructure', *Computer Methods in Applied Mechanics and Engineering*, 185, 433–454.

Kedar, S. (2006) 'Manufacturing and design methodology of hybrid fiber reinforced polymer (FRP)–autoclaved aerated concrete (AAC) panels and its response under low velocity impact', MS thesis, CCEE Department, The University of Alabama at Birmingham.

Khotpal, A. (2004) 'Structural characterization of hybrid fiber reinforced polymer (FRP)–autoclaved aerated concrete (AAC) panels'. MS thesis, CCEE Department, The University of Alabama at Birmingham.

Li, G. and Jones, V. (2007) 'Development of rubberized syntactic foam', *Composites Part A: Journal of Applied Science and Manufacturing*, 38, 1483–1492.

Li, G. and Venkata, D. (2008a) 'A cement based syntactic foam', *Journal of Materials Science and Engineering*, 478, 77–86.

Li, G. and Venkata, D. (2008b) 'Impact characterization of sandwich structures with an integrated orthogrid stiffened syntactic foam core', *Journal of Composites Science Technology*, 68, 2078–2084.

Mines, R.A.W., Worrall, C.M. and Gibson, A.G. (1998) 'Low velocity perforation behaviour of polymer composite sandwich panels', *Int. J. Impact Engng.*, 21, 855–879.

Mousa, M. (2007) 'Optimization of structural panels for cost-effective panelized construction'. MS thesis, CCEE Department, The University of Alabama at Birmingham.

Mousa, M. and Uddin, N. (2009) 'Experimental and analytical study of carbon fiber-reinforced polymer (FRP)/autoclaved aerated concrete (AAC) sandwich panels', *Journal of Engineering Structures*, 31, 2337–2344.

Olsson, R. (2000) 'Mass criterion for wave controlled impact response of composites', *Composites Part A: Journal of Applied Science and Manufacturing*, 31, 879–887.
Perez, J. (2003) 'Implementation of vacuum assisted resin transfer molding in the repair of reinforced concrete structures', MS thesis, ME Department, The University of Alabama at Birmingham.
Rhodes, M.D. (1975) 'Impact fracture of composite sandwich structures', *Proc. 16th ASME/AIAA/SAE Structural, Structural Dynamics Material Conference*, 311–316.
Shi, C. and Fouad, H.F. (2005) *Autoclaved Aerated Concrete – Properties and Structural Design*, Special Publication 226, American Concrete Institute.
Sika Corporation (2002) *Construction Products Catalog*, Sikadur, http://www.sikausa.com.
Sun, C.T. and Wu, C.L. (1991) 'Low velocity impact response of composite sandwich panels', *Proc. 32nd AIAA/ASME/ASCE/AHS/ASC Structural, Structural Dynamics Material Conference*, 1123–1129.
Uddin, N. and Fouad, H. (2007) 'Structural behavior of FRP reinforced polymer–autoclaved aerated concrete panels', *ACI Structural Journal*, 104(6), 722–730.
Yang, S.H. and Sun, C.T. (1982) 'Indentation law for composite laminates', *Composite Materials: Testing and Design (sixth conference), ASTM STP 787.1.* American Society for Testing and Materials, pp. 425–449.
Zenkert, D. (1995) *An Introduction to Sandwich Construction*, Engineering Materials Advisory Service Ltd, West Midlands, United Kingdom.

10.11 Appendix: symbols

E_{impact} = impact energy absorbed by FRP/AAC panel
E_{bs} = energy absorbed in bending and shear
E_c = energy absorbed by contact effect
m = mass of impact load
v = impact velocity
α = indentation value
C, n = parameters of Meyer's Indentation law
b = width of FRP/AAC panel
E_f = modulus of elasticity of FRP
D = flexural rigidity
h = thickness of the core
t = thickness of the facesheets
G = shear modulus of AAC
A = shear area
L = span of the specimen

11
Innovative fiber-reinforced polymer (FRP) composites for disaster-resistant buildings

N. UDDIN and M. A. MOUSA,
The University of Alabama at Birmingham, USA

DOI: 10.1533/9780857098955.2.272

Abstract: The chapter begins by discussing a new type of sandwich panel called composite structural insulated panels (CSIPs) intended to replace the traditional SIPs that are made of wood-based materials. A detailed analytical modeling procedure is presented in order to determine the global buckling, interfacial tensile stress at facesheet/core debonding, critical wrinkling stress at facesheet/core debonding, equivalent stiffness, and deflection for CSIPs. The proposed models were validated using experimental results that have been conducted on full-scale CSIP walls and floor panels. In order to be used as a hazard-resistant material, a detailed section was presented to show the resistance of CSIP elements to the different types of hazard effects, including impact loading, floodwater effect, fire effect, and windstorm loading.

Key words: composite structural insulated panels, SIP, floor and wall sandwich panels, flood, impact, windstorm.

11.1 Introduction

This chapter presents a new composite building panel system called composite structural insulated panels (CSIPs) for housing applications. This panel is intended to replace the traditional building materials. It is obvious for the US housing market that new building materials are required for higher sustainability and durability. Building materials and labor to construct the structural design are the largest cost components in housing construction, so the need for lower-cost and time-efficient technology becomes urgent. Also, to address the susceptibility of housing built in hurricane-prone communities, solutions are widely being sought to boost resiliency without straining budgets. The hurricanes that have devastated the Gulf States over the past few years, Ivan through Katrina and Rita, have clearly demonstrated the need for improved construction systems and building technologies that are resistant to high winds, storm surges, and flooding.

The current residential building system is based on the concept of panelized construction, where the building is subdivided into basic planar elements (i.e., wall and floor elements) then shipped directly to the construction site and assembled into the finished structure. Panelized construction has the ability to bring the benefits of mass production into the highly customized residential market through the pre-production of components and systems. There are many advantages of panelized structures, including cost reductions, possible through mass production; ease of assembly; a lower skill set required for field construction; and quality control and worker safety [1].

11.2 Traditional and advanced panelized construction

Most of the existing panelized products are traditional 'wood-based' structures. Incremental thinking and engineering refinements have led to better versions and what is commonly known as structural insulated panels (SIPs, Fig. 11.1). SIPs are made of foam core sandwiched between sheets of oriented strand board (OSB) or cement concrete board (CCB). SIPs have been found to be a good alternative to stick framed walls in which they can reduce onsite labor requirements and the amount of material needed. Further, SIPs have demonstrated the capabilities for energy-efficient, affordable housing and are stated to be stronger than traditional 'stick-built' construction. However, traditional constructions are often subjected to termite attack, mold buildups as they are subjected to harmful weather conditions, and also have poor penetration resistance against wind-borne debris in the event of hurricane, windstorm, or tornadoes. Further, their wind and storm surge resistance have not been sufficiently evaluated and could be a problem. Moreover, resistance to flood damage is unknown. Other traditional constructions made of steel and reinforced cement

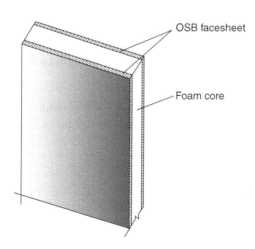

11.1 Illustration of a SIP panel.

concrete (RCC) also have some disadvantages. Steel is heavy and prone to corrosion damage. RCC, on the other hand, entails laborious and time-consuming construction techniques and the internal steel reinforcements are prone to corrosion.

Flood and windstorm hazards have led to significant damage to structures, especially those that are made of wood-based materials. They cause damage to the lower stories of all structures close to the shoreline. An evacuation of the residential areas is then required. The most common flood damage include, but are not limited to: direct damage as a result of inundation or flood-borne debris, degradation of building materials, contamination of the building due to flood-borne substances, and rotting of traditional structures (wood-based) and therefore the tendency for mold buildups (Fig. 11.2). The impact of these damages can be minimized through the use of new building materials that can withstand the direct contact with floodwater with or without little degradation in their strength and also can resist the rotting and mold buildups that result from flooding.

The vision for advanced panelized construction is to develop common building panels that perform multiple functions and integrate multiple tasks using new materials that deliver consistent levels or grades of performance from basic to high performance, and are easy to order, deliver, assemble, and integrate with the building process. During the last two decades, the use of composite materials in the construction industry has become more common [2]. Due to the superior properties of composites compared to the traditional construction materials, our vision is focused on using composites in panelized construction systems in the form of planar sheets to replace the OSB sheet used for producing traditional SIPs, thus producing a new panel system that can serve as a disaster-resistant building material. Furthermore, we aim to develop a design approach for structures against multiple natural hazards.

11.3 Innovative composite structural insulated panels (CSIPs)

CSIPs are made of low-cost orthotropic thermoplastic glass/polypropylene (glass–PP) laminate as facesheets and expanded polystyrene (EPS) foam as a core (Fig. 11.3). The developed CSIPs have a very high facesheet/core moduli ratio (E_f/E_c = 12,500) compared to the ordinary sandwich construction that has a ratio limited to 1,000 [3]. Glass–PP laminates provide high strength-to-weight ratio, excellent impact resistance, and high durability [4]. EPS, on the other hand, are characterized by light weight, thermal insulation, excellent impact properties for both low and high velocity impact, and fire resistance, which make them more competitive to use as core materials. All these properties lead finally to high performance of the panels which

Innovative FRP composites for disaster-resistant buildings 275

11.2 Flood effects on buildings: (a) degradation of building materials; (b) building contamination; and (c) tendency of traditional construction to absorb moisture and rot (mold buildups).

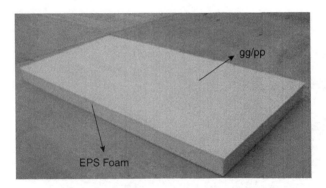

11.3 Proposed composite structural insulated sandwich panels (CSIPs).

can reduce the maintenance cost, enhance the structure's strength and increase service life of the structures. These panels can be used for different elements in the structure, including structural elements (e.g., floors, roofs, and load-bearing walls) and non-structural elements (e.g., non-load-bearing walls, lintels, and partitions).

The concept of CSIPs is based on that of the sandwich structure, in which a soft lightweight thicker core is sandwiched between two strong, thin facings. The facesheets carry the bending stresses while the core resists the shear loads and stabilizes the faces against bulking or wrinkling [3]. The core also increases the stiffness of the structure by holding the facesheets apart. Core materials normally have lower mechanical properties compared to those of facesheets. In addition, it usually has a low unit weight, which leads to a reduction in the overall weight of the building. The low unit weight makes the sandwich panels easy to manufacture and assemble into the structure. The stiffer the core material is, the closer the sandwich panel behavior gets to an isotopic plate. Despite the high strength resulting from this combination, the strength is not the only criterion governing the design of the panel [5]; the deflection is another aspect that controls the design of the sandwich panels.

11.3.1 Materials for CSIPs: thermoplastic facesheets

Thermoplastic composite facesheets used for CSIPs consist of 70% bidirectional glass fibers impregnated with polypropylene (PP) resin. Thermoplastic composites are produced using a hot-melt impregnation process (also called a DRIFT process). The hot-melt impregnation process enables continuous impregnation of fiber tows (e.g., glass, Kevlar/aramid, and carbon) with a wide range of low-cost thermoplastic polymers, such as polypropylene, nylon, polycarbonate, polyurethane, and others. In hot-melt impregnation, the fiber is pulled from a creel and can be heated to remove surface moisture or other surface contaminants. Heating of the fiber can also aid the wetting process that occurs during impregnation and can improve other aspects of the impregnation process. A single or twin-screw extruder compounds the polymer. The heated roving is impregnated in a die with the molten polymer by opening the fiber bundles over specifically designed spreader surfaces. After that, the impregnated product is passed through a chiller to cool the pre-preg, and then passed through a puller that controls the line speed. After the excess polymer is stripped to achieve the final fiber content, the impregnated material can be formed as continuous tapes. These tapes are then laid up in a 0/90 degree fiber orientation to achieve the bidirectional form. Furthermore, thermoplastic facesheets are produced by consolidating a number of woven glass/PP tapes using double-belt pressing under heating. The thickness of the final sheets depends on the amount of both pressure and heat. The pelletized form of the impregnated materials is referred to as long-fiber thermoplastics (LFTs). The fiber contents of the hot-melt impregnated materials typically range from 10 to 70% (by weight) with a control of ±2% [6]. Figure 11.4 shows the steps for producing thermoplastic sheets.

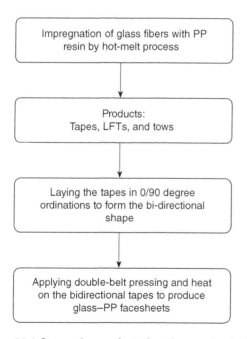

11.4 Steps of manufacturing thermoplastic facesheets.

Thermoplastics offer promise in terms of short processing time, extended shelf life and low-cost raw material. They can be reshaped and recycled by reheating, and also they have superior impact properties. The melt processing of thermoplastic polymers produces low-cost intermediate composite products and fabricates these products into complex, high-performance structures using flow molding and other low-cost manufacturing techniques. Superior impact resistance and large volume production potential make thermoplastic composites attractive as a structural member. The ability to maintain integrity after impact is unique to thermoplastics.

11.3.2 Materials for CSIPs: expanded polystyrene foam (EPS)

In general, the word 'foam' means a mass of bubbles of air or gas in a matrix of liquid film, especially an accumulation of fine, frothy bubbles formed in or on the surface of a liquid, as from agitation or fermentation. Foam is a material characterized by low cost and low weight, which reduces the structure weight. It also has good fire and thermal resistance. Because of these properties, it works very well as an insulation material. There are many types of foams, such as polystyrene, polyethylene, and polyurethane foam. These types vary in both properties and cost. Because of the lower cost, expanded polystyrene (EPS) foam was selected for use as the core of the proposed CSIP panels.

Unlike polyurethane and formaldehyde foams, which use unstable gases in their manufacture, EPS contains only stabilized air, thus its R-value will not decrease, as others do, with age. EPS can withstand the thermal shock of extreme freeze-thaw cycling without loss of insulation value or structural integrity. EPS is an inert, plastic insulation material with no future chemical activity. It has no nutritive value for plants, insects, or other animals. It will not rot and is extremely resistant to mildew. Expanded polystyrene is a rigid closed cell/cellular plastic resin made from petrochemicals derived from crude oil. The resin is incorporated with the blowing agent pentane. During pre-expansion, steam softens the plastic resin, causing the pentane to expand the plastic into beads at least 100 times their original size. As long as the plastic is exposed to steam, the pentane expands, therefore determining the expanded size of the bead. Before all of the pentane is displaced by air, the pre-expanded bead is then placed into a mold of any shape or block. Blocks are then cut to the desired size with hot wires, wrapped, and shipped.

Further, EPS foam has a compressive strength from 10 to 60 psf. It is also ideal for most construction applications. In the 1950s, contractors and carpenters began to realize expanded polystyrene's potential as an efficient foam insulation for commercial, residential, and civil engineering projects.

It did not take long for the thermal plastic material to become the standard in the industry. Insulation is often characterized by its R-value. The higher the R-value, the better resistance to the flow of heat. Compared to other rigid insulation boards, EPS provides the largest R-value for each dollar spent.

11.3.3 Manufacturing of CSIPs

The glass–PP facesheets are bonded to the EPS core using a hot-melt thermoplastic spray adhesive. This method of manufacturing is fast and less labor-intensive than manufacturing of traditional SIPs. The actual fabrication time for CSIPs is two hours, as against 24 hours required for traditional SIPs. Also high-quality and attractive panels can be produced in a short duration by adopting this manufacturing technique. To insure quality of processing, CSIP panels are tested at a casting and molding facility (Fig. 11.5).

11.4 Designing composite structural insulated panels (CSIPs) for building applications under static loading

This section presents an analytical modeling of the behavior of CSIPs under different types of loading. Formulas to obtain global buckling under concentric and eccentric loadings, equivalent wall stiffness, interfacial stress and critical wrinkling stress at the debonding, flexural strength, in-plane load capacity, effective core depth for floor panels, and deflection were developed. These equations were validated using full-scale experimental testing. The good agreement between the proposed models and the experimental results strongly verified the analytical models for calculating the strength and deflection of CSIPs. The explanations of how to obtain all of these formulas are provided elsewhere [7–10].

11.4.1 Global buckling

The global buckling load for CSIPs under concentric and eccentric loading scenarios can be calculated according to Eqs [11.1] and [11.2], respectively [9]:

$$P = \frac{\pi^2}{L^2} \frac{E_f I}{\left(1 + \frac{\pi^2 E_f I \left(1 - v_{xy}^2\right)}{L^2 \cdot b \left(\frac{d+c}{2}\right) \cdot G}\right)} \left(1 - v_{xy}^2\right) \quad [11.1]$$

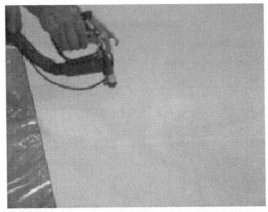

(a)

(b)

(c)

11.5 Manufacturing of full-scale CSIPs: (a) spraying hot-melt adhesive on the facesheet; (b) placing foam core on the sprayed facesheet; (c) ready panels stacked on the floor.

$$P = \frac{\pi^2}{\left(1 + \frac{6 \cdot e}{d}\right)L^2} \frac{E_f I}{\left(1 + \frac{\pi^2 E_f I(1 - v_{xy}^2)}{L^2 \cdot b\left(\frac{d+c}{2}\right) \cdot G}\right)}(1 - v_{xy}^2) \qquad [11.2]$$

Both equations were developed based on the Euler formula that is valid only for homogeneous sections and the general global buckling formula for sandwich panel with thin isotropic faces according to Allen [11]. As seen in Eqs [11.1] and [11.2], they considered the orthotropic nature of the glass–PP facesheet by using the term '$1 - v_{xy}^2$', where v_{xy} is the in-plane Poisson's ratio of the orthotropic facesheets in the xy-plane. This will consider the through-thickness anisotropy effect due to the orthotropic facesheets. Further, the formulas also consider the effect of core deformation in terms of shear area and core shear modulus.

According to the Euler formula, the critical buckling load (P_E) for a compression member is given by:

$$P_E = \frac{\pi^2}{L^2} \cdot D \qquad [11.3]$$

Comparing Eq. [11.1] with Eq. [11.3], the equivalent stiffness for a sandwich panel with orthotropic facesheets can be expressed as:

$$D_{equiv.} = \frac{E_f I}{\left(1 + \frac{\pi^2 E_f I(1 - v_{xy}^2)}{L^2 \cdot b\left(\frac{d+c}{2}\right) \cdot G}\right)}(1 - v_{xy}^2) \qquad [11.4]$$

It should be noted that the equivalent stiffness given by Eq. [11.4] is also used to obtain the lateral deflection of a sandwich wall panel subjected to in-plane loading.

11.4.2 Debonding of facesheet/core

The facesheet/core debonding is mainly caused by the wrinkling of the facesheet in compression when the panel is subjected to compressive loading. Debonding could be pre-existing due to manufacturing defects and in this case is called 'disbond' [7]. Almost all of the previous studies have focused on the modeling of pre-debonded sandwich panels. In the case of CSIPs, the focus is on the debonding due to compressive loading. Further, there is an apparent gap for modeling facesheet/core debonding for sandwich panels with very high facesheet/core moduli ratio, a characteristic of CSIPs. The developed debonding models also considered this criterion.

Wrinkling is caused by sudden localized short-wavelength buckling of facesheets in compression. In the case of wrinkling, the core acts as an elastic foundation to the facesheets [11, 12]. It can be either outward or downward. If the buckling is outward, it is known as debonding, while if it is downward, it is known as core crushing (Fig. 11.6). The former occurs in the case of sandwich panel with solid cores (e.g., EPS foam), while the latter happens in the case of sandwich panels with open cell cores (e.g., honeycomb core) [13].

11.4.3 Debonding modeling

During loading of the sandwich panel, two types of stresses are developed at the facesheet in the compression side; the first is a tensile stress at the facesheet/core interface, 'σ_z', while the other is a compressive critical wrinkling stress in the facesheet of the deboned part, 'σ_{cr}' (Fig. 11.7). The

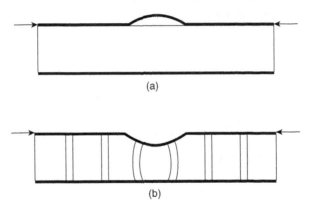

11.6 Upward and downward wrinkling: (a) upward wrinkling (debonding, case of solid cores), (b) downward wrinkling (core crushing, case of open cell cores).

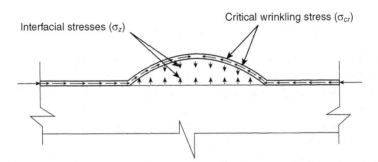

11.7 Types of stresses at the compressive facesheet for CSIP due to debonding.

debonding occurs when the tensile stress at the facesheet/core interface exceeds the tensile strength of the core material. It is a common mode of failure of sandwich panels, leading to the loss of panel stiffness. The facesheets under compression can be modeled as a strut or beam supported by an elastic foundation represented by the core.

The glass–PP facesheets under compression can be modeled as a strut or beam supported by an elastic foundation represented by the EPS foam core. In other words, CSIP wrinkling can be modeled as a Winkler foundation. In the analysis of the behavior of a long strut or beam supported by a continuous elastic medium, the medium can be replaced by a set of closed-spaced springs (Fig. 11.8); this phenomenon is normally known as the Winkler hypothesis, and the facesheet in this case is called the Winkler beam, while the core is known as the Winkler foundation. For a beam supported by a Winkler foundation, the governing differential equation of the beam is given as:

$$D_f \frac{d^4 w}{dx^4} + P \frac{d^2 w}{dx^2} + b\sigma_z = 0 \qquad [11.5]$$

where D_f is the flexural stiffness of the beam (facesheet), P is axial load developed in the facesheet due to loading, w is the displacement of the debonded part in the z-direction, σ_z is the interfacial tensile stress at the facesheet/core interface, and b is the width of the facesheet. The formulas developed for debonding were validated using full-scale CSIP testing and the results found were in good correlation [7].

Interfacial tensile stress (σ_z)

The interfacial tensile stress model was validated by demonstrating the close proximity to the experimental results. The results proved that the predicted interfacial stress is higher than the core tensile strength and, therefore, debonding was the general mode of failure. This validates the criteria that

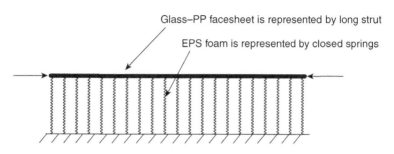

11.8 Winkler foundation model.

the interfacial stress is independent of loading and boundary conditions and depends only on the core properties [7]. This model is given by:

$$\sigma_z = 0.07\pi^2 f(\theta) E_c \quad [11.6]$$

where θ is a function of the core thickness and half-wavelength of l and is given by $\pi c/l$.

$f(\theta)$ is a function of the Poisson's ratio of the core and θ and has a different equation for each case of wrinkling [11] as follows:

$$\text{Case (I):} \quad f(\theta) = \frac{2}{\theta} \frac{(3-\upsilon_c)\sinh\theta\cosh\theta + (1+\upsilon_c)\theta}{(1+\upsilon_c)(3-\upsilon_c)^2 \sinh^2\theta - (1+\upsilon_c)^3 \theta^2} \quad [11.7]$$

$$\text{Case (II):} \quad f(\theta) = \frac{2}{\theta} \frac{\cosh\theta - 1}{(1+\upsilon_c)(3-\upsilon_c)\sinh\theta + (1+\upsilon_c)^2 \theta} \quad [11.8]$$

$$\text{Case (III):} \quad f(\theta) = \frac{\cosh\theta + 1}{3\sinh\theta - \theta} \quad (\upsilon_c = 0) \quad [11.9]$$

Critical wrinkling stress in the facesheet (σ_{cr})

The second stress that is associated with the debonding is the critical wrinkling stress in the facesheet in compression (σ_{cr}). This is a compressive in-plane stress developed in the facesheet due to loading. This stress was the focus of most of the previous studies that were conducted on the wrinkling of facesheets in compression, whereas the stress at the interface (σ_z) was not addressed appropriately. The main reason for that is all of these studies were investigating sandwich panels with pre-existing disbond that occur during the manufacturing of the panels. In this case, the interfacial stresses vanish at this region.

Based on the Winkler foundation model, a theoretical formula for the critical wrinkling stress was developed:

$$\sigma_{cr} = \frac{E_c}{t} cf(\theta) + \frac{E_f}{12}\left(\frac{\theta}{c}\right)^2 t^2 (1 - \upsilon_{xy}^2) \quad [11.10]$$

Equation [11.10] represents the general theoretical formula for wrinkling compressive stress in the facesheet for sandwich panel with orthotropic facesheets and solid core (referring to CSIP). As can be noticed, the wrinkling stress (σ_{cr}) is a function of the properties and thicknesses of facesheet and core, unlike the interfacial tensile stress ((σ_z) which is independent of the facesheet properties and mainly depends on the core material. The proposed theoretical model for the critical wrinkling stress less conservatively predicted the actual wrinkling stress. Accordingly, the following empirical formula was proposed to predict the critical wrinkling stress at the debonding for CSIP panels considering the orthotropic facesheets [7]:

$$\sigma_{cr} = 0.25(E_f E_c G_c)^{1/3}(1 - v_{xy}^2) \qquad [11.11]$$

11.4.4 Strength

CSIP floor

The strength of the CSIP floor member is expressed in terms of the flexural capacity. The strength in sandwich panels is developed due to the internal force couple in the facesheets. The core works as a separator between the two forces, as shown in Fig. 11.3, and carries the shear stresses [8, 14]. Based on this approach, the following formula was developed:

$$M_n = b \cdot t_{face} \cdot \varepsilon_{face} \cdot E_{face} \cdot d \qquad [11.12]$$

It should be noticed that in Eq. [11.12], the strain is taken as the rupture strain of the facesheets when the sandwich panel failed by facesheet rupture. This equation was validated using full-scale CSIP floor testing and the compassion showed a good agreement. It can also be noticed that the strain at the debonding represents 1.13% of the rupture strain of the glass–PP facesheets. In other words, the strain at failure (debonding) can be expressed as 0.0113 $\varepsilon_{rupture}$.

CSIP wall

Based on the critical wrinkling stress, the nominal load capacity (P_n) can be determined including the effect of the eccentricity. From the theory of mechanics, the stresses under combination of compressive normal force and moment are given by:

$$\sigma = \frac{P}{A} \pm \frac{M \cdot y}{I} \qquad [11.13]$$

The above equation can be rewritten in terms of nominal capacity load (P_n), facesheet thickness (t), panel thickness (d), face thickness, and eccentricity ($e = d/6$) as follows:

$$\sigma = \frac{P_n}{2 \cdot b \cdot t}\left[1 \pm \frac{6 \cdot e}{d}\right] \qquad [11.14]$$

By substituting in the above equation for $e = d/6$, as in the experiment, the maximum compressive force, which corresponds to the critical wrinkling stress, can be analytically obtained as follows:

$$\sigma_{cr} = \frac{P_n}{2 \cdot b \cdot t}\left[1 + \frac{6 \cdot (d/6)}{d}\right] = \frac{P_n}{b \cdot t} \qquad [11.15]$$

Therefore, the nominal load capacity is given by:

$$P_n = \sigma_{cr} \cdot b \cdot t \qquad [11.16]$$

11.4.5 Deflection

CSIP floor

The total central defection of a sandwich panel under out-of-plan loading is composed of bending deflection and shear deflection. The general formula for the deflection of a sandwich panel under out-of-plan loading is given by:

$$\Delta = \frac{k_b P L^3}{D} + \frac{k_s P L}{U} \qquad [11.17]$$

In the above equation, the first term on the right-hand side of the equation is the deflection due to bending, and the second term is the deflection due to shear. k_b and k_s are the bending and shear deflection coefficients, respectively. The values of both k_b and k_s depend on the loading and boundary conditions. D and U are flexural and shear rigidities of the sandwich panel, respectively. According to ASTM C-393 [15], D is determined from Eq. [11.18], whereas U can be determined from Eq. [11.19].

$$D = E_f I = \frac{E_f(d^3 - c^3)b}{12}(1 - v_{xy}^2) \qquad [11.18]$$

$$U = \frac{G_{core}(d+c)^2 b}{4c} \qquad [11.19]$$

It is generally recognized that the ordinary theory of bending and resulting deflection can be applied to a homogeneous panel (e.g., panel that is made from one material). For a sandwich panel, the behavior is different and this can be demonstrated by considering two extreme cases. First, when the core is rigid in shear, the sandwich panel is subjected to the same argument as those applied to a homogeneous panel (except for the difference in the flexural rigidity) and the deflections are expected to be small. Second, when the core is weak in shear, the faces act as two independent plates and the resulting deflections are expected to be much higher than in the first case. It was demonstrated by Allen [11], that the parameter λ represents the transition from one extreme to the other (known also as effective depth coefficient), varying from (−t/c) when the core is weak to (+1) when the core is rigid in shear. Thus, an empirical formula was developed by Allen [11] to define the effective thickness of a sandwich panel which varies from $2t$ ($G = 0$) to d ($G = \infty$):

$$d_{eff} = t\left(1+\frac{c}{d}\right)+\frac{c^2}{d}\lambda \qquad [11.20]$$

d_{eff} should be used instead of d in Eqs [11.18] and [11.19] when determining the flexural and shear rigidities. Effective depth coefficient (λ) is determined experimentally based on the deflection. In this study, λ was determined for CSIP floor panels based on the experimental deflection and was found to be 0.3. Thus, Eq. [11.20] can be rewritten as:

$$d_{eff} = t\left(1+\frac{c}{d}\right)+0.3\frac{c^2}{d} \qquad [11.21]$$

CSIP wall

Deflection of the sandwich wall is calculated based on the equivalent stiffness to consider the core shear deformations. Thus, the equivalent stiffness (D_{equiv}) of a CSIP sandwich panel should be determined first. As mentioned before, failures modes of sandwich walls include global buckling and local buckling or 'wrinkling' which can be debonding or core crushing. The full-scale CSIP wall panels failed by facesheet/core debonding. As for the global buckling, a theoretical formula was developed by the authors for the global buckling of CSIP walls, considering the core deformations (Eq. [11.1]). This equation was then led to an equivalent stiffness (Eq. [11.4]). Wall deflection at the mid-height is determined according to ACI-318 [16] as:

$$\Delta = \frac{(5M)L^2}{48(D)_{equiv}} \qquad [11.22]$$

where $M = \dfrac{M_{sa}}{1-\dfrac{5P_sL^2}{48(D)_{equiv}}} \qquad [11.23]$

Equation [11.23] takes into consideration the second order deflection or P-delta effect due to the compressive load, where P_s is applied in-plane eccentric loading and M_{sa} is the applied moment at the mid-height of the panel due to that load (P_s). D_{equiv} is determined using Eq. [11.4]. It should be noted that Eq. [11.22] is applicable only up to the linear stage. Therefore, loads used to determine the moments were those at the initiation of debonding.

11.5 Composite structural insulated panels (CSIPs) as a disaster-resistant building panel

11.5.1 Impact resistance

The response of CSIP was investigated under low and high velocity impact loading [17, 18]. Floor CSIPs are always subjected to tool drops which reflect low velocity impact loading, while wall CSIPs are always subjected to wind-borne debris in hurricane-prone areas which reflect high velocity impact loading. Accordingly, both low velocity impact (LVI) and high velocity impact (HVI) testing were conducted.

Low velocity impact (LVI)

Both traditional SIPs and CSIPs were tested for LVI loading to compare their responses [17, 18]. The impact energy was 68 J. The specimen sizes were 101.6 mm × 101.6 mm (4 in. × 4 in.). The typical load versus time curve for both panels is shown in Fig. 11.9, whereas deformations after impact loading are shown in Fig. 11.10. As seen in Fig. 11.9, the maximum load recorded by CSIP is much higher than that recorded by traditional SIPs. As shown in Fig. 11.10(a), the OSB facesheet, being weak in shear, could not resist the applied impact energy, and the impact energy was used up in penetrating the top OSB facesheet. In the case of CSIP (Fig. 11.10(b)), on the other hand, the damage was in the form of an indentation of the top facesheet, as against that of OSB facesheets in traditional SIPs, where the damage was highly localized.

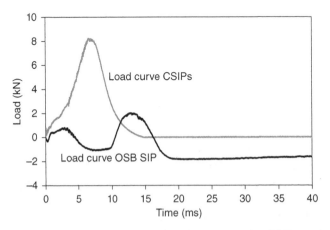

11.9 Typical impact load versus time curves for CSIPs and traditional SIPs.

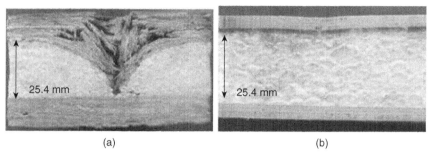

11.10 Damage after LVI for (a) OSB SIP, (b) CSIP.

High velocity impact (HVI)

Since the traditional SIPs failed completely under LVI, only CSIPs were tested under HVI [17, 18]. HVI was performed on full-scale CSIP cross section with 139.7 mm (5.5 in.) core and 3.04 mm (0.12 in.) facesheet. The energy of impact was calculated according to FEMA specifications [19]. A total energy impact of 1300 J was applied to CSIPs using a laboratory scale gas gun at the University of Alabama at Birmingham (UAB). It was observed that at the impact energy of 1300 J, the top face of the CSIP was completely damaged (Fig. 11.11(b)). The damage to the impacted facesheet was in the form of fiber breakage and some degree of foam crushing under the point of impact. Once the impactor hit the top face, the impact energy was transferred from the impactor to the top face. There was no delamination between the facesheet and the core and the energy was transferred effectively to the core. The foam core was seen to have cracked due to the applied impact energy. This implied that the strength of the adhesive used for bonding the facesheets and the foam was higher than that of the cohesive strength of the foam cells. The back face was seen to be intact (Fig. 11.11(c)). The impactor could not penetrate through the panel but was deflected away from the panel due to the large degree of flexing by the composite facesheets. This test thus proved that the full-scale CSIP could be used effectively in hurricane-prone areas and can protect the occupants inside the structure from wind-borne debris.

11.5.2 Fire resistance

Fire resistance is a very important issue for any element of the structure. The structural elements in particular must have adequate resistance to overcome flames and spreading of fire. For the composite structures in the proposed panels, high-composite fiber content promotes inherent flame retardant characteristics. In addition, additives and coating may enhance

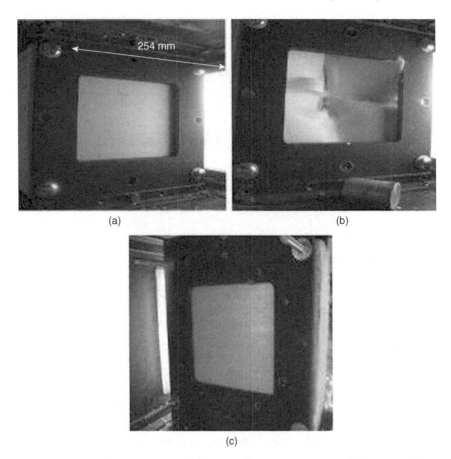

11.11 HVI on full-scale CSIP: (a) CSIP before impact; (b) impacted face of the CSIP; (c) intact back face of the impacted CSIP.

those characteristics. Generally, there are two different approaches to resist the spreading of fire for composite structures. The first is compounding a flame retardant into the polymer matrix of the composite. The second is to apply flame retardant as a top coat using a brush during or even after manufacturing of the component. The second approach is better than the first because the addition of flame retardant changes the impact characteristics of the component and makes it behave as a brittle material. On the other hand, the coating method does not affect the mechanical performance of the composite material [4].

Some fire tests were carried out by UAB investigators on 40 wt% glass polypropylene composite structures to be used in mass transit applications. The composite specimens were coated by a Flame Seal FX-PL as flame retardant. These materials were applied using a brush to the polypropylene panels. The flame retardant was selected particularly to be a surface finish after drying. Then the panels were tested after being cured at 60°C, and the evaluations were carried out with respect to flame spread and smoke density according to ASTM E 162-95 [20] and E 662-95 [21] for the fire safety requirements for mass transit applications. The results showed that both flame spread and smoke density were far lower than the limits [4]. The glass–PP facesheets used in CSIPs have passed the flame spread test according to ASTM E-84 [22] that tests the surface burning characteristics of building materials.

11.5.3 Flood resistance

Degradation of structural material due to flood is one of the major damaging effects during flooding events. Because of that consequence, thousands of homes in the coastal states, especially those that are constructed from wood-based materials, have been destroyed after each severe hurricane. One of the weaknesses of the current testing standards is that there is no specific standard to study the degradation of the building materials after exposure to floodwater. There are some ASTM standards such ASTM C 1601 [23], ASTM C 1403 [24], and ASTM E 514 [25] that investigate the moisture penetration of masonry walls and water absorption of masonry mortars due to rain effect. All of these standards use spray rack to simulate the rain conditions on one side of the wall specimen for 2 hours before structural tests under wet conditions are performed. Thus, none of them accurately represents the flood testing in which the wall or the building panel can be fully submerged in water and accordingly this will be a much more severe condition than spraying one side of the wall with water. Despite the limited exposure to water that these standards recommend, applying them to traditional construction materials such as wood, CMU and adobe, showed significant structural degradation and aesthetic problems [26]. Thus, this raises the need for an alternative building material that can be characterized with high resistance against degradation and yet in more aggressive conditions.

As we are proposing this new type of composite building panel, one goal was to investigate the structural behavior and performance of CSIPs after exposure to floodwater. The behavior of CSIPs under out-of-plane loading (Fig. 11.12) was investigated before and after exposure to floodwater for different periods to determine how much the panel's strength degraded due to floodwater. To have a more destructive situation, CSIPs were fully submerged in the floodwater for 3 and 7 days. In addition, a control group, not subjected to the floodwater, was tested initially to establish baseline properties. The performance was expressed in terms of flexural strength degradation due to floodwater effect as well as the moisture content throughout

11.12 Test setup with test instrumentation ASTM C 393 (2000).

Table 11.1 Specimens details

Group name	Exposure period (days)	No. of specimens	Dimension (mm)	Notes	Test time
G1: CSIP	–	3	1219.2 × 304.8 × 145.8	Control	–
G2: CSIP 3	3	3	1219.2 × 304.8 × 145.8	Flooded	After drying
G3: CSIP 7	7	3	1219.2 × 304.8 × 145.8	Flooded	After drying
G4: CSIP 7	7	3	1219.2 × 304.8 × 145.8	Flooded	After flood

the drying period. The structural performance was analyzed in terms of mode of failure, load–deflection and load–strain curves.

Four CSIP groups were tested and are summarized in Table 11.1. Each group consisted of three specimens. The digit at the end of the notation represents the exposure period in days. The CSIP specimens consisted of 140 mm (5.5 in.) thick EPS foam sandwiched between two 3.04 mm (0.12 in.) thick glass/polypropylene (glass/PP) composite facesheets. The length and width of each specimen were 1,219.2 mm × 304.8 mm (4 ft × 1 ft), respectively. A schematic of the test panel with the dimensions is shown in Fig. 11.13. As seen in Table 11.1, a control group that was not subjected to the floodwater was tested initially to establish baseline properties. Further,

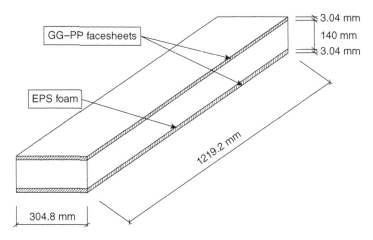

11.13 Schematic for CSIP test sample for flood testing.

some panels were subjected to 3 days of floodwater and others were exposed for 7 days. The intention here was to investigate the impact of exposure period on the strength degradation of the CSIPs. Further, G4 was tested directly after flooding to study the effect of the drying period on the strength degradation.

Flood testing results

The average results for deflections and strains were obtained for each group and then plotted against load to compare the behavior with respect to the control group that was not subjected to floodwater. All panels failed due to facesheet/core debonding (Fig. 11.14). Table 11.2 illustrates the average values for strength, deflection, stiffness, and degradation. The degradation was calculated for strength and stiffness. As seen in Table 11.2, the G2 group recorded strength and stiffness degradations of 5% and 10.5%, respectively. The G3 group recorded about 10% degradation in strength and about 25% in stiffness. The G4 group that was tested directly after flooding recorded the highest degradations in which its strength degraded by 16% whereas its stiffness degraded by about 33%. Figures 11.15 and 11.16 show the average capacity and stiffness for each group. It should be mentioned that the degradation in stiffness is higher than strength because stiffness incorporates both degradation in strength and deflection. The stiffness was considered as the slope of the average load–deflection curve.

The average load–deflection and average load–strain curves are shown in Plates VI and VII (between pages 240 and 241), respectively. As seen from Plate VI, the G1 group recorded the highest stiffness as expected. The

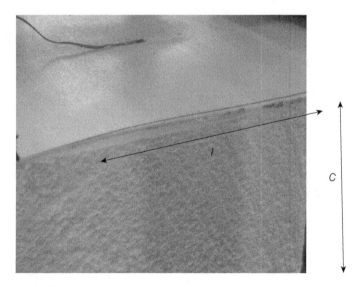

11.14 Debonding is the common failure mode of control and flooded CSIPs.

Table 11.2 Summary of the average results for the four groups

Group ID	Average failure load (N)	Average deflection (mm)	Average stiffness (N/mm)	Degradation % Strength	Stiffness
G1	3,004.40	24.89	120.71	N/A	N/A
G2	2,856.40	26.46	107.96	4.93	10.56
G3	2,708.40	30.11	89.95	9.85	25.48
G4	2,516.00	31.08	80.95	16.26	32.94

behavior of the G2 group was similar to that of G1 until a load of 2.5 kN (kips). After that load, G2 showed little stiffness and more strength degradations. Furthermore, G3 showed lower strength and stiffness than G2 (5% degraded) due to longer submersion in the floodwater (3 days for G2 versus 7 days in case of G3). As can be seen, both groups G3 and G4 showed similar behavior until a load of 1.25 kN (0.28 kips). After that load, G4 degraded more and the degradation increased as the load increased until failure with a degradation difference of 6% in strength and 7% in stiffness than G3. In addition, and as seen in Plate VII, although G3 and G4 showed similar tensile strain behavior, their compressive strain was slightly different

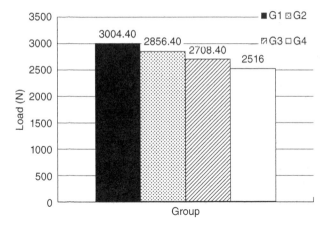

11.15 Average failure load for the four groups.

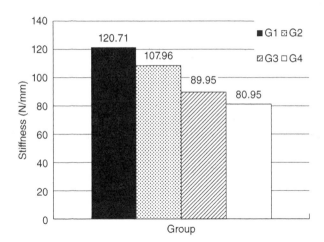

11.16 Average stiffness for the four groups.

in that G4 had lower stiffness in the compressive part than G3. This means that degradation is more likely to happen in the compressive part than in the tensile one due to the fact that the debonding occurs in the compressive portion of the sandwich panel.

As noticed from these results, the maximum degradation recorded by CSIPs was only 16% immediately after removal from the water (i.e., wet condition). However, this degradation can be decreased to 10% if the panels are left to dry. This small degradation in strength is highly acceptable when compared to the traditional wood structures, which can lose their strength significantly and therefore can lead to the total failure of the structure in the case of extreme floods. The findings of this study therefore recommend use of CSIP in construction in flood-prone areas.

11.5.4 Windstorm resistance

CSIPs were also tested under windstorm loading. The testing was conducted at The University of Florida (UF) and was led by Dr. Masters (co-principal investigator (Co-PI) of the NSF project). The purpose was to evaluate their resistance to wind load action resulting from tropical cyclones, thunderstorms, and other extreme wind events in order to optimize the material and geometric properties of the wall sections as well as their connection details. Further, the advantages of windstorm full-scale testing on structural systems include, but are not limited to, the following: aerodynamic effects of hurricane wind loads on structural components, enabling study of progressive damage to failure to analyze failure mode, and eliminate difficult scaling issues and permit testing to realistic dimension panels.

Pressure loading was generated by a custom-designed high airflow pressure loading actuator (HAPLA) system, which is based on the pressure loading actuator system developed by the University of Western Ontario for its Three Little Pigs project [27]. This system can replicate the temporal wind-induced pressures on the surfaces of buildings resulting from upwind turbulence and the flow distortion around the building. The HAPLA consists of two 75 HP centrifugal backward inclined Class IV SWSI blowers that supply air to a valve that modulates the airflow in and out of a test chamber called an airbox. The test specimen is integrated into the airbox such that it is acted on by the internal pressure.

Failure mode

Full-scale CSIPs were tested. The panels were simply supported from top and bottom to best simulate the actual condition and also to represent the most severe condition. Figure 11.17 shows the CSIP panel under windstorm testing. All panels approximately exhibited the same mode of failure in which the failure started by the facesheet/core debonding in the compression side followed by connection failure at the top and bottom (Fig. 11.18). The behavior of CSIPs under windstorm loading, from the beginning of loading till failure, is characterized by two main stages: bending stage and catenary action stage. The first stage is bending, which occurs from the beginning of loading till the initiation of facesheet/core debonding. During this stage, the panel behaves as a simple beam and the internal moment developed due to the couple force action in the facesheets. In this stage also, both facesheets recorded almost the same strains, with the compressive

Innovative FRP composites for disaster-resistant buildings 297

11.17 Back view of the CSIP panel under windstorm testing.

strain being a little higher than the tensile one. After the initiation of debonding, the catenary action stage starts. In this stage, the panel starts acting as a cable supporting the distributed load (windstorm loading). In this stage, the compressive strain starts decreasing until it becomes tensile prior to failure. The panel continues in the catenary action until connection failure occurred at both ends. Figure 11.19 illustrates the consequence of the CSIP failure under windstorm loading.

Capacity

The average capacity of a CSIP panel at the bending stage, with 5.5″ core thickness (3 pcf density) and 0.12″ facesheet thickness, was 54.28 psf, which translates to a wind velocity of 162 mph, whereas the inherent reserve capacity due to the catenary action was recorded as 99 psf, which corresponds to a wind velocity of 220 mph. These results proved that CSIPs, of the mentioned dimensions, can withstand up to a Category 5 hurricane, which is defined by a wind velocity greater than 155 mph. Therefore, CSIPs can be used in hurricane-prone areas.

298   Developments in FRP composites for civil engineering

11.18 General failure mode of CSIPs (debonding followed by connection failure).

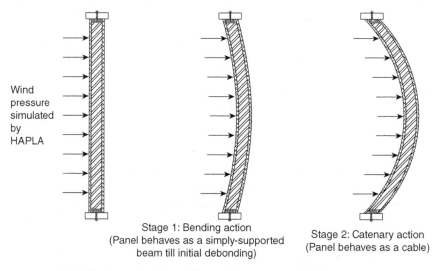

11.19 Schematic of the failure consequence of CSIP under simulated windstorm loading.

11.6 Conclusion

In this chapter, a new type of sandwich panel called composite structural insulated panels (CSIPs) was developed for structural floor and wall applications. This new hybrid panel is intended to replace the traditional SIPs that are made of wood-based materials. CSIPs are made of bidirectional glass–PP facesheets and EPS foam core resulting in a very stiff panel with a very high face/core moduli ratio. A comprehensive explanation of the CSIP concept, materials characteristics and manufacturing techniques was provided.

A detailed analytical modeling procedure was developed in order to determine the global buckling, interfacial tensile stress at facesheet/core debonding, critical wrinkling stress at facesheet/core debonding, equivalent stiffness, and deflection for CSIPs. The proposed models were validated using experimental results that have been conducted on full-scale CSIP wall and floor panels. The good correlation between the analytical and experimental results demonstrated the accuracy of the developed formulas for modeling the behavior of CSIPs under different types of loading.

In order to be used as a hazard-resistance material, a detailed section was presented to show the resistance of CSIP elements to the different types of hazard effects that a building can experience during its lifetime. The hazards included impact loading, floodwater effect, fire effect, and windstorm loading. As noticed from the impact testing, CSIPs have excellent performance in terms of strength and stiffness compared to traditional SIPs. Further, their impact resistance is much stronger than SIPs, which recommends CSIPs for areas that are prone to windstorm debris. Further, flood testing showed that CSIPs had insignificant strength and stiffness degradations and accordingly proved to be a hazard-resistant building material that can survive during a flood event. Finally, the windstorm testing on full-scale CSIPs proved that they can withstand high wind loading (up to hurricane Category 5), which therefore recommends the use of CSIPs as a building material in severe windstorm locations.

11.7 Acknowledgment

The authors gratefully acknowledge funding and support by National Science Foundation research project NSF (NSF-CMMI-825938).

11.8 References

1. PATH (Partnership for Advancing Technology in Housing), One Year Progress Report. (2002) US Department of Housing, Office of Policy Development and Research, Washington, DC.

2. Karbhari, V. and Zhao, L. (2000) 'Use of composites for 21st century civil infrastructure', *Computer Methods in Applied Mechanics and Engineering*, 185, 433–454.
3. Zenkert, D. (1995) *An Introduction to Sandwich Construction*, Engineering Materials Advisory Service Ltd, West Midlands, United Kingdom.
4. Center for Composites Manufacturing (2003) Final Report, FTA-AL-26-7001-2003.1, June.
5. Mousa, M. (2007) 'Optimization of structural panels for cost-effective panelized construction'. MS Thesis, CCEE Department, The University of Alabama at Birmingham.
6. Hartness, T., Husman, G., Koening, J. and Dyksterhouse, J. (2001) 'The characterization of low cost fiber reinforced thermoplastic composites produced by DRIFT process', *Composite: Part A*, 32, 1155–1160.
7. Mousa, M. and Uddin, N. (2010) 'Debonding of composites structural insulated sandwich panels', *Journal of Reinforced Plastics and Composites*, 29(22), 3380–3391.
8. Mousa, M. and Uddin, N. (2010) 'Experimental and analytical study of composite structural insulated floor panels', *Earth and Space 2010: Engineering, Science, Construction, and Operations in Challenging Environments, Proceedings of the 12th International Conference on Engineering*, ASCE, Honolulu, Hawaii.
9. Mousa, M. and Uddin, N. (2011) 'Global buckling of composite structural insulated wall panels', *Journal of Materials and Design*, 32(2), 766–772.
10. Mousa, M. (2011) 'Composites structural insulated panels (CSIPs) for hazards resistant structures'. PhD dissertation, CCEE Department, The University of Alabama at Birmingham.
11. Allen, H.G. (1969) *Analysis and Design of Structural Sandwich Panels*, Pergamon Press, London.
12. Kardomateas, G.A. (2005) 'Global buckling of wide sandwich panels with orthotropic phases', *Proceedings of the 7th International Conference on Sandwich Structures*, Aalborg University.
13. Galletti, G.G., Vinquist, C. and Es-Said, O.S. (2007) 'Theoretical design and analysis of a honeycomb panel sandwich structure loaded in pure bending', *Journal of Engineering Failure Analysis*, 15, 555–562.
14. Mousa, M.A. and Uddin, N. (2009) 'Experimental and analytical study of carbon fiber-reinforced polymer (FRP)/autoclaved aerated concrete (AAC) sandwich panels', *Journal of Engineering Structures*, 31, 2337–2344.
15. Annual Book of ASTM Standards C 393-00 (2000) Standard Test Method for Flexural Properties of Sandwich Construction, West Conshohocken, PA.
16. Building Code Requirements for Structural Concrete (ACI 318-02) and Commentary (ACI 318 R-02), American Concrete Institute.
17. Vaidya, S.A. (2009) 'Lightweight composites for modular panelized construction'. PhD dissertation, CCEE Department, The University of Alabama at Birmingham.
18. Vaidya, A., Uddin, N. and Vaidya, U. (2010) 'Structural characterization of composite structural insulated panels for exterior wall applications', *Journal of Composite for Construction*, 41, 464–469.
19. Federal Emergency Management Agency (FEMA) (2008) *Design and Construction Guidance for Community Safe Rooms*. FEMA 361, 1st edn, Washington, DC.

20. Annual Book of ASTM Standards E 162-95 (1995) Standard Test Method for Surface Flammability of Materials Using a Radiant Heat Energy Source, West Conshohocken, PA.
21. Annual Book of ASTM Standards E 662-95 (1995) Standard Test Method for Specific Optical Density of Smoke Generated by Solid Materials, West Conshohocken, PA.
22. Annual Book of ASTM Standards E 84-10 (2010) Standard Test Method for Surface Burning Characteristics of Building Materials, West Conshohocken, PA.
23. Annual Book of ASTM Standards ASTM C 1601-10 (2010) Standard Test Method for Field Determination of Water Penetration of Masonry Wall Surfaces, West Conshohocken, PA.
24. Annual Book of ASTM Standards ASTM C 1403-06 (2006) Standard Test Method for Rate of Water Absorption of Masonry Mortars, West Conshohocken, PA.
25. Annual Book of ASTM Standards ASTM E 514-09 (2009) Standard Test Method for Water Penetration and Leakage Through Masonry, West Conshohocken, PA.
26. Galitz, C.L. and Whitlock, A.R. (1998) 'The application of local data to the simulation of wind-driven rain', in Kudder, R.J. and Erdly, J.L. (eds), *Water Leakage through Building Facades, ASTM STP 1314*, West Conshohocken, PA: American Society for Testing and Materials, pp. 17–32.
27. Kopp, G.A., Morrison, M.J., Gavanski, E., Henderson, D.J. and Hong, H.P. (2010) 'The Three Little Pigs' Project: Hurricane risk mitigation by integrated wind tunnel and full-scale laboratory tests', *Natural Hazards Review*, 11(4), 151–161.

12
Thermoplastic composite structural insulated panels (CSIPs) for modular panelized construction

N. UDDIN, A. VAIDYA, U. VAIDYA and S. PILLAY,
The University of Alabama at Birmingham, USA

DOI: 10.1533/9780857098955.2.302

Abstract: Modular panelized construction is a modern form of construction technique in which precast multifunctional structural panels are used. In this technique, precast panels are fabricated in the manufacturing facility and are transported to the construction site. Traditional structural insulated panels (SIPs) consist of oriented strand boards (OSB) as facesheets and expanded polystyrene (EPS) foam as the core. These panels are highly energy efficient but have issues in terms of poor impact resistance and higher life cycle costs. Proposed panels consist of E-glass/polypropylene (PP) laminates as facesheets and EPS foam as core and are called composite structural insulated panels (CSIPs). Proposed CSIPs overcome the issues of traditional SIPs and retain all the energy-saving benefits of the traditional SIPs. This chapter describes manufacturing techniques developed for CSIPs and connection details for bonding CSIPs on the construction site. Based on the experimental investigation, ultrasonic welding was found to be the most suitable technique for joining the proposed CSIPs.

Key words: panelized construction, ultrasonic welding.

12.1 Introduction

Structural systems used in housing have historically developed into a single-purpose system. Structural elements of a building such as beams, columns, floors, and roof perform different functions in a traditional construction system. It is possible that structural design and materials in the housing industry can be dramatically improved through the development and application of innovative designs and new materials that capitalize on multifunctional components.

Modular building system is a rapidly growing form of construction, gaining recognition for its increased efficiency and ability to apply modern technology to the needs of the marketplace. In the modular construction technique, a single structural panel is manufactured which can perform a number of functions such as providing thermal insulation, sound and vibration damping, along with providing structural strength and hence it is termed as multifunctional. These multifunctional panels can be prefabricated in a manufacturing facility and then transported to the construction site. A system that uses prefabricated panels for construction is termed a 'panelized construction system'.

Structural insulated panels (SIPs) are high-quality, precast, sandwich composite panels, widely used in the modular panelized construction industry. In SIPs, lightweight foam is sandwiched between two oriented strand board (OSB) facesheets. The core of SIPs can be made from a number of materials, including molded expanded polystyrene (EPS), extruded polystyrene (XPS), and urethane foam [1]. Generally EPS foam is used as the core due to its excellent thermal and sound insulation properties and its low-cost benefits [1]. The proven superiority in transverse and axial loading over conventional stick-built framing systems make SIPs a stronger and safer alternative for modular building structures [2].

OSB is a wood-based composite laminate. As wood is organic in nature, it is prone to termite and molds. The mold buildups can result in the loss of millions of dollars. The OSB facesheets are also prone to swelling in the presence of moisture and can result in disintegration, hence loss of structural strength of the SIP panel. These issues ultimately increase the life cycle cost of the structures built using traditional SIPs. Also, SIPs have poor penetration resistance against low and high velocity impacts. Earlier studies [3] on OSB-based SIPs revealed that the SIPs failed at the low impact velocity of 50 J when impacted with blunt object impactor. This 50 J impact energy typically represented events such as a tool drop or a furniture drop on the structural floor panel.

To overcome these issues of the traditional SIPs, the OSB facesheets of the traditional SIPs are replaced with thermoplastic (TP) composite facesheets in this study, and the panels thus formed are termed as composite structural insulated panels (CSIPs). The facesheets of CSIPs comprise E-glass fibers impregnated with polypropylene (PP) matrix. Greater strength and stiffness, better impact resistance, and weight savings of approximately 180% (per unit area basis) can be obtained by using the proposed panels [3].

A detailed description of the manufacturing process for the proposed CSIPs is included in this chapter. Structural design and characterization of the proposed CSIPs can be obtained from Refs [3] and [4]. Section 12.2 describes the brief history of panelized construction, also covering the manufacture of traditional SIPs and the cost benefits of using precast SIPs in panelized construction. Based on the experimental results, joining techniques are proposed in this chapter for connecting the CSIPs.

12.2 Traditional structural insulated panel (SIP) construction

The concept of precast sandwich panels in the panelized construction industry was first introduced by Frank Lloyd in the 1930s. Some of the earliest houses designed by Frank Lloyd in the 1930s were designed using sandwich composite panels. These innovative panels were the result of Lloyd's attempts to build relatively low-cost houses. Some of the walls of these low-cost houses consisted of three layers of plywood and two layers of tar paper as structural elements. As the prototypes of these sandwich panels lacked the desired insulation, these panels were not produced on a large scale [1].

Alden Dow experimented further with the concept proposed by Lloyd in 1950, and solved the problem of insulation in Lloyd's panels [1]. Dow developed structural panels with insulating foam as the core and was credited for producing the first SIP. The load-bearing walls of these houses were made of 40 mm (1 5/8") Styrofoam core and 7 mm (5/18") plywood facesheets.

The first large-scale SIP manufacturing effort came in 1959 when Koopers Company converted an automotive production plant in Detroit, MI, into a SIP production facility. Koopers' method of producing SIPs involved blowing pre-expanded Styrofoam beads between two sheets of plywood and bonding them to the facings, which were already glued to a supporting framework. The manufacturing method proposed by Koopers was slow and hence was not competitive in the marketplace in the 1950s.

Alside® in the 1960s proposed significant changes in Koopers' SIP manufacturing process. Their new methods reduced the manufacturing time of SIPs from several hours to 20 minutes [1]. In the mid-1980s a significant number of manufacturers began producing SIPs on a large scale.

12.2.1 Manufacturing of traditional SIPs

Manufacturing of traditional SIPs is generally a three-step process. The OSB facesheets are adhesively bonded to the EPS foam core during the manufacturing process of traditional SIPs. OSB is a laminate consisting of a number of OSB chips glued together under high temperature and pressure. In the first step, the bottom facesheets are laid out in the assembly area. The core pieces are run through the glue spreading machine, where the adhesive is applied to both sides of the core pieces. These core sections are placed on the bottom facesheets, and then the top facesheet is positioned. The facesheet and the core are aligned before being moved into the compression molding press. After removal from the press, the SIPs are cured in place for 24 hours in order to achieve the full strength of the adhesive before moving them to a storage area. Once fully cured, the SIPs are transported to the construction site and are joined with each other using adhesives and mechanical connectors.

12.2.2 Economic analysis of traditional SIPs

Cost reductions in terms of energy efficiency

The R-value is a unit of thermal resistance used for comparing insulating properties of different materials. A higher R-value indicates greater insulating property of the material. The R-value of a material significantly varies depending on its type and thickness. Generally thicker materials have higher R-values than the thinner materials of the same type. Oak Ridge National Laboratory (ORNL) compared the R-values of a wall built using panels with EPS foam as the core and OSB laminates as facesheets [5]. According to this study at ORNL, the R-value of a wall with a 90 mm (3-1/2″) thick EPS core is R-14 compared to R-9.8 for a 50 mm × 100 mm (2″ × 4″) wood-framed wall insulated with fiberglass insulation. Thus higher energy efficiency can be achieved using SIPs as compared to traditional stick-built construction. This higher energy efficiency is reflected in lower utility costs of a structure.

Cost reduction in terms of construction time

Lesser construction time of a modular structure can reduce the total construction time from one to four weeks depending on the size of a building. This translates into huge cost savings. Earlier studies [1] showed that 34% lesser onsite construction time is required for constructing SIP-based structures than traditional stick-built structures.

Cost reduction in terms of labor cost

Unlike traditional stick-built construction, modular buildings can be constructed using unskilled labor, which translates into reduction of total labor cost. As the lightweight panels are manufactured in the factory, this saves 25–30% of the man-hours in building a structure [6].

12.3 Joining of precast panels in modular buildings

The SIP wall panels are typically connected with each other with the help of SIP splines. The SIP spline consists of EPS foam sandwiched between OSB facesheets. The height of the spline is equal to the height of the foam core of the panels to be connected. These splines are adhesively bonded to the panels to achieve the desired strength of the connection. At some places, instead of SIP splines, wooden splines are used to create a bond between the adjacent panels. Specially designed panel screws are used to connect the exterior wall panels. All these connections are designed in such a way that a bond is created between the OSB facesheet laminates. EPS foam is not involved in forming a bond between the panel connections due to its low shear strength properties.

Joining of precast composite panels poses challenges at the micromechanical level, fiber–matrix interface, macromechanical level, structural level, and at the interface between two or more separate components [7]. Three of the most commonly used techniques for joining composite panels are discussed here:

- adhesive bonding
- mechanical fasteners
- fusion bonding.

12.3.1 Adhesive bonding

The main function of the adhesive in an adhesively bonded joint is to transfer the load efficiently between the adherents. The bonding can also be used to increase the structural efficiency of a laminated structure. According to Smith and Pattison [8], the joint efficiency is only 30% when a single lap joint is used in connections. This value can be increased to 60% with a butt strap joint, 70% with a scarf joint, and 90% with a stepped joint. Issues with adhesive joints include: localized flaws which greatly affect the strength of the joint, requirement of surface preparation, and edge effects due to higher stress concentration.

12.3.2 Mechanical fasteners

Mechanical fasteners are preferred over adhesive bonding for on-site assembly of the precast composite panels [7]. Design methods that have been established for structural joints in metals are mainly applicable for composites. Each composite system has to be joined independently due to the anisotropic nature of the composites. Advantages of mechanical fasteners over adhesive joining include repeatability, absence of environmental effects on polymers, ease of inspection, and no specific surface preparation.

12.3.3 Fusion bonding

Traditional technologies such as mechanical fastening and adhesive bonding have limitations because of the stress concentrations resulting from hole drilling in mechanical fastening or requiring extensive surface preparation for adhesive bonding. Additional shortcomings of using mechanical fastening for composites include, but are not limited to, delamination originating from localized wear occurring due to drilling, differential thermal expansion of fasteners relative to composites, water intrusion between fastener and composite, possible galvanic corrosion at fastened joints, and extensive time and labor required for drilling holes [9].

Fusion bonding or welding is a long-established technology in the thermoplastic (TP) industry, where the efficiency of the welded joint can approach the bulk properties of the adherents. This technique eliminates the stress concentrations created by holes required for mechanical fasteners. One of the fusion bonding techniques tried for bonding CSIPs was ultrasonic welding, which is discussed in detail in Section 12.4.

12.4 Manufacturing of composite structural insulated panels (CSIPs)

Adhesive used for bonding the facesheets to the core is the most important component in a sandwich composite. Use of proper adhesive ensures effective load transfer between facesheets and the core. It is well known that delamination between the core and the facesheet is the predominant mode of failure for sandwich composite laminates. Hence, choosing a suitable adhesive is key to achieving the desired strength of the sandwich composite.

Three different adhesives were tried in this study for bonding the facesheets to the core of the CSIPs. These included:

1. 3M water-based contact adhesive
2. Hot-melt spray adhesive
3. TP film adhesive.

These adhesives were chosen based on their ease of handling and processing. The adhesive application techniques were different for the three adhesives, e.g. 3M was a water-based paintable adhesive, hot-melt adhesive was an adhesive that can be sprayed on to a substrate, while the TP film adhesive was in a film format. The candidate adhesive was chosen based on the bond strength between the facesheet and the core.

Dyna Z-16 apparatus was used for checking the bond strength between the facesheet and the foam core. Test discs of 50 mm diameter were bonded to the facesheet with 3M scotch weld adhesive as recommended in the Dyna-Z-16 operation manual [10]. According to ASTM 1583-04 [11], a core depth of 10 mm was chosen for the CSIPs. Dyna Z-16 apparatus is capable of applying a uniaxial tensile load on the test discs which creates a concentrated stress beneath the test discs. The shear strength at the interface is displaced on the circular disc attached to the base frame. Figure 12.1 shows the Dyna-Z-16 apparatus used for this study.

Results obtained from the pull-off test are plotted in Fig. 12.2 for the three candidate adhesives. From this figure it can be seen that the film adhesive and the spray adhesive were comparable with each other in terms of the shear strength. For film and spray adhesives, cohesive failure of the foam core was observed, while for the 3M water-based adhesive, failure between core and the facesheet was observed. Based on these results, two candidate adhesives were chosen for manufacturing CSIPs. Detailed descriptions of manufacturing processes for both the adhesives are discussed in the following sections.

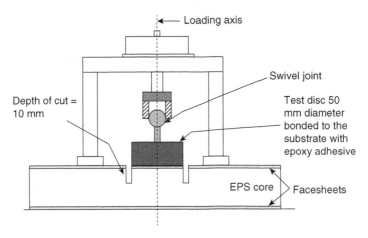

12.1 Schematic of pull off strength testing (ASTM-C-1583-04) (not to scale).

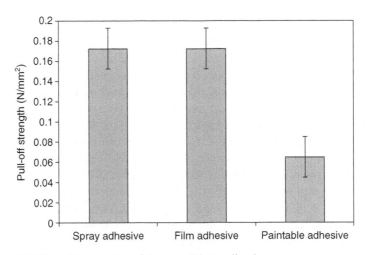

12.2 Pull-off strengths of the candidate adhesives.

Traditional SIPs are manufactured in sizes of 1,219.2 mm × 2,438.4 mm (4′ × 8′), based on this, the proposed CSIPs were manufactured to the same size. The full-scale manufacturing of the panels was undertaken using a heated press of 1220 mm × 3050 mm (4′ × 10′) available at Portage Casting and Molding at Portage, Wisconsin. Up to eight ceramic heaters were used in this press to heat up the platens. Temperature and pressure of this press were monitored with digital meters fitted to the press.

12.4.1 Manufacturing of CSIPs with film adhesive

Figure 12.3 shows the step-by-step manufacture of the CSIPs using film adhesive. As a first step, the facesheets were cleaned with the help of pressurized air to remove any loose dust present on the facesheet. The facesheets were then cleaned with acetone to dissolve sizings present on the laminates. Film adhesive was cut to the size of the facesheet and was laid on top of the facesheet prior to keeping them in the oven.

The melting point of the film adhesive was 65–74°C (150–165°F). After the adhesive was melted, the foam core was placed on top of the facesheet ensuring proper alignment. The whole assembly was then inverted and the bottom facesheet was adhered to the foam core in a similar manner. The sandwich was allowed to cool down in the oven for 15 minutes and then was taken out and stacked on the floor. The total time to fabricate one CSIP was 30 minutes using a crew of four.

During the initial trials of manufacturing, dry spots were observed on the surface of the facesheets as seen from Fig. 12.3(b). Air trapped between the facesheet and the film expanded when the facesheets were heated. This expansion of trapped air caused the thin film adhesive to bulge and rupture,

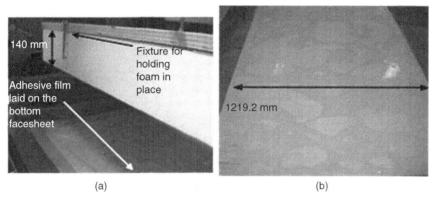

12.3 CSIP being manufactured using film adhesive: (a) position of the adhesive film and the bottom facesheet, (b) dry spots created on the facesheet after heating.

and this created dry spots on the facesheet where the bulging occurred. This problem was overcome by pre-melting the film adhesive with the help of a heat gun in order to remove the trapped air beneath the adhesive film. The film, though, did not melt completely, formed a weak bond with the facesheet which ensured lesser amount of air between the film and the facesheet. Pre-melting of the adhesive film avoided the dry spots formed on the facesheet surface which ensured proper bonding between the core and the facesheet.

Manufacturing the CSIPs with film adhesives had issues in terms of poor wet-out for larger spans. The dry spots acted as weak areas in the bonding and reduced the load-carrying capacity of the panels. Pre-melting the film adhesive on the facesheet increased the time of manufacturing which in turn increased the cost of manufacturing. Also, the material cost of film adhesive was approximately three times (per square foot) that of the hot-melt spray adhesive.

12.4.2 Manufacturing of CSIPs with TP spray adhesive

TP hot-melt adhesive was used to bond the facesheets to the core of the CSIPs. This adhesive was available in bead format. Beads of these adhesive were melted in the portable oven to achieve the desired viscosity for spraying. The temperature of the oven was maintained at the melting point of the adhesive. A spray gun and a spraying hose were connected to the oven. The adhesive was sprayed using the spray gun on the facesheets. Once both the facesheets were sprayed with the adhesive, EPS foam core was sandwiched between the facesheets. The sandwich thus obtained was subjected to constant dead weight for 24 hours. Figure 12.4 shows manufacturing of the CSIPs in a sequential manner.

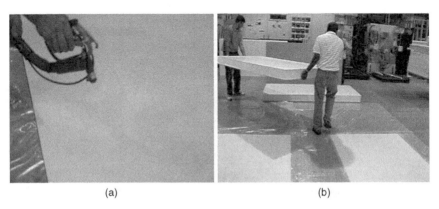

12.4 CSIP being manufactured using hot-melt impregnated spray adhesive: (a) TP spray adhesive being spread on the facesheets, (b) ready panels being stacked.

Table 12.1 Comparison of time required to manufacture CSIP panels using adhesive film and spray systems

Tasks	Time required to finish the task for adhesive film system (min.)	Time required to finish the task for adhesive spray system (min.)
Pressurized air cleaning	1	1
Cleaning with thinner	2	2
Pre-melting of adhesive	10	2
Application of adhesive	1	2
Application of pressure for curing of adhesive	120	120
Total	134	127

Table 12.1 summarizes the time required for each task for manufacturing the CSIPs using the spray and film adhesives.

12.5 Connections for composite structural insulated panels (CSIPs)

The connections between the CSIPs were designed in such a way that the TP facesheets would be the load-carrying members in a connection. Due to the poor shear strength and low surface energy of EPS foam, special techniques are required for putting inserts in the EPS foam [9]. Thus facesheets were considered as the two adherents for joining the proposed panels. The choice of a particular technique for joining the proposed panels was decided based on the single lap shear test performed on the facesheets of these panels. Specimens for this test were prepared according to ASTM-D-3163-01 [12].

12.5.1 Adhesives

Four different adhesives were verified for connecting the TP facesheets. These adhesives included 3M epoxy adhesive, 3M VHB acrylic foam tape 1, 3M VHB acrylic foam tape 2, and 3M water-based contact (brush-on) adhesive. For each adhesive, three specimens were tested for single lap shear tests and the average values are plotted in Fig. 12.5. Results of these tests were compared with the results obtained from ultrasonically welded specimens.

Adhesive failure was observed between the specimens bonded with 3M water-based adhesive and VHB tapes. The specimens bonded with ultrasonic welding failed in a cohesive manner. As seen from Fig. 12.5

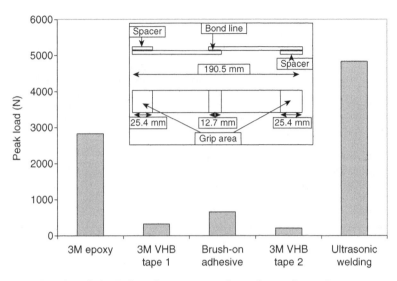

12.5 Results of single lap shear tests and specimen dimensions.

the peak load attained by the specimens bonded ultrasonically achieved highest peak load and hence ultrasonic welding was proposed for joining the CSIPs. Cost savings could be achieved using ultrasonic welding as this process was quick and no surface preparation was needed for ultrasonically welded connections.

12.5.2 Ultrasonic welding

The main components of ultrasonic welding apparatus consist of power supply, converter, booster, and horn. High-frequency electrical energy was supplied to the converter that transformed it to mechanical vibrations at ultrasonic frequencies. The mechanical vibrations were then transmitted through the booster to the horn. The horn amplified and transferred this vibration energy directly to the parts to be joined. The parts to be joined were held together under pressure and subjected to ultrasonic vibrations perpendicular to the contact area. The high-frequency stresses produced heat in the material and this heat was utilized at the joint interface through a combination of friction and hysteresis [13].

The welded joint was obtained by melting the polymer at the edges. For CSIPs, the reinforcing E-glass fibers acted as a skeleton for the matrix and avoided burning the matrix. Figure 12.6 illustrates the interaction between the fibers and the polymers during various stages of ultrasonic welding. In this study, the specimens were welded ultrasonically at Branson Ultrasonic Corp. in Atlanta, GA.

Polyvinyl chloride (PVC) pultruded shapes were used to connect the CSIPs. The connections were designed in such a way that the facesheets of the CSIPs were ultrasonically welded to the pultruded shapes. Figure 12.7 shows a hand-held ultrasonic welding machine. Due to its light weight and portable nature, this equipment would be ideal for joining the CSIPs on a construction site. Different horn sizes can be fitted to the booster in order to achieve a weld spot of desired diameter. The vibration energy melted the polymer at the edge of the panel and the pultruded shape. This created a strong bond between the pultruded shape and the CSIP. The time required to weld the PVC shape and the CSIP facesheet was 15 seconds for the total weld length of 305 mm.

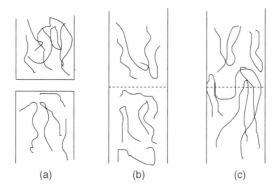

12.6 Healing of a polymer–polymer interface during ultrasonic welding: (a) two distinct composite interfaces, (b) achievement of intimate contact, (c) collapse of the interface through inter-diffusion [14].

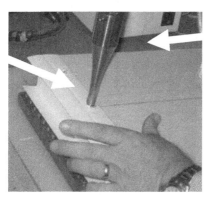

12.7 Proposed CSIP being ultrasonically welded to PVC extruded shape.

12.5.3 Connection details for CSIPs

Figure 12.8(a) shows a typical cross section of a structure built using modular panelized construction. Two representative locations were selected in this cross section to illustrate the details of the proposed connections for the CSIPs. These included exterior walls to floor connection (Fig. 12.8(b)) and connections at the junction of roof, wall, and the floor (Fig. 12.8(c)). Initially

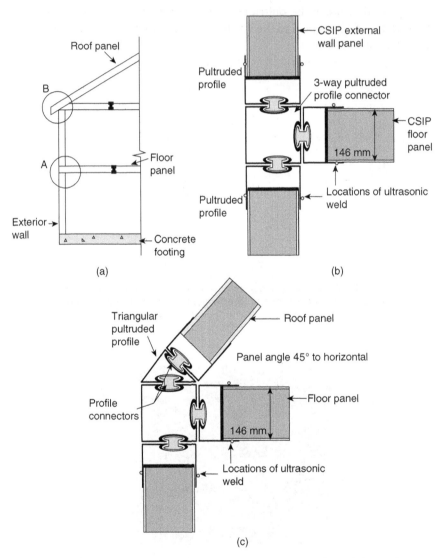

12.8 (a) Typical cross section of a building with CSIPs; (b) Details at A; (c) Details at B.

CSIPs were ultrasonically welded to the pultruded profiles. The pultruded shapes consisted of PVC connectors which created interlocking joints between the adjacent panels. The pultruded profiles were strong, lightweight and easy to install. For the connection between the roof panel and the exterior wall panel, a 45° connector was used, as shown in Fig. 12.8(b).

12.6 Conclusion

Manufacturing and connection details for joining the CSIPs were studied in this chapter. The findings from this study can be summarized as follows:

- Proposed CSIPs retained the advantages of the traditional SIPs in terms of energy efficiency, reduced construction time and cost, and reduced labor cost. Additional cost savings in terms of reduced transportation costs are possible with CSIPs, as these panels are 180% lighter than the traditional SIPs.
- Film adhesive used for bonding the facesheets to the core resulted in dry spots at the interface of the core and the facesheet due to expansion of trapped air. Pre-melting of the film adhesive on the facesheets solved this issue.
- TP spray adhesive used for bonding the facesheets to the core yielded the best results in terms of uniform distribution of the adhesive. The spray adhesive also proved more cost-effective than film adhesive. For these reasons, TP spray adhesive was chosen over film adhesive.
- Single lap shear test was used as a guideline for deciding the technique for connecting the CSIPs. During the single lap shear test, the specimens welded ultrasonically achieved the highest peak load, and hence ultrasonic welding was proposed as the best suited technique for joining the CSIPs.
- Interlocking assembly proposed for joining the CSIPs was less time-consuming and less labor-intensive.

12.7 Acknowledgment

The authors gratefully acknowledge funding and support provided by National Science Foundation research project NSF (NSF-CMMI-825938).

12.8 References

1. Morley, M. (2000) *Building with Structural Insulated Panels*, The Taunton Press, Newtown, CT.
2. Emerson, R. (2004) Moment resistant connections in prefabricated wood frame construction. *Proceedings of the NSF Housing Research Agenda Workshop*, February 12–14, Orlando, Florida.

3. Mousa, M. and Uddin, N. (2011) Flexural behavior of full-scale composite structural insulated floor panels. *Journal of Advanced Composite Materials*, 20, 547–567.
4. Vaidya, A.S., Uddin, N. and Vaidya, U.K. (2010) 'Design and analysis of composite structural insulated panels (CSIPs) for exterior wall applications', *Journal of Composites for Construction*, 14(4), 1313–1322.
5. Kosny, J. and Christian, J.E. *Whole Wall Thermal Performance*, Oak Ridge National Laboratory (ORNL), P.O. Box 2008 Oak Ridge, TN 37831.
6. Structural Insulated Panels for New Home and Commercial Building Construction, http://www.siphomesystems.com/pagemaster~PAGE_user_op~view_printable~PAGE_id~114~lay_quiet~1.html?b1c68fa39951e117fcf85297d112d3e7=a1566a120aefa4dc326f15c2e1c2bbb4 (accessed on March 27, 2008).
7. Hollaway, L. (1993) *Polymer composites for civil and structural engineering*, Chapman & Hall, London.
8. Smith, C.S. and Pattison, D. (1977) 'Design with fiber reinforced materials', *I. Mech. Eng. Conf.*, Paper C230/77.
9. Green, A.K. and Phillips, L.N. (1982) 'Crimp-bonded end fittings for use on pultruded composite sections'. *Composites* 13(3), 219–224.
10. Dyna-Z-16, *Operation Manual*, Proceq USA Inc., Aliquippa, PA.
11. Annual Book of ASTM Standards C1583-04, Standard Test Method for Tensile Strength of Concrete Surfaces and the Bond Strength or Tensile Strength of Concrete Repair and Overlay Materials by Direct Tension (Pull-off Method), West Conshohocken, PA.
12. Annual Book of ASTM Standards D-3163-02, Standard Test Method for Single Lap Shear Test, West Conshohocken, PA.
13. Zhang, J. (2005) 'Joining of polypropylene/polypropylene and glass fiber reinforced polypropylene composites', Dissertation, The University of Alabama at Birmingham.
14. Prager S. (1981) 'The healing process at polymer–polymer interfaces', *J. Chem. Phys*, 75, 5194–5198.

13
Thermoplastic composites for bridge structures

N. UDDIN, A. M. ABRO, J. D. PURDUE, and U. VAIDYA,
The University of Alabama at Birmingham, USA

DOI: 10.1533/9780857098955.2.317

Abstract: The primary objective of this chapter is first to introduce and demonstrate the application of thermoplastic (woven glass reinforced polypropylene) in the design of modular fiber-reinforced bridge decks, and next the development of jackets for confining concrete columns against compression and impact loading. The design concept and manufacturing processes of the thermoplastic bridge deck composite structural system are presented by recognizing the structural demands required to support highway traffic. Then the results of the small-scale static cylinder tests and the impact tests of concrete columns are presented, demonstrating that thermoplastic reinforcement jackets act to restrain the lateral expansion of the concrete that accompanies the onset of crushing, maintaining the integrity of the core concrete, and enabling much higher compression strains (compared to CFRP composite wraps) to be sustained by the compression zone before failure occurs.

Key words: thermoplastic composite, bridge deck, column jacket, dynamic load, impact, design, manufacturing.

13.1 Introduction

In the last few years, the presence of composite materials in the construction industry has become more common. Today, fiber-reinforced composite materials are used in a wide array of civil infrastructure applications [1]. Most of these applications utilize prepreg thermosetting composites, the most common of which is carbon fiber-reinforced polymer (CFRP). Thermoplastic composites are relatively new materials in civil engineering applications and lack the history of use in civil infrastructure. Limited time has been spent investigating the usage of thermoplastic materials. These materials offer comparable material characteristics to thermosetting composites. The ability to readily form these materials using epoxy makes them much more desirable. Thermoplastic polymers have several advantages over thermosets: they can be reshaped by reheating, are recyclable, are cost-effective and possess superior impact properties. They have comparable mechanical properties, higher notched impact strength, reduced creep tendency, and very good stability at elevated temperatures in humid conditions [2]. Long fiber reinforcement thermoplastics have significantly higher heat deflection temperature and better heat aging properties than the corresponding short fiber-reinforced matrix materials. Because of their good heat aging properties, thermoplastic composites like glass reinforced polypropylene components are suitable for continuous service temperatures up to 266°F (130°C) [2]. Fatigue strength is critical for designing girder components that are subjected to fluctuating stress. No significant loss of rigidity was recorded on a prototype glass reinforced thermoplastic composite slab during two million cycles of load at 2×24.8 kips (110.25 kN). The mode of failure is punching in the loading area and this mode of failure does not represent a catastrophic failure [3].

Thermoplastic composites typically comprise a commodity matrix such as polypropylene (PP), polyethylene (PE) or polyamide (PA) reinforced with glass, carbon, or aramid fibers. Progress in low-cost thermoplastic materials and fabrication technologies offer new solutions for very lightweight, cost-efficient composite structures with enhanced damage resistance and sustainable designs [3]. The primary objective of this chapter is to introduce and demonstrate the application of thermoplastic (woven glass reinforced polypropylene) first, in the design of modular fiber reinforced bridge decks, and next on the development of jackets for confining concrete columns against compression and impact loading.

13.2 Manufacturing process for thermoplastic composites

The process used to make long fiber thermoplastic products was generally very expensive. Thus, the market for long fiber thermoplastic composites was very limited due to the high cost of producing these products. Recently, a novel hot-melt impregnation technology has been developed that allows complete impregnation of long fibers with thermoplastic polymers at very high production rates, producing high-quality, low-cost thermoplastic composites. This technology, called DRIFT (Direct ReInforcement Fabrication Technology) [3], yields products that can be made as continuous rods, tapes and pultruded shapes, or they can be chopped into pellets of any length for injection or compression molding [2]. The process has been shown to work well with glass, carbon, aramid, and other polymer fibers and also with a wide variety of thermoplastic polymers.

E-glass/PP tapes of 0.5" (12 mm) width and an average layer thickness of 0.024" (0.6 mm) were produced using the DRIFT process [4]. The unidirectional E-glass/PP tape material with a fiber content of 67 wt% (42 vol%) has the tensile strength of 87.6 ksi (604 MPa), tensile modulus of 4,300 ksi (29,648 MPa) and density of 99 lb/ft^3 (15.5 kN/m^3) [4]. The hot-melt impregnated E-glass/PP tape can be woven into broad goods with various weaving patterns appropriate to the application. The unidirectional E-glass/PP tape material can be woven into a plain weave architecture fabric form (Fig. 13.1) through textile weaving operation.

Thermoforming is being used to produce large sized plastic components with varying wall thickness (greater than 0.04" (1 mm)), formed under low molding pressures (less than 50 psi (0.345 MPa)), with molds made of aluminum alloy, wood, or polymer composites. A simplistic overview of the single sheet thermoforming process consists of heating a plastic (or composite) sheet and forming the sheet over a male mold or into a female one. The operation deforms the sheets of the material into curvilinear shapes with the help of tools or molds. The process uses various configurations such as vacuum forming, drape forming, matched mold forming, etc. Basic vacuum forming represents the conventional technology; a vacuum is created between a female mold and a heated plastic sheet, which is forced to comply with the mold walls. The components can be produced with increasing thickness from the center to the edges. The process involves heating of polymer sheet that is firmly constrained along its perimeter above its transition temperature or the melt temperature, forming in a mold through vacuum and cooling by conduction in the case of thin films or through fans in the case of thick walls [4].

13.1 Close-up view of woven E-glass/polypropylene fabric.

13.3 Bridge deck designs

An integral modular fiber thermoplastic composite bridge structural system is described. To demonstrate the design concept, two bridge deck systems with different spans are modeled. The design concept of both decks presents a unique approach for a structurally efficient and low-cost bridge deck system. A modular fiber-reinforced thermoplastic panel with hat-sine rib stiffened shape is used as a bridge deck system. It consists of two/three components, i.e. top flat face, hat-sine rib, and/or bottom flat face. The other parameters are the interface contact length between shells of flat face and sine rib, wavelength of sine rib, depth of deck, and thickness of each deck component. All these parameters can be determined by considering the deck stiffness criteria set by the AASHTO code.

The deck shape based on the hat-sine rib stiffened design concept is selected by considering various issues such as the processability of the E-glass/PP woven tape, and the practical issues such as tooling, and design flexibility for the prototype studies. The glass/PP woven tape is relatively stiff, unlike the typical thermoset pre-pregs (such as glass/epoxy or carbon/epoxy); the material cannot be molded into tight radii/corners. The hat-stiffened rib design is shown to be structurally efficient in several studies [5–7]. A deck system as shown in Fig. 13.2 features E-glass/PP woven tape hat-sine shape ribbed profile bonded to a flat E-glass/PP woven face. A three-step concept was pursued for manufacturing the glass/PP thermoplastic composite floor segment type: (a) manufacture the flat face, (b) manufacture the hat-sine rib, and (c) adhesively bond the face to the hat-sine rib. The face and the rib portions of the deck floor can be processed through a number of choices, which include thermoforming, double belt press consolidation of the tape forms, reaction

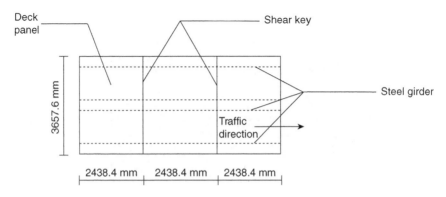

13.2 Plan of single-lane bridge deck.

injection molding, and/or extrusion. The contact area of the ribs to the face could be bonded adhesively and/or by a combination of adhesive bonding and fasteners [4].

13.3.1 Design criteria

The design criteria are set by following the loading conditions and performance limitations described in the AASHTO LRFD Bridge Design Specifications [8]. The dead load and the vehicular live load must be applied in different combinations to obtain the maximum effect. The dead load, DC, includes the weight of the structural system, wearing surface, and all attachments. The loads are taken as 15 psf (0.72 kPa) (self-weight of the deck) and 5 psf (0.24 kPa) (polymeric wearing surface) [9] applied as a uniformly distributed load over the surface of the bridge. The three specified types of vehicular loading, LL, are:

1. *Design truck load*: three axles with loads 32 kips (142 kN), 32 kips (142 kN) and 8 kips (35.6 kN). The spacing between the 32 kips (142 kN) axles varies from 14 ft (4.26 m) to 30 ft (9.14 m), and is chosen by the designer to produce the maximum effect for shear, moment, and deflection.
2. *Design tandem*: a pair of 25 kips (111 kN) axles spaced 4 ft (1.22 m) apart with transverse spacing of 6 ft (1.83 m).
3. *Design lane load:* a uniformly distributed load of 640 psf (30.64 kPa) applied over a 10 ft (3 m) wide strip.

The AASHTO category strength I load combination is used to compute the ultimate capacity of the bridge, i.e.

$$Q = 1.25DC + 1.75(LL + IM) \qquad [13.1]$$

The live load should include either a design truck load combined with a lane load, or a tandem design load combined with a lane load for every lane in the bridge. The AASHTO service I loading combination is used for checking the deflection of the bridge design, i.e.

$$Q = LL + IM \qquad [13.2]$$

For maximum deflection, the truck or tandem is placed such that the center of gravity of the truck or tandem is on the center of the bridge, i.e. AASHTO arrangement I. The shear stresses are checked by using arrangement II (with the rear axle of the truck or tandem at one end of the bridge) of the truck or tandem load.

The AASHTO specifications 3.6.1.3.2 and 2.5.2.6.2 are used to adopt the deflection limit of L/800 (where L is the span of the bridge). The deflection resulting from the design truck/tandem alone or that resulting from 25%

of the design truck/tandem taken together with the design lane load should not be greater than the maximum allowed limit.

The maximum work theory of Tsai-Hill is used to determine the failure of the structure which can be defined by the following equation:

$$(\sigma_X/\sigma_{X(ULT)})^2 - (\sigma_Y/\sigma_{Y(ULT)})(\sigma_X/\sigma_{X(ULT)}) + (\sigma_Y/\sigma_{Y(ULT)})^2 \\ + (\tau_{XY}/\sigma_{XY(ULT)})^2 < 1.0 \qquad [13.3]$$

where σ_X, σ_Y, and τ_{XY} are longitudinal, transverse, and shear stresses due to applied load, and $\sigma_{X(ULT)}$, $\sigma_{Y(ULT)}$, and $\sigma_{XY(ULT)}$ are the ultimate stresses in the longitudinal, transverse, and shear directions. These ultimate strength values in checking ply failure using the Tsai-Hill approach are adopted from the literature using experimental results whenever possible.

13.3.2 Analysis and design procedure

E-glass/PP is used in the design of the bridge structure. The ply properties, i.e. E(fiber), E(matrix), G(fiber), G(matrix) are based on experimental results mentioned in Vaidya et al. [4]. The elastic properties of the laminate for a specific volume fraction of fibers are analytically evaluated using laminate theory; these elastic constants used for analysis are Young's modulus in the longitudinal and lateral/transverse directions (E_X, E_Y, E_Z), Poisson's ratio in each direction (v_{XY}, v_{XZ}, v_{YZ}), and shear modulus (G_{XY}, G_{XZ}, G_{YZ}). Table 13.1 lists the elastic properties for the composite laminate.

The finite element analysis to model the bridge deck is carried out on Ansys 8.0 software, the composite face and the hat-sine ribs are modeled

Table 13.1 Material properties of E-glass/PP woven tape composite

Property	E-glass/PP woven tape composite 40% fiber content by volume
E_X	1,437 ksi (9,900 MPa)
E_Y	1,437 ksi (9,900 MPa)
E_Z	149 ksi (1,027 MPa)
v_{XY}	0.11
v_{YZ}	0.22
v_{XZ}	0.22
G_{XY}	184.16 ksi (1,270 MPa)
G_{YZ}	108.75 ksi (750 MPa)
G_{XZ}	108.75 ksi (750 MPa)
E_{FIBER}	10,150 ksi (69,982 MPa)
E_{MATRIX}	149 ksi (1,027 MPa)
G_{FIBER}	4,350 ksi (29,992 MPa)
G_{MATRIX}	108.75 ksi (750 MPa)

using the Shell 99 elements. The Shell 99 element used has six degrees of freedom at each node constituting the x, y, and z direction nodal translations and rotations. Each element is defined by eight nodes (the mid-plane and the corner nodes), average or corner layer thickness, orthotropic material properties, and ply orientations [10]. The contact region between the face panel and the hat-sine stiffened ribs is developed by merging the common nodes and key points. The hat-sine ribs were subjected to parametric studies, which included the amplitude of the hat-sine, the wavelength, and the contact width between the face panel and the ribs. Based on the combination of least deflection and corresponding stresses, the right combination of the sine amplitude with the other deck components (i.e., wavelength and contact width) is determined and is described in detail elsewhere [4]. Following the selection of the right combination of parameters, further analysis on the bridge deck was conducted as summarized in the following sections.

13.4 Design case studies

13.4.1 Single-lane bridge deck system

A typical single-lane bridge deck is modeled having width of 12 ft (3.65 m) and total length of 24 ft (7.30 m). The deck is supported on three steel girders having a span of 6 ft (1.83 m) and is divided into three panels, each having 8 ft (2.44 m) length and 12 ft (3.65 m) width with hat-sine rib direction perpendicular to the direction of traffic (Fig. 13.2). Typically, connection between deck and girder consists of shear studs cast into the cell of an FRP sandwich deck. Steel spirals are then positioned around each shear stud to aid in grout confinement.

Section with top flat face and hat-sine rib

In this case, we used a hat-sine shape ribbed profile bonded to a flat face (Figs 13.3 and 13.4), and each component was made of glass/PP woven tape ply which can be modeled as a 0/90° layer. Using the material properties as

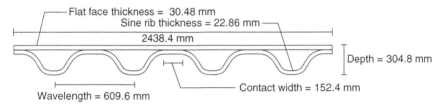

13.3 Single-lane bridge deck parameters for case 4.

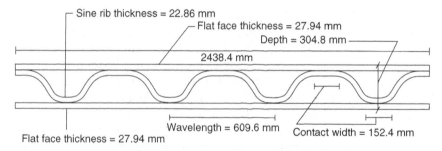

13.4 Single-lane bridge deck parameters for case 5.

Table 13.2 (a) Single-lane deck cross-sectional shape parameters (cases 1 to 4); (b) single-lane deck cross-sectional shape parameters (cases 5 to 7)

(a)

Case no.	Deck depth (amplitude) (mm)	Wavelength (mm)	Contact width (mm)	Top layer thickness (mm)	Sine curve thickness (mm)
1	152.4	304.8	76.20	22.86	15.24
2	304.8	609.6	152.4	25.40	17.78
3	304.8	609.6	152.4	27.94	20.32
4	304.8	609.6	152.4	30.48	22.86

(b)

Case no.	Deck depth (amplitude) (mm)	Wavelength (mm)	Contact width (mm)	Top layer thickness (mm)	Sine curve thickness (mm)	Bottom layer thickness (mm)
5	304.8	609.6	152.4	27.94	22.86	27.94
6	152.4	304.8	76.2	25.40	17.78	7.62
7	152.4	304.8	76.2	25.40	20.32	12.70

Note: The material for all cases is glass/PP except for cases 6 and 7, where the bottom layer material is carbon/PP.

in Table 13.1 with the loads mentioned, the bridge decks are subjected to optimization studies. Several deck model case simulations were carried out on ANSYS with appropriate amplitude, wavelength, and contact width as shown in Table 13.2(a). The simply supported boundary condition was chosen since the design was stiffness-based (deflection limited). It should be noted that the optimization of hat-sine rib is controlled by stiffness criteria satisfying the deflection criteria of AASHTO for bridge decks. The optimized section was checked for the Tsai-Hill failure limit and critical buckling load analyses were performed to obtain the thickness for the sine ribs [8]. Based on the combination of the least deflection (as shown in Fig. 13.5) and adequate stresses (not shown here for the sake of brevity), the bridge deck dimensions were determined to be optimal at 12 in. depth, 24 in. wavelength and 6 in. contact width. The corresponding layer thicknesses were 1.2 in. for top flat face and 0.9 in. for hat-sine rib (case 4 in Table 13.2(a) and Fig. 13.5). The 6 in. (150 mm) contact width was particularly chosen to have adequate bonding area of the hat-sine rib section to the flat face.

Section with top and bottom flat face with hat-sine rib

At the similar combination of wavelength and contact width, the deck system is also compared by adding an additional bottom layer, i.e., cases 5–7 as in Table 13.2(b), with the optimized case 5 as shown in Fig. 13.4. By comparing both shapes (cases 4 and 5), it is concluded that the most efficient and cost-effective section could be one which has lesser cross-sectional area while maintaining sufficient stiffness to control the deflection. Based on this criterion, the section shown by case 4 is still the optimized section (Fig. 13.5). It has 26.6% less cross-sectional area with a similar moment of inertia as case 5, but will consume lower manufacturing cost. It should be noted that the material for all cases is glass/PP except for cases 6 and 7, where the bottom layer material is carbon/PP, and that glass is cheaper than carbon fiber. For the optimized section (i.e., case 4), the maximum deflection of 0.09 in. (2.25 mm) is exactly at the point of contact wheel load (Plate VIII between pages 240 and 241). The maximum ultimate tensile stress of 6,392 psi (44 MPa) developed at the intermediate support is much less than the failure stress of 28,000 psi (193 MPa) for E-glass/PP composite laminate. The shear stresses σ_{yz}, σ_{xz} due to ultimate load are 3,339 psi (23 MPa) and 1,669 psi (11.5 MPa), respectively (Plate IX between pages 240 and 241).

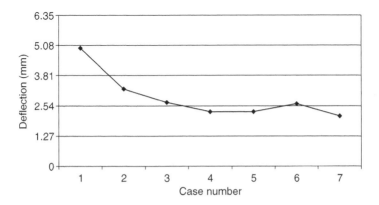

13.5 Maximum deflection for single-lane deck models.

13.4.2 Double-lane bridge deck system

Further demonstration of the performance of the deck conceptual design is presented by studying a typical two-lane traffic, 60 ft (18.3 m) span bridge having a width of 24 ft (7.3 m) and depth of 36 in. (900 mm) (Fig. 13.6). The length-to-depth aspect ratio of the deck system is 20:1, which is reasonable for highway bridges. Similar to the single-lane bridge deck system, connection between deck and girder can be assumed to consist of shear studs cast into the cell of an FRP sandwich deck. Steel spirals are then positioned around each shear stud to aid in grout confinement.

Section with top flat face and hat-sine rib

Following the similar analysis process as in Section 13.4.1, the bridge deck dimensions were determined to be optimal at 36 in. depth, 48 in. wavelength and 16 in. contact width. The top flat face thickness was 3 in., and the sine rib component thickness was 2.5 in. (Fig. 13.7). The corresponding deflection and the maximum stresses were also within the allowable limits [11].

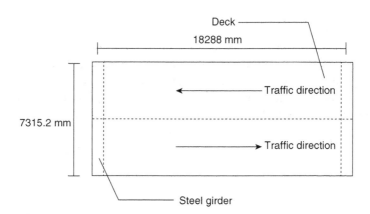

13.6 Plan of double-lane bridge deck.

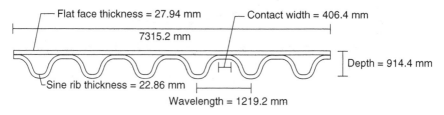

13.7 Double-lane bridge deck parameters for case 4 using glass/PP.

Top and bottom flat face with hat-sine rib

The analysis was carried out by varying sine rib parameters (Table 13.3); the sections having required stiffness to control the deflection within allowable limits are shown. In cases 1 and 2, carbon/PP was used in the bottom flat face of the deck system having a composite thickness between 3 and 3.5 in. Based on the combination of the least deflection and adequate stresses (not shown here for the sake of brevity) the hat-sine rib dimensions were determined to be optimal for case 3 compared to cases 1 and 2. Case 3, as shown in Fig. 13.8, provides a better section because it is cost effective and composed of glass/PP only with a component thickness between 2.0 and 2.4 in. Once again it should be noted that glass is cheaper than carbon fiber. Overall, the section shown by case 4 is the optimized section for this double-lane deck as it has 10.33% less cross-sectional area with comparable moments of inertia, and will consume less manufacturing cost compared to case 3 (Fig. 13.9). The deflection and ultimate shear stresses for the optimized section are 0.9 in. (22.5 mm), 6,366 psi (44 MPa), and 1,710 psi (11.8 MPa), respectively, and are shown in Plates X and XI (between pages 240 and 241).

Table 13.3 Double-lane deck cross-sectional shape parameters

Case no.	Deck depth (amplitude) (mm)	Wavelength (mm)	Contact width (mm)	Top layer thickness (mm)	Sine curve thickness (mm)	Bottom layer thickness (mm)
1	609.6	914.4	304.8	88.9	76.2	88.9
2	762.0	1016.0	381.0	76.2	76.2	76.2
3	914.4	1219.2	406.4	60.96	50.8	60.96
4	914.4	1219.2	406.4	76.2	63.5	–

Note: The material for all cases is glass/PP except for cases 1 and 2, where the bottom layer material is carbon/PP.

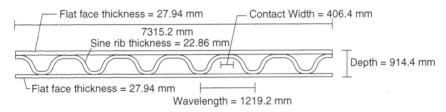

13.8 Double-lane bridge deck parameters for case 4 using glass/PP.

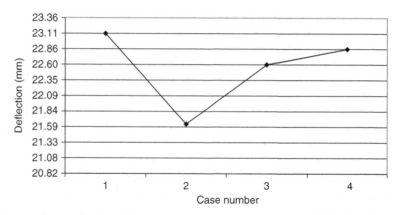

13.9 Maximum deflection for double-lane deck models.

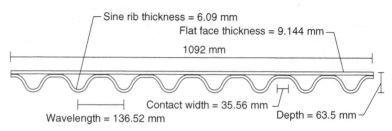

13.10 Panel shape and dimensional parameters used in an experiment.

13.4.3 Design verification

Design verification and analysis accuracy are compared by using the results of an experiment in which a panel made from E-glass/PP woven tape was tested under point loads (500 lbs (2.22 kN) to 2,000 lbs (8.90 kN)). The panel was simply supported and had a length of 43 in. (1075 mm) and a width of 29.5 in. (737.5 mm). Its shape, as shown by Fig. 13.10, consisted of 0.36 in. (9 mm) thick top flat face and 0.24 in. (6 mm) thick sine curve. The panel has the material properties as defined in Table 13.1. The experimental setup and details are mentioned in Ref. [4].

To check the accuracy of the finite element analysis, the panel is modeled on ANSYS using Shell 99 elements. The support boundary conditions are defined according to the experimental setup, and the process of finite element analysis is as defined. The experimental and FE analysis results are shown in Table 13.4. By comparing the experimental deflection with the Ansys analysis (Fig. 13.11), the analysis was found to underpredict the deflection by 10–15%. The difference between the analysis and the model

Thermoplastic composites for bridge structures

Table 13.4 Experimental and analytical deflection comparison

Concentrated load lbs (kN)	Maximum deflection inch (mm) (experimental)	Maximum deflection inch (mm) (finite element analysis)	Difference (%)
500 (2.22)	0.026 (0.65)	0.023 (0.65)	10.76
1,000 (4.45)	0.052 (1.30)	0.046 (1.15)	11.53
1,500 (6.67)	0.070 (1.75)	0.060 (1.50)	14.28
2,000 (8.90)	0.100 (2.50)	0.085 (2.12)	15.00

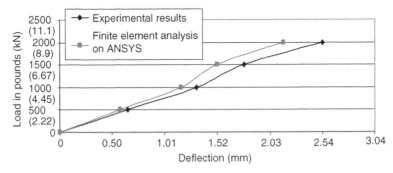

13.11 Panel load deflection comparison (experiment versus finite element analysis).

has been attributed to the perfect contact between the hat-sine rib profile and the face panel assumed in the model, while in the experiment there is a relative deformation between the two. The relative deformation can be explained because, in the tested panel, the face panel was bonded to the hat-sine rib profile using hot-melt glue. In an industrial setting, this step may adopt an ultrasonic bonding method, which would provide higher bond strength and eliminate any relative displacement between the face and the hat-sine rib profile. However, the analysis captured the trend observed in the experiments adequately for verification purposes. The results from some preliminary analysis using interface elements between the top plate and the ribs are reported elsewhere [11] (not reported here for the brevity), which also confirmed the observation.

13.5 Comparing bridge deck designs

The performance of the proposed bridge deck system is compared to two other designs: the Lockheed-Martin bridge [12] and the bridge system proposed by Aref and Parsons [13].

13.5.1 The Lockheed-Martin bridge

A 30 ft (9.14 m) span composite bridge built by Lockheed is described by Dumlao et al. [12]. A schematic of the cross-section of the Lockheed bridge is shown in Fig. 13.12 together with the proposed bridge deck design. The Lockheed design was loaded with a pair of 32 kips axles with measured deflections less than Span/800. A similar loading condition is imposed for the proposed deck system by using two 32 kips (142 kN) axles with 14 ft (4.26 m) spacing; also the design is checked for tandem loading condition with 25 kips (111 kN) axles spaced at 4 ft (1.22 m) distance from each other. The maximum vertical displacement of the proposed design is 0.45 in. (11.25 mm) which is equal to the AASHTO limit, and it occurs under the tandem loading condition. The maximum Tsai-Hill failure index is 0.48, and it occurs due to factored loads (strength 1 load combination) using a pair of 32 kips (142 kN) axles with 14 ft (4.26 m) spacing. The interface shear stresses between the outer (top flat face) and inner (sine ribs) shell contact area are σ_{yz} = 120 psi (0.82 MPa) and σ_{xz} = 366.5 psi (2.52 MPa). The weight of the proposed design is 28.1 kips and the dead-to-live load ratio is 0.439. The weight of the Lockheed bridge is 23 kips (102 kN), giving a dead-to-live load ratio of 0.36.

13.5.2 The bridge proposed by Aref and Parson [13]

The performance of the proposed double-lane bridge deck system (as discussed in Section 13.4.2) is compared to the bridge system proposed by Aref

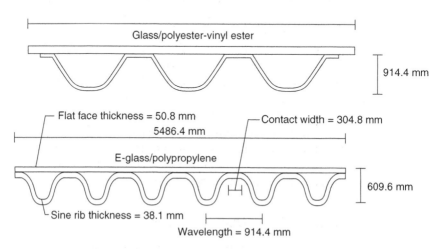

13.12 Cross sections of the Lockheed (thermoset glass/polyester-vinylester) and the comparison (proposed thermoplastic E-glass/PP) bridges.

and Parsons [13]. A schematic of the cross section of the bridge system proposed by Aref and Parsons [13] consists of seven inner cells encased in an outer shell and is shown in Fig. 13.13 together with our deck system. The performance comparison of both systems is summarized in Table 13.5, based on maximum deflection, failure indices, interface shear stresses, and the self weight of the deck system. The comparison of the proposed design with the modular fiber (S-glass/epoxy) deck system proposed by Aref and Parson [13] shows similar margin of safety with a Tsai-Hill index of 0.28 (proposed design) and 0.24 (Aref and Parsons [13]). Moreover, in the present design, a significant factor of safety is achieved in interface shear stresses between the outer (top flat face) and inner (sine ribs) shell contact area. The weight of our design is 121.5 kips (540 kN) which yields a dead-to-live load ratio of 0.84, and the weight of the S-glass/epoxy deck system is 67 kips (298 kN), giving a dead-to-live load ratio of 0.46.

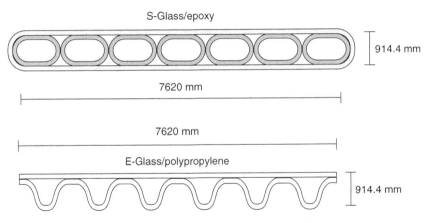

13.13 Cross sections of the Aref and Parsons [13] and the comparison thermoplastic (E-glass/PP) bridges.

Table 13.5 Performance comparison between S-glass/epoxy (Aref and Parsons [13]) and E-glass/PP (proposed design) deck

Material	S-glass/epoxy	E-glass/polypropylene
Deflection, inch(mm)	0.9 (22.5)	0.9 (22.5)
Tsai-hill failure index	0.24	0.28
Interface σ_{yz}, psi (MPa)	504 (3.48)	234 (1.6)
Interface σ_{xz}, psi (MPa)	484 (3.34)	175 (1.2)
Deck self-weight, lbs (kN)	67,000 (298)	121,500 (540)
Dead load:live load	0.46	0.84

13.6 Prefabricated wraps for bridge columns

As roadways and waterways become more congested, the risk of accidental collision with bridge piers remains a cause for concern. Bridge piers are designed for a variety of loading conditions, mainly compression, but often these structures fail when subjected to out-of-plane eccentric loading. Moreover, in engineering practice, there are many situations in which structures undergo impact or dynamic loading, such as during explosion, impact of ice load on pier structures, accidental falling loads, tornado-generated projectiles (i.e., objects picked up and converted to missiles by tornados), etc. For these reasons, along with the increased threat of terrorism, the need to find a way to protect these structures is critical. Over the last few years, the presence of composite materials in the construction industry has become more common. Most of these applications utilize prepreg thermosetting composites, the most common of which is carbon fiber-reinforced polymer, and so far most of the research conducted has concentrated on static and pseudo dynamic loading.

This study explored thermoplastic composite material produced in continuous pultruded form to produce a cost-effective split product form of directionally oriented glass fiber in polyurethane (or polypropylene) thermoplastic matrix for a representative bridge column. Two split halves will encapsulate the column with on-site mounting feasibility. The advantage of using pre-fabricated thermoplastic forms is they can be thicker than conventional thermoset wraps (such as presently used in bridge structures, only from a standpoint of enhancing stiffness/tensile strength). It is envisioned that under impact from unknown threats, such as collisions from trucks/trailers or blasts, the structure will have progressive failure potential, in place of catastrophic fracture presently witnessed. The tape and pultruded thermoplastic form has flexibility to accommodate curvatures encountered as part of the structure, and can be used either alone (only to suppress catastrophic failure) or in conjunction with conventional thermoset wraps if ductility improvement is also needed. An example of the concept is shown in Fig. 13.14. The split halves can be connected by a combination of thermoplastic tape jackets around the halves and a secondary mechanical reinforcement. Furthermore, the rate of strain induced on the structure is severe. The polypropylene alone or polypropylene/glass is cost-effective [14] and is expected to enhance the failure strain of concrete structures by several orders of magnitude.

The work presented here will compare the effects of dynamic loading of this type of confinement with the most common composite strengthening technique to date, CFRP composite wrapping. Two series of tests will be performed in this research: uniaxial compression testing of cylinders and impact loading of columns. An explanation of the specimen designation

Thermoplastic composites for bridge structures 333

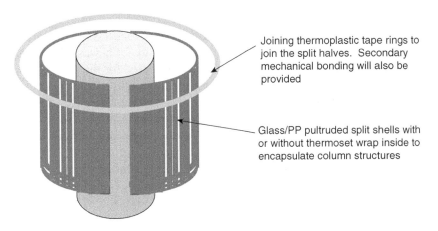

13.14 Schematic of the concept.

system is given first. The first letters are used to denote the type of specimen, 'Cy' for cylinder and 'Co' for columns. The second letter establishes the confinement type, 'N' for plain samples, 'C' for FRP, and 'P' for polypropylene jacket-confined cylinders. The next letter denotes the type of concrete, 'B' for high strength. The first number in the scheme is for the confinement thickness (mm) or number of plies. Finally, the last number represents the sample number.

13.7 Compression loading of bridge columns

The purpose of the uniaxial compression tests of the concrete cylinders was to evaluate static loading phenomena such as the effect on axial strength and strain capabilities. For this study, two variables were investigated: confinement material (polypropylene and CFRP composites (for reference)) and thickness of the polypropylene confinement. Strains were recorded using unidirectional electrical resistance strain gages. Compression loading was conducted using a Tinius-Olsen Universal Testing Machine, with the load applied manually at a constant rate of 103 kPa/s. A MegaDAC data acquisition system was used to record both the load and strain data.

13.7.1 Specimen details

The average compressive strength of the concrete was 58.6 MPa. The cylinders were grouped as follows: three control specimens; three 3 mm and three 6 mm thick polypropylene; and, finally, three single-ply unidirectional CFRP composites. Unidirectional SikaWrap Hex 103C was used for the CFRP, with Sikadur 300 used for the bonding agent to the concrete surface.

The fibers were oriented such that they provided reinforcement in the hoop direction (perpendicular to the applied compression load). After preparing the concrete surface and the CFRP composites, the material was rolled onto the cylinders. George Fischer beta (β)-PP [15] was used for the polypropylene and came in the following dimensions: 140 mm outer diameter, 13 mm wall thickness, and 5 m in length. This material was chosen since it had many desirable characteristics, including high impact strength, abrasion resistance, low weight, and a sizable operating temperature range, making it ideal for load-bearing applications. Since the polypropylene reinforcement is meant to act as passive reinforcement, the material was machined down from its original 13 mm wall thickness to the two thicknesses previously mentioned. Table 13.6 gives the material properties for the two types of confinement.

13.7.2 Stress–strain response

Comparison of the compression test data is presented in Fig. 13.15. Stress–strain data for the individual PP confinement materials can be found in

Table 13.6 Mechanical data for reinforcing materials

	Tensile strength (MPa)	Tensile modulus (GPa)	Elongation (%)	Nominal thickness (mm)
SikaWrap Hex 103C	958	73	1.33	1
β-PP	30	2.0	120	3 and 6

Note: β-PP = beta-nucleated polypropylene.

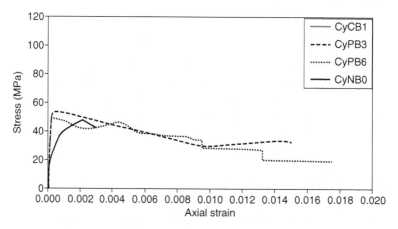

13.15 Comparison of stress versus strain for cylinders (averaged values).

Figs. 13.16 and 13.17. A review of the response curves in Fig. 13.15 demonstrates the polypropylene confinement produces a significant increase in the deformability of the concrete. However, it was unable to achieve similar compressive strength levels as that of the CFRP wrap concrete. For example,

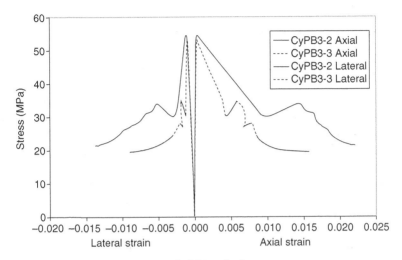

13.16 Stress versus strain for CyPB3 cylinders.

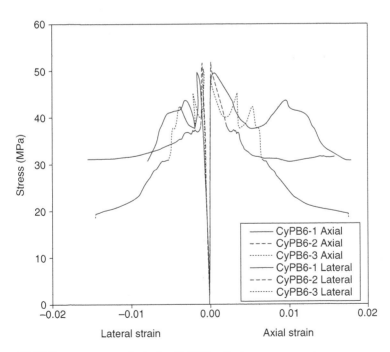

13.17 Stress versus strain for CyPB6 cylinders.

specimen CyCB1 reached a maximum axial stress of 98 MPa, whereas PP jacket specimens CyPB3 and CyPB6 could reach maximum 52 and 54 MPa, respectively. It was also obvious that isotropic CFRP wrapped around the circumference of the concrete did not affect the stiffness of the concrete up to peak concrete unconfined strength f'_c. Therefore, in this range, stiffness of the unconfined concrete was very similar to the CFRP wrap concrete. However, that was not the case for the PP jacket concrete. The PP, being more-or-less orthotropic, increased the stiffness to a considerable degree. Moreover, both CyPB3 and CyPB6 specimens achieved a similar stiffness.

CFRP wrap specimen CyCB1 achieved an axial strain at peak stress of 0.0056 mm/mm which is 3.5 times the average axial strain of the unjacketed cylinders. PP jacket specimen CyPB3 achieved a maximum strain of 0.0166 mm/mm, which is 9.2 times the average axial strain of the unjacketed cylinders. Specimen CyPB6 achieved a maximum strain of 0.0181 mm/mm, which is 10.1 times the average axial strain of the unjacketed cylinders. Both PP jacket specimens therefore recorded a maximum strain more than three times the CFRP wrap specimen. On the other hand, a maximum transverse strain at failure of 0.012 mm/mm (Fig. 13.16) and 0.015 mm/mm (Fig. 13.17) was recorded across the middle gage length for CyPB3 and CyPB6, respectively.

Initial matrix cracking formed in the wrap of specimen CyCB1 at an axial stress of 75 MPa. Rupture and debonding of strands of fibers from the matrix material of the CFRP wrap first occurred at a stress of 90 MPa. The specimen failed after a large number of strands of fibers ruptured and debonded from the wrap. Once ruptured, the fibers within these strands no longer contribute strength to the CFRP and thus confinement to the concrete. When CFRP wrap can no longer provide confinement, the specimen unloads. The rupture debonding of strands was concentrated along the middle gage length, though some also occurred within the top and bottom regions to a much lesser extent.

On the other hand, indication of initial concrete cracking was observed in the jacket of CyPB3 and CyPB6 at an axial stress of 54 MPa and 52 MPa, respectively. Once confinement was engaged, behavior of the concrete was a function of the circumferential stiffness of the confinement. Typically, three stages are seen in the stress–strain curves of fiber-reinforced polymer confined specimens. In the first region, it is the concrete that carries the axial load, due to minimal lateral expansion of the concrete core. Second, a non-linear transition range begins when the concrete starts to expand, generating greater lateral strain. Finally, the confinement takes effect and the stiffness is shown to remain at a constant rate. No debonding occurred along the PP. However, the PP jacket column showed a very small gain in compressive strength, an almost insignificant increase in strength. This can be explained by the fact that the bulging of PP affected the post-f'_c (post-peak) behavior. The polypropylene allowed the concrete to compact and dilate the confinement material. The effect of dilation can be seen in Fig. 13.18, since this phenomenon allowed the stress–strain curve to 'flat-line' (Fig. 13.15). This reflects a highly deformable mode of specimen failure relative to a specimen, for example, CFRP wrap which had failed by sudden rupture of all the fibers in a region of the wrap. The bulging of the jacket, however, was concentrated along the upper gage length, though some also occurred within the bottom regions to a much lesser extent.

Normalized strain data given by the ratio $\varepsilon_{tu}/\varepsilon_{to}$ (strain at failure of the confined cylinder by strain of the unconfined cylinder) are shown in Table 13.7. The 3 mm PP jacket produced an average ratio of 8.4, and the 6 mm

13.18 Comparison of failure among CyPB6 cylinders.

Table 13.7 Summary of compression test results

Specimen ID	Maximum strength (MPa)	Maximum strain	$\varepsilon_{tu}/\varepsilon_{to}$	Change in strain (%)
CyNB0*	50	0.0018	–	–
CyCB1-1	98	0.0056	3.1	211
CyCB1-3	106	0.0032	1.8	78
CyPB3-2	54	0.0135	7.5	650
CyPB3-3	53	0.0166	9.2	822
CyPB6-1	50	0.0181	10.1	906
CyPB6-2	52	0.0159	8.8	783
CyPB6-3	50	0.0176	9.8	878

*Average of the three specimens.

PP jacket yielded an average ratio of 9.6, which is impressive when compared with the 15.6 average ratio of the glass/PP confinement reported in earlier research [16].

13.7.3 Failure modes

The modes of failure observed during these tests varied depending on the confinement. The polypropylene confined samples exhibited a barreling effect as shown in Fig. 13.18. The ability to dilate considerably allowed the confined concrete to crush and compact inside the PP jacket. While this dilation was drastic, yielding of the polypropylene is evident in only a few places on the samples, and only one sample showed signs of material failure. Failure of the CFRP wrapped cylinders occurred due to fiber rupture near mid-height (Fig. 13.19).

13.8 Impact loading of bridge columns

The system used for impact testing was an Instron Model 8250 drop-weight impact machine with an instrumented striker (tup) assembly (Fig. 13.20). A flat striker was used for this test and had an impact area of 76 mm × 102 mm. For this study, the impact weight was 246 N. The hammer (tup) contained an internal load cell, which was used to record the contact load between the falling assembly and the column during the impact event. The load cell was rated for a maximum load of just over 44 kN. A drop height of 30 cm was used for all tests, since the combination of this height and the weight of the striker assembly produced loading close to that of the maximum allowed by the load cell. In previous studies using this machine, load–time

13.19 Comparison of failure among CyCB1 cylinders.

13.20 Instrumented drop weight low velocity impact test.

plots reported peak loads several times the expected value. Conclusions from similar tests, drawn by Suaris and Shah [17], illustrated that this loading was not indicative of the material properties but instead was a result of inertial effects of the samples. Though these effects have been accepted and calculated before testing metals, concrete creates a more complex problem due to the relatively small fracture strain and increased size of test specimen [18]. As in previous testing, a rubber pad was added to the striker to eliminate the inertial effects or 'ringing' being generated from the impact of the steel hammer and concrete specimens.

As mentioned earlier, impact testing was conducted using a total of four concrete columns: one control specimen, one CFRP confined, and two confined by a PP jacket of 3 mm and 6 mm thicknesses. For the CFRP confined, the fibers were oriented along the length of the column such that they provided reinforcement perpendicular to the axis of impact. The concrete used was from the same batch that produced the high strength cylinders tested under uniaxial compression. All columns tested were 152 mm × 914 mm. In an effort to illustrate a similar loading situation as would be seen in reality, the columns were placed horizontally inside a testing jig and subjected to axial compression. Deflection, velocity, and energy absorption were recorded using the DynaTup software that accompanied the Instron drop-tower. Strain data were recorded separately from the rest of the impact data. Strain was measured using unidirectional strain gages and recorded using the DATAQ recorder. After testing and subsequently failing the control column (CoNB0), it was clear that it would be impossible to fail the confined columns. The PP columns (CoPB3 and CoPB6) were impacted an average of five times. This was done to see if the material would exhibit any signs of weakening. Since the CFRP confined sample was assumed to be stiffer than the PP confined columns, only two tests were conducted. All samples were impacted from a height of 30 cm in an effort to keep the load cell free from damage. Due to the limitations of the drop-tower machine, the impact loading can be classified as low velocity impact or velocity less than 10 m/s [19]. Average impact velocity for these tests was 2.4 m/s. A summary of test data is given in Table 13.8.

Figure 13.21 shows the load versus time plot. The inertial effects have been reduced and are not visible in this plot due to the addition of the rubber pad to the striker. This phenomenon was evident in tests conducted by Erki and Meier [20]. Several observations can be made from the figure. First, the initial peak was the actual peak load. The subsequent peaks of smaller amplitude are simply rebounds of the tup. Since it was the stiffest, CoCB1 had the largest peak load of all the specimens with a value of 45 kN. As expected, CoPB3 and CoPB6 have desirable lower peak loads of 36 kN and 34 kN, respectively, since these specimens were less stiff than CoCB1. Since CoNB0 cracked under the loading, it seems not to have the second peak, only a third after a little rest due to a delayed rebound of the tup for the cracking. The dynamic bending load increase, on the other hand, was highest for the less stiff columns (CoPB3 and CoPB6) and gradually declined in value with increased stiffness. This trend was also observed by Jerome and Ross [21].

Based on the energy balance approach, the initial kinetic energy of the impactor deformed the structure during impact. Since the specimens used for this study were short columns and subjected to a compression sufficient

Table 13.8 Summary of impact test results

Specimen ID	Peak load (kN)	Maximum deflection at mid-span (mm)	Maximum strain	Change in strain (%)
CoNB0	38	3.38	0.0024	–
CoCB1	45	2.72	0.0057	143
CoPB3	36	4.52	0.0047	101
CoPB6	34	5.00	0.0058	148

Thermoplastic composites for bridge structures 341

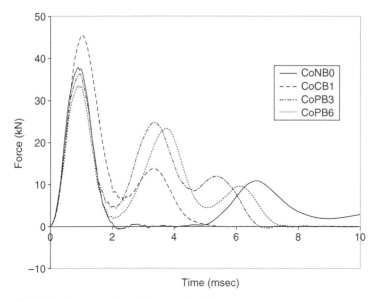

13.21 Load versus time for tested columns.

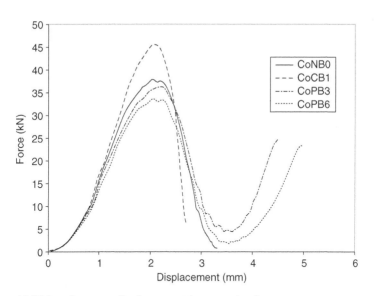

13.22 Load versus displacement for tested columns.

(corresponding to 1/8 P_o, where $P_o = 0.85\, A_g f'_c$) to provide fixity at the end against the impact, the stiffness was far too great to allow failure. It has been shown that the energy dissipated during vibration of composite structures is negligible [22]. From the load versus displacement curve (Fig. 13.22) however, CoPB6 deflected more than 5 mm and CoPB3 deflected about 4.5 mm. These

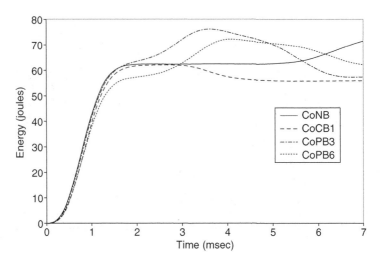

13.23 Energy versus time for tested columns.

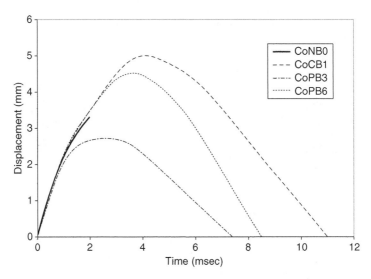

13.24 Deflection versus time for tested columns.

deflections are much higher than CFRP wrap column CoCB1, which deflected about 2.6 mm, about half that of the PP jacket confined columns.

From the energy versus time curve (Fig. 13.23), it appears that energy absorption of the polypropylene was higher than that of the CFRP composites confinement. Energy absorption for the 3 mm PP jacket was higher than that of the 6 mm jacket. Specimen CoCB1 absorbed 62 J, whereas specimens CoPB3 and CoPB6 absorbed about 78 and 70 J, respectively. The displacement versus time curve (Fig. 13.24) indicates that the polypropylene jacket confined specimens produced higher deflection than the CFRP confined column and unconfined column. This effect was due to the ability of the material to compress and further absorb the energy from impact. Transverse flexural strain values recorded across the middle gage length showed that CoPB6 and CoCB1 have equivalent strain values at the point of impact, with an increase in capability of approximately 145% over the unconfined concrete. The increase in deformability of sample CoPB3 was slightly less, with an increase of about 100%.

13.9 Conclusion

13.9.1 Bridge decks

The design concept and manufacturing processes of thermoplastic bridge deck composite structural systems are presented by recognizing the structural demands required to support highway traffic. The deck system is carefully engineered by considering the structural efficiency and manufacturing ease of deck components. Glass/PP woven tape material is used based on its effective utilization to produce structural deck components with flat geometries and gradual radii/curvatures.

The structural system presented possesses several special features that contribute to its effectiveness, including the use of curved panels (sine ribs) which provide the nonplanar core configurations to increase the performance of the bridge deck system. In all deck design cases, the stiffness (i.e. serviceability) is the main governing factor which controls the design. Once the stiffness requirement has been satisfied, the strength of the structure proved to be sufficient. In both deck systems (single-lane and double-lane), the outer shell (top flat face) with sine ribs offers a more efficient and economical section. The proposed design is compared to two published composite bridge concepts proposed by Dumalao *et al.* [12] and Aref and Parsons [13]. Although the present design has higher self-weight, which results in higher dead-to-live load ratio than both the glass/polyester-vinylester [12] and S-glass/epoxy deck systems [13], it could result in a better low-cost deck section based on the manufacturing and material cost comparison, as E-Glass/PP is much less expensive and the manufacturing process associated with it yields very good cost-effective results at a higher production rate [14].

13.9.2 Column wraps

Bridge columns are expected to sustain breaching and large inelastic rotation in plastic hinges during impact loading, a prime concern for the retrofit design to enhance the breaching and ductility capacity. Ductility will normally be provided by column plastic hinges. It is the plastic rotation of potential plastic hinges that is of greatest interest. The available plastic rotation capacity, and hence the ductility capacity, depends on the distribution of transverse reinforcement within the plastic hinge region. Transverse reinforcement provides the dual function of confining the core concrete, thus enhancing its breaching strength and enabling it to sustain higher compression strains, and restraining the longitudinal compression reinforcement against buckling. Most current bridge retrofitting applications utilize wet layup thermosetting composites, the most common of which is carbon fiber-reinforced polymer. However, FRPs possess a limited strain capacity relative to conventional materials such as steel. Finally FRP materials are relatively expensive.

This chapter discusses the results of the small-scale static cylinder tests and the impact tests on concrete columns. As summarized in the following, thermoplastic reinforcement jackets act to restrain the lateral expansion of the concrete that accompanies the onset of crushing, maintaining the integrity of the core concrete, and enabling much higher compression strains (compared to CFRP composite wraps) to be sustained by the compression zone before failure occurs. The mode of failure of PP wrapped specimens reflects a ductile mode of specimen failure relative to a specimen, for example, CFRP wrap, which had failed by sudden rupture of all the fibers in a region of the wrap. The bulging of the PP jacket, on the other hand, was concentrated along the upper gage length, though some also occurred within the bottom regions to a much lesser extent. No separation or debonding from the concrete surface occurred along the PP.

The impact tests were conducted to assess the energy absorption capacity of three concrete columns strengthened by PP confinement, a carbon/epoxy confined and one unconfined control specimen. All the results conclusively demonstrated the superior impact resistance properties of PP wrapped specimens over the CFRP. From the test results, the following conclusions were drawn.

1. Peak loading of the columns varied based on the stiffness of the confinement. Since the PP confined columns were the least stiff, they also exhibited the desirable least peak loading from the impact resistance design perspective.
2. Deflection of the PP confined columns was greater than the unconfined and CFRP confined columns. This is very favorable given that the time this displacement occurred was nearly six times greater than that of the plain specimen.
3. Transverse flexural strain values recorded across the middle gage length showed that the PP jacket strain demonstrates an increase in capability of approximately 145% over the unconfined concrete. The increase in

ductility of the PP jacket confined specimens was about 100% over CFRP wrap columns.
4. Energy absorption of the 3 and 6 mm PP was significantly higher than that of the single-ply CFRP composites confinement. Energy absorption for the 3 mm PP jacket was higher than that of the 6 mm jacket, which can be attributed to the lower stiffness of the former.
5. The PP jacket confined columns produced higher deflection than the CFRP confined column and unconfined column. This effect was due to the ability of the material to compress and further absorb the energy from impact. Though failure was not possible for the PP wrapped specimens, the above results demonstrated that the usage of a thermoplastic prefabricated jacket can be a potential solution to the threat of impact.

13.10 Acknowledgment

The authors gratefully acknowledge funding and support provided by University Transportation Center of Alabama UTCA research project (UTCA 3229 and 4210).

13.11 References

[1] ACI 440R-96 (1996). *State-of-the-Art Report on Fiber Reinforced Plastic (FRP) Reinforcement for Concrete Structures*, Committee 440, Farmington Hills, Michigan, 2000.
[2] Hartness, T., Husman, G., Koeing, J. and Dyksterhouse, J. (2001). 'The characterization of low cost fiber reinforced thermoplastic composites produced by the DRIFT process.' *Composites: Part A*, 32: 1155–1160.
[3] Center for Composites Manufacturing (2003). FTA Report Number FTA-AL026-7001-2003.1.
[4] Vaidya, U., Samalot, F., Pillay, S., Janowski, G., Husman, G. and Gleich, K. (2004). 'Design and manufacture of woven reinforced glass/polypropylene composites for mass transit floor system.' *Journal of Composite Materials*, 38(21): 1949–1972.
[5] Budiansky, B. (1999). 'On the minimum weights of compression structures.' *International Journal of Solids and Structures*, 39: 3677–3708.
[6] Christos, K. (1997). 'Simultaneous cost and weight minimization of composite-stiffened panels under compression and shear.' *Composites Part A*, 28A: 419–435.
[7] Swanson, G.D., Gurdal, Z. and Starnes Jr., J.H. (1990). 'Structural efficiency of graphite/epoxy aircraft rib structures.' *Journal of Aircraft*, 27(12): 1011–1020.
[8] AASHTO (2000). *Standard Specifications for Highway Bridges*, 18th edition. Washington, DC.
[9] Alampali, S., O'Connor, J. and Yannoti, A. (2000). 'Design, Fabrication, Construction, and Testing of an FRP Superstructure.' Report FHWA/NY/SR-00/134.
[10] Kohnke, P. (1994). *ANSYS Theory Reference*, 11th edn, SAS IP, Inc, Canonsburg, PA.

[11] Abro, A.M. (2006). 'Design and analysis of thermoplastic composite bridge superstructures,' MS thesis, CCEE Department, The University of Alabama at Birmingham, Birmingham, AL.
[12] Dumlao, C., Lauraitis, K., Abrahamson, E., Hurlbut, B., Jacoby, M., Miller, A., and Thomas, A. (1996). 'Demonstration low-cost modular composite highway bridge.' *Proceedings of the First International Conference on Composites in Infrastructure*, 1141–1155.
[13] Aref, A.J. and Parsons, I.D. (2000). 'Design and performance of a modular fiber reinforced plastic bridge.' *Journal of Composites Part B*, 31: 619–628.
[14] Wang, E. and Gutowski, T.G. (1990). 'Cost comparison between thermoplastic and thermoset composites,' *SAMPE Journal*, 26(6): 287–300.
[15] George Fischer (2004). 'Beta (β)-PP polypropylene piping system.' Products, <http://www.us.piping.georgefischer.com> (May 1999).
[16] Uddin, N. (2005). 'Vulnerability reduction for bridges.' University Transportation Center for Alabama, UTCA Project Report No. 032229, August.
[17] Suaris, W. and Shah, S.P. (1983). 'Properties of concrete subjected to impact.' *J. Struct. Eng.*, ASCE, 109(7): 1727–1741.
[18] Server, W.L., Wullaert, R.A. and Sheckhard, J.W. (1977). 'Evaluation of current procedures for dynamic fracture toughness testing.' *Flaw Growth and Fracture*, American Standards for Testing and Materials, STP 631: 448–461.
[19] Bartus, S.D. (2003). 'Long-fiber-reinforced thermoplastic: process modeling and resistance to impact.' MS thesis, The University of Alabama at Birmingham, Birmingham, AL.
[20] Erki, M.A. and Meier, U. (1999). 'Impact loading of concrete beams externally strengthened with CFRP laminates.' *J. Comp. Constr., ASCE*, 3(3): 117–124.
[21] Jerome, D.M. and Ross, C.A. (1996). 'Dynamic response of concrete beams externally reinforced with carbon fiber reinforced plastic (CFRP) subjected to impulsive loads.' *Structures Under Extreme Loading Conditions*, ASME, PVP 325: 83–94.
[22] Caprino, G., Lopresto, V., Scarponi, C. and Briotti, G. (1999). 'Influence of material thickness on the response of carbon fabric/epoxy panel to low velocity impact.' *Comp. Sci. Technol.*, 59: 2279–2286.

14
Fiber-reinforced polymer (FRP) composites for bridge superstructures

Y. KITANE, Nagoya University, Japan and A. J. AREF, University at Buffalo – The State University of New York, USA

DOI: 10.1533/9780857098955.2.347

Abstract: This chapter first reviews current structural applications of fiber-reinforced polymer (FRP) composites in bridge structures, and describes advantages of FRP in bridge applications. This chapter then introduces the design of a hybrid FRP–concrete bridge superstructure, which has been developed at The University at Buffalo for the past ten years, and discusses structural performance of the superstructure based on extensive experimental and analytical studies.

Key words: fiber reinforced polymer (FRP), bridge, superstructure, hybrid structure.

14.1 Introduction

14.1.1 Bridge conditions in the USA

Civil infrastructure systems play a crucial role in social and economic activities in any societies. Considering their importance, it is not surprising that the total investment in civil infrastructure is massive. The Federal Highway Administration (FHWA) estimates that the federal government alone has invested over $1 trillion in the US highway system (Wu, 2005). However, the US is now facing a major challenge to keep the nation's infrastructure systems in usable condition.

In 1967, the collapse of the Silver Bridge turned the bridge engineering community's attentions toward safety of bridges. State Departments of Transportation and FHWA collaborated to establish a systematic evaluation of structural safety, which resulted in the National Bridge Inspection Standards (NBIS) issued in 1971. These standards provide uniform procedures for the collection and maintenance of inventory and inspection data, minimum qualifications for bridge inspection personnel, and standardized methods for evaluating bridge conditions. The data collected by each state are submitted annually to FHWA, and FHWA maintains the data in the National Bridge Inventory (NBI) database. This NBI database contains only data for structures with a span of more than 6.1 m (Dunker and Rabbat, 1995; Small and Cooper, 1998).

A bridge is classified as deficient or not deficient by the inspection. There are two categories for deficient bridges: structurally deficient and functionally obsolete. The former indicates a deficiency in the health of a structure; the latter indicates a deficiency in the performance. Their definitions are as follows: bridges are structurally deficient if they have significant deterioration and have been restricted to light vehicles, and require immediate rehabilitation to remain open, or are closed; and bridges are functionally obsolete if they have deck geometry, load-carrying capacity, clearance or approach roadway alignment that no longer meet the criteria for the system of which the bridge is an integral part.

The number of deficient bridges is the most common indicator of the overall condition of bridges in the US, and the data for the last 20 years are plotted in Fig. 14.1. As of 2011, nearly 24% of 600,000 public bridges are either structurally deficient or functionally obsolete. Figure 14.2 shows a histogram of ages of US public bridges. The average age is about 40 years, and, as can be seen in the figure, a ratio of structurally deficient bridges to the number of bridges increases with age, implying that there will be an increasing need for maintenance work in coming years. According to a study by Brailsford *et al.* (1995), bridge decks are ranked the No. 1 bridge

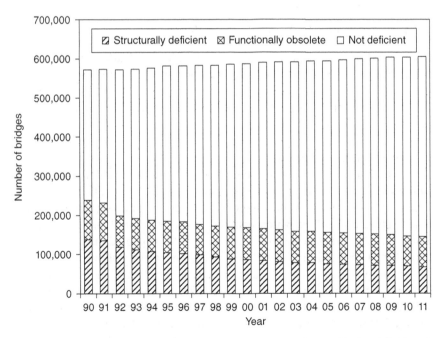

14.1 Number of deficient bridges in the US (source: National Bridge Inventory).

Fiber-reinforced polymer composites for bridge superstructures

maintenance item by State Department of Transportation (DOT) agencies. Tables 14.1 and 14.2 show the leading bridge maintenance priorities and the typical sources of bridge deterioration, respectively.

Although a quarter of US bridges are classified as deficient, the number of deficient bridges has been decreasing for the past few years as can be

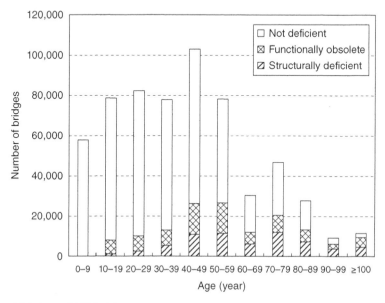

14.2 Age of US bridges.

Table 14.1 Bridge maintenance priorities at state DOTs (Brailsford et al., 1995)

Rank	Maintenance items
1	Bridge decking
2	Expansion joints
3	Steel trusses/connections
4	Painting
5	Concrete beams/columns
6	Steel bearings
7	Bridge railings
8	Timber piling
9	Drainage systems
10	Abutments/pier caps
11	Channel protection
12	Impacts
13	Electrical/mechanical

Table 14.2 Typical sources of bridge deterioration (Brailsford *et al.*, 1995)

Rank	Deterioration mechanism
1	Road deicing salts
2	Salt-water environment
3	Impact
4	Fatigue
5	Other: General aging Calcium chloride Freeze-thaw ASR of concrete Timber decay

seen in Fig. 14.1. There are a few reasons for this decreasing trend. First of all, the funding for preservation of the highway system has steadily increased to maintain deteriorating bridges. Secondly, the periodic and uniform inspection of bridges has played an important role in this decrease of deficient bridges. Not only does the inspection help State DOTs to locate deficient bridges, but the NBI database helps bridge experts and engineers point out critical problems with current bridge design practices. Thirdly, new materials and structural designs that require less maintenance have been developed and implemented in recent years.

However, this decreasing trend in the number of deficient bridges may not continue without better maintenance strategies. A large volume of the bridges were built in the 1960s (the Interstate era), and they will need more maintenance, major rehabilitation, or replacement in the near future. The US DOT also reported that it would require a significant increase in bridge funding in the coming years to fix all current and new deficiencies caused by further deterioration, since the investment that has been committed is only sufficient to preserve the bridge conditions and keep them from getting worse (US Department of Transportation, 2000).

14.1.2 Advantages of FRP composites

It is imperative to build bridge systems that have long-term durability and low maintenance requirements. As mentioned in the previous section, about 11% of US bridges are structurally deficient as of the year 2011, and billions of dollars will have to be spent in maintaining and improving conditions of the deficient bridges. A solution to this challenge may be to use new materials or to implement new structural systems. Among new structural materials, fiber reinforced polymer (FRP) composites have recently gained much attention in the civil engineering community due to their superior material properties, such as high specific stiffness, high specific strength, and high corrosion resistance.

FRP composites were applied in industry for the first time about 80 years ago. Since then, FRP composites have been developed mostly in the defense industry, particularly aerospace and naval applications. In the last two decades, considerable efforts have been made to apply FRP composites in the construction industry, and recently, structural applications of FRP composites started to appear in civil infrastructure systems. State-of-the-art reviews on FRP composites for construction can be found in papers by Karbhari and Zhao (2000) and Bakis *et al.* (2002). Some of the advantages of FRP composites over conventional materials in civil infrastructure applications are: high specific strength and stiffness, corrosion resistance, enhanced fatigue life, tailored properties, ease of installation, lower life-cycle costs, etc.

With these advantages, FRP composites have a great potential to be successfully applied in bridge structures. The high corrosion resistance of FRP composites makes them ideal alternative materials to resolve a number of persistent problems that the US highway system is now facing. Moreover, FRP composite structures can be much lighter than those built with conventional structural materials, which leads to reduction of dead load and ease of erection. Some of the research efforts and applications of FRP composites in bridge structures can be found in papers by Zureick *et al.* (1995), Seible *et al.* (1998), Tang and Hooks (2001), and Mufti *et al.* (2002).

14.2 Fiber-reinforced polymer (FRP) applications in bridge structures

Structural applications of FRP composites can be categorized into two large groups: rehabilitation and new construction. In a rehabilitation program, deficiencies of a structure are eliminated so that the structure can meet requirements of design codes as well as social and economic demands, and depending on its objective, the structural rehabilitation can be further divided into three groups: repair, strengthening, and retrofitting. Repairing a structure is to fix deficiencies by using FRP composites, so that the structure can regain its originally designed performance level. Strengthening a structure by applying FRP composites is to enhance the designed performance level. The retrofitting of a structure is to upgrade the seismic capacity of a structure by the use of FRP composites. For new construction, FRP composites can be used as reinforcing bars and tendons in concrete bridges, structural members of pedestrian bridges, deck and superstructure of vehicular bridges, bridge accessories such as inspection decks and manhole covers, etc. FRP applications in bridge structures are reviewed briefly in this section.

14.2.1 Rehabilitation

Repairing or strengthening beams, slabs, or bridge decks with FRP composite laminates is one of the most popular applications of FRP composites in bridge systems. There are many bridges where strengthening of the soffit of beams, slabs, or bridge decks is required. In some cases, concrete beams, slabs, and decks have deteriorated due to corrosion of steel reinforcement and freeze-thaw action. In other cases, structures have to be upgraded to bear higher load levels. Conventionally, either the external post-tensioning or the addition of epoxy bonded steel plates to the soffit is used for the repair and strengthening of concrete beams, slabs, and decks. Recently, externally-bonded FRP laminates have gained more popularity in this repair and strengthening procedure than the use of steel plates because FRP is corrosion resistant and very easy to handle.

A number of steel truss bridges in the US need either rehabilitation or replacement. As shown in Table 14.1, steel trusses are ranked No. 3 in the leading bridge maintenance priorities. Because a bridge replacement project can be very costly, a rehabilitation project is usually preferred if it is a viable option. In 2000, rehabilitation of a steel truss bridge was successfully completed by replacing the reinforced concrete deck with an FRP composite deck in the village of Wellsburg, NY (Alampalli and Kunin, 2001). This is the first project using this type of rehabilitation method. In this project, the application of an FRP composite deck reduced dead load of the bridge by 240 metric tons (the dead load became about one-fifth of the original dead load), and resulting in doubling the load ratings. The total cost of the project was $0.8 million, while it would have been $2.2 million to replace the entire bridge. In addition, a significant time reduction for the entire rehabilitation project was demonstrated.

In addition, FRP sheet bonding repair can be applied to repair corrosion-damaged steel bridges. In 2007, corrosion-damaged lower chord members of Asari Bridge, a three-span continuous steel truss bridge in Japan, was successfully repaired by bonding carbon fiber-reinforced polymer (CFRP) strand sheets.

For seismic retrofit, the most popular application of FRP composites is FRP composite jacketing of concrete columns. Recent earthquakes have shown the vulnerability of existing older concrete columns in bridges and buildings. Particularly, reinforced concrete bridge piers, designed before the 1971 San Fernando earthquake, are vulnerable because the transverse reinforcement is inadequate. Lessons learned from the 1971 San Fernando earthquake and the 1989 Loma Prieta earthquake led to the development of a steel jacket system for retrofitting those RC columns with substandard details. Although steel jackets work very well, they are costly because the installation is time-consuming, and they require constant maintenance. To speed up the installation of column jackets and reduce maintenance work, various column-jacketing techniques using FRP composites have been developed and successfully used in seismic retrofit projects.

14.2.2 Reinforcing bars and tendons

FRP reinforcing bars and tendons have been used for new concrete structures. Feasibility studies of FRP reinforcing bars began in the 1950s, and their applications gained much attention to prevent corrosion of steel reinforcing bars in the late 1980s (Bank, 2006). Currently, there is a design guide published by the American Concrete Institute (2006).

FRP composite tendons were implemented in a highway bridge deck in Germany in 1986, for the first time in the world. It has been over two decades since the first application, and the FRP tendons have widely been accepted in bridge construction. Pre- or post-tension tendons and stay cables made of FRP composites have an advantage over steel tendons and cables in their corrosion resistance. For these applications, aramid fiber-reinforced polymer (AFRP) and CFRP composites are often used due to their superior creep resistance. Moreover, FRP tendons and cables are lightweight and much easier to handle than those of steel.

14.2.3 Pedestrian bridges

The first pedestrian FRP composite bridge was built by the Israelis in 1975 (Tang, 1997). The first all-composite pedestrian bridge was installed in 1992 in Aberfeldy, Scotland (American Composites Manufacturers Association, 2004). It is a cable-stayed bridge with a total length of 113 m. Many others have been built in Asia, Europe, and North America since then. The development of pultruded structural shapes contributed to this increase in the number of FRP composite pedestrian bridges. Advantages of FRP composite pedestrian bridges over those made of conventional materials are lightweight, ease in installation, and low maintenance. These advantages made it possible to easily install pedestrian bridges in areas that are inaccessible by heavy construction equipment and environmentally restrictive. For this reason, many FRP composite pedestrian bridges can be found in State and National Parks in the US.

14.2.4 Decks

As shown in Table 14.1, bridge decks require the most maintenance of all elements in a bridge superstructure because the wearing surface deteriorates or wears out, and the deck system itself deteriorates due to de-icing salt and other causes. There has been a need for bridge decks made of new materials that can offer long-term durability and low life-cycle costs. To resolve these issues, FRP composite bridge decks have been developed and installed in many bridges. FRP composite bridge decks have several advantages over conventional bridge decks. First of all, because FRP composites are corrosion resistant, bridge decks made of FRP composites have lower life-cycle costs. Secondly, they are very easy to install due to their light weight. Thirdly, they can be prefabricated. This makes products uniform in quality and the installation time much shorter. Lastly, the reduction of deck weight is a major benefit to the bridge system. By reducing dead load, the bridge system with an FRP deck can either be used for higher live load levels or longer spans than a bridge system with a conventional bridge deck system. The use of an FRP bridge deck is reported to reduce the weight of conventional construction by 70–80% (Tang, 1997).

Many deck systems have been developed and tested since the early 1990s. The first US all composite vehicular bridge deck was built in 1996 in Russell, Kansas (Tang and Podolny, 1998), where the FRP deck panel is a sandwich construction of composite honeycomb and two composite face sheets. It took only one day to install the bridge deck in Russell County, showing how easy it is to install this type of modular FRP composite bridge deck.

One of the most popular types of FRP composite bridge deck is the one that has a sandwich structure. It consists of a core and top and bottom face sheets, and the core can be foam, honeycomb cells, or made of hexagon and double-trapezoid profiles. Other types of FRP composite decks have also been developed by making use of pultruded structural shapes (see, for example, Hayes et al., 2000).

Based on experiences with different types of FRP decks, maintenance issues peculiar to each type have been found. To have standard guidelines to inspect and evaluate the condition of existing FRP bridge decks, NCHRP conducted a research project and published an inspection manual as NCHRP Report 564 (Telang et al., 2006).

14.2.5 Accessories (such as inspection walkway and manhole cover)

Recently, accessories such as inspection walkways and manhole covers made of FRP composites have begun to be used. Corrosion of conventional steel inspection walkways in the bridges located in corrosive environments is always a problem. The use of FRP inspection walkway requires much less maintenance in such severely corrosive environments than conventional ones. In addition, lightweightness of FRP composites makes installation work of FRP inspection walkways much simpler without any heavy equipment and FRP manhole covers much easier to handle.

14.2.6 Superstructure

With the knowledge obtained from constructions of pedestrian bridges and bridge decks, many researchers and engineers have been trying to develop cost-effective all FRP composite vehicular bridges. Advantages of FRP composite vehicular bridges over bridges of conventional materials include corrosion resistance, ease of installation, reduction of construction period, less maintenance requirements, less dead load, a capability to have a longer span, and lower life-cycle costs.

Early examples of all FRP vehicular bridges include Laurel Lick Bridge, Tom's Creek Bridge, TECH21 Bridge, Kings Stormwater Channel Bridge.

- The Laurel Lick Bridge, with a span of 6.1 m, was built in 1997. The bridge has modular glass fiber-reinforced polymer (GFRP) decks on pultruded GFRP wide flange beams (Creative Pultrusions, Inc., 2002).
- The Tom's Creek Bridge, with a span of 5.33 m, was built in 1997. The bridge utilized pultruded hybrid FRP double-web beams as superstructure, and the FRP beam was composed of E-glass and carbon fibers in a vinylester matrix (Neely *et al.*, 2004).
- The Tech 21 Bridge, with a length of 10.1 m, was built in 1997. The bridge has GFRP decks of sandwich construction on three GFRP beams with trapezoidal cross section (Farheyl, 2005). The GFRP beams were composed of E-glass fibers and polyester resin.
- The Kings Stormwater Channel Bridge, with a length of 20.1 m, was built in 2000. The bridge has modular GFRP decks supported by concrete-filled filament-wound CFRP tubes (Seible *et al.*, 1999).

In spite of various advantages, there are several developments required to realize a successful design of a cost-effective all FRP composite vehicular bridge. Some of the areas needing more developments are:

- Although FRP composites have high specific strength, the stiffness has controlled the design rather than strength in almost all the demonstration FRP vehicular bridge projects where GFRP is used. It has been reported that the maximum stress level rarely exceeds 10% of the ultimate strength of the materials in these bridges (Karbhari and Zhao, 2000). To use the strength of the materials efficiently, more improvements and developments are needed on FRP composite bridge systems. Examples are structural optimization of the cross-sectional geometry, hybrid use of glass and carbon fibers, and hybrid use of FRP and conventional materials.
- FRP composites are susceptible to stress concentration due to their anisotropic material properties. Because FRP composites are not as ductile as steel, catastrophic failure may occur once the stress reaches the ultimate stress level. Therefore, better ways to connect FRP members

have to be developed in order to avoid these high stress concentrations. Some connection details of FRP members developed in the aerospace industry should be examined and transferred for bridge applications.
- Pultruded FRP composite structural shapes should be standardized. This will lead to achieving the uniform quality of structural components and reducing the initial costs. In addition, the standardization of structural shapes is necessary to facilitate the development of design guidelines for FRP composite bridges.
- Environmental durability of FRP composite materials has to be thoroughly investigated and well understood. The experimental and field data on the long-tem durability of FRP composites will facilitate the development of a design philosophy based on life-cycle costs. In addition, long-term performance data of FRP bridges should be accumulated so that proper maintenance procedures can be standardized.
- Initial costs of FRP composite bridges are still too expensive to compete with bridges of other conventional materials. Standardization of structural shapes and design with efficient use of materials are required to reduce the initial costs.

14.3 Hybrid fiber-reinforced polymer (FRP)–concrete bridge superstructure

14.3.1 Hybrid FRP–concrete structural system

To resolve some issues that all FRP composite bridges have, combinations of FRP and conventional materials have recently been investigated by a number of researchers. A good review on the combined or hybrid construction can be found in the paper by Mirmiran (2001). The advantages of the hybrid structural systems include the cost effectiveness and the ability to optimize the cross section based on material properties of each component. According to Mirmiran (2001), the most effective use of FRP composites is in the form of hybrid construction with concrete, where FRP acts as a load-carrying constituent and a protective measure for concrete.

The innovative idea of a hybrid FRP–concrete structural system for flexural members was first proposed by Hillman and Murray (1990). They proposed the combination of pultruded FRP sections and concrete to form lightweight decks in order to reduce the dead load in steel frame buildings. The FRP section was designed to serve as both reinforcement and permanent formwork. Most of the concrete was located above the neutral axis of the hybrid section. The study concluded that the weight reduction would be more than 50% when compared with a common type of concrete slab system.

Bakeri and Sunder (1990) investigated hybrid FRP–concrete bridge deck systems. They proposed a deck system of a simply curved membrane of FRP composites filled with concrete that was intended to resist the compressive force. A finite element analysis was performed to evaluate the mechanical performance of the deck system under an HS20-44 truck loading. They concluded that the hybrid FRP–concrete system was promising, particularly from the viewpoint of cost.

Saiidi *et al.* (1994) conducted experimental and analytical studies on composite beams that consist of CFRP structural sections and reinforced concrete slabs. Although in their study an epoxy resin was used to provide the bond between the concrete deck and the CFRP sections, they concluded that the use of epoxy resin to bond concrete to CFRP was only partially efficient, and that mechanical connectors or other reliable means would be needed to develop a good composite action between CFRP elements and a concrete slab.

Deskovic *et al.* (1995a) presented an innovative design of a hybrid FRP–concrete beam. The proposed beam consists of a filament-wound GFRP box section combined with a layer of concrete in the compression zone and a thin CFRP laminate in the tension zone. Three-point bending test results confirmed that the proposed design was feasible for producing an efficient and cost-effective hybrid system. They also studied long-term behavior (creep and fatigue) of the proposed hybrid section experimentally and analytically (Deskovic *et al.*, 1995b). Their analytical model agreed closely with experimental results. They concluded that the proposed hybrid section had very good time-dependent response characteristics.

Seible *et al.* (1998) designed a two-span, 210 m long, two-lane highway bridge with lightweight concrete-filled circular CFRP composite tubes. Their study showed that the design is stiffness-driven for this modular beam and slab bridge system, and that significant strength reserve remains in the carbon shell. Their preliminary estimates indicate that two different bridge systems, the concrete-filled CFRP beams with RC deck and the concrete-filled CFRP beams with pultruded modular E-glass deck, are 20% and 100% more expensive, respectively, when compared to a conventional RC slab bridge.

Ribeiro *et al.* (2001) investigated the flexural performance of hybrid FRP–concrete beams. In their design, GFRP pultruded channel profiles were assembled with a layer of concrete. Four types of hybrid rectangular beams were tested in the four-point bending. Two of them had all the section filled with concrete, while the other two had concrete only in the upper part of the profile. To compare the flexural performance of each assembly, they used two indices: flexural specific rigidity and synergistic effect. Flexural specific rigidity is a ratio between the flexural rigidity and the specific weight, while synergistic effect is a ratio of the ultimate load of the assembly to the sum of the ultimate loads of its two elements (GFRP channel and concrete). Test results showed that the flexural specific rigidity was the highest for the specimen that has more GFRP in the tension side and concrete only in the compression side. The synergistic effect was the highest for the specimen that did not exhibit a bond failure at the FRP–concrete interface. No special measure to provide the bond at the interface was taken in any specimen.

Fam and Rizkalla (2002) studied flexural behavior of concrete-filled FRP circular tubes experimentally. This hybrid system was developed as an excellent alternative for structural components subjected to aggressive corrosive environments. The FRP tube provides a lightweight permanent formwork for fresh concrete and acts as a non-corrosive reinforcement. The concrete core has two main functions: (1) to provide internal support to the tube and, consequently, prevent the local buckling of the FRP tube; and (2) to provide the internal resistance force in the compression zone and increase the strength and stiffness of the member. A total of 20 beams including steel tubes were tested in four-point bending. GFRP tubes were either filament-wound or pultruded. Some of their conclusions were: (1) the higher the stiffness of the hollow tube, the lower the gain in flexural strength and stiffness resulting from concrete filling; (2) although concrete-filled pultruded GFRP tubes showed higher stiffness than concrete-filled filament-wound GFRP tubes of the same thickness, they failed prematurely by horizontal shear due to the lack of fibers in the hoop direction; (3) concrete-filled FRP tubes with thicker walls or a higher percentage of fibers in the axial direction tended to fail in compression; (4) a higher flexural strength-to-weight ratio was achieved by providing a central hole in the core; (5) a shear transfer mechanism is necessary at the interface between the concrete core and the GFRP tube in flexural members; and (6) experimental results did not show a significant effect of the confinement of concrete on flexural strength, while ductility was improved.

Van Erp and his group (2002, 2005) have been developing hybrid FRP–concrete beams for a bridge application. The proposed beam is very similar to the one proposed by Deskovic *et al.* (1995a). The beam consists of two parts: concrete and FRP box section. The FRP box section is made of GFRP, and additional CFRP is added to the tensile flange of the box section to increase the stiffness of the beam. A layer of concrete is bonded on the GFRP box section by a high-quality epoxy adhesive. The weight of the hybrid beam is claimed to be about one-third that of a reinforced concrete beam. The top GFRP flange was designed to sustain more compression force after concrete is crushed. Concrete crushing is a warning of failure in this design. A full-scale bridge superstructure with a span of 10 m and a width of 5 m was designed and built with these hybrid beams. The

Fiber-reinforced polymer composites for bridge superstructures

beams were assembled using a high-quality epoxy adhesive to make up the bridge. The bridge was installed in Toowoomba, Australia, in January 2002. Field test results demonstrated high load-carrying capacity, excellent fatigue behavior, outstanding durability, and ability to carry high concentrated loads.

14.3.2 Design concept of the hybrid FRP–concrete bridge superstructure

By utilizing the concept of the hybrid FRP–concrete structural system, an FRP bridge superstructure has been developed at the University at Buffalo for the past ten years. Although many different configurations of hybrid FRP–concrete bridge superstructure are possible, the structural configuration shown in Fig. 14.3(a), where supports are not shown, was chosen based on numerous finite element analyses leading to the particular geometrical parameters. In combination, glass fiber reinforcement, and vinyl ester matrix were selected as constituents for the GFRP composites, because glass fiber reinforcement is much cheaper than carbon or aramid fiber reinforcement, and vinyl ester has high resistance to corrosion. Figure 14.3(b) shows a cross section of the proposed bridge superstructure. In this figure, parts in black are made of GFRP laminates, while the shaded portion

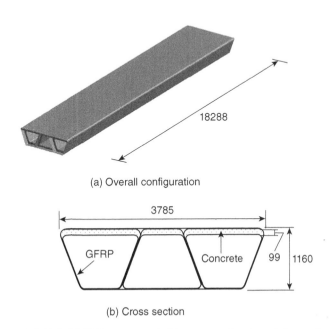

14.3 Hybrid FRP–concrete bridge superstructure (dimensions in mm).

is filled with concrete. In this design, three trapezoidal box sections are bonded together to form a one-lane superstructure. A trapezoidal shape was chosen so that forces could transfer between box sections efficiently. After being assembled, these trapezoidal sections are wrapped with a GFRP laminate, which ensures that three sections act together to produce an integral bridge superstructure.

Advantages of this bridge superstructure are summarized as follows:

- GFRP is corrosion-resistant, and concrete is not exposed to environmental conditions; therefore, the system is resistant to corrosion, and it requires less maintenance than conventional bridges.
- Concrete has high strength- and stiffness-to-cost ratio; thus, by using concrete efficiently, the total amount of FRP can be reduced, which leads to the reduction of initial costs.
- Concrete is designed to be always under compression in the longitudinal direction. The fact that concrete is not used in the tension side leads to significant weight reduction when compared to a concrete-filled FRP tube design.
- It has been reported that the local deformation under a loading point may become large for all-composite bridge decks (Bakeri and Sunder, 1990; Aref, 1997). A layer of concrete can reduce this local deformation of the top flange.
- Most parts can be fabricated in the factory; therefore, good quality control can be assured and the construction period can be shortened.

14.3.3 Design features

As the first trial design, a prototype bridge was designed as a simply supported single-span, one-lane bridge with a span of 18.3 m. Design parameters were determined to meet a set of design criteria by applying AASHTO design loads.

Design philosophy and assumptions

The design philosophy and assumptions used in the design process are as follows:

- In flexure, concrete should fail in compression first before GFRP laminates fail in either compression or tension.
- Under the service limit state condition, concrete does not crack due to the bending in the transverse direction.
- Under the strength limit condition, the top flange of the box section should resist the tire pressure with GFRP laminates only for the bending in the transverse direction if the concrete is already cracked.

- A perfect bonding between concrete and GFRP laminates is assumed.
- Confinement effect on the concrete strength is neglected.
- A knock-down factor or strength reduction factor for the GFRP is taken as 0.4. Seible *et al.* (1995) used 0.5 and Dumlao *et al.* (1996) used 0.25 in their designs. The value of 0.4 was chosen simply as a value between 0.25 and 0.5.
- Long-term degradation factor for FRP composites is not considered because the data are not available for the particular material used in this study. Typical values of degradation factors can be found in various sources. For example, see *ICE Design and Practice Guides* (Moy, 2001).

Box section

Box sections were chosen to be a basic structural form of the proposed design. According to Ashby (1991), thin-walled box sections are the most efficient structural forms for beams. However, the thin-walled box sections made of GFRP laminates have some disadvantages (Deskovic *et al.*, 1995a), which include:

- the compressive flange is considerably weaker than the tensile flange due to the local buckling failure;
- the failure of GFRP usually occurs in a catastrophic manner without giving much warning because the stress–strain curve of GFRP in the fiber direction does not show plastic deformation as much as conventional materials such as steel and concrete do;
- the design of a GFRP beam is usually governed by stiffness instead of strength; this is often resolved by introducing more material in the beam; therefore, the design tends to be uneconomical and the strength of the material is not used efficiently.

To overcome these disadvantages, a thin layer of concrete was decided to be placed in the compression zone of the section as shown in Fig. 14.3(b). The section was designed in such a way that concrete is surrounded with GFRP, and concrete will be protected from environmental exposure.

Web inclination

Another feature of the design is to have trapezoidal box sections, although it is more common to have a square or rectangular shape for a box section. When many rectangular box sections are put together to comprise a superstructure, the force will be transferred between the adjacent sections mainly through shear and bending. However, FRP composites usually have very low shear stiffness when compared to their normal stiffness in the fiber directions. By having the webs inclined, the structural shape of the cross section becomes more of a truss, and a greater portion of the applied load will be carried by axial force in FRP laminates within the cross section.

The inclination angle was determined based on finite element analyses of hybrid bridge superstructures with several different inclination angles subjected to live loads. The inclination of 3/8 (=21°) was chosen because that yielded the minimum deformation at the riding surface.

Stacking sequences

In the design process, it was assumed that the GFRP parts would be fabricated by a hand layup process. The inner trapezoidal tube is fabricated first, and the outer tube laminate is laid up onto the inner tube. After a certain number of the trapezoidal tubes are put together, they are wrapped with the outermost laminate. These three different laminates are shown in Fig. 14.4. Considering that the hand layup process is time-consuming, it is best to make the stacking sequences of these three laminates as simple as possible. A woven fabric was chosen as reinforcement; therefore, a layer of fabric has reinforcement fibers in the two orthogonal directions. The stacking sequence chosen for the outer tube and inner tube laminates was all 0°, while for the outer tube laminate, ±45° and 0° laminae were used for 67% and 33% of the thickness, respectively. The direction is measured from the longitudinal direction of the bridge. Because the reinforcement type is of a woven fabric, a 0° lamina has fibers in the 0° and 90° directions. By introducing ±45° laminae in the outer tube laminate, the shear stiffness of the hybrid bridge can be increased as the outer tube laminates make a large portion of the interior webs.

Thickness of the concrete layer

The thickness of the concrete layer is a key design parameter to optimize the hybrid FRP–concrete structural system because the increase in the thickness of concrete leads not only to a reduction in initial costs by reduc-

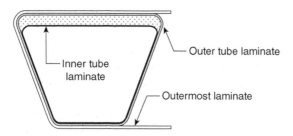

14.4 Three different laminates in the superstructure.

ing GFRP composites but also to an increase in the total weight. By examining flexural rigidity, it was found that the flexural rigidity rapidly increases with the concrete thickness until the concrete thickness is about 10% of the superstructure depth. After that point, additional concrete may not be used efficiently to increase the flexural rigidity, although the cracking and ultimate moments will still get benefits from the increase in the concrete thickness. Therefore, the maximum thickness of concrete that will increase flexural rigidity efficiently without adding too much weight was determined as 10% of the bridge depth for the proposed hybrid bridge. In the final design, the thickness was chosen as 99 mm, which is 8.5% of the total depth of the bridge. Thus, the ratio of the area of concrete to the total area in the cross section comes to 0.3.

Shear keys

To have good composite action between GFRP laminates and concrete, GFRP shear keys, as shown in Fig. 14.5, were designed. They are to be installed in staggered positions on the top and bottom laminates with a required interval.

14.3.4 Experimental study

To verify the feasibility of the proposed hybrid FRP–concrete bridge superstructure by examining the performance under live loads, a series of quasi-static loading tests and a fatigue loading test were performed on a test specimen of the proposed bridge superstructure.

Test specimen

The test specimen is a one-fifth scale model of the 18.3 m hybrid FRP–concrete bridge superstructure shown in Fig. 14.3. Its cross section is shown in Fig. 14.6. Each trapezoidal box section was fabricated individually by the

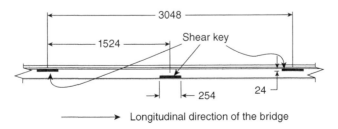

14.5 Shear keys (dimensions in mm).

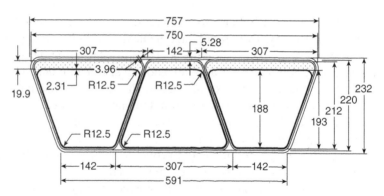

14.6 Cross section of the test specimen (dimensions in mm).

hand layup process. Each section consists of two layers of laminates: the inner tube laminate and the outer tube laminate. The inner tube laminate was first constructed with a $[0°_7]$ laminate construction, and the outer tube laminate was constructed over the inner tube laminate with a $[(\pm 45)_4 0°_4]$ laminate construction. At the same time, a cavity for concrete was created at the top of each section. Then, three trapezoidal sections were assembled together. A layer of glass fiber chopped strand mat wetted with vinyl ester resin was applied between box sections as a bonding material. After being assembled, the three sections were wrapped with the outermost laminate whose stacking sequence is $[0°_{16}]$.

Materials

E-glass woven fabric was chosen as reinforcement of FRP, and vinylester resin was chosen as the matrix. To determine the mechanical properties of an FRP laminate made of E-glass woven fabric and vinyl ester resin, tensile, compressive, and in-plane shear tests were conducted according to ASTM D3039-76, ASTM D3410-75, and ASTM D4255/D4255M-83, respectively. Tensile and shear coupons were cut from a 10-layer laminate, while compressive coupons were cut from a 20-layer laminate. These laminates were made by a hand layup process. Fiber volume fractions of the 10- and 20-layer laminates were 0.295 and 0.355, respectively.

The material properties based on testing are shown in Table 14.3. Modulus of elasticity and Poisson's ratio were determined for a strain range of 0.001 to 0.003, –0.005 to –0.007, and 0.001 to 0.005, for tensile, compressive, and shear tests, respectively. Stress–strain relationships of tension and compression were slightly nonlinear, while that of shear was highly nonlinear. Ultimate shear strain was not obtained because it was beyond the capacity of the strain gages used in the test.

Table 14.3 GFRP material properties

Test type	Direction	Modulus of elasticity (GPa)	Strength (MPa)	Ultimate strain
Tension	Fill	16.6	285	0.0218
	Warp	17.9	335	0.0235
	Average	17.3	310	0.0227
Compression	Fill	15.9	241	0.0177
	Warp	22.5	265	0.0158
	Average	19.2	253	0.0168
Shear	Fill	2.72	56.1	—
	Warp	2.45	63.8	—
	Average	2.59	60.0	—

As cavities for concrete in the test specimen were small, coarse aggregate was not used. The maximum aggregate size was 4.75 mm, and Type I Portland cement was used. The weight proportions of constituent materials in the concrete mix were: 9.4% water; 20% cement; 70% fine aggregate; 0.11% superplasticizer; and 0.39% shrinkage reducer. Four cylindrical specimens of 152.4 mm in diameter and 304.8 mm in length were prepared according to ASTM C192/C192M-98. Specimens were moist cured until an age of 28 days after the mixing of concrete. Compressive test of cylindrical specimens was performed after 28 days' cure according to ASTM C39-96. The obtained compressive Young's modulus and strength were 8.38 GPa and 37.9 MPa, respectively. The Young's modulus is about one-third that of normal concrete due to the lack of coarse aggregate.

Test setup

Figure 14.7 shows the test setup. Loads are applied vertically to the top surface of the test specimen by the actuator hanging from the top beam of a reaction frame. The specimen is instrumented with potentiometers and strain gages at various locations to measure displacement and strain, respectively. Spreader beams spread the load from the actuator to four contact points. The load configuration simulates the tandem load specified in the *AASHTO LRFD Bridge Design Specifications* (AASHTO, 1998). The design tandem load is a live load that has two axles of 110 kN, where one axle is 1,200 mm away from the other. Each axle has two tires that are 1,800 mm apart center-to-center, and each tire area is 510 mm long and 250 mm wide. For the one-fifth scale model, this design tandem load becomes two axles of 4.4 kN, 240 mm apart. Two tires of each axle are 360 mm apart and each tire area is 102 mm long and 50 mm wide.

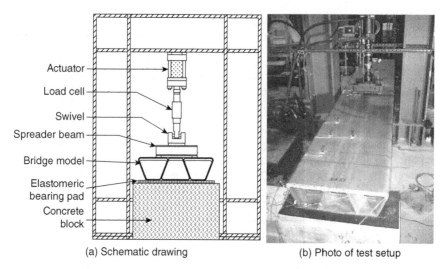

(a) Schematic drawing (b) Photo of test setup

14.7 Test setup.

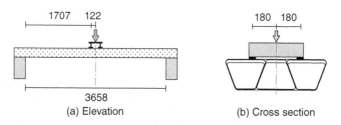

(a) Elevation (b) Cross section

14.8 Loading configuration (dimensions in mm).

The test specimen is supported by concrete blocks at its two ends. To protect the bottom surface of the specimen from damage and to allow rotation at the supports, elastomeric bearing pads are placed on the concrete blocks, and the specimen sits on the pads. Each elastomeric bearing pad is 13 mm thick and made of neoprene.

Test procedures

A series of non-destructive tests were first performed on the test specimen to obtain its structural characteristics in the elastic range under flexural loading. Then, the capacity of the specimen was investigated by destructive tests. Loading configuration for the static and fatigue loading is shown in Fig. 14.8. Displacement and strain measurement locations during the experiment are shown in Fig. 14.9.

Fiber-reinforced polymer composites for bridge superstructures

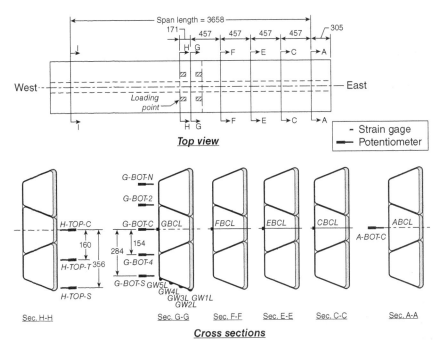

14.9 Instrumentation layout (dimensions in mm).

Before concrete was cast into the specimen, the FRP-only specimen was tested in flexure in order to examine the behavior without concrete. The test was performed by displacement control, and the maximum displacement applied was $L/480$ (7.6 mm), where L is a span length. The hybrid FRP–concrete bridge specimen was then tested in flexure up to the maximum displacement of $L/480$. Then the test specimen was subjected to 2×10^6 load cycles in flexure. This fatigue test was performed by force control with a load range of 0–17.6 kN and a frequency of 3.0 Hz. The maximum load of 17.6 kN is equivalent to twice the design tandem load for this test specimen. The test specimen was subjected to a static flexural loading every 2×10^5 cycles to obtain its stiffness and examine stiffness degradation.

After the fatigue test, the bridge model was tested in flexure to failure to examine its residual strength and failure modes. This test was performed by displacement control, and it was divided into two steps. In the first step (Step I), a cyclic displacement profile was used with the amplitude gradually increased. For each cycle, the displacement ranged from 0 to a predetermined amplitude, and three cycles were applied for each amplitude.

Although it was planned to repeat this process until failure of the bridge model, the test was stopped at the maximum displacement of 77.7 mm (17 × L/800) because the maximum force became close to the capacity of the load cell. By using a load cell with a higher capacity, the second step (Step II) of the test was performed. In this step, displacement was increased monotonically until the test specimen failed.

Test results

No sound of cracking of either the concrete or GFRP was heard during the non-destructive flexural loading where the applied maximum displacement is L/480. The obtained force–displacement responses were very much linear as shown in Fig. 14.10. Stiffness of the hybrid bridge specimen was 19% higher than that of the FRP-only specimen, which shows the effectiveness of concrete. For the prototype bridge, the stiffness increase resulting from the inclusion of concrete is expected to be as much as 40%, because Young's modulus of normal concrete to be used in the prototype bridge will be about three times as high as that of the concrete used in the test specimen. Figure 14.11 shows the deformed shapes of the top and bottom surfaces, respectively. It can be observed from the figure that the three box shapes deformed as a unit and there was not significant local deformation even near the loading points.

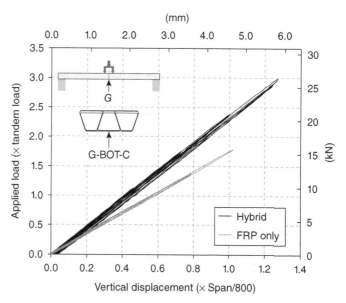

14.10 Force–displacement relationship at G-BOT-C.

Fiber-reinforced polymer composites for bridge superstructures

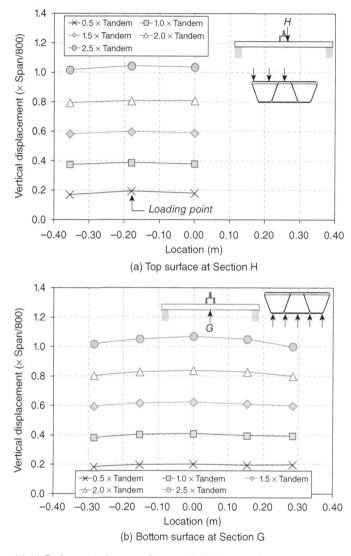

14.11 Deformed shapes of top and bottom surfaces.

It is recommended in the *AASHTO LRFD Bridge Design Specifications* that the maximum deflection under live loads be smaller than $L/800$ (L = span length). The live load for this check is $(1 + IM) \times$ truck load, where IM = dynamic load allowance, and has a value of 0.33 in this case. In this study, the condition was checked with the tandem load instead of the truck load. As the span length of the test specimen is 3.66 m, the maximum deflection under $(1 + IM) \times$ tandem load has to be smaller than $L/800 = 4.6$ mm.

The obtained deflection of the hybrid specimen due to $(1 + IM) \times$ tandem load was $0.547 \times L/800$. The hybrid specimen well satisfied the AASHTO live load deflection recommendation.

Variations of longitudinal strains over the height of the superstructure on the exterior web at the midspan of the specimen are shown in Fig. 14.12. Longitudinal strains on the web varied linearly, implying that the bending theory for small strain can apply, i.e., the plane section before deformation remains plane after deformation.

Figure 14.13 shows stiffness degradation over 2×10^6 cycles in the fatigue test. The stiffness degradation is defined as a ratio of the stiffness measured after a certain number of load cycles to the stiffness measured in the last cycle of the non-destructive flexural test. Although the trend of stiffness degradation is not very clear in this figure, it can be concluded that stiffness degradation was insignificant. The degradation was 5.9% after two million cycles.

Figure 14.14 shows force–displacement responses obtained at H-TOP-T and G-BOT-C during the destructive flexural test (Steps I and II). At the load of $19.1 \times$ tandem load in Step I, there was a loud cracking sound from concrete under the loading points. This is a local failure mode. The failure load was 8.2 times the AASHTO requirement of $1.75(1 + IM) \times$ tandem load for live loads in the Strength Limit I State. After concrete cracked under the loading points, the bottom flanges of the spreader beams came in contact with the top surface of the specimen, and the loading

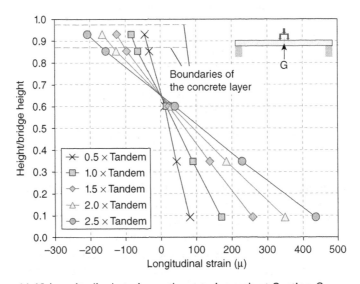

14.12 Longitudinal strain on the exterior web at Section G.

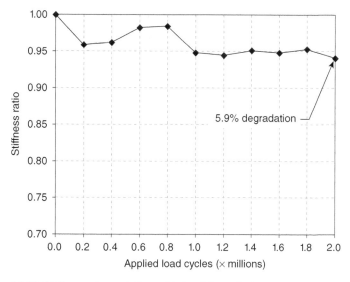

14.13 Stiffness degradation obtained from fatigue test.

configuration became two line loads instead of four point loads. Therefore, further local deformation was not observed at H-TOP-T.

The load path from Step II shows that the reloading path is very much linear until the maximum load experienced before. A global failure occurred at 35 × tandem load, which is 15 times the 1.75(1 + IM) × tandem load. A sequence of failure leading to global failure can be described as follows. Concrete failed in compression first, then the GFRP flanges took over the compressive force that had originally been carried by concrete. GFRP compression flanges failed in compression when the compressive stress reaches the compressive strength, followed by failure of the webs in compression. The bottom GFRP flange was found to be intact. Although the failure was sudden due to the nature of GFRP composites, the obtained global failure mode can be considered to be favorable because it did not lead to collapse of the entire bridge. At the failure section, crushing of concrete, compressive failure of the GFRP top flange, buckling failure of the GFRP interior flange, and significant delamination of the GFRP laminates were observed.

After the strength test, the test specimen was cut at a few sections that were away from the damaged section (Section H) to investigate the interface between the GFRP and concrete. Visual inspection did not find any trace of slippage between the GFRP and concrete.

The experimental study demonstrated excellent performance of the proposed hybrid FRP–concrete bridge superstructure. As is often the case

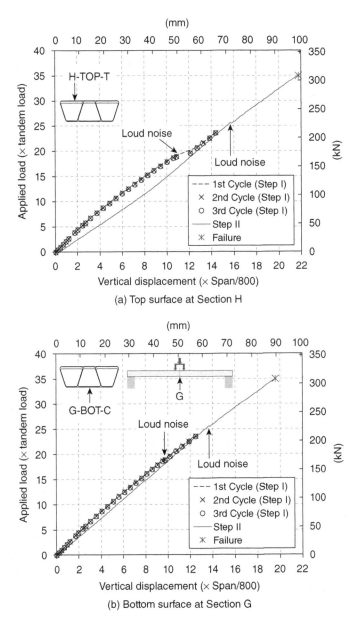

14.14 Force–displacement relationships in the destructive flexural test.

with many GFRP composite bridges, stiffness governs the design of the proposed bridge superstructure. Clearly, the proposed design can be designed to meet the AASHTO live load deflection recommendation. Also, the fatigue test revealed that the stiffness degradation was not significant over 2×10^6 cycles of the load equivalent to twice the tandem load. However, the test results showed that the proposed design has much higher strength than it is required to; i.e., the design was conservative. The depth of the bridge may be increased so that the same flexural rigidity can be achieved with less material and the strength can be used more efficiently.

14.3.5 Structural analysis

In this section, detailed finite element analysis (FEA) of the hybrid FRP–concrete bridge superstructure and simple methods of analysis are introduced, and their applicability for the proposed hybrid superstructure is discussed.

Linear finite element analysis

For the detailed FEA of the hybrid FRP–concrete bridge superstructure, the general purpose commercial finite element analysis software, ABAQUS ver. 6.1 (Hibbit, Karlsson & Sorensen, Inc., 2000), was used. A four-noded general shell element was used for GFRP laminates, while a general 3D eight-noded solid element was used for concrete. Figure 14.15 shows the finite element model for the one-lane hybrid FRP–concrete bridge superstructure. The longitudinal, transverse, and vertical directions are referred to as the x, y, and z directions, respectively. In this model, the total number of elements is 38,892 (22,764 shell elements and 16,128 solid elements), and the total number of nodes is 31,857.

The FEA model was supported by a line at each end. Boundary conditions were imposed to two lines of nodes of the bottom surface. Nodes at $y = 0$ and $z = 0$ were restrained in the y and z directions, and nodes at $y = $

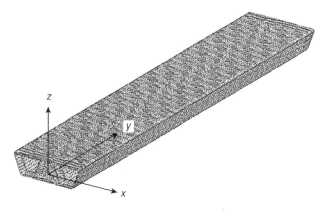

14.15 Finite element mesh.

L and $z = 0$ were restrained in the z direction. In addition, nodes at $x = 0$, $z = 0$, and $y = 0$ and L were restrained in the x direction, too.

Material properties of GFRP and concrete obtained from the experimental study presented in section 14.3.4 were used in the FEA. Although the obtained tensile and compressive properties are different to some degree, they are assumed to be the same in the analysis. The experimental study showed that GFRP almost behaves as a linear-elastic material until failure and that concrete cracking or compressive failure occurs at a much higher load level than the design load level. Therefore, only the linearly elastic behavior of the hybrid bridge superstructure is examined in this section, and only elastic properties are used in the FEA. A perfect bonding between concrete and GFRP laminates was assumed in the analysis.

Simple methods of analysis

Because a detailed finite element analysis for every trial design is very time-consuming and not cost-effective to select the bridge design parameters in a preliminary design phase, it is necessary to have a simple and quick method to analyze the bridge behavior under various loading conditions. The beam and orthotropic plate analyses are presented here as the simple methods of analysis. It is assumed that the estimation of deflection will be the primary objective of the simplified analysis because the design of the hybrid bridge superstructure is controlled by deflection. To model the bridge superstructure as a beam or a plate, cross-sectional properties are the key parameters. The procedures to evaluate the cross-sectional properties are presented based on the classical lamination theory.

It is assumed in the beam analysis that the cross section in the xz plane will not deform. The bridge superstructure is modeled as a beam with span length, L, effective flexural rigidity, EI_{eff}, and effective torsional rigidity, GJ_{eff}. The evaluation of EI_{eff} and GJ_{eff} is the main task of modeling the bridge superstructure as a beam. Equation [14.1] calculates the effective bending rigidity.

$$EI_{eff} = \int_{A_y} \bar{E}_y \bar{z}^2 dA \qquad [14.1]$$

where $\bar{E}_y$ is the effective modulus for laminates in the y direction or Young's modulus for the concrete; $\bar{z}$ is the vertical coordinate taken from the location of the neutral axis; and A_y is the cross-sectional area in the xz plane.

The effective engineering properties of a laminate were obtained based on the classical lamination theory. The effective engineering properties of different laminates are given in Table 14.4, and names of different laminates are found in Fig. 14.4.

The effective torsional rigidity, GJ_{eff}, is evaluated by considering the shear flow around the box section. Here, the following assumptions are made:

Table 14.4 Effective engineering properties of different laminates (units in GPa)

Laminate	$\bar{E}_x$	$\bar{E}_y$	$\bar{G}_{xy}$
Outermost	18.6	18.6	2.94
Outer tube	13.1	13.1	6.47
Inner tube	18.6	18.6	2.94

- the net shear flow through interior webs is negligible and the shear flow through the top and bottom flanges and exterior webs is of prime significance;
- the magnitude of the shear flow does not change along the top and bottom flanges and the exterior webs;
- the shear flow goes through the median line of each member;
- the top flange consists of the outermost laminate, the outer tube laminate, the concrete layer, and the inner tube laminate; and the top flange extends over the bridge width without any interruptions by the interior webs;
- the bottom flange consists of the outermost laminate, the outer tube laminate, and the inner tube laminate; and the bottom flange extends over the bridge width without any interruptions by the interior webs.

Based on these assumptions, the effective torsional rigidity may be approximated by using Bredt's formula:

$$GJ_{eff} = \frac{4 A_{encl}^2}{\int \frac{1}{\sum_k \bar{G}_{xy_k} t_k} ds} \qquad [14.2]$$

where A_{encl} is the area enclosed by the median lines of the top and bottom flanges and the exterior webs; $\bar{G}_{xy_k}$ is the effective shear modulus of the kth layer in a member; t_k is the thickness of the kth layer in a member; and s is the axis along the median line of each member.

In Eq. [14.2], the summation is carried out over layers in a member; for instance, there are four layers in the top flange: the outermost laminate, the outer tube laminate, the concrete layer, and the inner tube laminate. The integration is carried out along the median line of each member.

In the orthotropic plate analysis, the hybrid bridge is modeled as an orthotropic plate with two opposite edges simply-supported as shown in Fig. 14.16. By assuming some of the bending extension coupling stiffnesses are insignificant, the governing equations for the orthotropic plate model

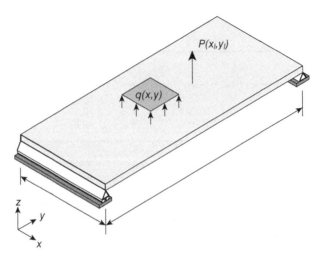

14.16 Bridge model for the orthotropic plate analysis.

of the hybrid bridge superstructure can be expressed as follows (Aref *et al.*, 2005):

$$D_x \frac{\partial^4 w_o}{\partial x^4} + 2D_t \frac{\partial^4 w_o}{\partial x^2 \partial y^2} + D_y \frac{\partial^4 w_o}{\partial y^4} = q(x, y) \quad [14.3]$$

where the following notations are used:

$$D_x = D_{11}^x, \quad D_y = D_{22}^y, \quad D_t = \frac{D_{xy} + D_{yx} + D_1 + D_2}{2},$$

$$D_1 = D_{12}^x, \quad D_2 = D_{12}^y, \quad D_{xy} = 2D_{66}^x, \quad \text{and} \quad D_{yx} = 2D_{66}^y \quad [14.4]$$

and w_o is vertical deflection, and D_{ij}^x and D_{ij}^y are 3 × 3 bending stiffness matrices evaluated by the classical lamination theory for the transverse direction and the longitudinal direction, respectively. These bending stiffness matrices can be evaluated by choosing representative units for the longitudinal and transverse directions as shown in Fig. 14.17. The solutions of Eq. [14.3] can be found in Cusens and Pama (1975).

Table 14.5 compares deflections obtained from the three different analytical methods and the experiment at the center of the test specimen of the bridge superstructure subjected to the flexural loading of twice the tandem load. Figure 14.18 compares deformed shapes from the three different analytical methods and the experiment at twice the tandem load. As can be seen in the table and figure, the predictions by the beam and plate analyses are very close to those by the detailed FEA with the maximum difference being only about 1.0%. Therefore, in this particular case, either the beam or the orthotropic plate analysis can be a very good tool to

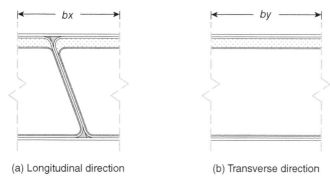

(a) Longitudinal direction   (b) Transverse direction

14.17 Representative units for the orthotropic plate analysis.

Table 14.5 Deflection at midspan under twice the tandem load

	Deflection		Difference from FEA
	(mm)	(L/800)	(%)
Linear FEA	3.89	0.850	−1.04
Beam analysis	3.85	0.842	0.687
Orthotropic plate analysis	3.92	0.856	0.706
Experiment	3.88	0.849	−0.135

estimate the overall bridge behavior under different loading conditions. However, when the bridge superstructure becomes wider, a variation of deflection in the transverse direction will get larger. To predict the maximum deflection of the hybrid bridge with more than three lanes, it is recommended to use the orthotropic plate analysis instead of the beam analysis.

14.3.6 Practical issues

Judging from results from the experimental and numerical study, the proposed hybrid FRP–concrete superstructure is highly feasible from a structural engineering point of view. To fully develop the superstructure and apply the concept in an actual bridge project, the following issues will have to be investigated in the future:

- automated fabrication process for trapezoidal sections,
- methods to expand traffic lanes,
- efficient methods to cast concrete in the field,
- design details to make better use of strength,
- design for the section to resist negative moments in a continuous girder,

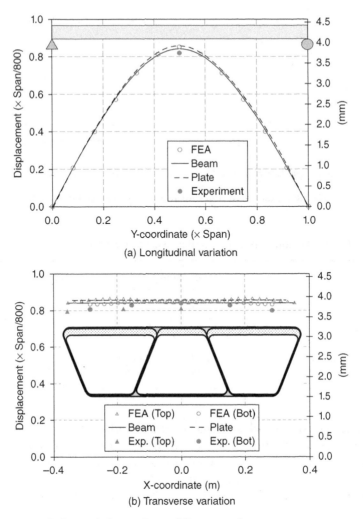

14.18 Deformed shapes from different analyses.

- installation method of guard fences,
- response to thermal effects,
- long-term performance degradation and maintenance requirements, and
- life cycle costs.

14.4 Conclusion

The concept of the hybrid design is a very prudent way to design a structure because different materials can be used efficiently where they perform best. In this chapter, the hybrid FRP–concrete bridge superstructure is introduced, where the novel concept of the hybrid FRP–concrete design is applied. The trial design of the bridge superstructure was proposed, and its structural performance was investigated through the detailed FEA and a series of static and fatigue loading tests. In addition, simple methods of analysis for this hybrid bridge superstructure were also proposed. Results from the research to date showed that the proposed hybrid bridge superstructure has excellent structural performance, and the hybrid FRP–concrete system for the bridge superstructure is highly feasible from a structural engineering point of view.

14.5 References

Alampalli, S. and Kunin, J. (2001), *Load Testing of an FRP Bridge Deck on a Truss Bridge*, Special Report 137, New York State Department of Transportation.

American Association of State Highway and Transportation Officials (AASHTO) (1998), *AASHTO LRFD Bridge Design Specifications*, 2nd edn, AASHTO, Washington, DC.

American Composites Manufacturers Association (2004), *Industry Overview: Overview of the FRP Composites Industry – A Historical Perspective*, http://www.mdacomposites.org/mda/overview.html (accessed April 30, 2012).

American Concrete Institute (2006), *Guide for the Design and Construction of Structural Concrete Reinforced with FRP Bars*, ACI 440.1R-06, American Concrete Institute, Farmington Hills, MI.

Aref, A. J. (1997), 'A novel fiber reinforced composite bridge structural system', PhD dissertation, the University of Illinois at Urbana-Champaign.

Aref, A. J., Kitane, Y., and Lee, G. C. (2005), 'Analysis of hybrid FRP–concrete multi-cell bridge superstructure,' *Composite Structures*, 69(3), 346–359.

Ashby, M. F. (1991), 'Overview No. 92 – materials and shape,' *Acta Metallurgica et Materialia*, 39(6), 1025–1039.

Bakeri, P. A. and Sunder, S. S. (1990), 'Concepts for hybrid FRP bridge deck systems,' *Serviceability and Durability of Construction Materials*, Proceedings of the First Materials Engineering Congress, Denver, CO, August 13–15, ASCE, Vol. 2, pp. 1006–1015.

Bakis, C. E., Bank, L. C., Brown, V. L., Consenza, E., Davalos, J. F., Lesko, J. J., Machida, A., Rizkalla, S. H., and Triantafillou, T. C. (2002), 'Fiber reinforced polymer composites for construction – state-of-the-art review,' *Journal of Composites for Construction*, 6(2), 73–87.

Bank, L.C. (2006), *Composites for Construction: Structural Design with FRP Materials*, John Wiley & Sons, New York.

Brailsford, B., Mikovich, S. M., and Hopwood, T. (1995), *Definition of Infrastructure Specific Markets for Composite Materials*, Technical Report, Project P93-121/A573, BIRL, Northwestern University, Evanston, IL.

Creative Pultrusions, Inc. (2002), *Superdeck Composite Bridge Deck Installation Profile – Laurel Lick Bridge*, http://www.creativepultrusions.com/library.html (accessed April 30, 2012).

Cusens, A. R. and Pama, R. P. (1975), *Bridge Deck Analysis*, John Wiley & Sons, London.

Deskovic, N., Triantafillou, T. C., and Meier, U. (1995a), 'Innovative design of FRP combined with concrete: short-term behavior,' *Journal of Structural Engineering*, 121(7), 1069–1078.

Deskovic, N., Meier, U., and Triantafillou, T. C. (1995b), 'Innovative design of FRP combined with concrete: long-term behavior,' *Journal of Structural Engineering*, 121(7), 1079–1089.

Dumlao, C., Lauraitis, K., Abrahamson, E., Hurlbut, B., Jacoby, M., Miller, A., and Thomas, A. (1996), 'Demonstration low-cost modular composite highway bridge,' *Fiber Composites in Infrastructure*, Proceedings of the First International Conference on Composites in Infrastructure, ICCI'96, Tucson, AZ, January 15–17, H. Saadatmanesh and M. R. Ehsani (eds.), pp. 1141–1155.

Dunker, K. F. and Rabbat, B. G. (1995), 'Assessing infrastructure deficiencies: the case of highway bridges,' *Journal of Infrastructure Systems*, 1(2), 100–119.

Fam, A. Z. and Rizkalla, S. H. (2002), 'Flexural behavior of concrete-filled fiber-reinforced polymer circular tubes,' *Journal of Composites for Construction*, 6(2), 123–132.

Farheyl, D. N. (2005), 'Long-term performance monitoring of the Tech 21 all-composite bridge,' *Journal of Composites for Construction*, 9(3), 255–262.

Hayes, M. D., Ohanehi, D., Lesko, J. J., Cousins, T. E., and Witcher, D. (2000), 'Performance of tube and plate fiberglass composite bridge deck,' *Journal of Composites for Construction*, 4(2), 48–55.

Hibbit, Karlsson & Sorensen, Inc. (2000), *ABAQUS/Standard User's Manual*, Version 6.1, Hibbit, Karlsson & Sorensen, Inc.

Hillman, J. R. and Murray, T. M. (1990), 'Innovative floor systems for steel framed buildings,' *Mixed Structures, Including New Materials*, Proceedings of IABSE Symposium, Brussels, Belgium, Vol. 60, IABSE, Zurich, pp. 672–675.

Karbhari, V. M. and Zhao, L. (2000), 'Use of composites for 21st century civil infrastructure,' *Computer Methods in Applied Mechanics and Engineering*, 185, 433–454.

Mirmiran, A. (2001), 'Innovative combinations of FRP and traditional materials,' *Proceedings of the International Conference on FRP Composites in Civil Engineering*, Vol. II, December 12–15, Hong Kong, China, J. G. Teng (ed.), Elsevier Science, New York, pp. 1289–1298.

Moy, S. S. J., (ed.) (2001), 'FRP composites – life extension and strengthening of metallic structures,' *ICE Design and Practice Guides*, The Institution of Civil Engineers, Thomas Telford Publishing, London.

Mufti, A. A., Labossière, P., and Neale, K. W. (2002), 'Recent bridge applications of FRPs in Canada,' *Structural Engineering International*, 12(2), 96–98.

Neely, W. D., Cousins, T. E., and Lesko, J. J. (2004), 'Evaluation of in-service performance of Tom's Creek Bridge fiber-reinforced polymer superstructure,' *Journal of Performance of Constructed Facilities*, 18(3), 147–158.

Ribeiro, M. C. S., Tavares, C. M. L., Ferreira, A. J. M., and Marques, A. T. (2001), 'Static flexural performance of GFRP–polymer concrete hybrid beams,' *Proceedings of the International Conference on FRP Composites in Civil Engineering*, Vol. II, December 12–15, Hong Kong, China, J. G. Teng (ed.), Elsevier Science, New York, pp. 1355–1362.

Saiidi, M., Gordaninejad, F., and Wehbe, N. (1994), 'Behavior of graphite/epoxy concrete composite beams,' *Journal of Structural Engineering*, 120(10), 2958–2976.

Seible, F., Burgueño, R., Abdallah, M. G., and Nuismer, R. (1995), 'Advanced composite carbon shell systems for bridge columns under seismic loads,' *Proceedings of the National Seismic Conference on Bridges and Highways – Progress in Research and Practice*, San Diego, CA, December 10–13, Section 10.

Seible, F., Karbhari, V. M., Burgueño, R., and Seaberg, E. (1998), 'Modular advanced composite bridge systems for short and medium span bridges,' *Developments in Short and Medium Span Bridge Engineering '98*, The Fifth International Conference on Short and Medium Span Bridges, Calgary, July 13–16, pp. 431–441.

Seible, F., Karbhari, V. M., and Burgueño, R. (1999), 'Kings Stormwater Channel and I-5/Guilman Bridges, USA,' *Structural Engineering International*, 9(4), 250–253.

Small, E. P. and Cooper, J. (1998), 'Condition of the nation's highway bridges – a look at the past, present, and future', *Transportation Research News*, No. 194, January–February, 3–8.

Tang, B. (1997), 'Fiber reinforced polymer composites applications in USA: DOT-Federal Highway Administration,' *Proceedings of the First Korea/U.S.A. Road Workshop*, January 28–29.

Tang, B. M. and Hooks, J. M. (2001), 'FRP composites technology is changing the American bridge building industry,' *Proceedings of the International Conference on FRP Composites in Civil Engineering*, Vol. II, December 12–15, Hong Kong, China, J. G. Teng (ed.), Elsevier Science, New York, pp. 1657–1663.

Tang, B. and Podolny, W. (1998), 'A successful beginning for fiber reinforced polymer (FRP) composite materials in bridge applications,' *Proceedings of the International Conference on Corrosion and Rehabilitation of Reinforced Concrete Structures*, December 7–11, Orlando, FL, Federal Highway Administration.

Telang, N. M., Dumlao, C., Mehrabi, A. B., Ciolko, A. T., and Gutierrez, J. (2006), *Field Inspection of In-Service FRP Bridge Decks*, NCHRP Report 564, Transportation Research Board.

US Department of Transportation, Federal Highway Administration, and Federal Transit Administration (2000), *1999 Status of the Nation's Highways, Bridges and Transit: Conditions and Performance – Report to Congress*, http://www.fhwa.dot.gov/policy/1999cpr/report.htm (accessed April 30, 2012).

Van Erp, G. (2002), 'Road bridge benefits from hybrid beams,' *Reinforced Plastics*, 46(6), 4.

Van Erp, G., Cattell, C., and Ayers, S. (2005), 'The Australian approach to composites in civil engineering,' *Reinforced Plastics*, 49(6), 20–26.

Wu, H. F. (2005), *Composites in Civil Applications*, National Institute of Standards and Technology, http://www-15.nist.gov/atp/focus/99wp-ci.htm (accessed April 30, 2012).

Zureick, A. H., Shih, B., and Munley, E. (1995), 'Fiber-reinforced polymeric bridge decks,' *Structural Engineering Review*, 7(3), 257–266.

15
Fiber-reinforced polymer (FRP) composites for strengthening steel structures

M. DAWOOD, University of Houston, USA

DOI: 10.1533/9780857098955.2.382

Abstract: This chapter summarizes the recent advances in the use of fiber-reinforced polymer (FRP) materials for repair, rehabilitation, and strengthening of steel structures. Conventional methods of strengthening and repairing steel structures are presented. The advantages and limitations of using FRP materials are summarized. Topics presented include strengthening of flexural members, strengthening with prestressed FRP materials, stress-based and fracture mechanics-based approaches to evaluating bond behavior, repair of cracked steel members, and strengthening of slender members subjected to compression forces. The chapter concludes with a brief discussion of future trends in this field and a summary of other resources for further information.

Key words: steel, steel–concrete composite members, carbon fiber-reinforced polymers (CFRP), stress-based approaches, fracture-mechanics based approaches.

15.1 Introduction

Steel has been a widely used construction material for building and bridge applications for over 70 years. Its high ductility, strength-to-weight ratio, stiffness-to-weight ratio, and ease of constructability make steel well suited for long-span bridges and in a wide range of building applications. However, steel structures are susceptible to several modes of degradation that can compromise their integrity and limit their serviceability. Exposure to moist environments, which is common for bridges, typically leads to corrosion and loss of cross section. This degradation can be localized or widespread and often leads to the reduction of the capacity of steel members such as bridge piles, stringers, and floor beams. Exposure to repeated cyclic loading can lead to cracking of fatigue-sensitive details. Welded details are particularly susceptible to fatigue-induced cracking. Cracking under the effect of cyclic loads can lead to progressive deterioration of a structure or possibly even catastrophic collapse. Steel members may also be overstressed due to changes in use of structures that lead to increased loading beyond the original design load. Oftentimes strengthening or repair of a structure are more viable options than replacement or major re-design. However, many conventional repair techniques are limited in their applicability or may be subjected to similar degradation modes as the original structure. Recently, the use of fiber-reinforced polymer (FRP) materials has emerged as a promising alternative to conventional methods of rehabilitation for steel structures. Initial investigations were based on the documented success of using FRP materials for rehabilitation and strengthening of reinforced concrete structures. More recently, efforts have focused on addressing the unique challenges associated with rehabilitation of high-strength, high-stiffness, or slender steel elements.

15.2 Conventional repair techniques and advantages of fiber-reinforced polymer (FRP) composites

Several methods have been adopted to repair or strengthen inadequate steel members and structures. Flexural members with inadequate strength are often retrofit by bolting or welding steel plates or auxiliary steel members to their tension flanges to increase their moment of inertia. Similarly, the stability of compression members can be increased by welding or bolting steel stiffening elements to corroded members or slender elements in the cross-section. However, this technique commonly requires heavy lifting and extensive onsite fabrication, making it time-consuming and costly. The installation of heavy steel plates or steel sections increases the self-weight of the structure, thereby reducing the efficiency of this strengthening technique. In aggressive environments the steel repair elements are susceptible to continued corrosion. Also, in some applications, such as in petroleum refineries, confined spaces, or hospitals, welding may be infeasible.

In bridge applications, non-composite steel stringers have been made composite with the bridge deck by installing welded shear studs during deck replacements. If deck replacements are planned, this can be an effective method to simultaneously increase the flexural capacity of the stringers. However, if the existing deck is in good condition, deck replacement may be an unnecessary expense. Post-installing shear connectors using grouts or structural adhesives has recently been investigated as an alternative to this technique (Kwon *et al.*, 2009). This study indicated that partial composite interaction between the steel stringers and the concrete deck can be achieved using this approach, resulting in an increase of the flexural capacity of the stringers of up to 50%. Similarly, slender steel compression members can be stabilized by jacketing them with concrete, thereby effectively turning them into composite columns. This approach can be effective, but site accessibility may limit its applicability and the increased dead load associated with the repair may limit the additional live load that can be carried by the structure. Also, continued corrosion of the embedded steel member may lead to premature splitting of the concrete jacket.

External post-tensioning using unbonded steel tendons can be used to reduce stress levels in simply supported and continuous steel beams (Klaiber et al., 1990). The reduction in stress level can be closely controlled using strategic placement of deviators to vary the eccentricity of the post-tensioning strands. The use of external post-tensioning can also help to enhance the flexural stiffness and strength of steel beams (Park et al., 2010). This approach has the added advantage that post-tensioning forces can be adjusted and tendons can be replaced to accommodate changing loads. However, the use of external post-tensioning typically requires installation of heavy duty, and often complex anchorage systems on the structure to provide self-reacting forces. Further, steel tendons are susceptible to corrosion that, if left unchecked, can compromise the integrity of the retrofit.

Repair of cracked members can be achieved using a number of techniques. Repair welding is a commonly used technique in which the cracked material is removed by arc gouging and the element is welded to re-join the material on either side of the crack. Alternatively, crack-stop holes can be drilled at the crack tips to reduce the stress concentration at these locations and slow or stop the crack propagation. Crack blunting can be accompanied by cold working of the material around the crack-stop hole to induce residual compressive stresses in the material and further slow crack propagation (Crain et al., 2010). These approaches are commonly used to repair so-called 'stress-induced' fatigue cracks. Alternatively, some details are susceptible to so-called 'distortion-induced' fatigue cracking. This type of cracking occurs at relatively flexible details that are subjected to loadings that induce distortion of the member. A classical example of this type of detail is the 'web-gap' detail that has been used to connect bracing members to steel bridge girders, as illustrated in Fig. 15.1.

Two repair alternatives for this type of detail include making the connection extremely compliant to minimize distortion-induced stresses, or increasing the fixity of the detail to minimize distortion. The compliance of the connection can be increased by significantly increasing the size of the web-gap and possibly even cutting large holes in the girder web (Zhao and Roddis, 2007). Alternatively, fixity can be introduced by extending the gusset plate and welding it to the tension flange of the girder or by installing bolted or welded connection elements such as 'tee' or angle sections (Cousins et al., 1998). All of these approaches require a considerable amount of field fabrication in confined areas and irreversible modification of the structure. Oftentimes the repaired details are susceptible to re-initiation of cracking and, in some cases, the fatigue resistance of the retrofitted detail may be worse than that of the original structure.

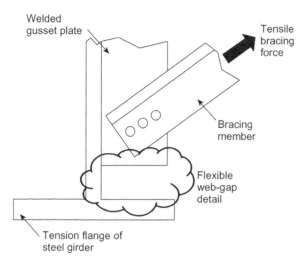

15.1 Schematic of typical flexible web-gap detail.

Compared to conventional repair techniques, the use of FRP materials provides several distinct advantages. The high strength-to-weight and stiffness-to-weight ratios of FRP materials facilitate rapid construction by small work crews with minimal use of heavy lifting equipment. These lightweight materials have a negligible effect on the self-weight of the structure, thereby maximizing the increase of the live load carrying capacity that can be achieved. Due to their light weight, FRP materials can be easily manoeuvred and installed in confined spaces with minimal disruption of service. Since FRP materials are typically bonded to steel structural members using structural adhesives, they represent a feasible alternative for repair applications in which welding is dangerous or not permitted. Further, adhesively bonded connection details typically have much higher fatigue resistances than welded details. The use of bonded FRP for repair applications does not result in the formation of 'locked-in' thermal stresses which commonly form in welded details. Additionally, the use of bonded details does not result in any fundamental change to the parent structure. Therefore, this repair technique does not degrade the condition or decrease the life of the existing member, and, in this regard, is 'reversible'.

The limitations of typical FRP repair techniques should also be noted. External bonding of FRP materials to steel surfaces requires thorough surface preparation of the parent structure to ensure the formation of a competent and durable bond. The performance of the strengthening system is sensitive to the bond performance. As such, careful detailing is essential to minimize bond stresses. In moist environments, special detailing may be required to minimize the potential for galvanic corrosion between steel and carbon FRP materials and to prevent ingress of moisture into the bond-line which can degrade the bond strength. Proper installation of FRP materials requires unique skills and specific environmental conditions. Therefore, qualified installers should be used and environmental conditions should be carefully monitored to ensure that they are within reasonable limits. Further, while many structural adhesives are rapid setting and rapid curing, bonded joints are sensitive to cyclic loads that are applied prior to complete curing of the adhesives. Finally, adhesives and FRP composites are susceptible to damage or degradation due to fire, UV exposure, and vandalism or unintentional impact. This can typically be accounted for by applying appropriate design constraints and using suitable protective coatings to the completed rehabilitation.

15.3 Flexural rehabilitation of steel and steel–concrete composite beams

Among the various topics related to the use of FRP materials for rehabilitation of steel structures, their use for flexural rehabilitation of steel and steel–concrete composite beams has received the most research attention to date. While the most common approach has been to bond unstressed FRP materials to the tension flange of the member, other researchers have used prestressed FRP materials to improve the efficiency of the strengthening system and to enable the FRP materials to participate in carrying the sustained dead loads acting on the structure. In both applications the bond behavior of the strengthening system has been identified as a critical factor in the system performance and the bond between steel surfaces and FRP materials has been studied in depth using various approaches.

15.3.1 Flexural behavior

The earliest studies on the use of FRP materials for rehabilitation of steel structures focused on repair of corroded steel bridge girders using externally bonded carbon FRP (CFRP) strips (Mertz and Gillespie, 1996). This research demonstrated that externally bonded CFRP strips could be used to restore the lost elastic stiffness and ultimate flexural strength of corroded steel girders to levels comparable to that of the original undamaged girder. All of the tested girders failed due to instability of the compression zone. Based on a moment-curvature analysis, this study further demonstrated that a more substantial increase of strength could be achieved in steel–concrete composite girders in which the concrete deck helps to stabilize the compression zone and provides additional compression capacity to balance the tensile forces developed in the CFRP strengthening materials. Subsequent studies demonstrated that the flexural strength of steel–concrete composite beams could be increased by up to 75% by bonding CFRP strips to the tension flanges of the girders (Sen et al., 2001; Tavakkolizadeh and Saadatmanesh, 2003c). However, since the elastic modulus of the CFRP materials used in these early studies was significantly lower than that of steel, in the range of 100–150 GPa, relatively thick plates were required to achieve a significant increase of the flexural capacity. More recent research has demonstrated that higher-modulus CFRP materials, with elastic moduli in the range of 230–460 GPa (so-called high modulus or ultra-high modulus CFRP), can be used to increase the elastic stiffness, yield moment, and ultimate strength of steel–concrete composite beams by up to 45%, 88%, and 66%, respectively (Dawood et al., 2007; Schnerch and Rizkalla, 2008; Fam et al., 2009). It has further been shown that the fatigue life of steel and steel–concrete beams strengthened or repaired with FRP materials is at least equal to that of unstrengthened beams and of common welded details that are typically used in steel construction (Tavakkolizadeh and Saadatmanesh, 2003a; Dawood et al., 2009).

The behavior of steel and steel–concrete composite girders, which are retrofit with FRP materials bonded to their tension flanges, can be accurately predicted using a sectional analysis that satisfies the principles of equilibrium and compatibility. This approach serves as the basis for the flexural analysis and design procedures recommended in several international design guidelines (Cadei et al., 2004; Schnerch et al., 2007; National Research Council, 2007). The moment-curvature analysis approach is illustrated schematically in Fig. 15.2. The strains in the deteriorated member immediately prior to strengthening can be calculated, typically based on a linear analysis and based on the properties of the transformed section. These pre-existing strains, due to sustained loads such as the self-weight of the structure, are 'locked in' prior to installation of the FRP strengthening

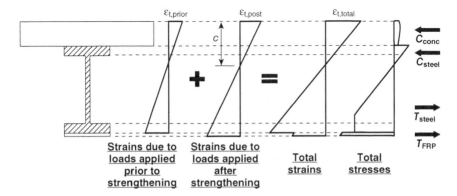

15.2 Schematic of moment-curvature analysis of FRP strengthened steel–concrete composite section (not to scale).

system. Therefore, the loads applied prior to strengthening do not induce any strain in the FRP materials. Any loads applied to the member after strengthening will induce strains in the FRP materials as shown. The incremental strains induced in the section due to the effect of loads applied after strengthening can be calculated by selecting a strain increment at the extreme compression surface of the member, $\varepsilon_{t,post}$. A neutral axis depth, c, can be assumed and the total strain profile can be determined by adding the incremental strains induced after strengthening to the 'locked-in' strains. Based on the total strains and the constitutive relationships of the different materials, concrete, steel and FRP, the distribution of stresses through the depth of the section can be determined. The total internal compression forces, C, and tension forces, T, in the cross-section can be calculated by integrating the strain profiles which is typically done using a numerical integration scheme by computer. The neutral axis depth can then be iterated until equilibrium of the internal compression and tension forces is achieved. The corresponding applied moment can be calculated by a summation of the internal moments and the curvature can be calculated as the slope of the strain profile. This analysis yields one point on the moment-curvature diagram of the strengthened cross section. The complete moment-curvature relationship of the section can be determined by incrementally increasing the applied strain at the top surface, $\varepsilon_{t,post}$, and repeating the above procedure until the ultimate strain of one of the materials is reached.

To illustrate the effect of various parameters on the response of steel and steel–concrete composite beams strengthened with FRP materials, several moment-curvature analyses were conducted. The basic beam configuration for the example calculations is shown schematically in Fig. 15.3. The section

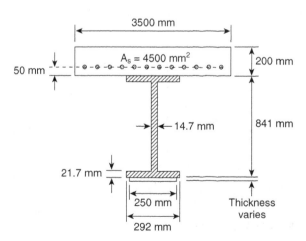

15.3 Schematic of the example beam studied in the parametric analysis (not to scale).

consisted of a US standard wide-flange section, W33 × 130, with a nominal depth of 841 mm, web thickness of 14.7 mm, flange width of 292 mm, and flange thickness of 21.7 mm. A 200 mm thick by 3000 mm wide composite concrete deck with a reinforcement ratio of 0.75% was also considered. The beams were strengthened with 250 mm wide CFRP strips with thicknesses ranging from 2 mm to 20 mm. The steel was treated as an elastic-perfectly plastic material with a yield strength of 248 MPa and an elastic modulus of 200 GPa. The yield strength of the reinforcing steel was taken as 400 MPa and the ultimate concrete compression strength, f'_c, was taken as 34 MPa with a parabolic stress–strain curve and an ultimate compression strain of 0.003. Two different types of CFRP materials were considered in the analysis; a high modulus CFRP with an elastic modulus of 450 GPa and an ultimate strain of 0.0037 and a standard modulus, high-strength CFRP with an elastic modulus of 165 GPa and an ultimate strain 0.017. The FRP materials were taken to be linearly elastic to failure. In this analysis, the sustained dead load acting on the structure immediately prior to strengthening was assumed to induce a strain at the outer surface of the tension flange equal to 30% of the yield strain. This corresponds to applied moments of 500 kN-m and 720 kN-m for the non-composite steel beams and the steel–concrete composite beams, respectively.

The predicted moment-curvature response of steel beams strengthened with different thicknesses of high-strength and high-modulus CFRP materials is presented in Fig. 15.4(a) and (b), respectively. Inspection of Fig. 15.4 reveals several interesting trends. First, it can be seen that for steel

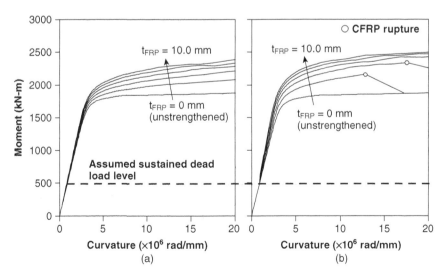

15.4 Moment-curvature response of steel beams strengthened with (a) high-strength CFRP and (b) high-modulus CFRP.

beams strengthened with high-strength CFRP materials, no significant increase of the elastic stiffness was achieved even for CFRP plates up to 10 mm thick (approximately half the thickness of the steel tension flange). In contrast, Fig. 15.4(b) indicates that a significant increase of the elastic stiffness can be achieved as indicated by the change of the slope of the moment-curvature relationship beyond the assumed sustained dead load level. Both sets of beams, strengthened with either high-strength or high-modulus CFRP materials, exhibit significant ductility and can achieve high levels of curvature without exhibiting any indication of brittle failure. This is primarily due to the fact that installing the CFRP strengthening materials lowers the neutral axis depth of the section. As such, yielding of the section initiates within the compression flange. This further lowers the neutral axis as yielding gradually propagates through the section. At high levels of curvature, although the steel tension flange may yield, the strain levels in the CFRP are typically not high enough to cause rupture of the CFRP except for beams strengthened with relatively thin high-modulus CFRP strips. Inspection of the figure further indicates that the flexural strength of steel beams can be increased by up to 30% using either high-strength or high-modulus CFRP materials. However, the efficiency of the strengthening system decreases as the amount of strengthening increases due to the shift of the neutral axis discussed previously.

The predicted moment-curvature response of steel–concrete composite beams strengthened with different thicknesses of high-strength and high-modulus CFRP materials is presented in Fig. 15.5(a) and (b), respectively. Comparison of Fig. 15.5 with Fig. 15.4 reveals a notable difference in the

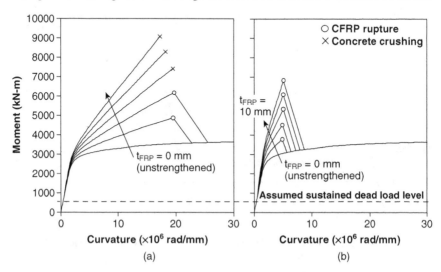

15.5 Moment-curvature response of steel–concrete composite beams strengthened with (a) high-strength CFRP and (b) high-modulus CFRP.

behavior of strengthened steel–concrete composite beams as compared to the behavior of strengthened plain steel beams. The strengthened steel–concrete composite beams exhibit a significant increase in the post-yield stiffness and a significant increase in ultimate strength. Failure of the strengthened beams is governed by one of two failure modes: rupture of the CFRP or crushing of the concrete. While rupture of the CFRP dominates in most cases, concrete crushing dominates for beams strengthened with large amounts of high-strength CFRP materials. Under displacement controlled loading, after rupture of the CFRP, the behavior of the member returns to that of the unstrengthened beam as illustrated in the figure. This behavior can be achieved because the large concrete deck provides adequate compression capacity to balance the increased tension capacity provided by the CFRP materials. Therefore, the strength of the section is dominated by failure of the materials rather than by instability of the cross section.

Comparison of Fig. 15.5(a) and (b) indicates that steel–concrete composite beams strengthened with high-strength CFRP materials can achieve significantly higher strengths than those strengthened with an equivalent thickness of high-modulus CFRP materials. This is mainly due to the higher strain capacity of the high-strength materials. Due to the high ultimate strain of the high-strength materials, the strengthened sections can also achieve relatively high values of curvature ductility, in the range of 8.5–10. This is beneficial in applications where energy dissipation is a major design consideration. Alternatively, due to the relatively low elastic modulus of the CFRP materials compared to steel, the elastic stiffness of the section is comparable to that of the unstrengthened beam. In contrast, examination of Fig. 15.5(b) indicates that, due to the relatively low rupture strain of the high-modulus CFRP, the ultimate flexural strength of the strengthened section is achieved at a much lower curvature. Therefore, members strengthened with high-modulus CFRP materials typically exhibit a much lower level of ductility, typically in the range of 2–3. On the other hand, due to the relatively high elastic modulus of the high-modulus CFRP materials, the strengthening significantly increases the elastic stiffness and yield moment of the strengthened members.

The effect of strengthening with high-strength and high-modulus CFRP materials on the behavior of plain steel and steel-concrete composite beams is summarized in Fig. 15.6. Figure 15.6(a) illustrates that high-modulus CFRP materials can be effectively used to increase the flexural stiffness, or transformed moment of inertia, of steel–concrete composite beams, and, to a lesser degree, plain steel beams. In contrast, high strength materials are significantly less effective at increasing the elastic flexural stiffness of steel beams with a maximum increase of only about 25% for the example beams when very thick CFRP plates are used. Similarly, Fig. 15.6(b) illustrates that,

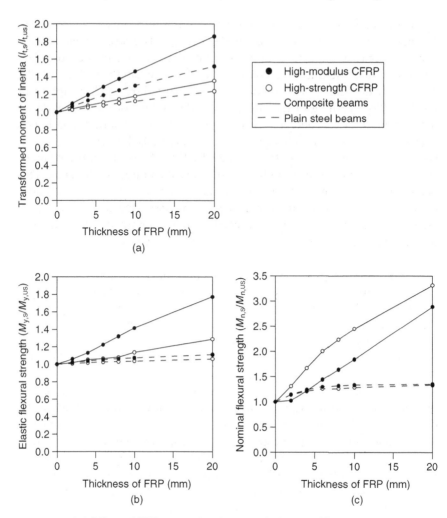

15.6 Effect of FRP strengthening on (a) elastic stiffness, (b) yield moment, and (c) ultimate strength of steel and steel–concrete composite sections.

for the example steel–concrete composite section, strengthening with 20 mm thick high-modulus CFRP plates resulted in an increase of the yield moment of the section of 80% compared to that of the unstrengthened beam. In contrast, use of a similar amount of high-strength CFRP resulted in a moderate increase of 25%. The significant increase of the elastic stiffness and yield moment that can be achieved using high-modulus CFRP materials make these materials well suited to enhance the serviceability and live-load-carrying capacity of sub-standard steel–concrete composite beams.

In contrast, for plain steel beams, the shift of the neutral axis results in initial yielding occurring at the extreme compression surface of the member. Therefore, regardless of the modulus of the CFRP materials, this strengthening approach does not significantly increase the flexural stiffness of plain steel beams.

Figure 15.6(c) illustrates that both high-modulus and high-strength CFRP materials can be effectively used to increase the ultimate strength of plain steel and steel–concrete composite beams. The presence of a large concrete deck provides balancing compressive forces and, therefore, makes significant increases in flexural strength, in excess of 150%, achievable with either high-strength or high-modulus CFRP materials. For plain steel beams, the lack of an integral concrete deck yields more modest, but still significant, increases in strength of up to 35% for both types of CFRP materials. If this level of strengthening is anticipated, it is important to note that the susceptibility of the strengthened member to other failure modes, such as shear, lateral-torsional buckling, or failure of shear connectors, should also be evaluated at the anticipated higher load levels.

15.3.2 Prestressed FRP strengthening

Several researchers have studied the possibility of prestressing CFRP materials as an alternative strengthening approach. In general, prestressing provides several advantages over bonding unstressed CFRP materials to the steel member, including:

- allowing the FRP materials to contribute to the dead load-carrying capacity of the member,
- reduced dead load deflections,
- reduction of tension stresses in tension flanges which can help slow or halt crack initiation and propagation, and
- maintaining the original ductility of the unstrengthened member by designing prestressing force to give essentially no increase of ultimate strength.

In general, the prestressing approach involves attaching an anchorage system to the tension flange at one end of the steel beam (the 'dead' end) (Schnerch and Rizkalla, 2008). The anchorage system can be bolted or welded to the tension flange. An anchor is typically bonded to the FRP strip and connected to the anchorage system. The adhesive is applied to the CFRP materials and the prestressing force is applied using threaded rods or a small jack through an abutment or fixture that is attached to the tension flange at the other end of the beam (the 'live' end). After the strengthening is completed and the adhesive has cured, the prestressing forces are transferred through the bond to the steel member. The anchorages can typically be left in place to provide some measure of redundancy in the highly stressed anchorage regions.

Analytical research results (Stratford and Cadei, 2006) indicate that significant bond stresses form near the end of the CFRP strengthening materials due to the influence of the prestressing forces (bond stresses are discussed in detail in the following section). The results suggest that the magnitude of the bond stresses induced by the prestressing forces can exceed the magnitude of the stresses induced due to applied loads and thermally induced stresses combined. Different approaches have been investigated to reduce the bond stress concentrations at the end of a prestressed CFRP strip. Most notably, a novel approach was developed at the Swiss Federal Laboratory (EMPA) for installing prestressed FRP materials on concrete structures which can, presumably, also be adopted for steel structures (Motavalli et al., 2011). Using this approach the prestressing force is gradually reduced from the maximum value to zero at the end of the prestressed strip in a step-wise manner over a short distance while simultaneously applying heat to the system. This approach requires the use of a specially designed installation device which eliminates the need for end anchorages. Other researchers have investigated the use of CFRP tendons for unbounded post-tensioning applications (Lee et al., 2005). This approach shares many of the same advantages of post-tensioning using steel tendons; however, the CFRP tendons have the added advantage of being non-corrosive. This post-tensioning technique was successfully used to strengthen portions of a three-span continuous steel bridge in central Iowa, USA.

15.4 Bond behavior

The integrity of a bonded FRP strengthening system for steel beams depends largely on the performance of the bond between the steel surface and the FRP materials. When strengthening concrete structures, the bond strength of the strengthening system is limited by the tensile strength of the substrate concrete. As such, in these applications it is sufficient to make the bond 'strong enough' to force the failure into the concrete layer. In contrast, when strengthening steel members, assuming that the steel surface is properly prepared, it is essentially impossible to force the bond failure into the steel substrate using modern adhesives that are conventionally used in structural applications. As such, a thorough understanding of the bond behavior is essential to ensure that the strengthening system performs as intended and to prevent a sudden, premature debonding failure. Two fundamental approaches have been adopted to evaluate the bond characteristics of FRP-strengthened steel members: the stress-based approach and the fracture mechanics-based approach. Each approach is summarized in the following sections.

15.4.1 Stress-based approaches

Several different stress-based approaches have been proposed to evaluate the nature of the bond stresses that develop in the adhesive layer between steel surfaces and FRP materials (Hart-Smith, 1974; Albat and Romily, 1999; Rabinovich and Frostig, 2000; Smith and Teng, 2001; Deng *et al.*, 2004; Stratford and Cadei, 2006; Al-Emrani and Kliger, 2006, Yang and Ye, 2010). While there are various differences between the different models, they all follow a fundamentally similar approach. These approaches are based on the principal that there is a mismatch between the strains and curvatures that would develop in the FRP and at the bottom surface of the strengthened beam if there was no composite interaction between the two and those that actually develop due to the presence of the adhesive. This mismatch must be accommodated by stresses that are induced in the adhesive layer. That is, it is the presence of the adhesive layer that forces the FRP to conform to the strain and curvature at the bottom of the strengthened member. The mismatch of axial force induces shear stresses in the adhesive layer and the mismatch of the curvature induces through-thickness or 'peeling' stresses in the adhesive layer perpendicular to the surface of the steel member. The distribution of these stresses can be obtained by considering equilibrium and compatibility of an infinitesimal segment of the strengthened member. Using this approach, the differential equations that dictate the stress distributions are formulated. The solution of the differential equations depends on the boundary conditions and loading acting on the strengthened member. The differences between the various models lie primarily in the different assumptions made to simplify the formulation and solution processes.

Regardless of the specific formulations that are considered, the behavior of the bond between FRP materials and steel surfaces exhibits several unique trends. Most notably, due to the abrupt change in geometry and material properties near the end of the strengthening plates, significant shear and peeling stress concentrations develop in the adhesive at these locations as illustrated in Fig. 15.7. This trend has been verified through a number of independent experimental studies (Miller *et al.*, 2001; Matta *et al.*, 2005; Colombi and Poggi, 2006; Dawood *et al.*, 2009). Inspection of the figure reveals several interesting trends. First, the through-thickness or peeling stress concentrations dissipates more quickly than the shear stress distribution. Additionally, the magnitude of the peak shear stress is greater than that of the peak peeling stress, approximately double in this specific example. It should be noted in this case, however, the peeling stresses are generally more detrimental to the performance of a bonded joint than shear stresses, as they place the adhesive into a state of direct tension. This is particularly critical for systems that employ brittle adhesives that can

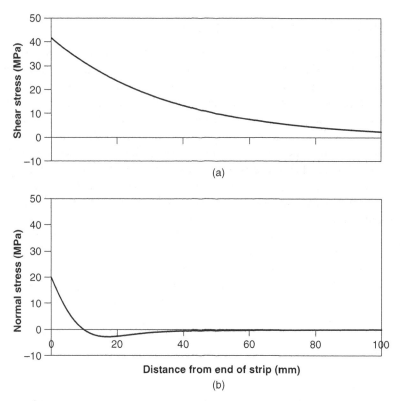

15.7 Predicted distribution of (a) shear and (b) peeling stresses in the adhesive layer near the end of a CFRP strip used to strengthen a steel beam (based on the formulation by Smith and Teng, 2001).

debond suddenly once the peak principal stress approaches the tensile strength of the adhesive.

Due to the nature of these stress concentrations, increasing the length of the bonded joint beyond a certain critical length will not increase the strength of the bonded joint. Alternatively, it has been recommended to modify the geometry or material properties of the strengthening plate and the adhesive locally near the plate end to help reduce the magnitude of the stress concentrations. These modifications, illustrated schematically in Fig. 15.8, may include providing an adhesive spew fillet or grading the elastic modulus of the adhesive near the plate end or tapering, 'reverse' tapering, or perforating the FRP plate itself. An extensive experimental and numerical study illustrated that providing a reverse tapered plate end detail with a 20° taper angle and a corresponding spew fillet helped to reduce the stress concentration factor in a bonded joint by a factor of two, thereby doubling the ultimate tensile strength of the bonded joint (Dawood *et al.*, 2009).

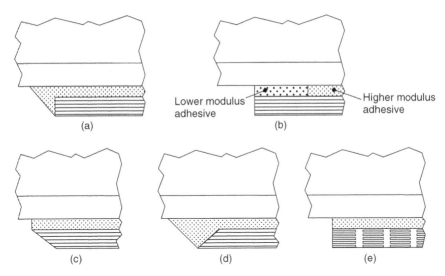

15.8 Various techniques to reduce local stress concentrations near the end of a bonded FRP plate: (a) adhesive spew fillet, (b) graded adhesive, (c) tapered plate end, (d) reverse tapered plate end with adhesive spew fillet, and (e) perforated plate end.

Recently, two interesting trends have been identified related to the bond behavior of FRP-strengthened steel beams. First, research indicates that the location of plate termination points must be carefully considered when strengthening indeterminate steel beams (Sebastian, 2003). Localized yielding of these indeterminate members causes load redistribution and migration of the points of inflection along the length of the member. To prevent premature debonding of FRP plates, it is common practice to locate the plate ends in regions of relatively low moment, near points of inflection. For heavily loaded indeterminate beams, the migration of the point of inflection can cause load reversals and significant stress concentrations near the plate ends. This effect is not a concern in the elastic service range. However, it may be more significant in structures that are expected to experience multiple excursions into the inelastic range, such as in seismic applications.

Another similar trend has been observed in beams that are expected to undergo significant inelastic deformations. At load levels above those required to cause yielding of the steel member, large shear stress concentrations form near the location of maximum moment, within the plastic hinge region of the beam (Sebastian, 2003; Linghoff *et al.*, 2009). These stress concentrations are due to the significant plastic flow that occurs in the tension flange of the beam at large strain values beyond yielding. Research indicates that the magnitude of these stress concentrations may exceed the magnitude of the stress concentrations that form near the plate end. The effect of these post-elastic stress concentrations on the behavior of beams strengthened with high-modulus CFRP materials has not been observed to be significant. This is primarily because the low rupture strain of the high-modulus materials does not permit the formation of large plastic strains in the beam flange prior to rupture of the CFRP. For beams strengthened with high-strength CFRP materials, the formation of these stress concentrations could reduce the ductility of the strengthened beam due to premature debonding, but is not expected to have a significant effect on the ultimate strength.

Recent studies have further shown that variations in temperature can induce shear and peeling stresses near plate ends that are comparable in magnitude to mechanically induced stresses (Stratford and Cadei, 2006). Experimental studies have demonstrated that the ultimate tensile strength and stiffness of bonded joints between steel and CFRP materials decrease dramatically as the ambient temperature approaches and exceeds the glass transition temperature of the adhesive (Nguyen et al., 2011). Conversely, at very low temperatures, many typical structural adhesives become stiffer and more brittle which may lead to abrupt and brittle debonding type failures. As such, due care should be taken to consider the influence of temperature variations on the adhesive properties and overall bond behavior when designing strengthening systems for applications in which temperature variations may be excessive, such as in bridge strengthening applications. Similarly, exposure to moisture may also degrade the adhesive properties and the bond strength of the system. While some concern has been raised regarding the potential for accelerated corrosion of steel structures due to galvanic corrosion with CFRP materials, research suggests that even a relatively thin layer of adhesive between the steel and the CFRP can minimize this risk (Tavakkolizadeh and Saadatmanesh, 2001). Other research suggests that the moist durability of the interface between the steel surface and the adhesive layer can be significantly enhanced by applying a properly selected silane-based primer to the steel surface prior to bonding the FRP materials (Dawood and Rizkalla, 2010).

15.4.2 Fracture mechanics-based approaches

The other main approach to evaluate the bond performance of steel members strengthened with FRP materials is based on the principles of fracture mechanics (Colombi, 2006; Lenwari et al., 2006). This approach recognizes that the plate end represents a point of stress singularity in the adhesive layer that is susceptible to sudden fracture. Failure of the adhesive layer and debonding are expected to occur when the stress intensity factor at the plate end, K, approaches the mode I fracture toughness of the adhesive, K_{IC}. Equivalently, failure can be defined when the fracture energy release rate of the system approaches the critical fracture energy of the adhesive, G_c. The stress intensity factor or fracture energy near the plate end are typically determined from a stress-based analysis, either analytically or numerically, and generally depend on the geometry near the plate end, the boundary conditions, and the loading on the beam. The fracture toughness and critical fracture energy of the adhesive are material properties which can be determined experimentally for different adhesives, much like the tension strength. This approach has been found to accurately predict the debonding strength of bonded joints between steel surfaces and FRP materials. The fracture mechanics-based approach has the added advantage that it can be relatively easily extended to evaluate the fatigue performance and crack propagation of bonded joints by considering the stress intensity factor range, ΔK (Liu et al., 2009).

15.5 Repair of cracked steel members

Repair of cracked members is a major challenge in the rehabilitation and maintenance of steel infrastructure. One of the primary challenges associated with conventional repair techniques is that the fatigue performance of the repair details is typically much lower than that of the virgin structure leading to rapid re-initiation of the crack. The application of an FRP patch can help reduce the stress intensity factor near the crack tip, which drives the crack propagation, through two primary mechanisms: (i) reduction of the stress range at the crack tip due to transfer of stresses along the FRP patch, and (ii) restraint of crack opening displacements. Research findings demonstrate that the installation of FRP patches can significantly affect the evolution of the mode I stress intensity factor, K_I near the crack tip. In unpatched members, as the crack length increases, K_I increases continuously leading to unstable propagation of the crack. In contrast, for FRP-patched members, as the crack length increases, K_I asymptotically approaches a limiting value leading to more stable crack growth (Rose, 1982; Paul et al., 1994; Wang et al., 1998). This reduction of the stress intensity factor can significantly improve the fatigue life of the repaired member.

Several research studies have highlighted the effectiveness of using FRP patches to repair cracked steel members (Tavakkolizadeh and Saadatmanesh, 2003b; Jones and Civjan, 2003; Nozaka et al., 2005; Shaat and Fam, 2008; Kim and Harries, 2011b). Stiffer patches are generally more effective for repairing cracked members as they attract more stress away from the crack tip region and are more effective in restraining the crack opening displacements. Therefore, thicker patches of higher modulus FRP materials are generally preferable for these types of repair. Research indicates that FRP patches are susceptible to two types of debonding.

Debonding that initiates near the crack location and propagates outwards is generally stable. While this type of cracking reduces the patch effectiveness, it does not generally lead to global failure of the patch. Debonding that initiates from the edges of the patches and propagates towards the crack is more critical and may lead to global failure of the patch. Thicker and stiffer patches are more susceptible to this latter type of debonding due to high bond stress concentrations that form near the ends of the patches. As such, the application of the remedial measures described previously to reduce stress concentrations near plate ends is advisable. Research also shows that double-sided crack repairs are more effective than single-sided repairs, since the stress intensity factor near the crack tip, on the unpatched side of a cracked member with a single-sided patch is higher than that on the patched side (Liu *et al.*, 2009; Lam *et al.*, 2010). Therefore, double-sided patching is recommended whenever both sides of a cracked member are accessible.

Research also indicates that a significant enhancement of the fatigue life can be achieved by applying prestressed FRP patches (Bassetti *et al.*, 2000; Täljsten *et al.*, 2009; Huawen *et al.*, 2010). Analytical studies indicate that applying moderate levels of prestressing in the FRP can reduce the stress intensity factor in cracked steel plates signficantly (Colombi *et al.*, 2003). In addition to the two primary mechanisms identified previously, prestressed patches also induce compressive stresses near the crack tip, thereby reducing the stress ratio that drives crack propagation. This can lead to a dramatic reduction of the crack growth rate and can even completely halt the crack propagation (Täljsten *et al.*, 2009; Huawen *et al.*, 2010).

15.6 Stabilizing slender steel members

Since steel members are typically very slender, they are particularly susceptible to local and global buckling failure mechanisms. Structural failure due to buckling can be avoided relatively easily in the design stages of a project. However, if members in service are found to be too slender, for example due to corrosion or design errors, or if a member needs to be reinforced to accommodate loads higher than the original design loads, addressing buckling concerns can be significantly more challenging. Researchers have shown that externally bonded FRP materials can be used to increase the global and local buckling resistance of different types of steel compression members including hollow structural sections (Shaat and Fam, 2006), tee-shaped compression braces (Kim and Harries, 2011a), and cold-formed steel struts (Silvestre *et al.*, 2008). These studies generally demonstrate that moderate increases in buckling strengths, on the order of 20%, can be achieved using externally bonded FRP materials. The presence of the FRP materials helps to increase the moment of inertia of slender elements and sections, thereby enhancing their buckling resistance. The effectiveness of externally bonded FRP strengthening in these applications is sensitive to the orientation and location of the fibers and the possible interaction between different buckling modes. Another approach that has proven to be effective is to stabilize slender compression elements with a cementitious core, made of either cast-in-place grout or precast cementitious blocks, which are subsequently confined by an FRP jacket (Liu *et al.*, 2005; El-Tawil and Ekiz, 2009). In this application the cementitious core provides stability to the slender member, while the FRP jacket primarily provides confinement of the cementitious core and protection from environmental exposure. With recent developments in underwater curing adhesives, the effectiveness of FRP jacketing for repairing corroded steel piles in bridge and marine applications is also being studied.

In another novel application, glass FRP (GFRP) pultruded sections have been used to stiffen the slender webs of plate girders (Okeil *et al.*, 2008). In this application, GFRP tee-shaped sections were bonded to the girder webs in the panel zones between existing welded stiffeners. This approach was found to increase the ultimate shear capacity of plate girders by up to 40%. The use of bonded GFRP stiffeners presents the interesting possibility of aligning the stiffeners with the direction of principal compression in the web, thereby maximizing the efficiency of the stiffener.

15.7 Case studies and field applications

Several steel structures have been retrofitted using CFRP materials to demonstrate the effectiveness of this strengthening technique. These early demonstration projects have focused primarily on the use of unstressed externally bonded CFRP materials. Externally bonded high-strength CFRP plates were used to retrofit a single steel girder on the Delaware Department of Transportation (DelDOT) 1-704 bridge. The bridge carries southbound traffic on Interstate Highway 95 over Christina Creek just south of Newark, Delaware, USA (Miller *et al.*, 2001). A single girder in the northern approach span was strengthened. The selected girder was designed to behave as a plain steel girder and, therefore, no shear connection was provided with the concrete deck. The girder was not in need of strengthening. Rather the retrofit was conducted to assess the long-term durability of the system. Load tests indicated that the presence of the CFRP helped to reduce the maximum strain in the steel tension flange by 12% as compared to tests conducted prior to strengthening. No deterioration or degradation of the strengthened girder has been reported to date.

In a similar application, high-modulus CFRP materials were used to strengthen several girders on a small bridge on Kentucky State Highway 32, west of Sadieville, Kentucky, USA. While the bridge was designed to be non-composite, composite interaction with the concrete deck was provided using post-installed shear anchors before installation of the CFRP materials. The installation of the CFRP strengthening is shown in Fig. 15.9. CFRP strips were bonded to the upper and lower surfaces of the tension flange in order to maximize the strength and stiffness increase that could be achieved using a single 2 mm thick layer of CFRP strips.

Several guidelines have been developed to facilitate the design and installation of externally bonded FRP materials for retrofit of steel flexural members. Specific guidance and a detailed design example, including flexural and bond considerations, are presented elsewhere (Schnerch et al., 2007).

15.8 Future trends

Significant advancements have been made to date on the use of FRP materials for retrofit and rehabilitation of steel structures. Based on the documented successes and advancement in this field, research has expanded to study several new and innovative solutions to complex challenges related to the performance of steel structures.

The majority of the research conducted to date has focused on the use of ambient temperature cure, thermosetting adhesives to bond pultruded FRP plates for flexural strengthening of steel beams. However, some researchers are investigating the use of vacuum bagging techniques, elevated cure temperatures, FRP prepreg materials, and thermoplastic adhesives for strengthening steel structures (Photiou et al., 2006; Hollaway et al., 2006). This approach can provide several distinct advantages over the current methods and materials. Namely, elevated temperature cure cycles produce cured adhesives with higher glass transition temperatures. Further, the use of FRP prepreg materials can be advantageous when rehabilitating structures with complex geometries and curved surfaces. In these applications, the use of vacuum bagging techniques can produce higher quality cured composites with higher fiber volume fractions than less advanced hand layup methods.

Integration of FRP materials with other types of advanced materials, including nanoparticles and shape memory alloys (SMA) could also expand their usefulness for repairing steel and other metallic structures. Current research efforts are assessing the benefits of using adhesives reinforced with carbon nanotubes or other types of nanoparticles (Faleh et al., 2011). This research highlights the importance of using advanced mixing and processing techniques to properly disburse the nanoparticles in the adhesives. The use of nanoparticle reinforced adhesives could lead to the development of the next generation of high-strength, multifunctional adhesives. These adhesives would be particularly well suited for steel strengthening applications

FRP composites for strengthening steel structures 403

(a)

(b)

15.9 Strengthening of steel bridge on Kentucky State Highway 32 using high-modulus CFRP materials: (a) installation of CFRP on upper surface of tension flange, (b) CFRP plates bonded to lower surface of tension flange (photos courtesy of Professor Issam Harik, University of Kentucky).

in which the bond characteristics and adhesive properties are particularly critical. On-going research is also focusing on integrating FRP with SMA materials with the objective of developing a new class of self-prestressing composite materials for crack repair applications.

The repair of cracked members with complex geometries, such as complex welded details, and gusset plates with closely spaced bolts, represents another challenge that may be well suited for the use of FRP-based repair alternatives (Kaan *et al.*, 2008; Nakamura *et al.*, 2009). FRP materials can be precisely fabricated to match the complex geometries at cover plate ends, weld fillets, gusset plates, and bolted connections. They can be easily bonded in confined spaces making their use very promising to repair fatigue-sensitive details in inaccessible locations that may otherwise require major retrofit of a significant portion of a structure or possibly even total replacement.

The use of FRP materials for repair of tubular steel structures and complex tubular joints such as those used in offshore and piping applications is another emerging field (Duell *et al.*, 2008; Alexander and Ochoa, 2010). Research to date has demonstrated the effectiveness of using wet layup FRP materials to repair welded joints in aluminum overhead sign structures (Pantelides *et al.*, 2003; Fam *et al.*, 2006). Other researchers are studying the bond behavior of FRP materials bonded to curved tubular members (Fawzia *et al.*, 2007). Modern advancements in adhesive joining technology have also led to the development of structural adhesives that can cure underwater and that are well suited for repairing underwater pipelines and offshore structures (Seica and Packer, 2007). Advancements in this field could prove increasingly beneficial based on growing global energy needs and advancements in both offshore petroleum and offshore wind energy technologies.

15.9 Sources of further information

Interested readers are directed to a number of relevant resources for more detailed information related to rehabilitation and retrofit of steel structures with FRP materials. Of significant note is the International Institute of FRP in Construction (IIFC) working group on Repair of Metallic Structures. IIFC's mission is to advance understanding related to the use of FRP materials in civil infrastructure applications for the advancement of the engineering profession and the betterment of society. The working group on Repair of Metallic Structures consists of active researchers from around the world who are conducting relevant, cutting-edge research. Among other activities, the Working Group publishes an annual list of publications, which, as of January 2011, included over 300 publications. IIFC also organizes two relevant biannual international conferences: the International Conference on FRP Composites in Civil Engineering (CICE), and the Asia Pacific Conference on FRP in Structures (APFIS). Similarly, the US Transportation Research Board committee AFF80 on Structural Fiber Reinforced Polymers has a subcommittee dedicated to studying advancements in FRP strengthened bridge girders.

A number of international design guidelines and specifications have been published that provide guidance to practitioners related to the proper design, installation, and monitoring of FRP-based systems for strengthening steel structures (Cadei et al., 2004, National Research Council, 2007). Interested readers are also directed to several summary papers and reports that provide a comprehensive overview of the state-of-the-art in the field (Hollaway and Cadei, 2002; Zhao and Zhang, 2007; Harries and El-Tawil, 2008).

15.10 References

Al-Emrani, M. and Kliger, R. (2006). Analysis of interfacial sear stresses in beams strengthened with bonded prestressed laminates. *Composites: Part B, 37*, 265–272.

Albat, A.M. and Romily, D.P. (1999). A direct linear-elastic analysis of double symmetric bonded joints and reinforcements. *Composites Science and Technology, 59*, 1127–1137.

Alexander, C. and Ochoa, O.O. (2010). Extending onshore pipeline repair to offshore steel risers with carbon-fiber reinforced composites. *Composite Structures, 92(2)*, 499–507.

Bassetti, A., Nussbaumer, A. and Hirt, M.A. (2000). Fatigue life extension of riveted bridge members using prestressed carbon fiber composite. In *Proceedings of the International Conference on Steel Structures of the 2000's ECCS*, 375–380.

Cadei, J.M.C., Stratford, T.J., Hollaway, L.C. and Duckett, W.G. (2004). *Strengthening Metallic Structures using Externally Bonded Fibre-Reinforced Polymers*. London: CIRIA.

Colombi, P. (2006). Reinforcement delamination of metallic beams strengthened by FRP strips: fracture mechanics based approach. *Engineering Fracture Mechanics, 73*, 1980–1995.

Colombi, P. and Poggi, C. (2006). An experimental, analytical and numerical study of the static behavior of steel beams reinforced by pultruded CFRP strips. *Composites: Part B, 37*, 64–73.

Colombi, P., Bassetti, A. and Nussbaumer, A. (2003). Analysis of cracked steel members reinforced by pre-stress composite patch. *Fatigue and Fracture of Engineering Materials and Structures, 26(1)*, 59–66.

Cousins, T.E., Stallings, J.M., Lower, D.A. and Stafford, T.E. (1998). Field evaluation of fatigue cracking in diaphragm–girder connections. *Journal of Performance of Constructed Facilities, 12(1)*, 25–32.

Crain, J.S., Simmons, G.G., Bennett, C.R., Barrett-Gonzalez, R., Matamoros, A.B. and Rolfe, S.T. (2010). Development of a technique to improve fatigue lives of crack-stop holes in steel bridges. *Transportation Research Record, 2200*, 69–77.

Dawood, M. and Rizkalla, S. (2010). Environmental durability of a CFRP system for strengthening steel structures. *Construction and Building Materials, 24(9)*, 1682–1689.

Dawood, M., Rizkalla, S. and Sumner, E. (2007). Fatigue and overloading behavior of steel–concrete composite flexural members strengthened with high modulus CFRP materials. *Journal of Composites for Construction, 11(6)*, 659–669.

Dawood, M., Guddati, M. and Rizkalla, S. (2009). Effective splices for a carbon fiber-reinforced polymer: strengthening system for steel bridges and structures. *Transportation Research Record, 2131*, 125–133.

Deng, J., Lee, M.M.K. and Moy, S.S.J. (2004). Stress analysis of steel beams reinforced with a bonded CFRP plate. *Composite Structures, 65*, 205–215.

Duell, J.M., Wilson, J.M. and Kessler, M.R. (2008). Analysis of a carbon composite overwrap pipeline repair system. *International Journal of Pressure Vessels and Piping, 85(11)*, 782–788.

El-Tawil, S. and Ekiz, E. (2009). Inhibiting steel brace buckling using carbon fiber polymers: large-scale tests. *Journal of Structural Engineering, 135(5)*, 530–538.

Faleh, H., Shen, L. and Al-Mahaidi, R. (2011). Fabrication and mechanical characterization of carbon nanotubes-enhanced epoxy. *Advances in Building Materials, 168–170*, 1102–1106.

Fam, A., Witt, S. and Rizkalla, S. (2006). Repair of damaged aluminum truss joints of highway overhead sign structures using FRP. *Construction and Building Materials, 20(10)*, 948–956.

Fam, A., MacDougall, C. and Shaat, A. (2009). Upgrading steel-concrete composite girders and repair of damaged steel beams using bonded CFRP laminates. *Thin-Walled Structures, 47(10)*, 1122–1135.

Fawzia, S., Al-Mahaidi, R., Zhao, X.L. and Rizkalla, S. (2007). Strengthening of circular hollow steel tubular sections using high modulus CFRP sheets. *Construction and Building Materials, 21(4)*, 839–845.

Harries, K.A. and El-Tawil, S. (2008). Steel-FRP composite structural systems. In *Composite Construction in Steel and Concrete VI – Proceedings of the 2008 Composite Construction in Steel and Concrete Conference*, 703–716.

Hart-Smith, L.J. (1974). *NASA CR2218: Analysis and Design of Advanced Composite Bonded Joints.* Washington, DC: National Aeronautics and Space Administration.

Hollaway, L.C. and Cadei, J. (2002). Progress in the technique of upgrading metallic structures with advanced polymer composites. *Progress in Structural Engineering and Materials, 4*, 131–148.

Hollaway, L.C., Zhang, L., Photiou, N.K., Teng, J.G. and Zhang, S.S. (2006). Advances in adhesive joining of carbon fibre/polymer composites to steel members for repair and rehabilitation of bridge structures. *Advances in Structural Engineering, 9(6)*, 791–803.

Huawen, Y., König, C., Ummenhofer, T., Shizhong, Q. and Plum, R. (2010). Fatigue performance of tension steel plates strengthened with prestressed CFRP laminates. *Journal of Composites for Construction, 14(5)*, 609–615.

Jones, S.C. and Civjan, S.A. (2003). Application of fiber reinforced polymer overlays to extend steel fatigue life. *Journal of Composites for Construction, 7(4)*, 331–338.

Kaan, B., Barrett, R., Bennett, C., Matamoros, A. and Rolfe, S. (2008). Fatigue enhancement of welded cover plates using carbon-fiber composites. In *Proceedings of the 2008 Structures Congress – Structures Congress 2008: Crossing the Borders, 314.*

Kim, Y.J. and Harries, K.A. (2011a). Behavior of tee-section bracing members retrofitted with CFRP strips subjected to axial compression. *Composites Part B: Engineering, 42(4)*, 789–800.

Kim, Y.J. and Harries, K.A. (2011b). Fatigue behavior of damaged steel beams repaired with CFRP strips. *Engineering Structures, 33(5)*, 1491–1502.

Klaiber, F.W., Dunker, K.F., Planck, S.M. and Sanders, W.W. Jr. (1990). *Strengthening of an Existing Continuous-Span, Steel-Beam, Concrete-Deck Bridge by Post-Tensioning (ISU-ERI-AMES-90210)*. Ames, IA: Iowa Department of Transportation.

Kwon, G., Englehardt, M. and Klingner, R. (2009). *Implementation Project: Strengthening of a Bridge near Hondo, Texas using Post-Installed Shear Connectors (FHWA/TX-09/5-4124-01-1)*. Austin, TX: Texas Department of Transportation.

Lam, A., Yam, M.C.H., Cheng, J.J.R. and Kennedy, G.D. (2010). A study of stress intensity factor of a cracked steel plate with a single-side CFRP composite patching. *Journal of Composites for Construction, 14(6)*, 791–803.

Lee, Y., Wipf, T.J., Phares, B.M. and Klaiber, F.W. (2005). Evaluation of steel girder bridge strengthened with carbon fiber-reinforced polymer posttension bars. *Transportation Research Record: Journal of the Transportation Research Board, 1928*, 233–244.

Lenwari, A., Thepchatri, T. and Albrecht, P. (2006). Debonding strength of steel beams strengthened with CFRP plates. *Journal of Composites for Construction, 10(1)*, 69–78.

Linghoff, D., Haghani, R. and Al-Emrani, M. (2009). Carbon-fibre composites for strengthening steel structures. *Thin-Walled Structures, 47(10)*, 1048–1058.

Liu, H., Al-Mahaidi, R. and Zhao, X. (2009). Experimental study of fatigue crack growth behavior in adhesively reinforced steel structures. *Composite Structures, 90(1)*, 12–20.

Liu, X., Nanni, A. and Silva, P. (2005). Rehabilitation of compression steel members using FRP pipes filled with non-expansive and expansive lightweight concrete. *Advances in Structural Engineering, 8(2)*, 129–142.

Matta, F., Karbhari, V.M. and Vitaliani, R. (2005). Tensile response of steel/CFRP adhesive bonds for the rehabilitation of civil structures. *Structural Engineering and Mechanics, 20(5)*, 589–608.

Mertz, D.R. and Gillespie, J.W. (1996). *Rehabilitation of Steel Bridge Girders through the Application of Advanced Composite Materials*. NCHRP-IDEA Program Project Final Report, Transportation Research Board, National Research Council, Washington, DC.

Miller, T.C., Chajes, M.J., Mertz, D.R. and Hastings, J.N. (2001). Strengthening of a steel bridge girder using CFRP plates. *Journal of Bridge Engineering, 6(6)*, 514–522.

Motavalli, M., Czaderski, C. and Pfyl-Lang, K. (2011). Prestressed CFRP for strengthening of reinforced concrete structures: recent developments at EMPA, Switzerland. *Journal of Composites for Construction, 15(2)*, 194–205.

Nakamura, H., Jiang, W., Suzuki, H., Maeda, K.I. and Irube, T. (2009). Experimental study on repair of fatigue cracks at welded web gusset joint using CFRP strips. *Thin-Walled Structures, 47(10)*, 1059–1068.

National Research Council (2007). *Guidelines for the Design and Construction of Externally Bonded FRP Systems for Strengthening Existing Structures: Metallic Structures*. Rome: CNR.

Nguyen, T.C., Bai, Y., Zhao, X.L. and Al-Mahaidi, R. (2011). Mechanical characterization of steel/CFRP double strap joints at elevated temperatures. *Composite Structures, 93(6)*, 1604–1612.

Nozaka, K., Shield, C.K. and Hajjar, J.F. (2005). Effective bond length of carbon-fiber-reinforced polymer strips bonded to fatigued steel bridge I-girders. *Journal of Bridge Engineering, 10(2)*, 195–205.

Okeil, A., Bingol, Y. and Ferdous, M.R. (2008). *A Novel Technique for Stiffening Steel Structures*. FHWA/LA.08/441.

Pantelides, C.P., Nadauld, J. and Cercone, L. (2003). Repair of cracked aluminum overhead sign structures with glass fiber reinforced polymer composites. *Journal of Composites for Construction, 7(2)*, 118–126.

Park, S., Kim, T. and Hong, S.N. (2010). Flexural behavior of continuous steel girder with external post-tensioning and section enhancement. *Journal of Constructional Steel Research, 66(2)*, 248–255.

Paul, J., Bartholomeusz, R.A., Jones, R. and Ekstrom, M. (1994). Bonded composite repair of cracked load-bearing holes. *Engineering Fracture Mechanics, 48(3)*, 455–461.

Photiou, N.K., Hollaway, L.C. and Chryssanthopoulos, M.K. (2006). Strengthening of an artificially degraded steel beam utilising a carbon/glass composite system. *Construction and Building Materials, 20*, 11–21.

Rabinovich, O. and Frostig, Y. (2000). Closed-form high-order analysis of RC beams strengthened with FRP strips. *Journal of Composites for Construction, 4(2)*, 65–74.

Rose, L.R.F. (1982). A cracked plate repaired by bonded reinforcements. *International Journal of Fracture, 18(2)*, 135–144.

Schnerch, D. and Rizkalla, S. (2008). Flexural strengthening of steel bridges with high modulus CFRP strips. *Journal of Bridge Engineering, 13(2)*, 192–201.

Schnerch, D., Dawood, M., Rizkalla, S. and Sumner, E. (2007). Proposed design guidelines for strengthening of steel bridges with FRP materials. *Construction and Building Materials, 21(5)*, 1001–1010.

Sebastian, W.M. (2003). Nonlinear influence of contraflexure migration on near-curtailment stresses in hyperstatic FRP-laminated steel members. *Computers and Structures, 81*, 1619–1632.

Seica, M.V. and Packer, J.A. (2007). FRP materials for the rehabilitation of tubular steel structures, for underwater applications. *Composite Structures, 80*, 440–450.

Sen, R., Liby, L. and Mullins, G. (2001). Strengthening steel bridge sections using CFRP laminates. *Composites: Part B, 32*, 309–322.

Shaat, A. and Fam, A.Z. (2006). Axial loading tests on short and long hollow structural steel columns retrofitted using carbon fibre reinforced polymers. *Canadian Journal of Civil Engineering, 33(4)*, 458–470.

Shaat, A. and Fam, A.Z. (2008). Repair of cracked steel girders connected to concrete slabs using carbon-fiber-reinforced polymer sheets. *Journal of Composites for Construction, 12(6)*, 650–659.

Silvestre, N., Young, B. and Camotim, D. (2008). Non-linear behavior and load-carrying capacity of CFRP-strengthened lipped channel steel columns. *Engineering Structures, 30(10)*, 2613–2630.

Smith, S.T. and Teng, J.G. (2001). Interfacial stresses in plated beams. *Engineering Structures, 23*, 857–871.

Stratford, T. and Cadei, J. (2006). Elastic analysis of adhesion stresses for the design of a strengthening plate bonded to a beam. *Construction and Building Materials, 20*, 34–45.

Täljsten, B., Hansen, C.S. and Schmidt, J.W. (2009). Strengthening of old metallic structures in fatigue with prestressed and non-prestressed CFRP laminates. *Construction and Building Materials, 23(4)*, 1665–1677.

Tavakkolizadeh, M. and Saadatmanesh, H. (2001). Galvanic corrosion of carbon and steel in aggressive environments. *Journal of Composites for Construction, 5(3)*, 200–210.

Tavakkolizadeh, M. and Saadatmanesh, H. (2003a). Fatigue strength of steel girders strengthened with carbon fiber reinforced polymer patch. *Journal of Structural Engineering, 129(2)*, 186–196.

Tavakkolizadeh, M. and Saadatmanesh, H. (2003b). Repair of damaged steel-concrete composite girders using carbon fiber-reinforced polymer sheets. *Journal of Composites for Construction, 7(4)*, 311–322.

Tavakkolizadeh, M., and Saadatmanesh, H. (2003c). Strengthening of steel-concrete composite girders using carbon fiber reinforced polymers sheets. *Journal of Structural Engineering, 129(1)*, 30–40.

Wang, C.H., Rose, L.R.F. and Callinan, R. (1998). Analysis of out-of-plane bending in one-sided bonded repair. *International Journal of Solids and Structures, 35(14)*, 1653–1675.

Yang, J. and Ye, J. (2010). An improved closed-form solution to interfacial stresses in plated beams using a two-stage approach. *International Journal of Mechanical Science, 1(52)*, 13–30.

Zhao, Y. and Roddis, W.M.K. (2007). Fatigue behavior and retrofit investigation of distortion-induced web gap cracking. *Journal of Bridge Engineering, 12(6)*, 737–745.

Zhao, X. and Zhang, L. (2007). State-of-the-art review on FRP strengthened steel structures. *Engineering Structures, 29*, 1808–1823.

16
Fiber-reinforced polymer (FRP) composites in environmental engineering applications

R. LIANG and G. HOTA, West Virginia University, USA

DOI: 10.1533/9780857098955.2.410

Abstract: This chapter presents dozens of select environmental engineering applications of fiber-reinforced polymer (FRP) composite materials with emphasis on their environmental benefits, followed by discussions on durability of composites. Significance of design codes and specifications in promoting and advancing the applications of FRP composites is addressed. With ever increasing attention toward a sustainable built environment, FRP composites have potential to be selected as a material of choice because of the performance and design advantages of FRPs.

Key words: fiber-reinforced polymer (FRP), durability, sustainable materials, infrastructural applications, recycling of composites, green composites.

16.1 Introduction

A composite material is a combination of two or more materials (reinforcing elements such as fibers, and binders such as polymer resins), differing in form or composition. The combination of these materials can be designed to result in a material that maximizes specific performance properties. For example, fiber reinforced polymer (FRP) composites are made of thermosetting or thermoplastic resins, and glass, carbon, or other types (e.g., Kevlar or natural fiber flax/kenaf) of fibers (rovings), mats, and/or fabrics. The fiber network is the primary load-bearing component, while the resin helps transfer loads including shear forces through fibers and fabrics and maintains fiber orientation. The resin primarily dictates the manufacturing process and processing conditions, and partially protects the fibers/fabrics from environmental damage, such as humidity, temperature fluctuations, and chemicals.

FRP composites are being promoted as the materials of the 21st century because of their superior corrosion resistance, excellent thermo-mechanical properties, and high strength-to-weight ratio. FRPs are also 'greener' in terms of embodied energy (the quantity of energy required to manufacture a product) than conventional materials such as steel and aluminum, on a per-unit-of-performance basis. The use of FRP composites in civil and military infrastructure can improve innovation, increase productivity, enhance performance, and provide longevity, resulting in reduced life-cycle costs and enhanced environmental protection. For over 25 years, the researchers at West Virginia University's Constructed Facilities Center (WVU-CFC) have been conducting research on fundamentals of engineering and material sciences, innovation and development including field implementation of FRP composite components and systems. The FRP implementation has touched upon a wide range of engineering applications with emphasis on enhancing performance, serviceability, and durability over conventional materials.

This chapter deals with FRP composites in environmental engineering to enhance environmental protection. Environmental engineering covers a range of applications involving science and engineering principles to protect and even improve the natural environment (air, water, and land resources), to provide healthy water, air, and land for human habitation (family dwellings or office spaces) including living organisms and if possible, to remediate environmental pollution. By no means will this chapter provide a complete coverage of the applications of FRP composites in environmental engineering.

After discussing the environmental benefits of composites, this chapter focuses on select FRP field applications related to environmental protection based on the authors' research and implementation experiences. These applications include:

- oil and gas storage tanks;
- FRP rebars;
- decking for ocean environment;
- sheet piling;
- cold water pipe for ocean thermal energy conversion power generation;
- chimney liners;
- cooling water tower;
- environmentally friendly utility poles;
- modular track panels;
- geosynthetics;
- corrosion-resistant pipelines;
- rock bolting for underground mining;
- modular buildings of environmental durability;
- FRP wraps for concrete and wooden structures;
- engineered recycled rail-road ties; and
- green composites.

The above applications are followed with discussions on durability of composites. Finally, design codes and specifications to promote and advance the applications of FRP composites are presented, in addition to their future research directions.

16.2 Advantages and environmental benefits of fiber-reinforced polymer (FRP) composites

FRP composites have been advanced over the years for mass production of chemical reactors, storage tanks, aerospace, automotive, offshore and highway structural applications, and many others. Such a wide range of applications can be attributed to the following favorable FRP material properties:

- higher specific (with reference to material density) strength and stiffness than steel or wood;
- higher fatigue strength and impact energy absorption capacity;
- better resistance to corrosion, rust, fire, hurricane, ice storm, acids, water intrusion, temperature changes, attacks from micro-organisms, insects, and woodpeckers;
- longer service life (over 80 years);
- lower installation, operation, and maintenance costs;
- non-conductivity;
- non-toxicity;
- reduced magnetic, acoustic, and infrared (IR) interferences;
- design flexibility including ease of modular construction; and
- consistent batch-to-batch performance.

The environmental benefits of FRP composites can be discussed in terms of:

- better durability;
- light weight;
- lower transportation costs;
- superior corrosion resistance, thus longer service life;
- ease of installation; and
- free of maintenance.

For example, corrosion-resistant FRP rebar in lieu of steel rebar in concrete makes roads and bridges last longer (Vijay and Hota, 1999; Chen et al., 2008). Modular FRP composite bridge decks are about ten times lighter and eight times stronger than the conventional concrete bridge decks requiring less erection time with only light equipment for installation (Shekar et al., 2005), thus reducing downtime and traffic tie-ups in the case of bridge deck replacement, increasing productivity, and reducing the life-cycle costs of construction. In addition, the low self-weight of FRP bridge decks can result in increased live-load-carrying capacity of old bridges after replacing deteriorated concrete decks with modern FRP decks.

FRP composites offer additional benefits in terms of energy efficiency and environmental performance. Several life cycle assessment analyses concluded that per unit of performance, GFRP composite products have lower embodied energy (Strongwell Corporation, 2009; Kara and Manmek, 2009). For example, per unit length of I-beam, GFRP I-beam consumes only 43% of the amount of embodied energy that is consumed to manufacture steel I-beam, while the greenhouse gas emission (CO_x) for GFRP I-beam is 75% less than that of steel I-beam, resulting in the total environmental impact index of GFRP I-beam being 20% that of steel I-beam (Kara and Manmek, 2009). It must be noted that the above stated relative embodied energy advantage of FRP over steel is based on specific application and specific performance. These embodied energy percentages vary depending upon the application type and material functionality. In addition, FRP composites, in lieu of wood for utility poles or cooling towers, will relieve the environmental concerns of chemically treated wood where preservatives could pollute the surrounding soil and damage the plant life or leach into and contaminate the water. According to Feldman and Shistar (1997), more than 667 million lbs of wood preservatives are consumed in the United States every year. Pultruded FRP cooling towers have already been accepted. As FRP utility poles penetrate into the wood pole market, wood preservative usage will be further reduced.

The development of natural ('green') composites will reduce the embodied energy of composite materials, improve environmental benefits, and promote sustainable developments (Mutnuri et al., 2010). The natural composites are made of natural fibers and bio-based resins. The natural fibers and bio-resins are from plants that grow by deriving energy from the sun, drawing carbon dioxide from the atmosphere, and releasing oxygen back into the atmosphere via photosynthesis. These plants can be continuously re-grown as feedstock for composite applications leading to a natural carbon sequestration approach. The potentially high performance and low cost of bio-composites not only provide carbon sequestration but also result in significant energy savings during manufacturing. For example, per unit weight, embodied energy of rough sawn timber is 23 times less than steel and 290 times less than aluminum; per ton of material, rough sawn timber stores 500 kg of carbon while steel releases 700 kg of carbon and aluminum releases 8,700 kg of carbon (Ferguson et al., 1996). Natural composites having embodied energy levels similar to timber offer ten-fold superiority in terms of mechanical properties. The development and implementation of innovative structural composites made of natural fibers and bio-based resins will allow us to reduce the use of traditional materials such as cement and fossil fuel-based composites, releasing large quantities of CO_x into the atmosphere. For example, one ton of cement releases one and a half tons of CO_x (Malvern, 2011), whereas one ton of natural fibers absorbs one ton of CO_x from the atmosphere. Noting that about 3 billion tons of cement is used every year, even partial replacement of cement-based concrete with natural composite structural elements in buildings will result in significant energy savings and reduction of carbon footprint.

16.3 Fiber-reinforced polymer (FRP) composites in chemical environmental applications

The advantages of FRPs in terms of both the chemical and corrosion resistances over conventional construction materials have been recognized since the initial development of FRPs following World War II. Their application in chemical process equipment dates back to the early 1950s when the chemical process and pulp bleaching industries began using FRPs to replace expensive materials such as alloys and rubber-lined steel (Kelley and Schneider, 2008). Currently, FRP composites are widely used in chemical industries, such as tanks, ducts, pipes, hoods, pumps, fans, grating, a range of other equipment for chemical processing, pulp and paper, oil and gas, water and wastewater treatment (ACMA, 2011). According to data from the American Composites Manufacturers Association, the above applications represent 22% of total FRP composites shipments during 2010 in the US, i.e. about 528 million lbs, as compared to a market share of approximately 10% in 1999.

16.3.1 Underground storage tank

Among other success stories is FRPs for underground storage tanks (UST) that store petroleum products. Gasoline underground storage tanks were made extensively from mild steel before the 1970s, leading to corrosion over time and eventually resulting in leakage of fuel into the surrounding environment and contamination of the soil and groundwater. FRP USTs came into use in the mid-1960s. These were made of glass fibers and isophthalic polyester resin and were proven to be a highly-engineered and cost-effective solution (Fig. 16.1). A lengthy process of third-party validation

(a)   (b)

16.1 The first generation FRP tank (a) before burial, May 15, 1963; (b) after burial, May 11, 1988.

and acceptance by major owners took place in the 1970s (Dorris, 2008). FRP UST designs were optimized in the late 1980s for cost-effectiveness. In the 1980s, epoxy vinyl ester resins were introduced to further increase the service life of FRP via improved chemical resistance and toughness.

USTs are regulated in the United States to prevent the release of petroleum and contamination of groundwater. In 1984, the US Congress established the UST Program to minimize and prevent environmental damage from petroleum contamination from USTs. Legislation requiring owners of USTs to locate, remove, upgrade, or replace underground storage tanks became effective on December 24, 1989. Through this program, many thousands of old underground tanks were replaced with FRP tanks. These tanks were constructed as double- or triple-walled tanks – first introduced in 1984 – to catch leaks from the inner tanks and to give an interstitial space to accommodate leak detection sensors. Piping was also replaced with FRP composite pipelines of multiple-wall construction. By the end of 2007, approximately 450,000 FRP tanks had been manufactured and sold (Dorris, 2008). According to the US Environmental Protection Agency (EPA) Report released in March 2012, the UST program had discarded 1,762,249 steel-based tanks and was regulating 587,517 active FRP USTs at approximately 212,000 sites across the country in 2011 (EPA, 2012).

The primary methods of manufacturing FRP USTs are chopped glass spray-up, rotating mandrel laydown, and filament winding (McConnell, 2007). The FRP UST models have evolved over the years, as new requirements were identified, from single-wall, to double-wall, to triple-wall construction. The first FRP tank shown in Fig. 16.1 is a single-wall UST with a capacity of 6,000 gallons and an 8-foot diameter. Since then, changes in the UST design and construction include:

- tank sizes are significantly larger, 8–10 ft diameter by 60–80 ft long, with a capacity of up to 50,000 gallons;
- the majority of FRP USTs have double-wall construction, ranging in wall thickness from 0.25 to 1.0 inch, including integral ribs;
- resin formulations have been improved so FRP USTs can contain aggressive fuels such as ethanol; and
- tanks now incorporate leak detection systems and containment sumps (McConnell, 2007).

Typically, a double-wall, multi-compartment FRP UST with a 10-ft diameter and 20,000 gallon capacity weighs about 8,000 lbs, which makes it easy to handle and install these tanks underground. Currently these FRP USTs have extensive applications beyond gas stations. Many other chemical industries have been using these tanks to store and transport chemicals (Wood, 2011). The water and wastewater markets are also recognizing the merits of factory-manufactured FRP tanks that meet high performance standards for containment and longevity (LeGault, 2011).

16.3.2 FRP rebars

Another outstanding example is FRP rebar in lieu of steel rebar in highly corrosive environments (Malnati, 2011). Concrete is a material that is very strong in compression, but relatively weak in tension. To compensate for this imbalance, steel reinforcing bars are embedded into concrete to carry the tensile loads. However, steel inherently corrodes (an electrochemical reaction) under salt exposure, leading to rusting. As rust takes up a greater volume than steel, its parent material, rust, causes severe internal pressure on the surrounding concrete, leading to cracking, spalling, and ultimately, concrete failure in tension due to rust-induced hoop stress. This is extremely serious when concrete is exposed to salt water, as in bridges where salt is applied to roadways in winter, or in marine applications.

FRP rebar appears to be the best solution to tackle this problem and offers a number of benefits to the construction of our nation's infrastructure, including bridges, highways, and buildings. It is light weight (a quarter the weight of steel), strong (about twice the strength of steel), impervious to chloride ion and chemical attack, free of corrosion, transparent to magnetic fields and radio frequencies, and nonconductive for electrical and thermal loads. FRP rebars are commercially available on the market and they are mostly made from unidirectional glass fiber-reinforced thermosetting resins.

WVU-CFC researchers started FRP rebar application in 1986 for a hospital building, but it was after eight years of research when the first vehicular bridge, McKinleyville Bridge, was built in the US to use FRP rebars in a concrete deck (Fig.16.2). McKinleyville Bridge is located in Brooke County, the northern panhandle of West Virginia. It is a 180-feet long, three-span, continuous integral abutment bridge accommodating two lanes of traffic. The selection of constituent materials and the manufacturing processes for FRP rebars were given careful consideration. Screening of several types of resins and fibers under harsh environments was extensively researched at the WVU-CFC (Vijay, 1999). The GFRP rebars were placed in the concrete deck as top and bottom layers in the transverse and longitudinal directions as shown in the insert of Fig. 16.2. McKinleyville Bridge deck with FRP rebars has been in service for 16 years.

With support from the sponsors and contractors, in 2007, WVU-CFC researchers completed the nation's first continuously reinforced concrete pavement (CRCP) test section with GFRP rebars, along with steel rebar–CRCP test segment for comparison (Fig. 16.3). These test segments are located on Route 9 in Martinsburg, in the northeastern corner of West

FRP composites in environmental engineering applications 417

16.2 McKinleyville Bridge with concrete deck reinforced with FRP rebars, Brooke County, WV, built in 1996.

(a) (b)

16.3 Pavement with FRP rebar, WV Route 9, Martinsburg, WV.

Virginia and are being studied for their performance. Field studies show that GFRP rebar offers a low life-cycle cost option for reinforcement in concrete pavements (Chen *et al.*, 2008). It is anticipated that FRP rebar reinforced pavements will offer many years of additional service life as compared to steel rebar reinforced pavements. There have been many other successful field implementations, in particular using GFRP rebars in bridge deck applications in WV and many other states (FHWA, 2001; Gremel, 2007). WVU-CFC has been working with Federal and State agencies and private industries for over 25 years in promoting and implementing FRP composite products on the US highway system, including construction or deck replacement of over 100 bridges with FRP bridge deck, FRP bridge superstructure, FRP reinforcing bar as well as FRP dowel bar for concrete pavements.

16.4 Fiber-reinforced polymer (FRP) composites in sea-water environment

Few materials can survive long under the aggressive sea-waterfront environment, i.e. onslaught of sea waves, impact from vessels, corrosive salts, sand and pebble erosion, high atmospheric humidity, inter-tidal wetting and drying, sun, marine borers, and immense storm forces. Historically, steel has been the primary structural material used for ships and submarines. Steel structures make up the largest weight group of any ship, typically contributing 35–45% of the overall vehicle weight (Beach and Cavallaro, 2002). This fact implies that ship structures have a major influence on the overall characteristics such as displacement, payload, signatures, combat system effectiveness, and life-cycle cost. Currently, 52% of a ship's manpower is focused on maintenance, because the existing primary construction material – steel – requires constant maintenance to avoid rapid rate of corrosion (Greene, 2003). Costs of spare parts and associated downtime needed to repair corroded structures and hardware severely hamper the operational readiness of a ship.

16.4.1 All-composite deckhouse

FRP composites offer a potential solution to improve performance, survivability, and reliability of future naval ships and submarines. The US Navy is currently expanding the use of composites in the first of a new family of advanced, multi-mission destroyers, known as the DDG-1000 *Zumwalt* class (LeGault, 2010). The DDG-1000 destroyer is designed to support both sea-based and land-based missions. It features a 'tumblehome' wave piercing hull and an upper section deckhouse made predominantly of fiber-reinforced sandwich composites. Both the hull shape and composite deckhouse are intended to reduce the ship's radar footprint.

The all-composite deckhouse superstructure of DDG-1000 is illustrated in Fig. 16.4. It is approximately 130 ft long by 60 ft wide by 40 ft high (39.6 m by 18.3 m by 12.2 m), divided into four levels, and is made of balsa-cored glass and/or carbon/vinyl ester sandwich panels (LeGault, 2010). Each superstructure will use approximately 200,000 square feet of flat composite sandwich panels. The all-composite superstructure not only reduces infrared and radar signatures but also reduces topside weight and total ship tonnage, and lowers construction and maintenance costs.

A sandwich panel can be defined as a three-layer construction, i.e., two thin facesheets (skin) and a thick core (Marshall, 1998). The skin is thin and stiff with high strength, while the core is thick and lightweight. A good sandwich construction requires the core to be strongly bonded to the skin

FRP composites in environmental engineering applications

(a)

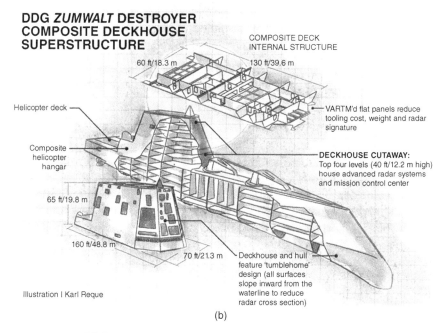

(b)

16.4 DDG-1000 *Zumwalt*: stealth warship.

so that the core can transfer loads from one facesheet to another; thus, the core and skins will act in unison offering greater stiffness than the facesheets alone. Typically, the thickness ratio of core to skin of a composite is in the range of 10 to 20.

16.4.2 Vacuum-assisted resin transfer molding (VARTM) versus pultrusion

A vacuum-assisted resin transfer molding (VARTM) process has been selected as the manufacturing method used to produce the sandwich panels for the first DDG-1000. In a VARTM process, dry reinforcements in the form of mats, rovings, or fabrics are pre-shaped and manually oriented into a skeleton of the actual part, known as the preform. After the preform is inserted into a tool (typically comprises one mold surface and one bag surface), the resin is injected at low pressures into the closed mold. During resin injection, a vacuum is applied to reduce voids and assist infusion of the resin into the fabric, before allowing the resin to cure at room temperature for 12 to 24 hours. Its advantages include low tooling cost, low volatile emission, low void content, and design flexibility for large and complex parts, but the process is labor-intensive and joining of VARTM panels is a challenge.

To overcome the limitations of VARTM panels, researchers at WVU-CFC, in close collaboration with composites industry personnel and government researchers, have been developing an automated pultrusion process for producing composite sandwich panels (Fig. 16.5; Liang *et al.*, 2005). Many pultruded profile shapes such as angles, beams, channels, tubes, bars, rods, plates, and sheets are commercially available, but to the authors' knowledge, the process had not been used to produce thick sandwich panels (such as the target 2.25–3.5″ panel). The pultruded sandwich panels have resulted in improved mechanical performance and reduced production costs, thus producing less expensive and more durable deckhouse (Liang and Hota, 2012).

Pultrusion is a process where FRP composites are produced continuously at speeds ranging from a couple of inches to a couple of feet per minute, through a heated die of desired cross section, i.e., no part length limitation. The reinforcements are in continuous forms such as rolls of unidirectional roving, biaxial fabric, or multiaxial fabric, which are properly positioned by a set of creels and guides for subsequent feeding into the resin bath. As the reinforcements are saturated (wet-out) with the resin in the resin bath and pulled into the forming and curing die, the heat curing of the resin is initiated from the preheated die, leading to a rigid profile. The advantages of the pultrusion process include:

- high fiber content,
- high cure percent age,
- minimal kinking of fibers/fabrics,
- rapid processing,
- low material scrap rate, and
- good quality control.

FRP composites in environmental engineering applications 421

16.5 Pultrusion of 4′ GFRP composite sandwich panels with integral joint edges: (a) impregnated fabric entering into the forming die, (b) puller stretching sandwich panel out of the die, (c) female joining profile (groove) of panel, (d) sandwich panel passing through cut-off saw (Bedford Reinforced Plastics Inc.).

The process disadvantages are:

- sometimes local inadequate or non-uniform fiber wet-out,
- die jamming,
- die size/geometry limitation, and
- initial capital investment and die cost (Goldsworthy, 1982).

Hence, a glass fiber-reinforced vinyl ester composite sandwich panel composed of ¼ inch thick facesheets with a 3 in thick balsa core was pultruded and evaluated for its thermo-mechanical properties, including joining efficiency (Fig. 16.6). The study has concluded that the pultruded panels are about 15–20% stronger and stiffer and 50% cheaper than VARTM panels. In particular, 8-ft wide sandwich panels were made through adhesive bonding of two 4-ft wide modular panels (with tongue and groove joint profiles using three layers of 24 oz/sq biaxial glass fabric as external reinforcement) and had 100% joint efficiency under both shear and bending

16.6 Full-scale panels under four point bend loading: (a) 4′ × 10′ with span 100″, (b) joined panel of 4′ × 5′ with span 80″.

with failure in the balsa core and away from the joint (Fig. 16.6). The pultrusion process is proven to be a viable manufacturing technique to produce high-quality composite sandwich panels for shipbuilding applications, in lieu of labor-intensive and costly steel structures. The authors are currently working with Huntington Ingalls Industry, Ingalls Shipbuilding to evaluate three types of joint configurations of the pultruded sandwich panels to build most typical joint configurations, including flat, 'L', 'T', and cruciform. These joining configurations will enable the fabricators to assemble an actual ship structure, such as a deckhouse (Liang and Hota, 2012).

16.4.3 Sheet piling

FRP composites in the form of panels, pipes, and posts, in addition to pultruded standard shapes can find broad applications in every type of seawaterfront facilities. These applications include: decking, walkways, platforms, ship-to-shore bridges, fenders, docking systems, retaining walls, crosswalks, moorings, cables, piles, piers, underwater pipes, railings, ladders, handrails, and many others, as representatively shown in Fig. 16.7. For example, WVU-CFC has recently tested several composite sheet piles made by Creative Pultrusion Inc. as an application to protect soil erosion near sea-front homes (Hota and Skidmore, 2012). These FRP products can survive under constant exposure to saltwater and salt air, and will not corrode, rust, or create sparks.

16.4.4 Cold water pipe for ocean thermal energy conversion (OTEC)

For the past four years, the US Department of Energy (DOE) and Lockheed Martin have been collaborating to develop innovative technologies to

16.7 FRP handrail, ladder, platform, and walkway for sea waterfront facilities.

enable ocean thermal energy power generation (PRNewswire, 2008; Chiles, 2009). Ocean thermal energy conversion (OTEC) uses the ocean's thermal gradient to drive a heat engine (Fig. 16.8). Since the ocean's temperature difference is relatively small, large volumes of seawater must be used to generate commercial levels of power. The fabrication and installation of large diameter cold water pipe (CWP) that reaches depths of thousands of feet (~3000) represent one of the largest technical challenges to successfully install and operate an offshore OTEC system (Chiles, 2009). Figure 16.8 shows a section of 4 m (13 ft) diameter FRP CWP at the Lockheed Martin facility. The goal is to manufacture a 10 m (33 ft) diameter pipe that will reach depths downwards of 1,000 m (3300 ft) in the bottom of the ocean (Miller, 2011). On this project, WVU researchers provide technical consultation and conduct testing and evaluation of composite materials and FRP CWP sections with emphasis on FRP durability and service-life prediction in a seawater environment (Dittenber and Hota, 2010).

16.5 Fiber-reinforced polymer (FRP) composites in coal-fired plants

More than half of electricity in the United States is generated from coal. There are currently 14 coal-fired electricity generating facilities located in West Virginia. Environmental concerns and increasing demand for energy have created a situation where flue gas generated by coal-fired power plants, old or new, needs to be cleaned before being rejected into the atmosphere. Early in the 1970s, the US government's regulation of stack emissions from coal-fired plants led to the deployment of flue gas scrubber systems that remove sulfur dioxide and mercury from the flue gas produced when coal is burned (Fig. 16.9). These flue gas scrubber systems require corrosion-resistant chimney liners to resist corrosive chemicals.

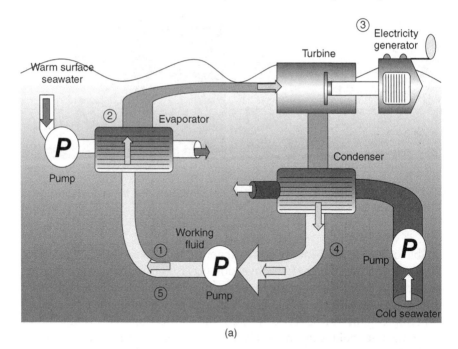

16.8 Ocean thermal energy conversion cold water pipe.

FRP composites in environmental engineering applications

16.9 Coal power plants: (a) no air pollution control devices; (b) with air pollution control devices.

16.5.1 Chimney liners

FRP composites have been successfully applied as chimney liners and gas ducts in power plants for many years due to their non-corrosive properties, ease of fabrication, and cost effectiveness. Recently, more stringent requirements have led to new greenhouse gas scrubber technologies such as jet bubbling reactors (JBR) and have further expanded the use of FRP composites. FRP composites have proven to be durable both structurally and chemically when used in the flue gas desulfurization (FGD) process of a coal-fired power plant (Southern Company Services, 2002). FRPs are also used in water piping, storage tanks, top ash and fly ash pipe, and cable trays in many coal-fired plants, while more than 70% of new and replacement field erected cooling towers in the United States are constructed with pultruded FRP structures. FRP composites enable coal-fired power plants to be operated under more environmentally friendly conditions.

Again, these applications involve the mass use of FRP composites for large diameter FRP structures. A recently built coal power plant in Springfield, IL has a 440-ft (134-m) tall concrete chimney with a FRP composite liner that is made of 27 segments that were fabricated using a filament winding method (Lucintel, 2008). Each segment is 13.5 ft tall by 15 ft in diameter. For joining in the field, the segments were aligned to within 0.25-inch (6.35-mm) tolerance. After surface preparation, the segments were then bonded together with interior and exterior laminates of 0.375-inch thick fiberglass/resin composite. Figure 16.10 shows a module of FRP liner and connection elbow on an International Chimney worksite near Morgantown, WV, while Fig. 16.11 shows a view of stack liners installed inside a

16.10 Large diameter FRP chimney flue liner: (a) module liner section; (b) connection elbow.

16.11 View of stack liner installation inside a power plant chimney.

concrete power plant chimney (Kelley and Schneider, 2008). International Chimney installed the FRP liner at Fort Martin Power Plant, Maidsville, WV in 2009. The plant has a 60 ft diameter × 529 ft high reinforced concrete chimney that took about two rows of 400 ft tall FRP liners that are 25 ft in diameter, in addition to an interconnecting elbow. Each liner segment is 25 ft in diameter and 31 ft high, as shown in Fig. 16.10.

16.5.2 Cooling towers

American Electric Power (AEP) is one of the largest electricity suppliers in the United States, delivering electricity to more than 5 million customers in 11 states. AEP ranks among the nation's largest generators of electricity, owning more than 38,000 megawatts (MW) of generating capacity in the

US, with individual unit ratings ranging from 25 MW to 1,300 MW. The authors have been providing advice and services to AEP Engineering on the aging and durability of GFRP composites for JBR and FGD applications since 2009.

There are a total of 15 hyperbolic and 30 mechanical draft cooling towers in the AEP system. These towers utilize a cross-flow or counter-flow thermal transfer design, and almost all of the cross-flow towers are treated wood structures. AEP started using pultruded glass fiber reinforced (GFRP) composite structures in cooling towers in 2008. AEP replaced four cross-flow mechanical draft towers during a period from 2008 through May 2010, and a counter-flow mechanical draft tower was built for a new unit in 2009. All five of these new towers were constructed using GFRP structures. They are the first batch of in-service FRP composite cooling towers on the AEP system (Cashner, 2011).

The AEP engineering team has been working on converting a hyperbolic tower structure from a cross-flow to a counter-flow design as part of Cardinal Plant Unit 3 FGD Project that is owned by Buckeye Power Inc. (BPI). The cold water basin is roughly 385 ft in diameter with the bottom of the concrete shell measuring 262 ft in diameter. Fig. 16.12(a) shows a general view of the tower. The heat transfer area, distribution pipes and drift eliminators are located about 35 ft above the cold water basin and are supported by a structure composed of pultruded glass fiber-reinforced vinyl ester columns. The columns measure 5.2 inches square, 3/8 inch thick, and were supplied in full lengths (e.g., no splices). There are roughly 464 columns which are laid out on a 12 ft by 12 ft grid. Figure 16.12(b) shows the duct work entering the tower and Fig. 16.12(c) the FRP beam-column latticed structure in the basin.

FRP composites have become the structural material of choice in industrial cooling towers in view of their superior performance in hostile environments (such as high temperature, wet, corrosive, abrasive, and sustained loading) and other beneficial properties. The design flexibility of FRPs has allowed new types of cooling tower to be developed which are more efficient and cost-effective than previous designs with conventional materials. The modular construction systems provide structures of high integrity that can be rapidly installed. The desirable environmental properties of FRP composites also aid compliance with the increasingly stringent legislation.

16.5.3 Utility poles

Utility companies rely on transmission and distribution poles (Fig. 16.13) to connect to end users. Currently, there are 130 million utility poles in-service in the United States, with about 97% of them being creosote treated wood poles, less than 2% steel poles and less than 1% composite poles.

16.12 Buckeye Power Inc. Cardinal Plant Unit 3 FGD Project, Brilliant, OH.

More than 70% of the utility poles in use are distribution poles in class 4 or class 5 ranging to 40 ft or less in height. Both new installation and replacement markets are worth about $4 billion per year (Hiel, 2001).

Wood poles require treatment with environmentally unfriendly toxic preservatives (e.g. creosote, copper chromium arsenate (CCA), penta-chloro-phenol) to resist rot, decay, etc., in order to yield a service life of about 30–35 years. However, these preservatives have been found to be hazardous to humans through leaching to the surrounding soil and water. This led to petitions for the EPA to ban the use of CCA, penta and creosote (Feldman and Shistar, 1997).

Utility companies are searching for alternatives to treated wood poles. FRP poles (Fig. 16.13), which represent one of three alternatives along with

16.13 FRP utility poles (courtesy of Duratel).

steel poles and concrete poles, are beginning to penetrate into both the distribution and the transmission pole markets. FRP poles have advantages such as non-conductive and non-corrosive properties over steel poles, and lighter weight, easier installation and better ductility over prestressed concrete poles. Thus, FRP poles have been receiving greater attention from electrical utility and telecommunication companies. This is especially true now because mass production of FRP poles in a more cost-effective manner has been made possible in order to receive a greater market share, other than niche applications beyond the mountainous terrain or corrosive soils.

16.6 Fiber-reinforced polymer (FRP) composites in mining environments

FRP offers lightweight, high strength, corrosion resistance, humidity resistance, impact resistance, non-sparking, long-term durability, and other advantages that are essential for mining applications (Richter, 1999; Tusing, 2003). Many pultruded shapes have already found a wide range of applications in mining facilities, such as handrails, walkways, platforms, caged ladders, non-slip decking and grating (see Fig. 16.7).

16.6.1 Modular track panels

West Virginia is the second largest coal-producing state in the United States and has over 50 coal mines in production. Mine cars often derail (two or three times daily) in coal mines causing fatal injuries, fires initiated from sparks during derailment, and costly downtimes. To improve coal mining productivity and safety of miners during transportation, WVU-CFC researched GFRP modular track panels for the mining environment. The GFRP modular track panels comprised two box beams bonded together with epoxy resin (Fig. 16.14). The modular track panels were tested for bending behavior and load-sharing characteristics, between track modules and also between rails. Tests on the panels included static and fatigue loading at discrete locations to determine the response of the beams in a simulated mine foundation. Static tests of the modular panels included both vertical and horizontal loading. Fatigue tests were used to determine the change in stiffness of the modules. The study indicated that the FRP panels would distribute load more efficiently, possess good fatigue performance, and be able to sufficiently sustain the loading. FRP panels are well suited for field applications and the light weight of FRP panels would allow the miners to install these panels at ease within the mine. In addition, FRP panels would alleviate the derailment and other structural problems (Tusing, 2003).

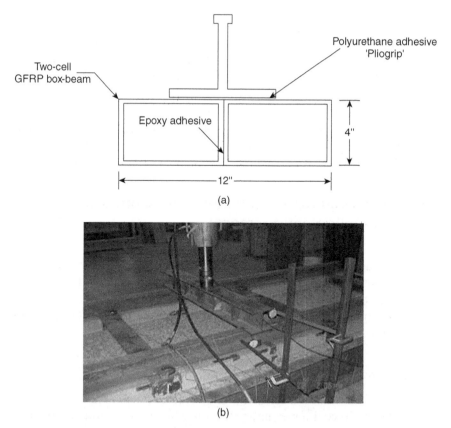

16.14 Cross section of GFRP track panel and test specimen (Tusing, 2003).

FRP composites in environmental engineering applications 431

16.6.2 Geosynthetics

Zijin Mining group is a leading gold, copper, and non-ferrous metals producer and refiner in China, with gold output of 69 tons in 2010. In July 2010 two leaks of waste acid copper solution at the Zijinshan Gold and Copper Mine polluted the Ting River in Fujian province and poisoned hundreds of tons of fish. The waste acid was found to have leaked from the cracking of a GSE high-density polyethylene geomembrane liner. The company believed that the cracking was caused by a shearing force created by the accumulation of water beneath the liner. Again in September 2010, there was a deadly tailing dam collapse at its Yinyan Tin Mine that killed four people. The accident was later attributed to a landslide triggered by heavy rains from Typhoon Fanapi. The above accidents have alerted engineers at Zijin Mining College of Fuzhou University, Fujian, China, to re-evaluate the material requirements under specific loadings.

Zijin engineers reached out to WVU-CFC researchers for possible collaboration. For example, Zijin engineers are looking into the potential use of lightweight, high strength, corrosion-resistant, durable FRP composite materials for construction of new dams or strengthening of existing dams. WVU researchers are proposing to use geosynthetics as soil reinforcements to construct a tailing dam as illustrated in Fig. 16.15 (Wu, 1994; Koerner, 2012). Geosynthetics are available in a wide range of forms and materials, including geotextiles, geogrids, geomembranes, geofoam, geocells, or a combination of the above. Each type of geosynthetic has at least one of the following functions:

- separation,
- reinforcement,
- filtration,
- drainage, and
- containment (Wikipedia, 2012).

The tailing dam needs to be strong but still allow for drainage without soil loss. The geosynthetic under consideration would be geocells that are made of strips and can be expanded into three-dimensional, stiff honeycombed cellular structures, resulting in a confinement system when infilled with compacted soil. The cellular confinement reduces the lateral movement of soil particles, thereby maintaining compaction, retaining the earth, and protecting the slope. Geogrids offer open, gridlike configurations and as reinforcement materials, play similar roles to those of geocells. In addition, geotextiles are flexible, porous, woven or knitted synthetic fibers/fabrics and can function as reinforcement or drainage or filtration or separation, depending on design, while geomembranes are thin, impervious sheets of polymeric material and function as containment. Geomembranes

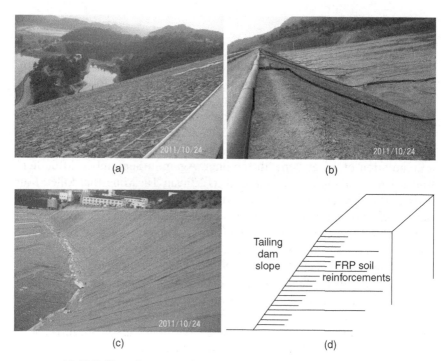

16.15 Tailing dam and design concept of a FRP strengthened mining tailing dam: (a) downward view of a Zijin tailing dam; (b) tailing dam to store mining waste slurry; (c) new pond with geomembrane to store waste acid solution; (d) concept of using FRP soil reinforcement for durable tailing dam.

are being extensively used as linings of waste acid copper solution reservoirs in Zijin Mines as shown Fig. 16.15.

16.6.3 Pipelines

Zijin has over 270 km of various pipelines across the mine, with a typical pipeline diameter of 22 inches. They are typically HDPE pipes of half-inch thickness. FRP pipes are only used when chemical and corrosion resistance is required. Figure 16.16 shows some venting systems made of FRP composites while extensive pipelines are also seen along a tailing dam. WVU-CFC has extensive experience with design, manufacturing, testing, and in-service monitoring of FRP pipelines. Figure 16.17 shows a 16 inch diameter FRP pipe being tested at WVU-CFC Laboratory.

Researchers at WVU-CFC are currently collaborating with Beijing Huade Creation Environmental Protection Equipment Corporation to develop FRP products for mining and environmental applications in China. Huade Corporation serves the coal mining and metal mining industries as

FRP composites in environmental engineering applications 433

16.16 Extensive uses of FRP vessels and pipelines at Zijin Mining Group, China.

16.17 A 16 inch FRP pipe being tested at WVU-CFC Laboratory.

well as coal-fired power plants. They manufacture and/or supply coal preparation equipment, fine coal preparation systems, tailing mine surface disposal and backfill treatment equipment, mill circuit clarification/de-slime/dewatering equipment, power plant flue gas desulfurization cyclones, and many others. The company has manufacturing capabilities in cast polyurethane, mold-pressed rubber, spray polymer, FRP composites, precise steel structure fabrication, ceramic liner equipment, and others. Huade is an associate member of NSF IUCRC Center for Integration of Composites into Infrastructures at WVU.

16.6.4 Rock bolting

The products under consideration by Huade include FRP rock bolt, vessels, pipelines, duct valves and fittings, cooling tower, spiral, safety helmet, electric fan, cable pipe, railing, grating, crane span, and others. Rock bolt (Fig. 16.18) is widely used in underground mining to provide support to the roof or sides of the cavity. It can be used in any excavation geometry and it is simple and quick to apply, and is relatively inexpensive. The installation can be fully mechanized. The length of the bolts and their spacing can be varied, depending on the reinforcement requirements. However, in aggressive environmental conditions such as in coal mines, steel bolts deteriorate in a matter of days rather than years. FRP rock bolt is particularly suitable in harsh chemical and alkaline environments because of its corrosion resistance. It is durable, lightweight, easy to install, non-conductive, dimensionally stable under thermal loading, anti-static rating, and can be cut without the danger of sparks. Currently, four production lines are being installed at the Beijing Huade facilities. Furthermore, spiral (Fig. 16.19) is widely used in coal preparation equipment and fine coal washing systems, and is being manufactured using a hand layup method. It is made of FRP composites with a wear-resistant coating.

(a)　　　　　　　　　　　　　　(b)

(c)

16.18 Rock bolt for underground mining (courtesy of Huade).

FRP composites in environmental engineering applications

16.19 Spirals in coal preparation process (courtesy of Huade).

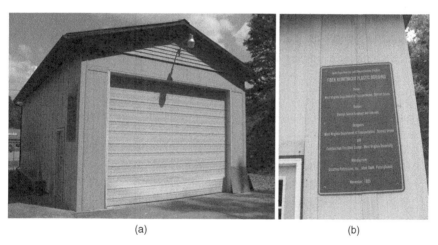

16.20 Multi-purpose FRP building, Weston, WV, built in November 1995.

16.7 Fiber-reinforced polymer (FRP) composites for modular building of environmental durability

Traditional housing focuses on uses of masonry, timber, steel, and concrete. FRP composites were initially used for small components, such as windows, canopies, doors, profiles, and other decorative features. The WVU-CFC team designed, manufactured, and constructed its first innovative FRP building in Weston, West Virginia, in November 1995 (Fig. 16.20). This experimental building was the result of a joint research and development effort among the US National Science Foundation, West Virginia Department of Transportation/Division of Highway and WVU-CFC. All multicellular panels were of standard size of 24 inches × 5.5 inches, with a wall thickness of 3/16 in. The cells of the panels contained polymeric insulation. The building is about 40 ft × 21 ft, with an inside height from slab to trusses of 14 ft. The weight of total FRP material used in the construction was approximately 9500 lbs.

All the components used for this project were installed without the use of mechanical equipment such as a crane because of its lightweight nature. FRP parts were delivered precut, thus speeding up the entire construction process. This building is being used as a multi-purpose facility requiring heat, insulation, electric power, and ventilation. A 2009 inspection revealed that the building has been performing excellently for the past 16 years and looks new, as shown in Fig. 16.20. The building with maintenance-free interior and exterior walls has demonstrated outstanding chemical and environmental durability. A newer modular unit is shown in Fig. 16.21. Currently, the WVU-CFC researchers are developing FRP modular dwellings using natural composites integrated with many green concepts towards future sustainable buildings.

FRP composite houses are now readily available on the markets (CBS, 2009; Stewart, 2011). Such modular FRP houses have the following features:

- no maintenance (no painting and do not deteriorate from weather, rot, or insect infestation);
- lower heating and cooling costs (the modular FRP panels have built-in insulations and the dome shape further increases energy efficiency);
- high structural strength (the curved surface of the panels reduces wind resistance, enabling the house to withstand hurricanes);

16.21 FRP composite house at Bedford Reinforced Plastics Inc. facility (2008) (courtesy of Bedford Reinforced Plastics Inc.).

FRP composites in environmental engineering applications 437

- quick construction (modular design for ease of erection);
- good earthquake resistance (the FRP panels flex instead of breaking);
- high water resistance (completely sealed from the ground up);
- portability (a FRP house can be easily disassembled and relocated to a new site).

These buildings serve as disaster-resistant shelters, military barracks especially at cold and high altitude locations, school buildings, industrial factory and warehouse buildings, and large dormitory settings for workers in remote locations, greenhouses, etc.

16.8 Fiber-reinforced polymer (FRP) wraps

The infrastructure in the United States is deteriorating and aging. The constructed facilities are being used far in excess of their original design life and are close to being obsolete; these structures are vital to support the economic, transportation, and societal functions. Therefore, the most attractive alternative would be to extend the physical life of the existing infrastructure in a cost-effective manner. By extending the service life of existing structures, the need to demolish, dispose of, and reconstruct existing structures is reduced, leading to lower life-cycle costs, as well as minimal energy use and construction waste minimization (Edwards *et al.*, 2009).

FRP retrofitting has been widely used to strengthen civil and military structures as an effective disaster prevention approach or to restore the damaged structures after disasters such as hurricanes and earthquakes (see, for example, Saadatmanesh *et al.*, 1997). Recently, FRP composites have been receiving more attention as a material of choice to strengthen existing structures to continue serving their functions (Mirmiran *et al.*, 2008; Belarbi *et al.*, 2011). In the United States, many of the existing highway or railroad bridges have either reached the end of their service life or require rehabilitation to continue in service. Due to decreased funding levels for new construction, government agencies are interested in utilizing GFRP wraps to rehabilitate structures at a fraction of the outright replacement cost, while extending the structural service life for a few more decades. The advantages of FRP wraps include minimal traffic disruption, efficient labor utilization, ease of rehabilitation, optimization of load transfer, and cost effectiveness.

16.8.1 Strengthening of concrete structures

WVU-CFC has been actively involved with advanced FRP wrapping technology development, including specific design methods, material selection, field installation procedures, performance requirements, and subsequent inspection techniques since 1988 (FHWA, 2001). The rehabilitation of cracked reinforced concrete beams of the floor of a San Antonio, TX library building, using Tonen Carbon Tow Sheet was completed in 1993 after testing and evaluating the steel reinforced concrete beams stiffened with Tonen Carbon Tow Sheet at the WVU-CFC laboratory. The creep behavior of such stiffened beams and the long-term performance of bond between the concrete surface and Tow Sheet were also evaluated before field implementation. Figure 16.22 shows the details of rehabilitation of T-shaped reinforced concrete superstructure girder of Muddy Creek Bridge, Preston County, WV using carbon FRP wraps.

16.8.2 Strengthening of timber structures

WVU-CFC's research and development on repair and rehabilitation of wood railroad bridges using FRP composites began in 1999. The laboratory

16.22 Rehabilitation of T-section reinforced concrete superstructure using carbon FRP wraps, Muddy Creek Bridge, Preston County, WV (October 2000): (a) blue primer coating and wrap application; (b) pressing of wrap for full bond with concrete; (c) carbon wrapped concrete beam; (d) gray paint application.

FRP composites in environmental engineering applications   439

testing revealed that compression testing of previously failed creosote-treated railroad stringer, after wrapping, regained 80% of strength using GFRP wraps (Petro et al., 2002). The fieldwork was conducted on two railroad bridges on South Branch Valley Railroad (SBVR) lines in Moorefield, WV in summer 2000. Field static and dynamic testing (5 mph, 10 mph, 15 mph) using GE 80 ton locomotive supplied by SBVR was conducted before and after applying wraps. IR thermography measurements were also carried out to assess the bond status of rehabilitated members. The results turned out to be a success and the rehabilitated members have been performing satisfactorily for the past 12 years. Figure 16.23 shows a group of

16.23 Retrofitting of railroad bridges using FRP wraps without interrupting railroad service, SBVR, Moorefield, WV (July 2010).

photos showing how damaged piles of timber railroad bridges on SBVR lines in Moorefield, WV, were rapidly rehabilitated and restored *in-situ* without affecting the rail traffic by using GFRP composites, in summer 2010. These timber bridges consisted of total span lengths varying from 75 ft to 1,200 ft with timber pile bents spaced 12–20 ft apart. The deteriorated piles were cracked, heart-rotted, and damaged to varying lengths. This rapid rehabilitation technique can be used on various other structural members including steel and reinforced concrete members in a cost-effective manner to extend the service life of structural systems. West Virginia Department of Transportation, Division of Highways is embarking on rehabilitating 400–500 concrete bridges using FRP composite wraps in the next five years (by 2018) because of their cost-effectiveness, minimal user inconvenience, and proven success.

16.8.3 Strengthening of waterway structures

The US Army Corps of Engineers maintains aging infrastructure along their navigable waterways and flood control facilities. For example, Tygart Dam near Grafton, WV was built in 1934 as the first of 16 flood control projects in the Pittsburgh District (Fig. 16.24(a)). Tygart Dam was constructed with the most concrete in any dam east of the Mississippi River with a staggering 324,000 cubic yards. This concrete gravity dam has an uncontrolled spillway and measures 1,921 ft long and 209 ft thick at the base. According to US Army Corps of Engineers, Tygart Dam protects areas from West Virginia to Pittsburgh, Pennsylvania and has prevented multi-billion dollars in flood damage to date. Fig. 16.24(b) shows Lock and Dam #52 with its wooden wicket gates in the up position, located on the Ohio River near Brookport, IL. The wooden wicket dam was completed in 1928

16.24 The aging infrastructure along the US waterways: (a) Tygart Lake Dam, Grafton, WV; (b) wicket gates at Lock and Dam #52, Ohio River (courtesy of Richard Lampo).

and the locking system consists of a main lock of 1,200 ft by 110 ft and auxiliary lock of 600 ft by 110 ft along the length of the river.

However, corrosion, materials degradation, and damage during operations are taking a toll on these aging facilities. In addition, the high costs associated with repair and replacement of critical components present many challenges in keeping open these waterways that are vital to the nation's economy and security. With recent material and processing advances, FRP composites offer the potential for repair, rehabilitation, and replacement of these critical structural components of the waterways at a reduced cost (Hota and Vijay, 2010). Additionally, greater durability can be attained with corrosion-resistant FRP components.

The researchers at WVU-CFC in cooperation with the US Army Corps of Engineers (USACE) are undertaking a significant effort to demonstrate the use of corrosion-resistant FRP composites for the repair and replacement of components on Civil Works navigation structures. FRP composites will be used to repair and replace select lock and dam components at Lake Washington Canal, Washington; Willow Island Lock and Dam, near Newport, OH; Heflin Dam near Gainesville, AL; Chickamauga Lock and Dam, near Chattanooga, TN; and Lock and Dam #52 on the Ohio River, near Brookport, IL. These components consist of miter-blocks, tainter-gates, recess filler panels, discharge ports, and wicket gates. For example, the concrete discharge ports on the Chickamauga Dam have been described to experience tension failures and need to be rebuilt. FRP composite sections will be used as an alternative to steel panels around the pier-like sections and grout will be pumped. These repairs will be completed underwater by divers, including necessary overlapping and adhesive bonding. In the case of wicket gates, current wicket gates used at lock and dams by USACE are fabricated from white oak and steel elements. These wooden wicket gates deteriorate quickly and the steel corrodes. The WVU-CFC team has been investigating FRP composites as an alternative to wood for direct replacement of wooden wicket gates. In addition, guidelines for the selection of FRP composite constituents, the design of the components, and installation of FRP composite components are being developed. A detailed economic analysis including life cycle assessment is carried out to compare the use of the demonstrated FRP composite materials with conventional repair or replacement using coated steel components and materials.

16.9 Recycling composites

16.9.1 Thermoplastics

Recycling is discussed more in the context of recycling of thermoplastic polymers than thermosetting polymers, because thermoplastics account for the majority of commercial usage. Americans consume approximately 60 billion pounds of plastics each year, while the annual thermosetting composites shipment is less than 4 billion pounds. Waste polymeric materials exist in three sources: domestic waste, industrial waste, and discarded plastics products. Unlike natural macromolecules, most synthetic polymers cannot be assimilated by micro-organisms. In the past, most of these plastic products were not reused and thus ended up in landfill. However, over the past two decades the disposal of plastics has become a serious concern to the environment, leading to government legislation for recycling.

Although there has been increasing public demand for recycling discarded thermoplastic products, in most cases the recycling of polymers is technically difficult and expensive. This is attributed primarily to the fact that post-consumer plastics are commingled. Mixed plastics have poor mechanical properties due to compatibility problems and thus have little value. Separating the chemically different plastics from each other, however, is expensive. Most of the implementation activities concentrate on recycling that is economically feasible, i.e., plastic wastes having slight contamination and higher resale value. These plastics can be recycled in the form of usable materials. For heavily contaminated plastic wastes which are difficult to sort into single polymer streams, a thermal recycling process is available. During this process, recovery of plastics is in the form of gaseous, liquid, or solid fuels (Liang, 2001).

16.9.2 Material recycling and thermal recycling

Material recycling (i.e. re-use) is usually implemented in the recovery of plastics when the process can acquire large quantities of reasonably clean polymers with very good purity. Recovered plastics can be reprocessed by formulating alloys, blends, or composites (known as ABC techniques) to upgrade their performance and bring them to desired levels of properties for applications (Liang, 2001). A material recycling process comprises collecting and sorting of waste plastics, shredding, washing, drying, and upgrading. The recovered plastics are generally reprocessed by extrusion into granules for normal plastics molding. Reuse of polypropylene recovered from automobile bumpers, for example, has shown good market potential. The recovered polypropylene was reinforced during reprocessing with waste cord-yarns from the tire industry, resulting in a grade of extrudable and injection-moldable, fiber-reinforced thermoplastics.

Thermal recycling plays a key role for unwashed waste plastics or rubber wastes (often heavily contaminated, multilayered, heavily pigmented, mixed, and unable to be recycled in the form of materials) to recover gaseous, liquid, or solid fuels, and sometimes oligomers or monomers. There are three principal thermal methods: pyrolysis, gasification, and hydrocracking. For example, the pyrolysis of acrylic polymers recovers 25–45% gas with a high heating value and 30–50% oil of rich aromatics. In addition, chemical processes such as methanolysis, glycolysis, hydrolysis, ammonolysis, and aminolysis can be employed for recycling of some types of plastics (typically polyesters, polyurethanes, and polyamides). This approach involves chemically decomposing macromolecules by chain cracking into monomers that can be reused for manufacturing new polymers.

As a classic example of reusability of thermoplastics, the recycled materials were reportedly used to make bridges that are strong enough to support a US Army tank (UPI, 2009). A pair of bridges were made entirely from more than 170,000 pounds of recycled consumer and industrial plastics using a patented materials technology integrated with a patent pending I-beam design. The bridges withstood several M1 Abrams tank crossings during the tests. The M1 Abrams weighs nearly 70 tons, making it too heavy for the vehicle to use most standard bridges and roads. The report (UPI, 2009) claims that the structures are less expensive to build than traditional timber, concrete, or steel bridges used on US military bases.

The construction industry can serve as an end-user of post-consumer plastics. Processes for recycling polymers into construction materials such as plastic lumber are being developed in many countries (Kibert, 1993). Polymer concrete can be made from PET beverage bottles or textile waste and fly ash wastes. The properties of the polymer concrete based on recycled PET were comparable to the polymer concrete made from virgin materials. Their good strength and durability properties make these polymer concretes suitable for effective uses in many construction applications such as utility, transportation, and building components, and the repair and overlay of pavements, bridges, and dams (Rebeiz, 1996; AbdelAzim, 1996). In the highway industry, it is reported that New Mexico has added recycled polymers as aggregate additives to asphalt in virgin and recycled hot-mix, in recycled cold-mix, seal coats, friction courses, and base courses. It is concluded that for restoration of asphalt roads, cold-mix recycling systems with polymer additives are superior and less expensive than any other system tested (Anon., 1987).

16.9.3 Material research

The material research and product development on plastics recycling at WVU-CFC have been continuing for 14 years, focusing on engineering plastics recovered from electronics shredder residue (ESR) (Liang and Gupta, 2000, 2001; Vijay et al., 2000). Polycarbonate (PC) and acrylonitrile-butadiene-styrene (ABS) are relatively expensive polymers that are used in significant quantities in the manufacture of computer, monitor, and printer housings. These products are discarded after being used for only a few years. Recycling these polymers after the end of their life is receiving greater attention and tremendous progress has been made in reusing these discarded plastics for a variety of applications (Vijay et al., 2000; Aditham, 2004; Kalligudd, 2010; Chada, 2012).

The objectives of the material research were to reuse these polymers in their original, high-value applications by blending with virgin polymers and/or produce acceptable, high-quality, low-cost, green products using as high recycle content as possible. To achieve these, each recycled polymer has to be characterized thermally, rheologically, and mechanically. Once the properties are defined, additives are introduced by mixing them to produce a specific compound. Since recycled polymers are recovered from unknown sources, approaches must be sought to minimize the batch-to-batch variations in properties in order to yield consistently high-quality compounds. Hence, four strategies were investigated by WVU-CFC researchers:

1. blending recycled polymers with chemically identical virgin resins,
2. blending recycled polymers with chemically different virgin resins,
3. adding short glass fibers to reinforce the recycled polymer blends, and
4. using molecular weight modifiers to adjust the average molecular weight distribution.

One of the findings was the '15% blending rule', i.e., to attain excellent properties, up to 15 wt% recycled polymer can safely be added to the virgin polymer without significantly altering properties of the virgin resin if the recycled polymer has a purity level of about 99% (Liang, 2001).

16.9.4 Product development

Both structural and non-structural applications of recycled PC and ABS polymers, with chopped or continuous glass fiber/fabric reinforcements, have been extensively investigated at WVU-CFC (Vijay et al., 2000; Aditham, 2004; Kalligudd, 2010; Chada, 2012). Some of the products developed include: guardrail post, offset spacer block, rectangular grids, rib-stiffened panels, sign posts and sign boards, dowel bars, window panels, and wood plastics composite (WPC). A couple of these products have been field-installed in the highway systems with the approval of West Virginia Department of Transportation (Aditham, 2004).

Recycled polymers are also engineered at WVU-CFC laboratory to manufacture full-scale railroad crossties that use end-of-life railroad wood ties as the core and recycled polymer composite as a shell (Kalligudd, 2010; Chada, 2012). Figure 16.25 shows a recycled GFRP composite railroad tie before and after demolding. These railroad ties have been extensively evaluated under static and fatigue loads in the laboratory (Fig. 16.26), followed by field installation in straight and curved locations and testing

FRP composites in environmental engineering applications 445

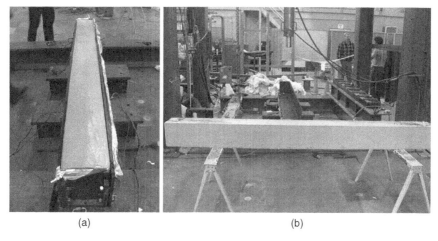

16.25 Manufacturing of recycled GFRP composite railroad ties (Chada, 2012): (a) before demolding; (b) after demolding.

16.26 Recycled GFRP composite railroad tie testing (Chada, 2012): (a) three-point bending test; (b) fatigue test in gravel bed.

under standard locomotive loads (Fig. 16.27). With over 12 million railroad ties being replaced annually in the United States, this green product is being negotiated for mass field implementation (Chada, 2012).

16.9.5 Thermosets

FRP composites with thermosets are generally believed to be more difficult to recycle than thermoplastics because liquid resin becomes rigid via chemical curing upon application of heat, and cured resins will not melt or flow upon reheating. Thermosetting polymers and FRP composites can

(a) (b)

16.27 Field testing of recycled GFRP composite railroad ties (256 Kips) on SBVR line in Moorefield, WV (Chada, 2012).

be recycled using the above-mentioned recycling processes. Automotive manufacturers have had good practice in recycling all the component materials used in their vehicles, including composites such as sheet molding compounds (SMC) and bulk molding compounds (BMC) (SPI, 1993). A material recycling of FRP composites can be viewed as grinding thermosetting materials into particles or powder, resulting in a particle size of about 1.5 mm, and using these particulates as fillers or aggregates. These scrap FRP fine powders can be used as aggregate for producing panels, cement paste and mortar, as well as in asphaltic concrete. As in the case of recycling of thermoplastics, the construction industry should play a critical role in the recycling of composite materials. In fact, the construction industry is one of the major end-users of FRP composites and consumes 45% of total FRP shipment in the United States (ACMA, 2011). Alternatively, the thermal recycling process can be used to recover reinforcing fibers for reuse, such as carbon fibers from cured fabric/epoxy composites (Allred *et al.*, 1996).

In addition, plastics including FRP composites can be collected with other combustible wastes and incinerated to recover heat value. The energy recovery of wastes by incineration could utilize all thermoplastic and thermosetting wastes when no reuse of the materials is possible. It is reported that incineration systems with energy recovery can recover about 8,000 thermies per ton of unsorted plastics (Dawans, 1992).

To facilitate the end-of-life recycling of polymers and composites, a 'design for recycling' concept should be incorporated into the development of next generation products. One of the general principles is to minimize the number of material types and select compatible polymeric materials when designing a new product. Another important consideration is to avoid contaminants such as labels, adhesives, nails, and metal plates. The success of composite recycling depends on various factors including government policies, industry commitment, technology, and public support.

16.10 Green composites

'Green' composites are biocomposites that are completely biodegradable and made of natural fiber-reinforced biodegradable resins, while the term 'natural fiber reinforced composite' (NFRP) generally refers to natural fibers in any sort of polymeric matrix (thermoplastic or thermoset; natural or synthetic). According to Patel and Narayan (2005), a sustainable development is a 'development that meets the needs of the present without compromising the needs of future generations to meet their own needs.' This definition implies that sustainable development must include environmental, economic, and social factors. Biocomposites are promising sustainable composite materials due to their substitution of renewable resources for fossil fuel-based polymers and synthetic fibers, lower greenhouse gas emissions, closure of the cyclical loop from raw material growth to biodegradation, potential for lower production costs, and opportunities for growth in agricultural and chemical industries, including new jobs (Dittenber and Hota, 2012; Baillie, 2004). While the production of cement and other building materials results in a large amount of carbon emissions, efficiently produced natural composites would provide a minimal carbon footprint due to their natural ability to absorb CO_2.

FRP composites are used in a wide range of non-structural and structural applications. FRPs are 'greener' in terms of embodied energy (overall energy required in a process to make a product) than conventional construction materials. In view of durability, sustainability, energy efficiency in manufacturing and other advantages, a growing number of architects and building owners are now choosing FRP-based building products over conventional materials. Several case studies have proven that FRP components for building construction are good thermal insulators, economical, strong, dent-resistant, scratchproof, have good acoustic barrier properties and are user friendly (Nadel, 2006). However, GFRP composites are made from fossil fuel-based polymers and synthetic fibers. Increasing awareness about LEED (Leadership in Energy and Environmental Design) ratings has created strong demand for innovative eco-friendly materials of low carbon footprint.

16.10.1 Bioresins and natural fibers

In response to growing demands to curb the usage of petroleum-based thermosetting resins due to environmental as well as economic and resource sustainability issues, researchers have been developing bio-based and sustainable resin systems for composite manufacturing. Upon processing, plant-based materials such as soy, crambe, linseed, and castor oil produce unsaturated triglycerides. The triglycerides constitute unsaturated and saturated fatty acids which can be polymerized to form an elastomeric network which can replace petroleum-based resins. With constant emphasis from markets pertaining to LEED, through the US Department of Agriculture's BioPreferred Program, Ashland Chemicals, Inc. initiated production of soy-based (bio-based) resins for commercial applications (CT, 2008).

Natural fibers used in composites are mostly derived from plant fibers. Among the natural fibers, stem-based fibers such as flax, kenaf, jute, hemp, and leaf-based fibers such as abaca rattan and sisal are considered important with respect to their specific properties and compatibility for composite manufacturing (Drazl et al., 2004). Amongst these fibers, high grade flax fiber's mechanical properties are nearly on a par with conventional E-glass fiber for composite reinforcement. Flax fibers are available on the market for half the cost of conventional E-glass fiber. A comparison between natural fibers and conventional fibers with respect to their mechanical properties and cost ratio are presented in Table 16.1.

16.10.2 Pultrusion of natural composites

Engineers at Bedford Reinforced Plastics (BRP) Inc. took a lead in utilizing plant-based naturally renewable fiber reinforcement with a commercially available bio-based resin system to manufacture 'green' composites on an industrial-scale pultrusion line (Fig. 16.28), resulting in more environmentally friendly composites having less embodied energy compared to conventional GFRP composites (Mutnuri et al., 2010). The natural fiber used was flax with two different densities, 225 gsm and 685 gsm, while Ashland

Table 16.1 Mechanical properties and cost ratio of natural fibers (Gaceva et al., 2007)

Fiber	Specific gravity (g/cm^3)	Tensile strength (ksi)	Tensile modulus (msi)	Specific strength (ksi/g cm^3)	Specific modulus (msi/g cm^3)	Cost ratio
Sisal	1.20	11.6–72.5	0.44–14.2	10–61	0.4–12	1
Kenaf	1.20	135	7.69	112	6.4	1
Flax	1.20	290	12.3	242	10.3	1.5
E-glass	2.60	508	10.4	195	4.0	3
Kevlar	1.44	566	19.0	393	13.2	18
Carbon (standard)	1.75	435	34.1	249	19.5	30

16.28 Pultrusion of natural fiber-reinforced bioresin composite panel (Mutnuri *et al.*, 2010).

ENVIREZ 70301 resin was used, which contains 22% bio-derived content and is compatible for the pultrusion process. The engineers at BRP explored processing variables and thermo-mechanical responses of environmentally benign natural composites. Moreover, the possibilities of using 'green' composites in various applications such as thermal insulators and sound transmission barriers are being investigated for their feasibility in FRP markets (Mutnuri *et al.*, 2010). For discussion purposes, the mechanical properties of pultruded natural composites are presented in Table 16.2 along with those of hand layup samples. Note that natural fibers used in pultrusion did not experience any surface treatment while the natural fibers used in hand layup samples went through treatment in alkali solution. As seen from Table 16.2, all natural composite samples resulted in poor strength and stiffness properties under tension as well as flexure. The alkali solution treatment did not improve the properties significantly. The pultruded natural composite sample had a high fiber volume fraction but that high fiber content did not translate to good mechanical properties.

Table 16.2 Tensile and flexural properties of NFRPs

Composite	Resin	Fiber	Natural fiber treatment	Fiber FVF	Manufacturing method	Tensile strength (ksi)	Tensile modulus (msi)	Flexural strength (ksi)	Flexural modulus (msi)	Source
NFRP 2008	Envirez 70301	Flax 225gsm	untreated	0.38	Hand layup	8.52	0.78	10.15	0.44	Mutnuri et al., 2010
NFRP 2008	Envirez 70301	Flax 225gsm	NaOH	0.34	Hand layup	9.54	0.84	11.53	0.47	Mutnuri et al., 2010
NFRP 2008	Envirez 70301	Flax 685gsm	untreated	0.43	Hand layup	8.00	0.70	9.47	0.44	Mutnuri et al., 2010
NFRP 2008	Envirez 70301	Flax 685gsm	NaOH	0.39	Hand layup	6.91	0.64	9.29	0.42	Mutnuri et al., 2010
Hybrid 2009	Envirez 70301	Flax + glass	untreated	0.46 + 0.08	Pultrusion	26.00	1.83	15.94	0.69	Mutnuri et al., 2010
NFRP 2009	Envirez 70301	Flax 685gasm	untreated	0.62	Pultrusion	12.31	1.45	16.35	1.01	Mutnuri et al., 2010
NFRP 2011	Derakane 510A	Kenaf	untreated	0.32	Compression	n/a	n/a	22.40	0.87	Dittenber 2012a
NFRP 2011	Derakane 510A	Kenaf	NaOH	0.42	Compression	n/a	n/a	30.25	1.13	Dittenber 2012a
NFRP 2011	Derakane 510A	Kenaf	Silane	0.33	Compression	n/a	n/a	24.27	1.06	Dittenber 2012a
NFRP 2011	Derakane 510A	Kenaf	Silane + NaOH	0.49	Compression	n/a	n/a	29.64	1.20	Dittenber 2012a
NFRP 2011	Derakane 8084	Kenaf	untreated	0.32	Compression	n/a	n/a	25.24	0.97	Dittenber 2012a
NFRP 2011	Derakane 8084	Kenaf	NaOH	0.39	Compression	n/a	n/a	30.76	1.22	Dittenber 2012a
NFRP 2012	Derakane 510A	Kenaf	untreated	0.35	Compression	25.57	3.29	26.57	1.10	Dittenber 2012b
NFRP 2012	Derakane 510A	Kenaf	NaOH	0.39	Compression	23.26	3.31	29.11	1.10	Dittenber 2012b

Natural fiber-reinforced composites offer improved sustainability and eco-friendly characteristics, and have the future potential to be lighter-weight and lower-cost than many synthetic composites, as well as being easier to handle. Natural composites are being evaluated for some automotive applications as interior paneling, but are not yet in use as primary structural elements due to their perceived lower mechanical properties and reduced environmental performance (Netravali *et al.*, 2007; Dittenber and Hota, 2012).

16.10.3 Hydrophilic nature of fibers

All natural fibers are hydrophilic in nature. This is the main drawback of natural fibers, as this causes them to have high water absorption and to be incompatible with hydrophobic polymer matrices (Mohanty *et al.*, 2001). Glass fibers, on the other hand, are essentially moisture resistant. The most common way to reduce the moisture absorption capability of a natural fiber is reported to be through the process of alkalization (also known as mercerization). Alkali treatment (usually with KOH or NaOH) reduces the hydrogen-bonding capacity of the cellulose, eliminating open hydroxyl groups that tend to bond with water molecules. Alkalization can also dissolve hemicellulose. The removal of hemicellulose, which is the most hydrophilic part of natural fiber structures, reduces the ability of the fibers to absorb moisture (Symington *et al.*, 2009).

Another process that shows promise for reducing the moisture content in natural fibers is the Duralin steam treatment process (Stamboulis *et al.*, 2000). This process depolymerizes the hemicellulose and lignin into aldehyde and phenolic functionalities, which are subsequently cured into a water-resistant resin. Untreated fibers reached a maximum moisture content of over 42% in 100% RH while the fibers treated by the Duralin process reached a maximum moisture content of only around 14% (Stamboulis *et al.*, 2000). The moisture diffusion time was also slowed for the fibers treated by the Duralin process. Some advantages of the Duralin process include the omitting of dew retting, increased fiber yield and quality, better dimensional and temperature stability, better resistance to fungal attack, and generally improved mechanical properties. On the other hand, the steaming process would add a significant amount of energy to the fibers, resulting in higher embodied energy of the fibers (Dittenber and Hota, 2012).

16.10.4 Incompatibility of fiber with resin

In addition to the data obtained by Mutnuri *et al.* (2010), shown in Table 16.2, Netravali *et al.* (2007) also noted that most 'semi-green' or 'green' composites have maximum tensile strength and stiffness in the ranges of

14.5–29 ksi (100–200 MPa) and 0.14–0.58 msi (1–4 GPa), too low to be used for primary, load-bearing components. The main reason leading to reduced mechanical properties in natural FRPs is the poor compatibility/adhesion between the hydrophilic fibers and the hydrophobic matrix materials. Many researchers have been developing treatments to modify the surface characteristics of the fibers to improve compatibility and adhesion.

The three factors affecting the bond between two materials are the mechanical interlocking, the molecular attractive forces, and the chemical bonds. Ideally, hydroxyl groups in a resin would bond with the hydroxyl groups that are available in all natural fibers, creating hydrogen bonds. The bond strength between resins and fibers is significantly lowered by the presence of moisture while curing due to the fact that H_2O molecules will bond with the available hydroxyl groups on the surface of the fiber, lessening the connections available for matrix bonding. When water evaporates, voids are left in a cured natural composite. If fibers are properly dried before a suitable matrix is introduced, then a better bond ought to result and future moisture uptake ought to be limited due to the lack of available hydroxyl bonding locations (Dittenber and Hota, 2012).

There are a number of different modifications that can be made in order to improve the interface between the fibers and the matrix. One of the most popular surface modifications for improving strength is the alkali treatment, where the reduction of moisture absorption capacity and the surface modification of the fibers work together to improve the mechanical properties of the composite. Most alkalization treatments intended to improve mechanical properties are conducted with a procedure similar to the treatment to reduce moisture absorption: soaking in 2–10% concentration NaOH for between 10 minutes and a few hours (Symington *et al.*, 2009; Ibrahim *et al.*, 2010). During alkalization, the cementitious materials in the fibers are removed, leading to rougher fiber surface. Even in low NaOH concentrations, fibers can fibrillate, increasing the aspect ratio and the bondable surface area. By improving the fiber/matrix adhesion in this manner, the tensile and flexural properties of the composite were improved up to 50%.

16.10.5 Durability and fatigue

A third major concern with natural composites is their long-term behavior when exposed to different environments, such as hygrothermal aging and dynamic loading as well as prediction of lifetimes. The general consensus from the literature is that very little fatigue work has been carried out on natural fiber composites. However, this will be necessary before natural fiber composites are accepted as primary structural components. More work has been going on to explore the response of natural composites exposed to moisture and weathering (Ray and Rout, 2005).

The growth of fungus and bacteria in natural composites due to biodegradation or moisture retention is a major concern in their development as structural materials. On samples of jute/phenolic composites exposed to humidity by Singh *et al.* (2000), some black spots and white patches appeared, which when viewed under a microscope were observed to be fungal hyphae. In another study by Stamboulis *et al.* (2000), moisture was found to cause fungus development on the surface of flax fibers after as little as 3 days of exposure. The Duralin treatment appeared to lessen this effect, improving the environmental durability.

Weathering studies showed that UV can cause color fading. After 2 years of exposure, jute/phenolic composites showed resin cracking, bulging, fibrillation, and black spots, with a tensile strength reduction of over 50%. Mehta *et al.* (2006) also found that weathering produced color change, weight loss, and surface roughening. In an effort to reduce weathering, some samples were coated in polyurethane, and exhibited very little surface deterioration (Singh and Gupta, 2005). Others found that mechanical properties deteriorate quickly with exposure to outdoor conditions for natural fiber/polyester composites. The strength was reduced by around 5–25%, with a more pronounced reduction in wetter conditions. The same composites exposed to indoor conditions for the same duration showed no significant changes in mechanical properties. Poor weathering performance is not only a problem for natural fibers; soy protein-based plastic lost strength and toughness and became stiff and brittle over time, possibly due to leaching of the plasticizer which is an additive in the resin system. The use of proper coatings and certain types of fiber modification (bleaching, alkalization, or silanes) seem to slow the effects of weathering (Dittenber and Hota, 2012).

16.10.6 Recent progress and discussion

The near-term goal of WVU-CFC research is to overcome the aforementioned barriers to advance the development of natural composites for interior structural applications in infrastructure with emphasis on green buildings, while the long-term goal is to evolve the natural composites as an alternative to GFRP for both interior and exterior structural applications. For the past five years, a variety of NFRP composites have been manufactured using different techniques including hand layup, compression, and pultrusion. In addition to flax-based composites, kenaf fibers are also being researched. Those NFRPs have been characterized for their thermomechanical properties. Recent data on kenaf-based composites including their tensile and flexural properties are also listed in Table 16.2 for comparison. As observed from the data in Table 16.2, certain improvements in mechanical properties have been made. A summary of the findings of WVU-CFC research on natural composites is given below and more details can be found in the paper by Dittenber (2012a).

A screening study of several resins was conducted to identify the best resin system that is the most compatible with the kenaf fibers. Among the resin systems tested, vinyl esters outperformed the polyurethane and phenolic resins in both mechanical properties and moisture absorption. The Derakane 8084 vinyl ester exhibited similar mechanical properties as Derakane 510A vinyl ester, but when the higher moisture absorption characteristics of Derakane 8084 are taken into consideration, the Derakane 510A is a better choice for a kenaf fiber-reinforced composite.

Individual kenaf fibers were tested for their mechanical properties. The kenaf fibers used in the study seemed to be of average to good quality when their tensile properties were compared to published results (Table 16.1). Alkalization with 5% NaOH for 20–40 minutes resulted in an optimum single fiber tensile strength of 116–145 ksi (800–1000 MPa) and stiffness of average 14.5 msi (100 GPa). Both the strength and stiffness of treated fibers were increased by 30–50% over untreated fibers.

The strength of an individual fiber does not directly correspond to the strength of the resulting composite, indicating that alkalization not only affects the fiber but also the interface between the fiber and the matrix, and has significant influence on mechanical properties. Alkalization with 5% NaOH for 40 minutes resulted in an optimum composite flexural strength of average 29.0 ksi (200 MPa) and stiffness of average 1.16 msi (8 GPa). This has shown an increase of 25–35% over composites manufactured with untreated fibers.

Higher stiffness (up to around 18%) can be achieved by the use of silanes in stiffness-critical situations, where the added cost can be justified. Although higher concentration alkali treatments may be able to marginally improve some of the flexural properties of the composites, they also lead to increased moisture absorption. Lower concentration treatments produce the best moisture absorption capabilities, with the use of silanes additionally reducing the amount of moisture absorbed by potentially as much as 50%.

Flax fibers are generally believed stronger and stiffer than kenaf fibers, as shown in Table 16.1. However, Table 16.2 appears to suggest that this flax-fiber advantage was not translated to the performance of flax fiber-reinforced polymer composites. The flax composites (Mutnuri et al., 2010) showed inferior properties when compared to some of the published literature (Gaceva et al., 2007). This might be due to the difference in grade of the flax fiber, inconsistency in processing, sizing (chemical treatment) of the fiber, fiber waviness caused during pultrusion, etc. (Mutnuri et al., 2010).

Research on flax and kenaf composites at WVU-CFC is ongoing. The most recent data on kenaf composites in Table 16.2 (Dittenber, 2012b) indicate that the effectiveness of alkali treatment is not satisfactory, increasing the flexural strength by 10% but decreasing the tensile strength by 10% and with no effect on either tensile or flexural modulus. A breakthrough regarding natural fiber treatment technique that will enable hydrophobic transformation but will not deteriorate fiber strength is being sought. Additional durability tests on NFRPs are underway at WVU-CFC under the US-NSF sponsored research to establish long-term degradation rates of natural fiber composites under limited accelerated aging conditions.

The WVU-CFC team is focusing on promising natural fiber composite systems with reference to glass fiber-reinforced composites for some structural applications, particularly once weight, cost, and environmental impact are taken into consideration. As the WVU-CFC team continues to upgrade the NFRP performance, future work will incorporate the evaluation of the long-term performance of NFRPs under different loading and aging conditions, including fatigue tests. In order for NFRPs to be suited for construction applications, many advances to deal with the challenges discussed in this section have to be achieved by the research community of NFRP composites.

16.11 Durability of composites

FRP composites have superior performance under harsh end-use environments over conventional construction materials (steel, concrete, and wood). Such design advantages of composites have to be based on sound understanding of the durability (long-term) response of the FRPs under the harsh environments and loading conditions. The durability response of composites is identified typically in terms of chemical, physical, and mechanical aging and their combinations, which depends primarily on pH level, temperature, creep/relaxation, UV radiation, and externally induced thermo-mechanical stress fluctuations. The durability response is further accelerated in the presence of water or salt solutions because of their expansion under freezing (ACI, 2006; Chin et al., 2001; Karbhari, 2006; McBagonluri et al., 1998; Antoon and Koening, 1980). In terms of the above parameters, fluid absorption in and out of composites under freeze-thaw conditions has the highest influence on durability (Lesko et al., 1998).

Long-term performance determination and modeling of composites due to aging, moisture and pH, freeze-thaw, fatigue, and creep are critical for high volume applications. In-service composite structural systems are exposed to various environmental factors such as moisture, thermal (freeze-thaw cycling or elevated temperature) and other weathering conditions that affect their overall performance. The physical weathering occurs when composites are subjected to mechanical loadings such as static, fatigue, and creep (sustained stress), while the chemical weathering occurs when the material is exposed to moisture and chemicals such as alkaline or acid solutions. Under those exposures, the mechanical and physical properties (e.g., strength, stiffness, creep, fatigue life, glass transition temperature, fiber–resin bond strength) change with exposure time.

The process of composite physical aging is dependent primarily on service temperature and sustained load, and could be substantial on composite performance. Creep of polymers involves deformation of the molecules with molecular segments changing their conformations and sliding past one another. If the deformations are large enough, molecular chain rupture (scission) may occur. As a result, the mechanical properties vary with time under load (Liao et al., 1999). The chemical aging of composites is related to: diffusion process, polymer chemical composition, temperature, stress corrosion, fluid ion exchange, and others. Polymer chain scissions and altered material chemistry are the direct result of chemical aging that lead to loss of constituent materials including physical and mechanical properties. Fluid sorption in and out of FRP composites is the most influential factor on thermo-mechanical properties when a material is exposed synergistically to environmental and mechanical loads.

The rate of degradation of polymer composites exposed to a fluid environment is related to the rate and quantity of fluid sorption (Bott and Barker, 1969), which are governed mostly by: chemical structure of the resin; degree and type of crosslinking; state of the material including void content; type, temperature, and concentration of fluid; and applied stress and hydrostatic pressure (Antoon and Koening, 1980). Exposure of composites to moisture and chemicals causes stress corrosion, which is driven by the exchange of alkali ions in glass fibers and hydrogen ions from a reactive fluid, leading to spontaneous fiber surface cracking and stress failure (Price, 1989). Diffusion of fluids through polymer can cause swelling due to hydrogen bond disruption and induce stresses within the composite. Freeze-thaw in the presence of salt can also result in accelerated degradation due to the formation and expansion of salt deposits in addition to the effects of moisture-induced swelling and drying (Karbhari et al., 2003).

Degradation of composites is compounded when exposed to other environmental factors along with freeze-thaw fatigue (McBagonluri et al., 1998, 2000). Karbhari (2006) provided excellent insights into the effect of moisture on E-glass/vinyl ester composites through dynamic mechanical analyses. Differential scanning calorimetry (Verghese et al., 1999) data revealed the nature and presence of freezable water for each constituent material within an E-glass/vinyl ester composite (matrix and interface). Heat flow measurements during thawing were taken for a single cycle (−150°C to +50°C) on saturated, unreinforced vinyl ester resin samples, which indicated the absence of freezable water. Since this free volume size within the resin is in the order of about 6–20 Å, these voids are thermodynamically too small for water to freeze. This is due in part to hydrogen bonding in addition to geometric space constraints, thus impeding the freezing process. However, the dimensions in a cracked composite system were large enough to facilitate water freezing and damage accumulation (Verghese *et al.*, 1999). Similarly, high damage accumulations were noted in pretensioned FRPs under marine environments (Sen *et al.*, 1998). Life prediction model(s) for the durability of composites over a range of chemo-thermo-mechanical environments have to be developed from accelerated test data and validated from field studies. (See, for example, the work done on FRP rebars for chemo-thermo-mechanical properties by Vijay and Hota, 1999.)

Fatigue damage in FRP composites is progressive and cumulative in nature at the micro level (e.g., Reifsnider, 1990 among many others). For metals, the damage due to fatigue is more localized at the macro level (Degrieck and Paepegem, 2001) and can be identified visually and by micrographs of fracture surfaces (Rakow and Pettinger, 2006). Unlike metals, composites accumulate damage at various locations under fatigue. However, the potential defects present in the material (matrix cracks, broken fiber, fiber wrinkling, inadequate cure, etc.) at the time of manufacturing are higher in FRP composites than in metals. When subjected to fatigue loading, these defects lead to crack initiation, crack growth and interconnection, and eventual failure. Figure 16.29 shows a sandwich panel under three-point bending fatigue where crack initiation occurs at locations of fiber wrinkling.

Due to matrix cracking or fiber debonding, the life prediction of composites under thermo-mechanical fatigue is challenging, except in terms of damage evaluation in an average sense, i.e., not size-specific. Fatigue studies on composites under combined thermal and mechanical loads are sketchy, and limited research has been performed by Lesko *et al.* (1998), Kellogg (2005), and a few others. The load rate has the greatest influence on fracture

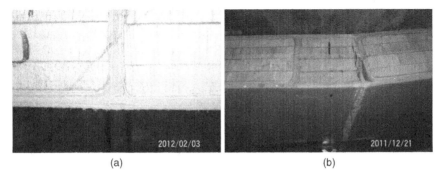

16.29 Cracking initiation and growth of a sandwich sample under three-point bending fatigue, leading to eventual failure at the corner.

sensitivity for a notched specimen where the rate of increase of loading results in increased mean notch toughness values for wide ranges of moisture content or temperature. Similar results, including stiffening under low temperatures, were reported by Dutta (1995). The greatest concern with temperature effects on composites is the laminate debonding under freeze–thaw cycling (thermal fatigue) due to moisture expansion upon freezing. The WVU-CFC team developed a strain energy model using strain energy release rate as the damage metric to predict fatigue life of a composite material (Dittenber and Hota, 2010; Natarajan et al., 2005). This model has been extensively verified using a large amount of WVU-CFC data and non-WVU-CFC data to be viable to predict fatigue life of both composite laminates and composite components. Additional technical work to develop a unified life prediction model for composite materials for infrastructural applications has been recently funded by NSF (Award No. IIP-1230351), so that FRPs can be considered for use in building a sustainable environment.

16.12 Design codes and specifications

The construction industry relies on design codes, specifications, and standards for any material to be used in construction. FRP design codes and standards are needed not only to provide credibility to FRP products and penetrate into the market against existing materials, but also to do business with the government. A good example would be Underwriters Laboratories (UL) 1316 (1966) 'Non-Metallic Tanks for Petroleum Products Only' for single-wall FRP tanks. This standard as a Performance Specification was revised for double-wall FRP tanks in 1984 under the same code number UL 1316: 'Glass-Fiber-Reinforced Plastic Underground Storage Tanks for Petroleum Products, Alcohols, and Alcohol-Gasoline Mixtures'. This standard is the only nationally recognized standard governing the design, manufacture, installation, and inspection of FRP tanks.

FRP composites as chimney liner for smoke stacks are governed by the ASTM Standard D5364 entitled 'Standard Guide for Design, Fabrication, and Erection of Fiberglass Reinforced Plastic (FRP) Chimney Liners with Coal-Fired Units'. This standard was first published in 1993 and was revised in 2002 and further expanded in 2008. This guide provides information, requirements and recommendations for design professionals, fabricators, installers, and end-users of FRP chimney liners. This guide provides uniformity and consistency to the design, material selection, fabrication, erection, inspection, confirmatory testing, quality control, and assurance of FRP liners for concrete chimneys with coal-fired units, while the Cooling Technology Institute (CTI) Code STD 137 'Fiberglass Pultruded Structural Products for Use in Cooling Towers' offers recommendations for classification, materials of construction, tolerances, defects, workmanship, inspection, physical, mechanical, and design properties of glass fiber-reinforced pultruded structural shapes intended for use as construction items in cooling tower applications. This specification was first published in 1994 and last revised in October 2009.

The research advances made through various sponsored programs have directly resulted in the development of FRP rebar design codes and specifications. Among others, Bank *et al.* (1998) studied the behaviors of pultruded shapes and full-sized structural frames under short-term and long-term loading as well as FRP reinforced concrete structures. Nanni *et al.* (1997) researched the use of FRP rebars in concrete decks and externally bonded FRP composite wraps of RC structures. Through NSF award, Vijay and Hota (1999) systemically investigated the durability responses of reinforced concrete members with FRP rebar. Based on accelerated aging test results calibrated with respect to naturally aged composites, the study concluded that the service life of the FRP rebar with durable low-viscosity urethane-modified vinyl ester resin is about 60 years as a minimum with 20% sustained stress on the bar. Concrete cover protection to the FRP bars would enhance the service life up to 120 years. All the above studies have found their way into the design and construction specifications for composite rebars for concrete structural elements and composite wraps to strengthen infrastructural systems (ACI 440.1R.03 and AASHTO LRFD Bridge Design Guide).

Recently, a new design code entitled, 'Pre-standard for Load and Resistance Factor Design (LRFD) of Pultruded Fiber Reinforced Polymer (FRP) Structures' is being developed through the American Composites Manufacturers Association (ACMA) and the American Society of Civil Engineers (ASCE). This code will allow architects and structural engineers to incorporate FRP composite materials to build stronger, safer, and better buildings. The introduction of these codes will help FRP composites to compete on a level playing field with other construction materials such as concrete, steel, wood, and aluminum. Performance criteria for design, specification, and installation will mean a higher degree of confidence for professional engineers and contractors to design and construct with FRP composites, in addition to instilling confidence in owners to field implement the advanced FRPs.

For example, the released draft LRFD prestandard design guide states that during analysis and design of pultruded FRP structural components and systems, the nominal strength will be determined by multiplying the reference strength by the adjustment factors for end-use conditions, as represented in the following formula:

$$R_n = R_0 C_1 C_2 C_3 \ldots C_n \qquad [16.1]$$

where R_0 is the reference strength and C_i represents the applicable adjustment factors for sustained end-use conditions that differ from the reference conditions. More specifically, the reference strength or stiffness is obtained from tests under short-term loading at ambient temperature of 73 ± 3°F and relative humidity of 50 ± 10%. The above-mentioned draft LRFD code specifies C_M (moisture condition factor) and C_T (temperature factor) as set out in Table 16.3 to account for sustained in-service moisture and temperature, respectively. Here C_T is valid for in-service temperatures higher than 90°F but less than T_g minus 40°F. For sustained temperatures in excess of 140°F, C_T shall be determined from tests.

For chemical environmental factor C_{CH} in high alkalinity or acidity, the adjustment factor shall be determined from interpolation or extrapolation of the results of ASTM C581 tests performed on the laminate exposed to the exposure chemical environment for a period of 1,000 hours. Also, the reference strength in Eq. [16.1] shall be obtained for single members or connections without load sharing or composite action. There are adjustment factors for member strength or stiffness in structural assemblies to account for the increase in strength of the assembly over the strength of an individual member or for the increase in assembly stiffness when the members are constrained to act in a composite fashion. In addition, fatigue shall be considered in the design of members and connections subjected to repeated loading.

Step-by-step design guidelines, along with all the computation details, for pultruded sections under given load conditions and FRP cooling tower using those pultruded sections, are provided in the reference by Qureshi (2012), where a 4″ × 4″ square section, a 4″ × 4″ wide-flange section, and a 4″ diameter round section are analysed as per given fiber architecture of a quarter inch thick GFRP composite laminate. The design capacities of these sections under bending, torsion, combined bending, and torsion are compared. This is followed with design details for design of girts,

Table 16.3 Adjustment factors for end-use conditions (extract from Draft LRFD code, Table 2.4-1)

Reference property	Moisture C_M	Temperature C_T for (90°F < T ≤ 140°F)
Vinyl ester material		
Strength	0.85	1.7–0.008T
Elastic modulus	0.95	1.5–0.006T
Polyester material		
Strength	0.80	1.9–0.010T
Elastic modulus	0.90	1.7–0.008T

connections with columns, and girt splices for a full supported structure in a cooling tower, using pultruded FRP sections and stainless steel bolts. Loads considered include:

- structural dead load,
- fill media dead load (15 psf),
- water held up by filled media (long-term load, 18 psf),
- axial force from wind load (860 lbs on bent-line joist and joist supporting mid-bay joists),
- temporary service live loading (20 psf, not operating tower),
- ice load (30 psf),
- fouling load (30 psf, applied to both operating and non-operating tower).

The serviceability requirements are to limit the deflection to maximum deflection = $L/180$ and maximum live load deflection = $L/240$, where L is the span. The environmental conditions include wet environment, 110°F temperature, and high salt content.

16.13 Future trends

With the world facing a crisis in terms of sustainable growth and environmental stability, the responsibility is on the engineering community to develop cost-effective and durable construction materials having lower embodied energy. The green building movement, science in energy and environmental design, innovation for sustainability, sustainable materials, and many other programs have been established to steer the United States in a greener direction. We can envision an eco-urban habitat of zero carbon footprint. As a mid-term goal, research and development efforts on game-changing technologies can aim at an urban habitat capable of breathing with ambient environment and minimizing energy and water usage. There are several projects being funded by the NSF that focus on buildings of net-zero energy operation. The goal of these projects is to maximize heating/cooling/lighting influence of solar energy and other natural resources to reduce energy consumption and yet maintain habitable conditions, including net-zero water design for buildings. With reference to FRP composites, durable, strong, and stiff composite panels made of natural fibers and natural resins with lower embodied energy and cost per unit performance than steel need to be developed and integrated with prefabricated modular sub-system design concepts for potentially zero construction waste.

FRP composites have the potential to help achieve a sustainable environment because of their advantages over conventional materials (Stewart, 2011). From a life cycle assessment perspective, the selection of FRPs would require characterization of its long-term durability and development of predictive models to assess the useful life of structural components or systems utilizing composites. Durability responses of composite materials and understanding the mechanisms have become very important topics for mass implementation of these advanced composite materials. Serviceability, durability, and cost-effectiveness are essential for widespread use of any material. Long-term responses of composite structures under environmental loads including moisture exposure have to be established so that the accelerated aging test methodology (ATM) can be used to predict long-term performance of FRP composites through life prediction models; these data have to be calibrated with field response data, monitored from the implementation works. Thus more durable, efficient, and safer FRP structures can be designed based on data collected using ATM and appropriate safety (knock-down) factors.

Furthermore, smart and multifunctional materials will be an important future focal area of FRP research. The topics of importance include: phase-changing materials for energy storage and release, conductive polymers for solar cells, protective self-cleaning and de-polluting coatings, self-deicing materials, self-assessing and self-healing materials, coatings that can be used as sensors, and many others. For example, advances in the use of carbon fibers for sensing and detecting damage through polymers housed in nano-fibers will be heavily researched. Coatings consisting of nano-fibers can be used as sensors to detect micro-cracks, fire, and hazardous chemicals. Electrically conductive coatings in conjunction with wireless networks will be developed to detect fire and other structural hazards.

In conclusion, FRP composites are found to be excellent in corrosive environments. For the past two decades, FRPs have been gradually accepted in infrastructural applications including structures for highway and waterway, utility poles, wind turbine blades, and pipelines. With the recent launching of new design codes, new high volume FRP composite markets will open up and existing markets will broaden further. With ever increasing attention toward a sustainable built environment, the selection of construction material is to be justified at a life cycle level accounting for environmental benefits, social benefits, and other factors. Based on research, development and field implementation, FRP composites have the potential to be selected as a material of choice because of FRPs' long-term performance and design advantages.

16.14 Acknowledgment

The authors would like to thank Al Dorris for providing information on FRP tanks; Bob Cashner for permission to present AEP's status of FRP application; Don Kelley for permission to use his photos; Bhyrav Mutnuri and David Dittenber for their materials on natural composites used in this chapter.

16.15 References

AbdelAzim A A A, 1996, Unsaturated polyester resins from polyethylene terephthalate waste for polymer concrete. *Polym Eng Sci*, 36(24), 2973–2977.

ACI Manual, 2006, Durability of Fiber Reinforced Polymer (FRP) Composites Used with Concrete, ACI Subcommittee 440-L.

ACMA, 2011, *American Composites Manufacturers Association 2010 Industry Report*.

Aditham R P, 2004, Manufacturing and evaluation of structural products with recycled polymers. MS thesis, West Virginia University.

Allred R E, Coons A B and Simonson R J, 1996, Properties of carbon fibers reclaimed from composite manufacturing scrap by tertiary recycling. *International SAMPE Technical Conference*, Vol. 28, pp. 139–150.

Anon, 1987, Tests mark progress with polymers in asphalt. *Highway & Heavy Construction*, 130(6), 48–49.

Antoon M K and Koening J L, 1980, The structural and moisture stability of the matrix phase in glass-reinforced epoxy composites. *Journal of Macromolecular Science – Review Macromolecular Chemistry*, C19(1), 135–173.

Baillie C, 2004, *Green Composites: Polymer Composites and the Environment*. Cambridge: Woodhead Publishing.

Bank L C, Puterman M and Katz A, 1998, The effect of material degradation on bond properties of FRP reinforcing bars in concrete. *ACI Materials Journal*, 95(3), 232–243.

Beach J and Cavallaro J, 2002, An overview of structures and materials work at the division, *Carderock Division Technical Digest*, NSWCCD, September.

Belarbi A, Bae S-W, Ayoub A, Kuchma D, Mirmiran A and Okeil A, 2011, *Design of FRP Systems for Strengthening Concrete Girders in Shear*, NCHRP Report 678, Transportation Research Board, Washington, DC.

Bott T R and Barker A J, 1969, The behavior of model composites in contact with different environments. *Transactions of the Institute of Chemical Engineers*, 47, T188-T193.

Cashner B, 2011, AEP experience with pultruded polyester fiberglass structure for cooling towers. Paper No. TP11-19, *Cooling Technology Institute Annual Conference*, San Antonio, TX, February 6–10.

CBS, 2009, Composite Dome Homes. http://www.compbldgsys.com/dome_applications.htm.

Chada V R, 2012, Manufacturing, evaluation and field implementation of recycled GFRP-composite railroad ties. MS thesis, West Virginia University.

Chen R H L, Choi J-H, Hota V G and Kopac P A, 2008, Steel versus GFRP rebars? *Public Roads*, 72(2), 2–9.

Chiles J R, 2009, The other renewable energy. *Invention & Technology*, Winter, 24–35.

Chin J W, Hughes W L and Signor A, 2001, Elevated temperature aging of GFRP vinyl ester and isophthalic polyester composites in water, salt water and concrete pore solution. *American Society of Composites 16th Tech Conference Proceedings*, September.

CT, 2008, Bio-Composites Update: Bio-Based Resins Begin to Grow. *Composites Technology*, 04/1/08, available at: http://www.compositesworld.com/articles/bio-composites-update-bio-based-resins-begin-to-grow.aspx.

Dawans F, 1992, Treatment of polymer wastes – chemical or energy upgrading. *Revue de l' Institut Français du Petrole*, 47(6), 837–867.

Degrieck J and Paepegem W V, 2001, Fatigue damage modeling of fiber-reinforced composite materials: review. *Appl Mech Rev*, 54(4), 279–300.

Dittenber D B, 2012a, Effect of alkalization on flexural properties and moisture absorption of kenaf fiber reinforced composites. *SAMPE*, Baltimore, MD, May 21–24.

Dittenber D B, 2012b, Internal communication, August 14, 2012.

Dittenber D B and Hota G V S, 2010, Evaluation of a life prediction model and environmental effects of fatigue for glass fiber composite materials. *Struct Eng Int*, 20(4), 379–384.

Dittenber D B and Hota G V S, 2012, Critical review of recent publications on use of natural composites in infrastructure. *Journal of Composites Part A: Applied Science and Manufacturing*, 43(8), 1419–1429.

Dorris A, 2008, Four decades (1964–2007) of FRP underground storage tank (UST) use in the USA. *Proceedings of the International Conference and Exhibition on Reinforced Plastics ICERP 2008*, February 7–9, 2008, Mumbai, India.

Drazl L T, Mohanty A K, Burgueno R and Mishra M, 2004, Biobased structural composite materials for housing and infrastructure applications: opportunities and challenges. *NSF-PATH Housing Research Agenda Workshop, Proceedings and Recommendations* (eds M G Syal, M Hastak and A A Mullens), pp. 129–140.

Dutta P K, 1995, Fatigue evaluation of composite bridge decks under extreme temperatures. *Proceedings, SAMPE Symposium*.

Edwards J, Lee L and Jain R, 2009, Assessment of FRP composites for sustainable construction. Paper No. 3, *Proceedings of Int. Conference on FRP Composites for Infrastructure Applications*, San Francisco, CA, November 4–6.

EPA, 2012, *FY 2011 Annual Report on Underground Storage Tank Program*, EPA 510-R-12-001, March 2012. Available at: http://www.epa.gov/oust/pubs/fy11_annual_ust_report_3-12.pdf.

Feldman J and Shistar T, 1997, *Poison Poles – A Report about Their Toxic Trail and Safer Alternatives*. Available at: http://www.beyondpesticides.org.

Ferguson I, La Fontaine B, Vinden P, Bren L, Hateley R and Hermesec B, 1996, *Environmental Properties of Timber*. Available at: http://www.timberbuilding.arch.utas.edu.au/environment/env_prop/env_prop.asp.

FHWA, 2001, *Fiber Reinforced Polymer Composite Bridges of West Virginia*. Publication No. FHWA-ERC-2-002.

Gaceva G B, Avella M, Malinconico M, Buzarovska A, Groxdano A, Gentile G, Errico M E, 2007, Natural fiber eco-composites. *Polymer Composites*, 28(1), 98–107.

Goldsworthy W B, 1982, Continuous manufacturing process, in *Handbook of Composites* (ed. G. Lubin), New York: Van Nostrand Reinhold, pp. 479–490.

Greene E, 2003, Over the bounding main: large composite structures in the US Navy. *Composite Fabrication*, July, p. 10.

Gremel D, 2007, Internal FRP reinforcing update: codes, standards, specifications and developments. *Polymer Composites IV*, Morgantown, WV, March 20–22.

Hiel C, 2001, Three examples of practical design & manufacturing ideas for the emerging composites infrastructure industry. *Polymer Composites II* (ed. R Creese and H GangaRao). Boca Roton, FL: CRC Press.

Hota G and Skidmore M, 2012, FRP Sheet Piling, February 15, West Virginia University, Morgantown, West Virginia; available at http://www.creativepultrusions.com/LitLibrary/sheetpile/brochure.pdf.

Hota G and Vijay P V, 2010, Feasibility review of FRP materials for structural applications. Technical Report submitted to US Army Corps of Engineers Engineering Research and Development Center and Construction Engineering Research Laboratory, Vicksburg, Mississippi.

Ibrahim N A, Hadithon K A and Abdan K, 2010, Effect of fiber treatment on mechanical properties of kenaf fiber-ecoflex composites. *J Reinf Plast Compos*, 29(14), 2192–2198.

Kalligudd S K, 2010, Characterization and durability evaluation of recycled FRP composites and sandwich specimens. MS thesis, West Virginia University.

Kara S and Manmek S, 2009, Composites: calculating their embodied energy. Life Cycle Engineering & Management Research Group, University of New South Wales. Available at: http://www.wagnerscft.com.au/documents/?5.

Karbhari V M, 2006, Dynamic mechanical analysis of the effect of water on E-glass–VE composites. *Journal of Reinforced Plastics and Composites*, 25(6), 631–644.

Karbhari V M, Chin J W, Hunston D, Benmokrane B, Juska T, Morgan R, Lesko J J, Sorathia U and Reynaud D, 2003, Durability gap analysis for fiber-reinforced polymer composites in civil infrastructure. *Journal of Composites for Construction*, 7(3), 238–247.

Kelley D and Schneider G, 2008, Epoxy vinyl ester in coal fired power plant applications: overview and case histories. *Proceedings of the International Conference and Exhibition on Reinforced Plastics, ICERP 2008*, February 7–9, Mumbai, India.

Kellogg K G, 2005, Effect of load rate on notch toughness of glass FRP subjected to moisture and low temperature. *Int J of Offshore and Polar Engineering*, 15(1), 54–61.

Kibert C, 1993, Construction materials from recycled polymers. *Proc Institution of Civil Engineers – Structures and Buildings*, 99(4), 455–464.

Koerner R M, 2012, *Designing with Geosynthetics*, 6th edn, Indianapolis, IN: Xlibris Publishing.

LeGault M R, 2010, DDG-1000 Zumwalt: stealth warship. *Composites Technology*, January 18.

LeGault M, 2011, Composite vs. corrosion: battling for market share. *Composites Technology*, October 10. Available at: http://www.compositesworld.com/articles/composite-vs-corrosion-battling-for-marketshare.

Lesko J J, Hayes M D, Garcia K and Verghese K N E, 1998, Environmental mechanical durability of e-glass/vinyl ester composites. *Proceedings of 3rd DURACOSYS*, Blacksburg, VA, pp. 173–179.

Liang R, 2001, Recycling polymer and polymer composite materials: a review. *Polymer Composites II 2001 – Composites Applications in Infrastructure Renewal and Economic Development* (November 14–16, Morgantown, WV), (eds R C Creese and G Hota) Boca Raton, FL: CRC Press, pp. 147–157.

Liang R F and Gupta R K, 2000, Rheological and mechanical properties of recycled polycarbonate. *SPE ANTEC*, May 7–11, Vol. 46, pp. 2903–2907.

Liang R F and Gupta R K, 2001, The effect of residual impurities on the rheological and mechanical properties of engineering polymers separated from mixed plastics. *SPE ANTEC*, May 6–10, Vol. 47, pp. 2753–2757.

Liang R F and Hota G, 2012, Experimental evaluation of pultruded panel joints: tensile, bending, and fatigue strengths. Technical Report submitted to Huntington Ingalls Industries (HII), Ingalls Shipbuilding, Pascagoula, Mississippi.

Liang R, Mutnuri B and Hota G R, 2005, Pultrusion and mechanical characterization of GFRP composite sandwich panels. *SPE 63th ANTEC*, May 1–5, Boston, MA, Vol. 1, pp. 1601–1605.

Liao K, Schultheisz C R and Hunstow D L, 1999, Effects of environmental aging on the properties of pultruded GFRP. *Composites Part B*, 30(5), 485–493.

LRFD, 2011, *Pre-Standard for Load and Resistance Factor Design (LRFD) of Pultruded Fiber Reinforced Polymer (FRP) Structures*. American Composites Manufacturers Association (ACMA), Arlington, VA, January 21.

Lucintel, 2008, New Power Plant Uses Composite Stack Liner. Available at: http://www.lucintel.com/newspage.aspx?sno=5245.

Malnati P, 2011, A hidden revolution: FRP rebar gains strength. *Composites Technology*, December 1. Available at: http://www.compositesworld.com/articles/a-hidden-revolution-frp-rebar-gains-strength.

Malvern, 2011, *Today's Cement Industry*. Available at: http://www.malvern.com/ProcessEng/industries/cement/overview.htm.

Marshall A C, 1998, Sandwich construction, in *Handbook of Composites*, 2nd Edition (ed. S T Peters), New York: Chapman & Hall.

McBagonluri F, Hayes M D, Garcia K and Lesko J J, 1998, Durability of E-glass/vinyl ester composites in an elevated temperature simulated sea water environment. *Proc 7th Int. Conf on Marine Applications of Composite Materials*, Melbourne, FL, Paper F10.

McBagonluri F, Garcia K, Hayes M D, Verghese K N E and Lesko J J, 2000, Characterization of fatigue and combined environment on durability performance of glass/vinyl ester composite for infrastructure application. *Int J Fatigue*, 22(1), 53–64.

McConnell V P, 2007, Global underground: the state of composite storage tanks. *Reinforced Plastics*, June 1. Available at: http://www.reinforcedplastics.com/view/2098/global-underground-the-state-of-composite-storage-tanks/.

Mehta G, Mohanty A K, Drzal L T, Kamdem D P and Misra M, 2006, Effect of accelerated weathering on biocomposites processed by SMC and compression molding. *J Polym Environ*, 14(4), 359–368.

Miller A K, 2011, The Lockheed Martin Ocean Thermal Energy Conversion Cold Water Pipe, Lockheed Martin, Palo Alto Colloquium, May 5.

Mirmiran A, Shahawy M, Nanni A, Karbhari V, Yalim B and Kalayci A S, 2008, *Recommended Construction Specifications and Process Control Manual for Repair and Retrofit of Concrete Structures Using Bonded FRP Composites*. NCHRP Report 609, Transportation Research Board, Washington, DC.

Mohanty A K, Misra M and Drzal L T, 2001, Surface modifications of natural fibers and performance of the resulting biocomposites: an overview. *Compos Interface*, 8(5), 313–343.

Mutnuri B, Aktas C, Marriott J, Bilec M and Hota G, 2010, Natural fiber reinforced pultruded composites. *ACMA COMPOSITES 2010*, Las Vegas, NV, February 9–11.

Nadel B A, 2006, *Fiberglass Fenestration: A Durable, Sustainable, and Economic Alternative for Windows and Doors*. McGraw-Hill Construction-Continuing Education Center, June.

Nanni A, Nenninger J, Ash K and Liu J, 1997, Experimental bond behavior of hybrid rods for concrete reinforcement. *Structural Engineering and Mechanics*, 5(4), 339–354.

Natarajan V, Hota G and Shekar V, 2005, Fatigue response of fabric-reinforced polymeric composites. *J Compos Mater*, 39(17), 1541–1559.

Netravali A N, Huang X and Mizuta K, 2007, Advanced 'green' composites. *Advanced Composite Materials*, 16(4), 269–282.

Patel M and Narayan R, 2005, How sustainable are biopolymers and biobased products? The hope, the doubt, and the reality. In *Natural Fibers, Biopolymers, and Biocomposites*. (eds A K Mohanty, M Misra, L T Drzal), Boca Raton, FL: Taylor & Francis, pp. 833–854.

Petro S H, Hota G, Halabe U B, Aluri S, Smith A and Vasudevan A, 2002, Fiber Reinforced Polymer Composites Used to Repair and Rehabilitate Wood Railroad Bridges, Phase II, Project Report, Constructed Facilities Center, West Virginia University, Morgantown WV.

Price J N, 1989, Stress corrosion cracking in glass reinforced composites. In *Fractography and Failure Mechanism of Polymers and Composites* (ed. A C Roulin-Moloney) New York: Elsevier Applied Science, pp. 495–531.

PRNewswire, 2008, US Department of Energy Awards Lockheed Martin Contract to Demonstrate Innovative Ocean Thermal Energy Conversion Subsystem, October 8.

Qureshi M A M, 2012, Failure behavior of pultruded GFRP members under combined bending and torsion. PhD dissertation, West Virginia University.

Rakow F J and Pettinger M A, 2006, Failure Analysis of Composite Structures in Aircraft Accidents, ISASI 2006 Annual Air Safety Seminar, Cancun, Mexico, September 11–14.

Ray D and Rout J, 2005, Thermoset biocomposites. In *Natural Fibers, Biopolymers, and Biocomposites* (eds A K Mohanty, M Misra and L T Drzal). Boca Raton, FL: Taylor & Francis, pp. 291–346.

Rebeiz K S, 1996, Strength and durability properties of polyester concrete using PET and fly ash wastes. *Advanced Performance Materials*, 3(2), 205–214.

Reifsnider K L, 1990, *Fatigue of Composite Materials*, Composite Materials Series 4, Elsevier Science Publishers, Amsterdam.

Richter P J, 1999, Developing cost effective polymer composite structural elements for mining facilities applications. *Polymer Composites I* (ed. R Creese and G Hota), Lancaster, PA: Technomic Publishing, pp. 21–30.

Saadatmanesh H, Ehsani M R and Jin L, 1997, Repair of earthquake-damaged RC columns with FRP wraps. *ACI Structural Journal*, 94(2), 206–214.

Sen R, Shahawy M, Rosas J and Sukumar S, 1998, Durability of aramid pretensioned elements in a marine environment. *ACI Structures Journal*, 95(5), 578–587.

Shekar V, Aluri S and Hota G H, 2005, Performance evaluation of FRP composite bridge decks. *Transportation Research Record: Journal of the Transportation Research Board*, CD 11-S, 465–472.

Singh B and Gupta M, 2005, Performance of pultruded jute fibre reinforced phenolic composites as building materials for door frame. *J Polym Environ*, 13(2), 127–137.

Singh B, Gupta M and Verma A, 2000, The durability of jute fibre-reinforced phenolic composites. *Compos Sci Technol*, 60, 581–589.

Southern Company Services, Inc., 2002, Demonstration of Innovative Applications of Technology for the CT-121 FGD Process, Clean Coal Technology Demonstration Program, August.

SPI, 1993, International composites recycling, in *Proceedings of 48th Annual Conference, Composites Institute*, The Society of the Plastics Industry Inc, February 8–11.

Stamboulis A, Baillie C A, Garkhail S K, Van Melick H G H and Peijs T, 2000, Environmental durability of flax fibres and their composites based on polypropylene matrix. *Appl Compos Mater*, 7, 273–294.

Stewart R, 2011, Composites in construction advance in new direction. *Reinforced Plastics*, September 6.

Strongwell Corporation, 2009, A Life Cycle Assessment Approach in Examining Composite. Raw Materials, Steel and Aluminum Materials Used in the Manufacturing of Structural Components, June 19. Available at: http://www.strongwell.com/pdffiles/green/Life-Cycle-Report.pdf.

Symington M C, Banks W M, West O D and Pethrick R A, 2009, Tensile testing of cellulose based natural fibers for structural composite applications. *J Compos Mater*, 43(9), 1083–1108.

Tusing D S, 2003, Modular track panels for improved safety in the mining transportation industry. MSCE thesis, West Virginia University, Morgantown, WV.

UPI, 2009, Bridges Built from Recycled Plastic, July 2.

Verghese K N E, Hayes M D, Garcia K, Carrier C, Wood J, Riffle J R and Lesko J J, 1999, Influence of matrix chemistry on the short term, hydrothermal aging of vinyl ester matrix composites under both isothermal and thermal spiking conditions. *J Compos Mater*, 33(20), 1918–1938.

Vijay P V, 1999, Aging and design of concrete members designed with GFRP bars. PhD dissertation, West Virginia University, Morgantown, WV.

Vijay P V and Hota G V S, 1999, Accelerated and natural weathering of glass FRP bars. ACI Special Publication, SP-188, *4th Int Symposium FRP Reinforcement for Reinforced Concrete Structures*, Baltimore, MD, November (eds C W Dolan, S H Rizkalla and A Nanni), pp. 605–614.

Vijay P V, Hota G V S and Bargo J M, 2000, Mechanical characterization of recycled thermoplastic polymers for infrastructure applications. Presented at the Conference on Advanced Composite Materials for Bridges and Structures (ACMBS-3), August.

Wikipedia, 2012, 'Geosynthetic'. Available at: http://en.wikipedia.org/wiki/Geosynthetic (accessed September 1, 2012).

Wood K, 2011, Tough resins for aggressive environments. *Composites Technology*, June 1. Available at: http://www.compositesworld.com/articles/tough-resins-for-aggressive-environments.

Wu J T H, 1994, *Design and Construction of Low Cost Retaining Walls: The Next Generation in Technology*. CTI-UCD-1-94, Colorado Transportation Institute, Denver, CO.

17
Design of all-composite structures using fiber-reinforced polymer (FRP) composites

P. QIAO, Washington State University, USA and
J. F. DAVALOS, The City College of New York, USA

DOI: 10.1533/9780857098955.2.469

Abstract: This chapter presents a systematic approach for material characterization, analysis, and design of all-fiber-reinforced polymer or plastic (FRP) composite structures. The suggested 'bottom-up' analysis concept is applied throughout the procedure, from materials/microstructures, to macro components, to structural members, and finally to structural systems, thus providing a systematic analysis methodology for all-FRP composite structures. The systematic approach described in this chapter can be used efficiently to analyze and design FRP shapes and bridge systems and also develop new design concepts for all composite structures.

Key words: FRP composites, micromechanics, FRP structural shapes, design of composite structures, design concepts.

17.1 Introduction

Fiber-reinforced polymer or plastic (FRP) composites have been increasingly used in civil engineering applications due to their high-strength, high-stiffness fibers (e.g., E-glass, carbon, and aramid), light weight, environmentally resistant matrices (e.g., polyester, vinylester, and epoxy resins), and high energy efficiency. Structures made of FRP composites have shown to provide efficient and economical applications in bridges and piers, retaining walls, airport facilities, storage structures exposed to salts and chemicals, and others (Qiao *et al.*, 1999). In addition to lightweight, noncorrosive, nonmagnetic, and nonconductive properties, FRP composites exhibit excellent energy absorption characteristics – suitable for seismic response; high strength, fatigue life, and durability; competitive costs based on load-capacity per unit weight; and ease of handling, transportation, and installation. FRP materials offer the inherent ability to alleviate or eliminate the following four construction-related problems adversely contributing to transportation deterioration worldwide (Head, 1996): corrosion of steel, high labor costs, energy consumption and environmental pollution, and devastating effects of natural hazards such as earthquakes.

With the increasing demand for infrastructure renewal and the decreasing costs of composites manufacture, FRP materials began to be extensively used in civil infrastructure from the 1980s and have continued to expand in recent years. Attention has been focused on FRP shapes as alternative bridge deck materials, because of their high specific stiffness and strength, corrosion resistance, light weight, and potential modular fabrication and installation that can lead to decreased field assembly time and traffic-routing costs. Most currently available commercial bridge decks are constructed using assemblies of adhesively bonded pultruded FRP shapes or honeycomb FRP sandwich panels.

17.2 Review on analysis

17.2.1 Design considerations

A critical obstacle to the widespread use and application of FRP structures in construction is the lack of simplified and practical design guidelines. Unlike standard materials (e.g., steel and concrete), FRP composites are typically orthotropic or anisotropic, and their analyses are much more complex. For example, while changes in the geometry of FRP shapes can be easily related to changes in stiffness, changes in the material constituents do not lead to such obvious results. In addition, shear deformations in pultruded FRP composite materials are usually significant, and therefore the modeling of FRP structural components should account for shear effects. For applications to pedestrian and vehicular FRP bridges, there is a need to develop simplified design equations and procedures, which should provide relatively accurate predictions of bridge behavior and be easily implemented by practising engineers.

17.2.2 Analytical techniques

Closed-form, mechanics-based methods for designing sectional stiffness properties of composite shapes were detailed by Barbero *et al.* (1993) and Davalos and Qiao (1999). These mechanics concepts combined with elastic equivalence analysis can be translated into approximate methods for estimating the equivalent orthotropic properties of members, such as plate behavior of cellular panels. In this way, cellular bridge deck configurations can be defined as assemblies of repetitive structurally efficient and easy-to-manufacture pultruded composite sections. A systematic analysis and design approach for single-span FRP deck-and-stringer bridges was presented by Qiao *et al.* (2000). While systematic methods for optimizing both geometry and lay-up of pultruded sections were given (Davalos *et al.*, 1996a; Qiao *et al.*, 1998), most developments of cellular deck geometries and material lay-ups have been derived by trial and error.

Similar to pultruded FRP structural shapes, honeycomb structures are broadly used in many structural applications in such industries as aerospace and automobile (Noor *et al.*, 1996) and civil infrastructure (Davalos *et al.*, 2001). As recognized by several feasibility studies (Plunkett, 1997), honeycomb FRP structures are found to be very efficient in providing high mechanical performance for minimum unit weight. In light of this, the concept of lightweight and heavy-duty FRP honeycomb panels with a sinusoidal wave core configuration in the plane and extending vertically between face laminates (Fig. 17.1) was introduced for highway bridge decks (Plunkett, 1997; Davalos *et al.*, 2001).

A series of studies for testing and field evaluations were successfully accomplished (Plunkett, 1997), and several analytical studies of sinusoidal sandwich construction were recently conducted (Davalos *et al.*, 2001; Xu *et al.*, 2001; Qiao and Wang, 2005a, b). The bending behavior of FRP sandwich specimens was analytically and experimentally evaluated, and an approximate mechanics of materials approach was used to determine the effective stiffness properties of the sinusoidal core (Davalos *et al.*, 2001; Qiao and Wang, 2005a). Using a homogenization theory, the transverse shear stiffness of a composite honeycomb core with a general configuration was studied, and explicit formulas were derived for the transverse shear properties of the several common cellular cores (Xu *et al.*, 2001). A further effort (Xu and Qiao, 2002; Qiao and Xu, 2005) was conducted to derive all the elastic tensor components for flexure, stretching, transverse shearing, in-plane shearing, and twisting of orthotropic hexagonal-core sandwich plates using a multipass homogenization technique. The effective transverse shear moduli of composite honeycomb cores are important material properties in analysis and design of sandwich structures. An analytical

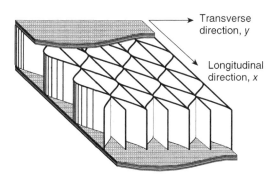

17.1 Honeycomb FRP panel geometry.

approach using a two-scale homogenization technique (Xu *et al.*, 2001) was presented to predict the effective transverse shear stiffnesses of thin-walled composite honeycomb cores with general configurations. To improve the performance of core transverse shear behavior, a nondimensional index, the so-called efficiency of material (EOM), was introduced to evaluate the optimal design of periodic cellular cores (Qiao and Wang, 2005b).

The vertical compression (crushing) behavior of the core was determined by two distinct tests to capture both material strength and buckling (Chen *et al.*, 2002), leading to the development of a closed-form solution and simplified practical formulas for buckling strength (Chen and Davalos, 2003a; Davalos and Chen, 2005). Bending tests were used to evaluate shear strength of the core and core–facesheet interface delamination (Chen and Davalos, 2003b), leading to an interesting study of a complex state of stress at the interface, due to shear warping and bending warping induced by the facesheet as predicted by analytical models, that in combination with a failure criteria proved useful for predicting onset of delamination (Chen and Davalos, 2004a, b, 2005, 2007). Finally, strength optimization of the facesheet was investigated, using a finite element (FE) progressive failure model (Chen and Davalos, 2004c). In a parallel study, the delamination of the core–facesheet interface was addressed by a rigorous approach based on fracture mechanics, to obtain Mode-I energy release rate (G_{Ic}) by a contoured double cantilever beam (CDCB) specimen (Wang *et al.*, 2002). A linear-exponential traction law was used to develop a cohesive zone model (CZM), which was incorporated into the FE program ABAQUS as a special 3D interface finite element, which effectively simulated the core–facesheet delamination propagation of HFRP panels under mixed-mode loading (Wang and Davalos, 2003a, b).

Most recently, Davalos *et al.* (2006) presented a comprehensive design approach for FRP composite structures. First, the material properties involving constituent materials, ply properties, and laminated panel engineering properties were analyzed and given in the format of Carpet plots for design convenience. Then the mechanics of laminated beam (MLB) theory was used to study the member stiffness, and several critical mechanical behaviors of FRP shapes including bending, buckling, and material failure were presented, followed by equivalence formulations for FRP cellular panels and macro-flexibility analysis for deck-and-stringer bridge systems. Finally, design guidelines and examples for FRP beams and bridge systems were given in detail to illustrate the applications and design procedures. In this chapter, the comprehensive design approach in Davalos *et al.* (2006) is reviewed and summarized, from which a 'bottom-up' analysis concept and systematic design methodology is formulated to develop design guidelines for FRP structural members and systems.

Design of all-composite structures using FRP composites 473

17.3 Systematic analysis and design methodology

In this section, a systematic methodology for analysis and design of all-FRP composite structures (Davalos *et al.*, 2006) is presented. As illustrated in Fig. 17.2, this methodology consists of analysis and design at micro level (material), macro level (structural component), and system level (structure) to design all-FRP composite structural systems.

First, based on information from manufacturers and material lay-up, ply properties are predicted by micromechanics/ply mechanics. Once the ply stiffnesses are obtained, macromechanics is applied to compute the panel mechanical properties which are then presented in the form of Carpet plots

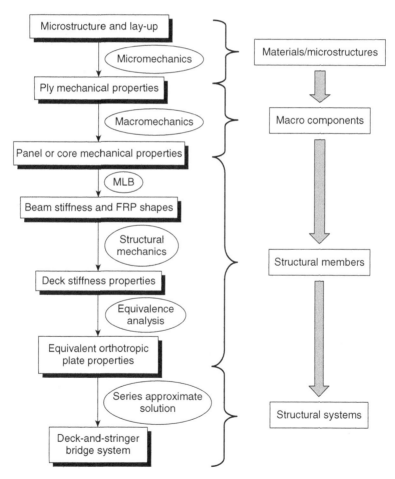

17.2 Systematic analysis methodology for all-FRP composite structures.

for design convenience. Effective material properties for honeycomb FRP cores are evaluated using homogenization techniques, and they are then used in sandwich theory to obtain the sandwich deck properties. Also, beam or stringer stiffness properties are then evaluated from mechanics of thin-walled laminated beams (MLB). The elastic deformation, including shear effects and maximum stress of FRP beams, are correspondingly predicted. Then, based on the stability analyses, formulas for global and local buckling loads of FRP shapes (columns and beams) are presented. From the panel strength data and Carpet plots, the material failure strengths of FRP beams are evaluated. Using elastic equivalence, apparent stiffnesses for composite cellular decks are formulated in terms of panel and single-cell beam stiffness properties, and their equivalent orthotropic material properties are further obtained. Similarly, for honeycomb FRP sandwich panels, the equivalent stiffnesses for cellular core are obtained using homogenization techniques, and combining with the face sheet laminate properties the sandwich panel stiffness coefficients are evaluated. For design analysis of FRP deck-and-stringer bridge systems, an approximate series solution for first-order shear deformation orthotropic plate theory is applied to develop simplified design equations, which account for load distribution factors for various load cases.

As shown in Fig. 17.2, the systematic design methodology, which accounts for the microstructure of composite materials and geometric orthotropy of a composite structural system, represents a 'bottom-up' analysis concept: 'Materials/microstructures,' → 'Macro components,' → 'Structural members' → 'Structural systems,' which can be employed to design and optimize efficient FRP structural components and systems. In the following, the analysis procedures at the aforementioned four levels are presented.

17.3.1 Materials and microstructures

FRP shapes are not laminated structures in a rigorous sense. However, they are produced with material architectures that can be simulated as laminated configurations (Davalos et al., 1996b). For a typical pultruded FRP section shown in Fig. 17.3, the panels in the thin-walled structure can be simulated as laminates consisting of combinations of the following four types of layers:

1. A thin layer of randomly-oriented chopped fibers (Nexus) placed on the surface of the composite, which is a resin-rich layer primarily used as a protective coating, and its contribution to the laminate response can be neglected.
2. Continuous or chopped strand mats (CSM) of different weights consisting of either continuous or chopped randomly-oriented fibers.
3. Stitched fabrics (SF) with different fiber orientation.
4. Roving layers that contain continuous unidirectional fiber bundles.

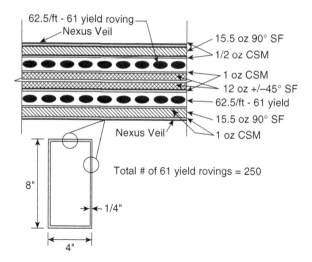

17.3 Lay-up of a typical FRP box-section.

Table 17.1 Material properties of the constituents

Material	E (psi)	G (psi)	ν	ρ (lb/in³)
E-glass fiber	10.5×10^6	4.1833×10^6	0.255	0.092
Vinyl ester resin[a]	7.336×10^5	2.3714×10^5	0.30	0.041

[a] Obtained experimentally (Tomblin, 1994).

In the following, we present constituent material properties, modeling of materials based on accurate estimation of volume fractions of constituents, and computation of ply stiffness by micromechanics and ply mechanics.

Constituent materials

Ply stiffnesses of pultruded panels cannot be readily evaluated experimentally, since the material is not produced by lamination lay-up, but they can be computed from micromechanics formulas for roving, CSM, and SF layers (Qiao, 1997). The ply stiffnesses can then be used in classical lamination theory (CLT) (Jones, 1999) to predict the laminate stiffnesses. The material properties of E-glass fiber and vinyl ester resins commonly used in pultruded products are given in Table 17.1.

Fiber volume fraction

The properties of a pultruded composite section are controlled by the relative volumes of fiber and matrix used. Each layer is modeled as a homogeneous, linearly elastic and generally orthotropic material (Davalos et al., 1996b), and to evaluate its properties, the information provided by the material producer and pultrusion manufacturer are used to compute the fiber volume fraction (V_f). The fiber volume fraction (V_f) is defined as the ratio of the volume of fibers present to the total volume of the layer. The relative volumes of fiber and matrix are first determined in order to evaluate the ply stiffnesses. Similarly, the fiber volume fraction (V_f) of the whole section can be defined. For the CSM and SF layers, which are, respectively, specified commercially in oz/ft² and oz/yd², V_f can be determined as follows:

$$(V_f)_{CSM/SF} = \frac{w}{\rho t_f} \qquad [17.1]$$

where w is the weight per unit area in lb/in² (i.e., convert the units of oz/ft² and oz/yd² to the one of lb/in²), ρ is the unconsolidated density of the CSM or SF fibers in lb/in³, and t_f is the 'as manufactured' thickness of the fabrics (inches) as provided by the material producer. For the roving layers, the fiber volume fraction (V_f) is defined as:

$$(V_f)_r = \frac{n_r A_r}{t_r} \qquad [17.2]$$

where n_r is the number of rovings per unit width (in⁻¹) provided by the manufacturer, t_r is the thickness of the composite layer, which can be evaluated by subtracting the CSM and SF layer thickness from the total panel thickness. The area of one roving is, $A_r = 1/Y\rho_r$, where Y is the yield specified in yard/lb and converted to in/lb, and ρ_r is the density of the fibers. Once the V_f for all the typical layers are computed, the ply stiffnesses are predicted using selected micromechanics formulas.

Micromechanics using periodic microstructure model

There are several micromechanics models available to predict the effective elastic properties of composite materials (Chamis, 1984). Because of its accuracy, it is recommended to compute the ply stiffnesses for the roving and SF layers using favorably the micromechanics model for composites with periodic microstructure (Luciano and Barbero, 1994). Detailed expressions for the computations of the elastic constants E_1, E_2, G_{12}, and v_{12} are given in the original paper along with experimental correlations. The SF and roving layers are usually modeled as unidirectional composites in two orthogonal directions. The CSM layer, however, is assumed to be isotropic in the plane and the properties can be obtained as (Harris and Barbero, 1998):

$$G_{CSM} = \frac{1}{8}\frac{E_1}{\Delta} - \frac{1}{4}\frac{v_{12}E_2}{\Delta} + \frac{1}{8}\frac{E_2}{\Delta} + \frac{1}{2}G_{12}$$

$$v_{CSM} = \frac{\dfrac{1}{8}\dfrac{E_1}{\Delta} + \dfrac{3}{4}\dfrac{v_{12}E_2}{\Delta} + \dfrac{1}{8}\dfrac{E_2}{\Delta} - \dfrac{1}{2}G_{12}}{\dfrac{3}{8}\dfrac{E_1}{\Delta} + \dfrac{1}{4}\dfrac{v_{12}E_2}{\Delta} + \dfrac{3}{8}\dfrac{E_2}{\Delta} + \dfrac{1}{2}G_{12}} \qquad [17.3]$$

$$E_{CSM} = 2G_{CSM}(v_{CSM} + 1)$$

$$\Delta = 1 - v_{12}v_{21} = 1 - v_{12}^2 \frac{E_2}{E_1}$$

17.3.2 Macro components

Macro components in FRP design include laminates and cores, and they are the subcomponents of FRP shapes (e.g., beams, columns, and cellular decks) and sandwich panels (e.g., laminates as facesheets). The analyses of laminates and cores and their engineering properties are provided in this section.

Engineering properties of laminated panels using macromechanics

Once the ply stiffnesses for each flat panel or wall section of an FRP shape are computed, the stiffness of a panel can be computed from CLT (Jones, 1999). For a laminated panel, the general constitutive relation between the resultant forces (N_x, N_y, N_{xy}) and moments (M_x, M_y, M_{xy}) and the midsurface strains (ε_x^0, ε_y^0, γ_{xy}^0) and curvatures (κ_x, κ_y, κ_{xy}) is defined by CLT (Jones, 1999) as:

$$\begin{Bmatrix} N_x \\ N_y \\ N_{xy} \\ M_x \\ M_y \\ M_{xy} \end{Bmatrix} = \begin{bmatrix} A_{11} & A_{12} & A_{16} & B_{11} & B_{12} & B_{16} \\ A_{12} & A_{22} & A_{26} & B_{12} & B_{22} & B_{26} \\ A_{16} & A_{26} & A_{66} & B_{16} & B_{26} & B_{66} \\ B_{11} & B_{12} & B_{16} & D_{11} & D_{12} & D_{16} \\ B_{12} & B_{22} & B_{26} & D_{12} & D_{22} & D_{26} \\ B_{16} & B_{26} & B_{66} & D_{16} & D_{26} & D_{66} \end{bmatrix} \begin{Bmatrix} \varepsilon_x^0 \\ \varepsilon_y^0 \\ \gamma_{xy}^0 \\ \kappa_x \\ \kappa_y \\ \kappa_{xy} \end{Bmatrix} \qquad [17.4]$$

where [A], [B], and [D] are the panel stiffness submatrices. By full inversion of the panel stiffness matrix, we can express the midsurface strains and curvatures in terms of the compliance coefficients and panel resultant forces as:

$$\begin{Bmatrix} \varepsilon_x^0 \\ \varepsilon_y^0 \\ \gamma_{xy}^0 \\ \kappa_x \\ \kappa_y \\ \kappa_{xy} \end{Bmatrix} = \begin{bmatrix} \alpha_{11} & \alpha_{12} & \alpha_{16} & \beta_{11} & \beta_{12} & \beta_{16} \\ \alpha_{12} & \alpha_{22} & \alpha_{26} & \beta_{12} & \beta_{22} & \beta_{26} \\ \alpha_{16} & \alpha_{26} & \alpha_{66} & \beta_{16} & \beta_{26} & \beta_{66} \\ \beta_{11} & \beta_{12} & \beta_{16} & \delta_{11} & \delta_{12} & \delta_{16} \\ \beta_{12} & \beta_{22} & \beta_{26} & \delta_{12} & \delta_{22} & \delta_{26} \\ \beta_{16} & \beta_{26} & \beta_{66} & \delta_{16} & \delta_{26} & \delta_{66} \end{bmatrix} \begin{Bmatrix} N_x \\ N_y \\ N_{xy} \\ M_x \\ M_y \\ M_{xy} \end{Bmatrix} \quad [17.5]$$

where $[\alpha]$, $[\beta]$, and $[\delta]$ are the panel compliance submatrices. In particular, the panel compliance matrix $[\alpha]$ is used to compute the equivalent elastic properties of the panel as:

$$E_x = 1/(t\alpha_{11}), \quad E_y = 1/(t\alpha_{22}), \quad v_{xy} = -\alpha_{12}/\alpha_{11}, \quad G_{xy} = 1/(t\alpha_{66}) \quad [17.6]$$

where t is the thickness of the panel. The panel moduli given in Eq. [17.6] are valid for in-plane loads only. The laminate stiffnesses of the panel as indicated in Eq. [17.6] can be directly obtained from tests of coupon samples, which are cut from the FRP sections and tested in tension and shear (Iosipescu) (see Table 17.2).

Strength of pultruded FRP panels

In design of FRP shapes, strength of individual panels is a critical design factor which determines the ultimate material failure load of the component. As mentioned above, most pultruded FRP panels consist of three types of layer: CSM, SF and rovings. Experimental data (Barbero *et al.*, 1999) indicate that the compressive strengths of CSM and SF layers are relatively lower than that of a roving layer. All the CSM and SF layers failed

Table 17.2 Panel stiffness properties of FRP shapes

FRP shapes	E_{xx} (×10⁶ psi)		G_{xy} (×10⁶ psi)	
	Tension test	Micro/macro-mechanics	Iosipescu test	Micro/macro-mechanics
WF-beam 6" × 6" × 3/8" (**WF6 x 6**)	4.155 (COV = 5.28%)	4.206	0.686 (COV = 8.39%)	0.682
I-beam 4" × 8" × 3/8" (**I4 x 8**)	5.037 (COV = 2.24%)	4.902	0.745 (COV = 9.79%)	0.794
WF-beam 4" × 4" × 1/4" (**WF4 x 4**)	4.391 (COV = 5.55%)	4.167	0.778 (COV = 11.28%)	0.676
Square tube 4" × 4" × 1/4" (**Box4 x 4**)	4.295 (COV = 10.70%)	3.604	0.548 (COV = 8.39%)	0.550

before the rovings reached their ultimate capacity, and therefore the roving layers sustain the applied load up to ultimate failure. Thus, the compressive strength of a panel laminate is approximately proportional to the fiber volume fraction of the roving layers and the percentage of roving layers in the laminate, and it is therefore possible to derive a simplified formula for prediction of compressive strength of FRP panels based on the assumption that the roving layers are the last to fail at ultimate load.

For pultruded FRP panels, the expression for the ultimate compressive load can be written as:

$$P = F_{1c}^* n_r A_r \qquad [17.7]$$

where F_{1c}^* is defined as the specific roving compressive strength and can be obtained from the roving layer compressive strength (F_{1c}) divided by the fiber volume fraction of roving layer (V_f)$_r$ as $F_{1c}^* = F_{1c}/(V_f)_r$, n_r is the number of rovings per unit width (in^{-1}), and A_r is the area of one bundle of roving. The panel compressive strength is thus obtained as:

$$F_{xc} = \frac{P}{1.0" \times t} = \frac{F_{1c}^* n_r A_r}{t} = \frac{F_{1c}^* n_r}{\rho_r Y t} \qquad [17.8]$$

where t is the thickness of the panel in inches. Hence, we can rewrite Eq. [17.8] as:

$$F_{xc} = \frac{F_{1c}^* n_r A_r}{t} = \frac{F_{1c}^* t_r (V_f)_r}{t} \qquad [17.9]$$

where t_r is the thickness of the roving layers.

By defining the percentage of roving layers in the panel as $\alpha = t_r/t$, we can simplify Eq. [17.9] as:

$$F_{xc} = F_{1c}^* \alpha (V_f)_r \qquad [17.10]$$

For typical rovings used for pultruded FRP shapes, the specific roving compressive strengths are given by Barbero *et al.* (1999). Based on Eq. [17.10], the compressive strength of the FRP panels can be predicted once the percentage of the roving layer in the panel and corresponding roving layer fiber volume fraction are known.

The classical approach for strength design of laminates includes the predictions of first-ply-failure (FPF) and fiber failure (FF) loads (Barbero, 1999). The FPF load can be obtained when failure first occurs in any layer of the laminate, and the FPF prediction can be based on any failure criteria commonly used (e.g., Tsai-Wu). Due to the weakness of transverse strength of polymer matrix composites, the FPF is usually associated with matrix cracking. Following the FPF, the material is degraded and the load is increased until a fiber failure occurs (called the FF load).

In FF, a single degradation factor (f_d) is commonly used to reduce all the stiffness values of a degraded layer, except for the stiffness in the fiber direction, which is assumed to be unaffected by FPF. Both the compressive and shear strength values for common pultruded FRP panels based on the FPF and FF load concepts are obtained and further used as baseline data for failure design of FRP shapes.

Carpet plots for FRP panel properties

Based on the analyses presented above for the panel stiffness and strength properties, Carpet plots are developed and used for simplified design of FRP panels. In Carpet plots, diagrams of apparent moduli and strength for various laminate configurations can be produced beforehand and effectively applied to design of FRP structures.

Most pultruded FRP shapes typically consist of three layers (unidirectional roving (0°), CSM, and ±45° angle-ply) and their lay-ups are usually balanced-symmetric and can be simply defined as:

$$[0_{\alpha/2}/CSM_{\beta/2}/(\pm 45°)_{\gamma/2}]_s \qquad [17.11]$$

where α, β, and γ represent the percentages of roving, CSM, and SF (±45°) layers in the laminated panel, and they should satisfy the following relationship:

$$\alpha + \beta + \gamma = 1 \qquad [17.12]$$

Subsequently, Carpet plots for panel moduli and strengths are produced using the micro/macromechanics and strength data introduced above. Based on Eq. [17.6], Carpet plots for panel stiffness properties (E_x, E_y, G_{xy}, v_{xy}) are produced for an E-glass/polyester composite panel with $V_f = 50\%$ (Fig. 17.4). A statistical study is performed (Li, 2000) to compare the corresponding stiffness coefficients with the same fiber lay-up and fiber percentage combinations but different fiber volume fractions. Due to the similar trends among Carpet plots for panel stiffness coefficients of different fiber volume fractions, it can be sufficient to produce only one set of master plots (e.g., with $V_f = 50\%$ of Fig. 17.4) and the corresponding ratio plots (Fig. 17.5), which can be used effectively to estimate stiffness properties for panels of various fiber volume fractions. For example, with respect to the one for $V_f = 50\%$, we can determine the equivalent panel moduli for different fiber volume fractions by multiplying the corresponding ratios by the equivalent panel moduli for $V_f = 50\%$. Using the concept described above, we define Carpet plots for $V_f = 50\%$ (Fig. 17.4) as master plots, and their corresponding ratio plots for E_x, E_y, G_{xy}, and v_{xy} with various fiber volume fractions as given in Fig. 17.5.

Design of all-composite structures using FRP composites

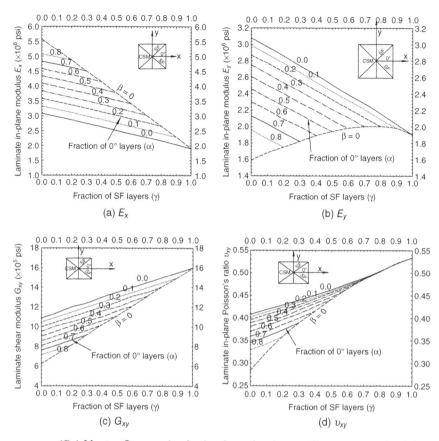

17.4 Master Carpet plot for laminate in-plane stiffness properties (V_f = 50%).

As mentioned above, the roving compressive strength dominates the compressive strength of the whole laminated panel. A graphical representation of panel compressive strength values for various fiber volume fractions is very useful for preliminary design. In Fig. 17.6, Carpet plots for compressive strength of FRP panels are produced based on Eq. [17.10], and the panel ultimate compressive strength depends solely on the properties of roving in the panel (e.g., specific roving compressive strength, roving fiber percentage in the panel, and roving fiber volume fraction). The specific roving compressive strengths for most common pultruded materials have similar values; therefore, an average (F^*_{1c} = 173 ksi, COV = 6.4%) is used in Eq. [17.10] to produce Carpet plots of Fig. 17.6. The plots can generally be applied to predict the panel compressive strength for most pultruded FRP shapes.

An alternative approach to produce Carpet plots for compressive strength F_{xc} is to use conventional failure criteria (e.g., Tsai-Wu criterion) by

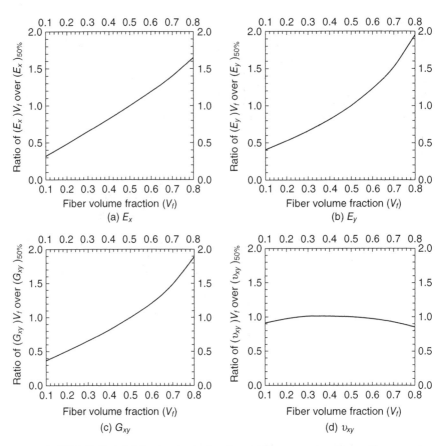

17.5 Ratio plot for laminate in-plane stiffness properties with respect to V_f.

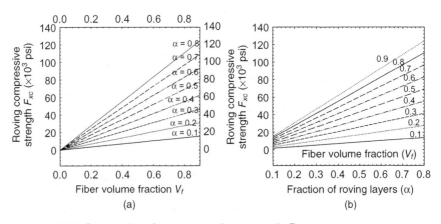

17.6 Carpet plots for compressive strength F_{xc}.

predicting the first ply failure (FPF) load and fiber failure load (FF) (Davalos et al., 2006). Similar to the panel compressive strength prediction, the approach to produce Carpet plots for shear strength F_{xy} is based on the Tsai-Wu criterion, and the FPF shear and FF shear strength ratios are predicted (Davalos et al., 2006). The ultimate shear strength of panels with various fiber volume fractions can be approximated by the plot given in Fig. 17.7.

Honeycomb cores and laminated facesheets

A similar procedure as for pultruded laminate analysis can be used for the analysis of honeycomb FRP composite cores and laminated facesheets in sandwich structures (Fig. 17.1). The core geometry typically consists of closed honeycomb-type FRP cells, and the constituent materials used for the honeycomb sandwich panel (both face laminates and core) consist of E-glass fibers and vinyl ester or polyester resins. It is noteworthy that the composite honeycomb cores are different from their metal counterparts (e.g., aluminum hexagonal or tubular cores) in both manufacturing and consequent corrugated shapes. Unlike traditional metal sandwich structures, the shape of the FRP corrugated cell wall is defined, for example, by a sinusoidal function in the plane (Figs 17.1 and 17.8). In this case, the combined flat and waved FRP cells are produced by sequentially bonding a flat sheet to a corrugated sheet, which is similar to the processing of corrugated cardboard. The assembled cellular core is then co-cured with the upper and bottom facesheet laminates to build a sandwich panel (Fig. 17.1). The waved core elements are

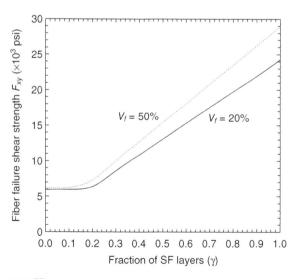

17.7 FF shear strength F_{xy} for $V_f = 20\%$ and $V_f = 50\%$.

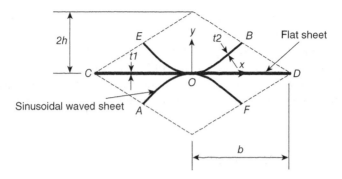

17.8 Representative sinusoidal core element.

Table 17.3 Equivalent material properties of sinusoidal honeycomb core

E_x^e, GPa ($\times 10^4$ psi)	E_y^e, GPa ($\times 10^4$ psi)	G_{xy}^e, GPa ($\times 10^4$ psi)	G_{xz}^e, GPa ($\times 10^4$ psi)	G_{yz}^e, GPa ($\times 10^4$ psi)	ρ_{cr} g/cm³ (lb/in³)
0.531 (7.702)	0.0449 (0.651)	0.0237 (0.344)	0.292 (4.235)	0.119 (1.726)	0.149 (0.00538)

produced by forming FRP sheets in a corrugated mold. The core wall materials (both the flat and waved sheets) are made either of chopped strand or continuous randomly oriented mat of E-glass fibers and polyester resin, and they can be modeled as an isotropic layer. On the other hand, the facesheets in the sandwich panels are in the form of laminate configurations. The stiffness of each ply in the laminated facesheet can be predicted from the micromechanics model with periodic microstructure (Luciano and Barbero, 1994), and the apparent engineering properties of facesheets can be predicted by a combined micro- and macro-mechanics approach (Davalos *et al.*, 1996b) or directly from Carpet plots of Figs 17.4–17.7.

A mechanics of materials approach, recently developed by Qiao and Wang (2005a), can be used to obtain the effective in-plane moduli of sinusoidal cores. For the transverse shear moduli (G_{xz}^e and G_{yz}^e), the formulas for general core configuration obtained from homogenization theory (Xu *et al.*, 2001) are applied for the case of sinusoidal core, and the explicit solutions are given as:

$$G_{xz}^e = \left(\frac{t_1}{2h} + \frac{bt_2}{2hS}\right) G_{12}^s; \quad G_{yz}^e = \frac{2ht_2}{bS} G_{12}^s \qquad [17.13]$$

where G_{12}^s is the shear modulus of solid walls, and S is the length for the curved segment, $S = \int_A^B ds$ in Fig. 17.8. Typical values for a sinusoidal core used in bridge deck panels are given in Table 17.3.

17.4 Structural members

FRP composite shapes are used as structural members (e.g., beams, columns, and deck panels) due to their high strength and stiffness in relation to their weight, corrosion resistance, and structural efficiency. Beams and columns are structural members that carry mainly transverse flexural and axial compressive loads, respectively. Deck panels produced as either cellular FRP shapes or sandwiches primarily carry flexural, transverse shear, and twisting loads. In this section, the analysis techniques for beams, columns, and deck panels are presented.

17.4.1 Beam stiffness properties

The stiffness properties of FRP members are evaluated using a formal engineering approach to the mechanics of thin-walled laminated beams (MLB) (Barbero *et al.*, 1993), based on kinematic assumptions consistent with Timoshenko beam theory. The MLB approach is adopted in this study to model pultruded structural shapes based on first-order shear deformation theory for thin- and moderately thick-walled laminated beams with open or closed cross sections. In this model, the stiffnesses of a beam are computed by adding the contributions of the stiffnesses of the component panels, which in turn are obtained from the effective beam moduli as given in Eq. [17.6]. The model accounts for membrane stiffness and flexure stiffness of the walls. Warping effects due to non-uniform bending (shear lag) are not included in this model. Therefore, this theory is more appropriate for moderately thick-walled laminated beams than for thin-walled laminated beams. The position of the neutral axis is defined in such a way that the behavior of a thin-walled beam-column with asymmetric material and/or cross-sectional shape is completely described by axial, bending, and shear stiffness coefficients (A_z, D_y, F_y) only.

The basic kinematic assumptions in MLB (Barbero *et al.*, 1993) are: (1) the contour does not deform in its own plane, and (2) a plane section originally normal to the beam axis remains plane, but not necessarily normal due to shear deformation. Straight FRP beams with at least one axis of geometric and material symmetry are considered. The pultruded sections are modeled as assemblies of flat walls. The compliance matrices $[\alpha]_{3 \times 3}$, $[\beta]_{3 \times 3}$, $[\delta]_{3 \times 3}$ of the individual panels (see Eq. [17.5]) are obtained from classical lamination theory (CLT) (Tsai, 1988). For each wall panel, the position of the middle surface is defined as (Fig. 17.9):

$$y(s_i) = s_i \sin \phi_i + \overline{y}_i \quad \text{for} \quad -\frac{b_i}{2} \leq s_i \leq \frac{b_i}{2} \qquad [17.14]$$

where b_i is the wall width and $\overline{y}_i$ is the position of the wall centroid.

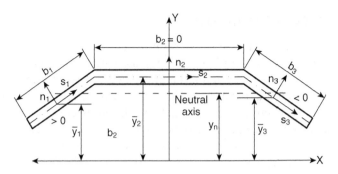

17.9 Global (beam) and local (panel) coordinator systems used in MLB.

By restricting the off-axis plies to be balanced-symmetric, the shear-extension (α_{16}) and shear-bending (β_{16}) coupling coefficients in Eq. [17.5] vanish ($\alpha_{16} = \beta_{16} = 0$). Then, using the beam theory assumptions ($N_y = M_y = 0$), the compliance equations (Eq. [17.5]) can be reduced to

$$\begin{Bmatrix} \bar{\varepsilon}_x \\ \bar{\kappa}_x \\ \bar{\gamma}_{xy} \\ \bar{\kappa}_{xy} \end{Bmatrix} = \begin{bmatrix} \alpha_{11} & \beta_{11} & 0 & 0 \\ \beta_{11} & \delta_{11} & 0 & 0 \\ 0 & 0 & \alpha_{66} & \beta_{66} \\ 0 & 0 & \beta_{66} & \delta_{66} \end{bmatrix} \begin{Bmatrix} \bar{N}_x \\ \bar{M}_x \\ \bar{N}_{xy} \\ \bar{M}_{xy} \end{Bmatrix} \qquad [17.15]$$

Inverting Eq. [17.15], the reduced constitutive equation for the ith panel is obtained as

$$\begin{Bmatrix} \bar{N}_x \\ \bar{M}_x \\ \bar{N}_{xy} \\ \bar{M}_{xy} \end{Bmatrix} = \begin{bmatrix} \bar{A}_i & \bar{B}_i & 0 & 0 \\ \bar{B}_i & \bar{D}_i & 0 & 0 \\ 0 & 0 & \bar{F}_i & \bar{C}_i \\ 0 & 0 & \bar{C}_i & \bar{H}_i \end{bmatrix} \begin{Bmatrix} \bar{\varepsilon}_x \\ \bar{\kappa}_x \\ \bar{\gamma}_{xy} \\ \bar{\kappa}_{xy} \end{Bmatrix} \qquad [17.16]$$

where $\bar{A}_i$, $\bar{B}_i$, $\bar{D}_i$, $\bar{F}_i$, $\bar{C}_i$ and $\bar{H}_i$ are, respectively, the ith panel extensional, bending-extension, bending, shear, twisting-shear and twisting stiffnesses; they can be expressed in terms of panel engineering properties (see Eq. [17.6]) and given as:

$$\bar{A}_i = (\delta_{11}\Delta_1^{-1})_i, \quad \bar{B}_i = (-\beta_{11}\Delta_1^{-1})_i, \quad \bar{D}_i = (\alpha_{11}\Delta_1^{-1})_i$$
$$\bar{F}_i = (\delta_{66}\Delta_2^{-1})_i, \quad \bar{H}_i = (\alpha_{66}\Delta_2^{-1})_i, \quad \bar{C}_i = (-\beta_{66}\Delta_2^{-1})_i, \qquad [17.17]$$
$$\Delta_1 = \alpha_{11}\delta_{11} - \beta_{11}^2, \quad \Delta_2 = \alpha_{66}\delta_{66} - \beta_{66}^2$$

General expressions for the axial, bending and transverse shear stiffness coefficients of FRP members are derived from the beam variational problem (Barbero et al., 1993); while the torsional stiffness is found using the energy balance between the work done by the external torque and the strain energy due to shear (Barbero, 1999). Hence, the axial (A_z), bending (D_x or D_y), shear (F_x or F_y), and torsional (D_t^{open} for open cross section and D_t^{close} for closed cross section) stiffnesses that account for the contribution of all the walls can be computed as:

$$A_z = \sum_{i=1}^{n} \bar{A}_i b_i$$

$$B_y = \sum_{i=1}^{n} [\bar{A}_i(\bar{y}_i - y_n) + \bar{B}_i \cos\phi_i] b_i$$

$$D_y = \sum_{i=1}^{n} \left[\bar{A}_i \left((\bar{y}_i - y_n)^2 + \frac{b_i^2}{12} \sin^2\phi_i \right) + 2\bar{B}_i(\bar{y}_i - y_n)\cos\phi_i + \bar{D}_i \cos^2\phi_i \right] b_i \quad [17.18]$$

$$F_y = \sum_{i=1}^{n} \bar{F}_i b_i \sin^2\phi_i$$

$$D_t^{open} = GJ = 4 \sum_{i=1}^{n} \bar{H}_i b_i$$

$$D_t^{close} = \frac{(2\Gamma_s)^2}{\sum_{i=1}^{n} b_i/\bar{F}_i} + \frac{3}{4}\left(4\sum_{i=1}^{n} \bar{H}_i b_i \right)$$

where Γ_s is the area enclosed by the contour of shear flow. The beam bending-extension coupling coefficient (B_x or B_y) can be eliminated by defining the location of the neutral axis of bending (x_n or y_n) as:

$$y_n = \frac{\sum_{i=1}^{n}(\bar{y}_i \bar{A}_i + \cos\phi_i \bar{B}_i) b_i}{A_z} \quad [17.19]$$

By introducing the coordinate $y' = y - y_n$, we are able to decouple the extensional and bending responses (i.e., $B_y = 0$). An explicit expression for the static shear correction factor (K_x or K_y) is derived from energy equivalence. As an approximation in design, the shear correction factor for pultruded sections can be taken as 1.0. General equations of the shear correction factor for various FRP sections are presented by Lopez-Anido (1994), and for laminates by Madabhusi-Raman and Davalos (1996).

For each laminated wall (e.g., a flange or a web), the stiffness values are obtained either by the micro/macromechanics approach (see Eq. [17.6]) or from Carpet plots. If we incorporate stress-resultant assumptions compatible with beam theory, and we assume that the off-axis plies of pultruded panels are balanced-symmetric, and no extension-shear and bending-twist couplings are present, the extensional, bending, shear, and twisting stiffnesses of the ith panel are expressed in terms of panel apparent moduli as:

$$\bar{A}_i = (E_x)_i t_i, \quad \bar{D}_i = (E_x)_i t_i^3/12, \quad \bar{F}_i = (G_{xy})_i t_i, \quad \bar{H}_i = \frac{(G_{xy})_i t_i^3}{12} \quad [17.20]$$

where the engineering properties (E_x, E_y, v_{xy}, and G_{xy}) of the panel are computed by using micro/macromechanics approach or obtained from Carpet plots.

For most common thin-walled structures (e.g., I- and Box-sections), the beam centroid is the neutral axis of bending (no beam bending-extension coupling), and the axial (A), bending (D), shear (F), and torsional (D_t) stiffnesses of the beam (that account for the contribution of all the panels) can be simplified as:

$$A = \sum_{i=1}^{n}(E_x)_i t_i b_i, \quad D = \sum_{i=1}^{n}\left[(E_x)_i t_i \left(\frac{b_i^2}{12}\sin^2\phi_i\right) + \frac{(E_x)_i t_i^3 \cos^2\phi_i}{12}\right]b_i,$$

$$F = \sum_{i=1}^{n}(G_{xy})_i t_i b_i \sin^2\phi_i \quad [17.21]$$

$$D_t^{open} = \frac{1}{3}\sum_{i=1}^{n}(G_{xy})_i t_i^3 b_i, \quad D_t^{close} = \frac{(2\Gamma_s)^2}{\sum_{i=1}^{n}b_i/((G_{xy})_i t_i)} + \frac{1}{4}\left(\sum_{i=1}^{n}(G_{xy})_i t_i^3 b_i\right)$$

where b_i is the panel width, and ϕ_i is the cross-sectional orientation of the ith panel with respect to the x-axis. The preceding stiffness formulas are readily applied in engineering design to various structural shapes. As an example, the bending and shear stiffnesses of four FRP beams are listed in Table 17.4.

17.4.2 Mechanical behaviors of FRP components

Most FRP shapes are thin-walled structures and made of E-glass fiber and polyester or vinyl ester resins. Due to the relatively low stiffness of FRP composites and thin-walled sectional geometry, problems with large deformation, including shear deformation and structural stability, need to be considered in design and analysis (Qiao et al., 1999).

Table 17.4 Beam bending and shear stiffness properties

FRP shapes	$D = EI$ ($\times 10^8$ psi-in^4)		$F = GA$ (10^6 psi-in^2)	
	Strong-axis	Weak-axis	Strong-axis	Weak-axis
WF-beam 6" × 6" × 3/8" (**WF6 × 6**)	1.776	0.570	1.292	3.066
I-beam 4" × 8" × 3/8" (**I4 × 8**)	2.558	0.199	1.772	2.379
WF-beam 4" × 4" × 1/4" (**WF4 × 4**)	0.334	0.111	0.585	1.351
Square tube 4" × 4" × 1/4" (**Box4 × 4**)	0.364	0.338	1.100	1.176

Design of all-composite structures using FRP composites

Requirements for service-load deflections are usually the dominant design limits for FRP shapes subjected to transverse loading. Critical global and local buckling due to the thin-walled structure and/or large slenderness ratios of component panels is another important criterion in design of FRP shapes. Further, potential material failure due to the relatively low compressive and shear strengths of composites should also be considered. Thus, the mechanical behaviors of FRP shapes associated with large deformation, elastic stability, and material failure are addressed in this section.

Elastic deflections

According to Timoshenko beam theory, the deflection of a beam is caused by bending and shear when subjected to transverse loading and can be obtained by solving the equilibrium equations. Deflections at discrete locations can be computed by employing energy methods that incorporate the beam bending and shear stiffnesses. For some conventional materials, the shear deformation can be neglected because of the higher shear modulus. But for FRP materials, the shear modulus is relatively low in relation to the axial modulus, and is in the range of $E/6$ to $E/10$ as compared to $E/2.5$ for metals. Therefore, the shear deformation must be considered, especially for relatively short spans. Formulas for maximum bending and shear deflections for typical beam loading and boundary conditions are given in Table 17.5 as:

$$\delta_{Total} = \delta_{Bending} + \delta_{Shear} \qquad [17.22]$$

Table 17.5 Bending and shear deflections of Timoshenko beams

Beam (Loading + B.C.)	$\delta_{Bending}$	δ_{Shear}
Simply supported, center point load P	$\dfrac{1}{48}\dfrac{PL^3}{D}$	$\dfrac{1}{4}\dfrac{PL}{KF}$
Simply supported, two point loads P	$\dfrac{23}{648}\dfrac{PL^3}{D}$	$\dfrac{1}{3}\dfrac{PL}{KF}$
Cantilever, end point load P	$\dfrac{1}{3}\dfrac{PL^3}{D}$	$\dfrac{PL}{KF}$
Simply supported, uniform load q	$\dfrac{5}{384}\dfrac{qL^4}{D}$	$\dfrac{1}{8}\dfrac{qL^2}{KF}$
Cantilever, uniform load q	$\dfrac{1}{8}\dfrac{qL^4}{D}$	$\dfrac{1}{2}\dfrac{qL^2}{KF}$
Simply supported, end moments M	$\dfrac{1}{8}\dfrac{ML^2}{D}$	0

where in Table 17.5, P is the concentrated load, q is the uniformly distributed load, M is the transverse bending moment, L is the span length, K is the shear correction factor ($K = 1.0$ can be assumed), and D and F are the beam bending and shear stiffnesses (see Eq. [17.21]).

Elastic strains and stresses

For the ith panel (Fig. 17.9), the mid-surface strains and curvatures in terms of the beam resultant forces and moments are calculated as:

$$\bar{\varepsilon}_x = \frac{N_Z}{A_Z} + (\bar{Y}_i - Y_n)\frac{M_X}{D_X}, \quad \bar{\kappa}_x = \frac{M_X}{D_X}\cos\phi_i, \quad \bar{\gamma}_{xy} = \frac{V_Y}{K_Y F_Y}\sin\phi_i \quad [17.23]$$

where N_Z, M_X, and V_Y are, respectively, the resultant internal axial force, bending moment, and transverse shear force acting on the beam in the global coordinate system. The subscripts x, y, and z refer to the local coordinate system of individual panels defined similarly as those in MLB approach. Then, applying Eq. [17.16], we can obtain the resultant forces and moments ($\bar{N}_x, \bar{M}_x$, and $\bar{N}_{xy}$) acting on the ith panel. Combining the constitutive relations of Eq. [17.5] with the assumptions of $\bar{N}_y = \bar{M}_y = 0$ and $\bar{M}_{xy} = 0$, the mid-surface strains and curvatures on the ith panel are obtained as:

$$\begin{Bmatrix} \bar{\varepsilon}_x^0 \\ \bar{\varepsilon}_y^0 \\ \bar{\gamma}_{xy}^0 \\ \bar{\kappa}_x \\ \bar{\kappa}_y \\ \bar{\kappa}_{xy} \end{Bmatrix} = \begin{bmatrix} \alpha_{11} & \alpha_{16} & \beta_{11} \\ \alpha_{12} & \alpha_{26} & \beta_{12} \\ \alpha_{16} & \alpha_{66} & \beta_{16} \\ \beta_{11} & \beta_{16} & \delta_{11} \\ \beta_{12} & \beta_{26} & \delta_{12} \\ \beta_{16} & \beta_{66} & \delta_{16} \end{bmatrix} \begin{Bmatrix} \bar{N}_x \\ \bar{N}_{xy} \\ \bar{M}_x \end{Bmatrix} \quad [17.24]$$

where the overbar identifies a panel quantity. Based on CLT, the ply strains (ε_x, ε_y, and ε_{xy}) and stresses (σ_x, σ_y, and σ_{xy}) can be correspondingly obtained through the thickness of each panel. Using coordinate transformations, the ply strains (ε_1, ε_2, and γ_{12}) and stresses (σ_1, σ_2, and τ_{12}) can be computed in principal material directions, and they can be used with failure criteria to predict the failure loads.

Euler buckling of FRP columns

For an elastic column with pin-pin boundaries at the ends and under axial load P, the Euler buckling load can be easily obtained by applying beam theory, and it is defined as:

$$P_E = \frac{n^2\pi^2 D}{L_e^2} \quad [17.25]$$

Design of all-composite structures using FRP composites

where D is the bending stiffness, n is the buckled wave number, and L_e is the effective span length. When the load exceeds the critical value P_E for $n = 1$, the column becomes unstable. For a column made of composite materials, the bending stiffness (D) is computed from beam theory using Eq. [17.21]. It is important to note that D is computed assuming that the cross section does not deform during bending. Eq. [17.25] can be generalized for various end-conditions by identifying the inflection points of the deflected shape (Fig. 17.10). The effective length is the distance between the inflection points or the end pinned supports, and it is represented by L_e in Eq. [17.25], where its value is conveniently computed as $L_e = \xi L$, in terms of tabulated end-restraint coefficient ξ (Table 17.6). The ideal boundary conditions in Fig. 17.10 cannot be realized in practice because of flexibility of the connections. Therefore, experimentally adjusted values of the end-restraint coefficient

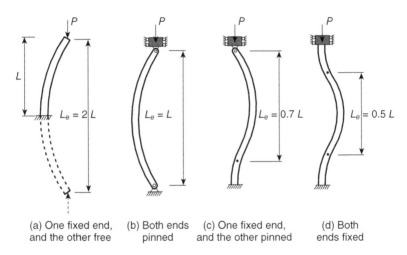

(a) One fixed end, and the other free (b) Both ends pinned (c) One fixed end, and the other pinned (d) Both ends fixed

17.10 Effective length of column for various end-conditions.

Table 17.6 End-restraint coefficients for long column buckling

End-restraint	ξ_{theory}	ξ_{steel}	ξ_{wood}
Pinned-pinned	1.0	1.0	1.0
Clamped-clamped	0.5	0.65	0.65
Pinned-clamped	0.7	0.8	0.8
Clamped-free	2.0	2.1	2.4

are used, as given in Table 17.6. Unfortunately, there is no experimental data for FRP composite columns. The values for wood columns are often used in design since the material properties of wood are relatively similar to those of FRP composites. Then, the Euler buckling load in Eq. [17.25] becomes:

$$P_E = \frac{\pi^2 D}{(\xi L)^2} \qquad [17.26]$$

If the end-conditions are the same with respect to the weak and strong axes of bending, buckling will occur with respect to the weak-axis of bending, because the stabilizing effect is weaker with respect to this axis. But if the end-conditions are different, not only the values of (D) but also the values of the end restraint condition will determine the direction of buckling. In those cases, the bending stiffness (D) and coefficient ξ with respect to both the weak and strong axes are needed.

The shear deformation increases the deflection of a member when undergoing buckling, and it can be interpreted as a reduction of the stabilizing effect caused by the low transverse shear stiffness of the column. Shear deformation beam theory can be used to derive the equation for critical load, and the result is a reduction of the critical load given in Eq. [17.26], which is modified as:

$$P_{ES} = \frac{P_E}{1 + P_E/F} \qquad [17.27]$$

The above correction to Eq. [17.26] is negligible for practical cases. The shear stiffness (F) affects the beam deflections only for short beams, and the same holds for columns. But short thin-walled columns usually fail by local buckling at loads lower than P_E. Therefore, the correction of P_E for shear has very limited application.

Flexural-torsional buckling of FRP beams

The flexural-torsional buckling of pultruded FRP composite I- and C-section beams was analyzed using the second variational total potential energy principle and Rayleigh-Ritz method (Qiao et al., 2003; Shan and Qiao, 2005). By applying the Rayleigh-Ritz method and solving for the eigenvalues of the potential energy equilibrium equation, the flexural-torsional buckling load, P_{cr}, for a cantilever I-section with a point load applied at the centroid of the free end is obtained as (Qiao et al., 2003):

$$P_{cr} = \Psi_1 \cdot \left\{ b_w L \Psi_2 + \left(\sqrt{\Psi_3 + \Psi_4 + \Psi_5 + \Psi_6 + \Psi_7} \right) / b_w \right\} \qquad [17.28]$$

where $\Psi_1 = (6b_f + b_w)/[2L^3 \cdot (76.5b_f^2 - 6.96b_f b_w + 0.16b_w^2)]$

$$\Psi_2 = (123b_f - 5.6b_w)D_{16}$$

$$\Psi_3 = a_{11}b_f^3(279.5b_f^2 - 25.5b_f b_w + 0.6b_w^2)$$

$$\Psi_4 = b_f b_w^5(62.7L^2 d_{66}D_{11} - 305.4b_w^2 D_{11}^2 - 1377.4L^2 D_{16}^2 - 5511L^2 D_{11}D_{66})$$

$$\Psi_5 = b_w^6(7b_w^2 D_{11}^2 + 31.4L^2 D_{16}^2 + 125.5L^2 D_{11}D_{66})$$

$$\Psi_6 = a_{11}b_f^3 b_w^2(1118b_f^5 d_{11} - 101.8b_f^4 b_w d_{11} + 2.3b_f^3 b_w^2 d_{11} + 5043.5b_f^3 L^2 d_{66} + 4.64b_w^5 D_{11} + 20.9b_w^3 L^2 D_{66})$$

$$\Psi_7 = a_{11}b_f^4 b_w^3 [b_w(-203.6b_w^2 D_{11} + 10.5L^2 d_{66} - 918.5L^2 D_{66}) + b_f(2235.8b_w^2 D_{11} - 459.5L^2 d_{66} + 10087L^2 D_{66})]$$

and the following material parameters are defined as:

$$a_{11} = 1/\alpha_{11}, a_{66} = 1/\alpha_{66}, d_{11} = 1/\delta_{11}, d_{66} = 1/\delta_{66} \quad [17.29]$$

For design purposes, the simplified engineering equations for flexural-torsional buckling of I-beams developed by Pandey *et al.* (1995) for several commonly used loading conditions can be adopted in practice, and these formulas are derived based on Vlasov's theory. In a similar fashion, the solution for global buckling load of cantilever C-section beams was obtained by Shan and Qiao (2005).

Local buckling of FRP shapes

For short-span FRP shapes, local buckling occurs more readily. In general, the local buckling analyses of FRP shapes are accomplished by modeling the flanges and webs individually and considering the flexibility of the flange-web connections (Qiao *et al.*, 2001). In this type of simulation, each component of FRP shapes (Fig. 17.11) is modeled as a composite plate subjected to elastic restraints (R) along the unloaded edges (i.e., the flange-web connections) (Qiao *et al.*, 2001). The explicit solution for rotationally restrained (R) plates is given in Qiao and Shan (2005). For the RR (restrained on both unloaded edges) plate when $k_L = k_R = k$, the local buckling stress resultant is simplified to:

$$N_{cr}^{RR} = \frac{24}{b^2}\left\{1.871\sqrt{\frac{\tau_2}{\tau_1}}\sqrt{D_{11}D_{22}} + \frac{\tau_3}{\tau_1}(D_{12} + 2D_{66})\right\} \quad [17.30]$$

where the coefficients τ_1, τ_2, and τ_3 are functions of the rotational restraint stiffness k, and defined as:

$$\tau_1 = 124 + 22\frac{kb}{D_{22}} + \frac{k^2 b^2}{D_{22}^2}, \quad \tau_2 = 24 + 14\frac{kb}{D_{22}} + \frac{k^2 b^2}{D_{22}^2}, \quad \tau_3 = 102 + 18\frac{kb}{D_{22}} + \frac{k^2 b^2}{D_{22}^2}$$

and the resulting critical aspect ratio for the RR plate is given as:

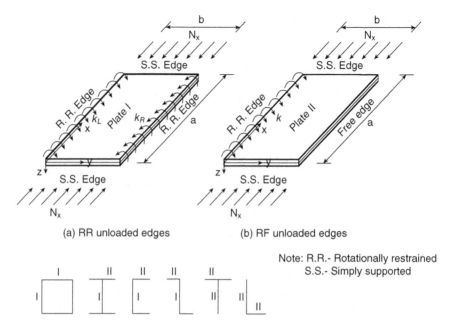

17.11 Geometry of orthotropic plate elements of FRP shapes.

$$\gamma_{cr}^{RR} = 0.663 \left\{ \frac{m^4 \left(\eta_1 k^2 b^2 + \eta_2 D_{22} kb + 4\eta_3 D_{22}^2 \right) D_{11}}{\left(\eta_{12} k^2 b^2 + \eta_{13} D_{22} kb + 36\eta_{14} D_{22}^2 \right) D_{22}} \right\}^{\frac{1}{4}} \quad [17.31]$$

where $\eta_1 = \dfrac{76 D_{22}^2 + 17 D_{22} kb + k^2 b^2}{D_{22}^2}$, $\eta_2 = \dfrac{1{,}140 D_{22}^2 + 272 D_{22} kb + 17 k^2 b^2}{D_{22}^2}$,

$\eta_3 = \dfrac{1{,}116 D_{22}^2 + 285 D_{22} kb + 19 k^2 b^2}{D_{22}^2}$, $\eta_{12} = \dfrac{36 D_{22}^2 + 13 D_{22} kb + k^2 b^2}{D_{22}^2}$,

$\eta_{13} = \dfrac{396 D_{22}^2 + 156 D_{22} kb + 13 k^2 b^2}{D_{22}^2}$, $\eta_{14} = \dfrac{24 D_{22}^2 + 11 D_{22} kb + k^2 b^2}{D_{22}^2}$

For the *RF* (restrained on one unloaded edge and free on the other) plate, the local buckling stress resultant and the critical aspect ratio are obtained, respectively, as:

$$N_{cr}^{RF} = \frac{112 \left(15 D_{22}^2 + 10 D_{22} kb + 2 k^2 b^2 \right) D_{66} - 28 \left(5 D_{22} kb + k^2 b^2 \right) D_{12}}{b^2 \left(140 D_{22}^2 + 77 D_{22} kb + 11 k^2 b^2 \right)}$$
$$+ \frac{4 \sqrt{35 D_{11} D_{22} kb \left(3 D_{22} + kb \right)}}{b^2 \sqrt{140 D_{22}^2 + 77 D_{22} kb + 11 k^2 b^2}} \quad [17.32]$$

Design of all-composite structures using FRP composites

$$\gamma_{cr}^{RF} = 0.9133m \left\{ \frac{(140D_{22}^2 + 77D_{22}kb + 11k^2b^2)D_{11}}{(3D_{22} + kb)kbD_{22}} \right\}^{\frac{1}{4}} \quad [17.33]$$

Design formulas of critical local buckling load (N_{cr}) for applications to several common orthotropic plate cases of applications (see Fig. 17.12) and their related critical aspect ratios (γ_{cr}) are given in Table 17.7, and they can be applied in local buckling analysis of FRP plates.

Once the explicit solutions for elastically restrained plates (Fig. 17.12) are obtained, they can be applied to predict the local buckling of FRP shapes (Fig. 17.11). In the discrete plate analysis of FRP shapes, the rotational restraint stiffness (k) is needed to determine the critical buckling strength. Based on the studies by Bleich (1952) for isotropic materials and Qiao et al. (2001) and Qiao and Zou (2002, 2003) for composite materials, the rotational restraint stiffness coefficients (k) for local buckling of different FRP shapes were examined by Qiao and Shan (2005) and Shan (2007). The explicit formulas for local buckling stress resultants (N_{cr}) and rotational restraint stiffness (k) are summarized in Table 17.8, and they can be used to predict the local buckling of several common FRP profiles.

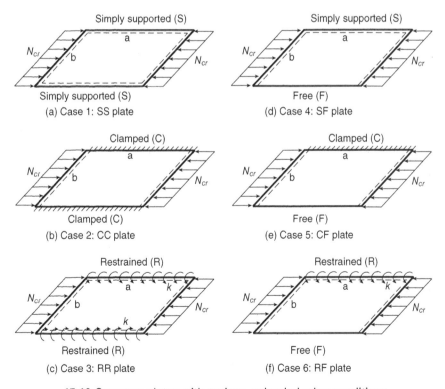

17.12 Common plates with various unloaded edge conditions.

Table 17.7 Local buckling loads and critical aspect ratios for FRP plates

Plate type	Local buckling stress resultant (N/cm)	Critical aspect ratio
SS plate (Fig. 17.12(a))	$N_{cr}^{SS} = \dfrac{2\pi^2}{b^2}\{\sqrt{D_{11}D_{22}} + (D_{12} + 2D_{66})\}$	$\gamma_{cr}^{SS} = \left(\dfrac{m^4 D_{11}}{D_{22}}\right)^{1/4}$
CC plate (Fig. 17.12(b))	$N_{cr}^{CC} = \dfrac{24}{b^2}\{1.871\sqrt{D_{11}D_{22}} + (D_{12} + 2D_{66})\}$	$\gamma_{cr}^{CC} = 0.663\left(\dfrac{m^4 D_{11}}{D_{22}}\right)^{1/4}$
RR plate (Fig. 17.12(c))	Eq. [17.30]	Eq. [17.31]
SF plate (Fig. 17.12(d))	$N_{cr}^{SF} = \dfrac{12 D_{66}}{b^2} + \dfrac{\pi^2 D_{11}}{a^2}$	—
CF plate (Fig. 17.12(e))	$N_{cr}^{CF} = \dfrac{-28 D_{12} + 4\sqrt{385 D_{11} D_{22}} + 224 D_{66}}{11 b^2}$	$\gamma_{cr}^{CF} = 1.6633 m \left(\dfrac{D_{11}}{D_{22}}\right)^{1/4}$
RF plate (Fig. 17.12(f))	Eq. [17.32]	Eq. [17.33]

Ultimate bending and shear failure

Due to the relatively low compressive and shear strength properties of FRP composites, the material failure needs to be evaluated as well. Similar to beam deflection and buckling, the beam bending and shear strengths (ultimate failure loads) for three-point bending can be expressed in terms of panel strength properties as:

$$\text{Bending: } P_{fail}^{bending} = \frac{8 F_c D}{(E_x)_f (b_w - t_f) L} \quad [17.34]$$

$$\text{Shear: } P_{fail}^{shear} = F_{xy} b_w t_w \quad [17.35]$$

where F_c and F_{xy} are the compressive and shear strengths of FRP panels, and their values can be obtained from Carpet plots (see Figs 17.6 and 17.7).

17.4.3 Equivalent analysis of FRP decks

Each of the honeycomb sandwich and multicellular FRP composite bridge decks can be modeled as orthotropic plates, with equivalent stiffnesses that account for the size, shape, and constituent materials of the deck. Thus, the complexity of material anisotropy of the panels and structural orthotropy of the deck system can be reduced to an equivalent orthotropic plate with global elastic properties in two orthogonal directions: parallel and transverse to the longitudinal axis of the deck. These equivalent orthotropic plate properties can be directly used in analysis and design of deck-and-stringer

Table 17.8 Rotational restraint stiffness (k) and critical local buckling stress resultant (N_{cr}) for different FRP shapes[a]

FRP section	Buckled plate[b]	Critical local buckling stress resultant N_{cr}	Rotational restraint stiffness k
Box-	Flange	Eq. [17.30] with D_{ij}^f	$k = \dfrac{D_{22}^{w*}}{b_f \rho_1} \left(1 - \dfrac{b_w^2 \sqrt{D_{11}^f D_{22}^f} + D_{12}^f + 2D_{66}^f}{b_f^2 \sqrt{D_{11}^w D_{22}^w} + D_{12}^w + 2D_{66}^w}\right)$[c]
	Web	Eq. [17.30] with D_{ij}^w	$k = \dfrac{D_{22}^{f*}}{b_w \rho_1} \left(1 - \dfrac{b_f^2 \sqrt{D_{22}^w D_{11}^w} + D_{12}^w + 2D_{66}^w}{b_w^2 \sqrt{D_{11}^f D_{22}^f} + D_{12}^f + 2D_{66}^f}\right)$[c]
I-	Flange	Eq. [17.32] with D_{ij}^f	$k = \dfrac{D_{22}^{w*}}{b_w}\left(1 - \dfrac{6b_w^2}{\pi^2 b_f^2}\dfrac{D_{66}^f}{\sqrt{D_{11}^w D_{22}^w} + D_{12}^w + 2D_{66}^w}\right)$
	Web	Eq. [17.30] with D_{ij}^w	$k = \dfrac{D_{22}^{f*}}{b_w \rho_2}\left(1 - \dfrac{\pi^2 b_f^2 \sqrt{D_{11}^w D_{22}^w} + D_{12}^w + 2D_{66}^w}{6 b_w^2 D_{66}^f}\right)$[c]
Channel and Z-	Flange	Eq. [17.32] with D_{ij}^f	$k = \dfrac{2D_{22}^{w*}}{b_w}\left(1 - \dfrac{6 b_w^2}{\pi^2 b_f^2}\dfrac{D_{66}^f}{\sqrt{D_{11}^w D_{22}^w} + D_{12}^w + 2D_{66}^w}\right)$
	Web	Eq. [17.30] with D_{ij}^w	$k = \dfrac{D_{22}^{f*}}{b_w \rho_2}\left(1 - \dfrac{\pi^2 b_f^2 \sqrt{D_{11}^w D_{22}^w} + D_{12}^w + 2D_{66}^w}{6 b_w^2 D_{66}^f}\right)$[c]

Table 17.8 Continued

FRP section	Buckled plate[b]	Critical local buckling stress resultant N_{cr}	Rotational restraint stiffness k
T-	Flange	Eq. [17.32] with D_{ij}^f	$k = \dfrac{D_{22}^{w*}}{1.9 b_f} e^{-\frac{1}{2}\left(\frac{b_w - \frac{b_f}{2}}{4.5}\right)^2}$
	Web	Eq. [17.32] with D_{ij}^w	$k = \dfrac{D_{22}^{f*}}{1.9 b_w} e^{-\frac{1}{2}\left(\frac{b_f - \frac{b_w}{2}}{4.5}\right)^2}$
L-	Flange	$N_{cr} = \dfrac{12 D_{66}^f}{(b^f)^2} + \dfrac{\pi^2 D_{11}^f}{a^2}$	$k = 0$[d]
	web	$N_{cr} = \dfrac{12 D_{66}^w}{(b^w)^2} + \dfrac{\pi^2 D_{11}^w}{a^2}$	$k = 0$[d]

[a] D_{ij} ($i, j = 1, 2, 6$) are the bending stiffness per unit length and D_{22}^* is the transverse bending stiffness of a unit length.
[b] Buckled plate refers to the first buckled discrete element (either flange or web) in the FRP shapes.
[c]
$$\rho_1\left(\frac{b_i}{b_j}\right) = \frac{1}{2\pi} \tanh\frac{\pi b_i}{2 b_j} \left\{ 1 + \frac{\frac{\pi b_i}{b_j}}{\sinh\left(\frac{\pi b_i}{b_j}\right)} \right\}, \quad \rho_2\left(\frac{b_i}{b_j}\right) = \frac{1}{4\pi} \frac{3\cosh^2\left(\frac{\pi b_i}{b_j}\right) + \left(\frac{\pi b_i}{b_j}\right)^2 + 1}{\frac{\pi b_i}{b_j} + 3\sinh\left(\frac{\pi b_i}{b_j}\right)\cosh\left(\frac{\pi b_i}{b_j}\right)}$$
, where $b_l(b_j)(l = f \text{ or } w)$ is the width of flange or web, respectively.
[d] In the L-section, only the case of equal flange and web legs is herein given.

bridge systems, and they can also serve to simplify modeling procedures either in explicit or numerical formulations.

For the honeycomb sandwich deck, the material properties obtained for the face laminates using micro/macromechanics and the core using homogenization theory at macro component level can be used with sandwich theory to predict the equivalent properties of the sandwich panel. For brevity, the derivation of the sandwich theory (Vinson, 1999) is not presented here.

The development of equivalent stiffness for cellular decks consisting of multiple FRP box beams was presented by Qiao *et al.* (2000). Multicell box sections are commonly used in deck construction because of their light weight, efficient geometry, and inherent stiffness in flexure and torsion. Also, this type of deck has the advantage of being relatively easy to build. It can be either assembled from individual box-beams or manufactured as a complete section by pultrusion or vacuum assisted resin transfer molding (VARTM). The elastic equivalence concept used (Troitsky, 1987) accounted for out-of-plane shear effects and the results for a multicell box section were verified experimentally and by finite element analyses (Qiao *et al.*, 2000).

The bending stiffness of the deck in the longitudinal direction, or x-axis in Fig. 17.13, is expressed as the sum of the bending stiffness of individual box beams (D_b can be obtained by MLB):

$$D_x = n_c D_b \qquad [17.36]$$

where n_c is the number of cells. For the section shown in Fig. 17.13, b is the width of a cell, h is the height of a cell, t_f and t_w are the thicknesses of the

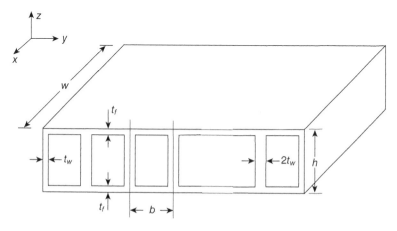

17.13 Geometric notation of multi-cell box deck.

flange and web, respectively. If all panels have identical material lay-up and $t_f = t_w = t$, Eq. [17.36] becomes:

$$D_x = n_c E_x \left(h^2 + t^2 + 3(h)b \right) \frac{(h)(t)}{6}$$ [17.37]

where E_x is the modulus of elasticity of a panel in the x-direction computed by micro/macromechanics, or obtained experimentally, or evaluated from Carpet plots (Fig. 17.4).

The out-of-plane shear stiffness of the deck in the longitudinal direction, F_x, is expressed as a function of the stiffness for the individual beams (F_b):

$$F_x = n_c F_b$$ [17.38]

where F_b is the shear stiffness obtained by MLB, and n_c is the number of cells. This expression can be further approximated in terms of the in-plane shear modulus of the component panel, G_{xy}, and cross-sectional area of the beam webs:

$$F_x = n_c G_{xy} (2t) h$$ [17.39]

where G_{xy} is the in-plane shear modulus of the panel walls, and can be evaluated from Carpet plots (Fig. 17.4).

An approximate value for the deck bending stiffness in the transverse direction, D_y, may be obtained by neglecting the effect of the transverse diaphragms and the second moment of area of the flanges about their own centroids. For a deck as shown in Fig. 17.13 with $t_f = t$:

$$D_y = \frac{1}{2} E_y (w)(t) h^2$$ [17.40]

where w is the length of the deck in the longitudinal direction, and E_y is the modulus of elasticity of the panel in the y-direction, which can be obtained from Carpet plots (Fig. 17.4).

For multiple box sections, the simplest way to obtain the decks out-of-plane transverse shear stiffness is to treat the structure as a Vierendeel frame in the transverse direction (Cusen and Pama, 1975). For the Vierendeel frame, the inflection points are assumed at the midway of top and bottom flanges between the webs. The shear stiffness in the transverse direction, F_y, for the cross section shown in Fig. 17.13 may be written as:

$$F_y = \frac{V}{\theta} = \frac{12 E_y}{b \left(\frac{h}{I_w} + \frac{b}{2 I_f} \right)}$$ [17.41]

where the moments of inertia I are defined as:

$$I_f = \frac{wt_f^3}{12}; I_w = \frac{w(2t_w)^3}{12} \quad [17.42]$$

For $t_f = t_w = t$, Eq. [17.41] can be simplified as:

$$F_y = \frac{2E_y wt^3}{b\left(b+\dfrac{h}{4}\right)} \quad [17.43]$$

where E_y is the modulus of elasticity of a panel in the y-direction.

The torsional rigidity of a multi-cell section, GJ, is evaluated by considering the shear flow around the cross section of a multi-cell deck. For a structure where the webs and flanges are small compared with the overall dimensions of the section, Cusen and Pama (1975) have shown that the torsional rigidity may be written as:

$$GJ = \frac{4A^2 G_{xy}}{\sum \dfrac{ds}{t}} + \sum G_{xy}(ds)\frac{t^3}{3} \quad [17.44]$$

where A is the area of the deck section including the void area and is defined as $A = n_c bh$, and $\sum ds/t$ represents the summation of the length-to-thickness ratio taken around the median line of the outside contour of the deck cross section. For a constant panel thickness t, the torsional rigidity can be simplified as:

$$GJ = \frac{2(n_c bh)^2 G_{xy} t}{(n_c b + h)} + \frac{2}{3}(n_c b + h) G_{xy} t^3 \quad [17.45]$$

The above approximate equation is justified by the fact that for a multi-cell deck, the net shear flow through interior webs is negligible, and only the shear flow around the outer webs and top and bottom flanges is significant. The second term in Eq. [17.45] is relatively small compared to the first term and can be ignored.

If the deck is treated as an equivalent orthotropic plate, its torsional rigidities depend upon the twist in two orthogonal directions. Thus the torsional stiffness D_{xy} may be taken as one-half of the total torsional rigidity given by Eq. [17.45] divided by the total width of the deck:

$$D_{xy} = \frac{GJ}{2n_c b} \quad \text{or} \quad D_{xy} = \frac{n_c G_{xy} bh^2 t}{(n_c b + h)} \quad [17.46]$$

where D_{xy} is the torsional stiffness per unit width (lb-in^4/in).

Once the stiffness properties of an actual deck are obtained, it is relatively simple to calculate the equivalent orthotropic plate material properties, which can further simplify the design analysis of deck and deck-and-stringer bridge systems. The equivalent orthotropic plate material properties of the cellular box deck are given as (Qiao et al., 2000):

$$(E_x)_p = 12 \frac{D_x}{t_p^3 b_p}(1 - v_{xy}v_{yx}) \quad [17.47a]$$

$$(E_y)_p = 12 \frac{D_y}{t_p^3 l_p}(1 - v_{xy}v_{yx}) \quad [17.47b]$$

$$(G_{xz})_p = \frac{F_x}{t_p b_p} \quad [17.47c]$$

$$(G_{yz})_p = \frac{F_y}{t_p l_p} \quad [17.47d]$$

$$(G_{xy})_p = 6 \frac{D_{xy}}{t_p^3} \quad [17.47e]$$

where the subscript 'p' indicates property related to the equivalent orthotropic plate; t_p is the thickness of the plate (= h for the actual deck, Fig. 17.13), b_p is the width of the plate (= $n_c b$ for the actual deck), and l_p is the length of the plate (= w for the actual deck).

17.5 Structural systems

The equivalent properties for panels and stiffnesses for FRP beams can then be efficiently used to analyze and design structural systems (e.g., a deck-and-stringer bridge system as in Fig. 17.14). For the deck-and-stringer

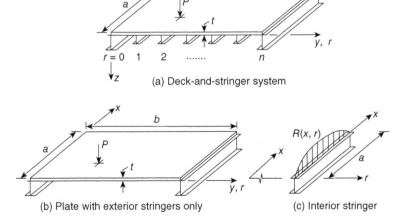

17.14 Deck-and-stringer bridge system.

bridge system, a first-order shear deformation macro-flexibility analysis (Brown, 1998; Qiao et al., 2000) is conducted, and the solutions for symmetric and antisymmetric load cases are used to obtain the solution for asymmetric loading. Based on deck-stringer transverse interaction force functions, wheel load distribution factors are derived, which are used later to provide design guidelines for deck-and-stringer bridge systems.

The general expressions of load distribution factors in terms of the number of stringers m (where $m = n + 1$) for symmetric and asymmetric loads (Brown, 1998) are, respectively:

$$W_f^{Sym}(r) = \frac{\sin\frac{r-1}{m-1}\pi + W_o}{\frac{2}{\pi}(m-1) + mW_o} \quad [17.48a]$$

$$W_f^{Asym}(r) = \frac{\left[R_{11}\left(\sin\pi\frac{r-1}{m-1} + W_o\right) + R_{12}\left(\sin 2\pi\frac{r-1}{m-1} + W_1\left(1 - 2\frac{r-1}{m-1}\right)\right)\right]}{\sum_{r=1}^{m}\left[R_{11}\left(\sin\pi\frac{r-1}{m-1} + W_o\right) + R_{12}\left(\sin 2\pi\frac{r-1}{m-1} + W_1\left(1 - 2\frac{r-1}{m-1}\right)\right)\right]} \quad [17.48b]$$

The maximum wheel-load distribution factor under symmetric loading occurs when $(r-1)/(m-1) = 1/2$; i.e., $\sin(\pi/2) = 1$. Therefore, the maximum load distribution factors for symmetric and antisymmetric loading are given as:

$$\left(W_f^{Sym}\right)_{max} = \frac{1 + W_o}{mW_o + \frac{2}{\pi}(m-1)} \quad [17.49a]$$

$$\left(W_f^{Antisym}\right)_{max} = \frac{1 + 0.5W_1}{(m-1)\left(\frac{1}{\pi} + \frac{W_1}{4}\right)} \quad [17.49b]$$

The maximum wheel load distribution factors given in Eq. [17.49] can be used in analysis and design of deck-and-stringer bridge systems, as illustrated in design examples by Davalos et al. (2006).

17.6 Design guidelines

Based on the aforementioned analytical studies, design guidelines for FRP structures including structural members and deck-and-stringer bridge systems are provided, and the proposed step-by-step design analysis procedures follow closely the flowchart illustrated in Fig. 17.2. The following steps for analysis and design of FRP shapes are suggested:

(1) Based on the manufacturer's information and material lay-up, the panel stiffness and strength properties are obtained by using either Carpet plots or conducting experimental coupon tests.
(2) Using homogenization theory, the effective stiffness properties of honeycomb core are obtained, and combining with the face laminate properties obtained in Step (1), they can be used to compute the elastic properties of honeycomb sandwiches.
(3) Based on the panel material properties obtained in Step (1), the elastic stiffness properties of FRP beam members and bridge deck are calculated by MLB and equivalent analysis, respectively.
(4) Using Timoshenko's beam theory, the deflections of FRP beams with different boundary and loading conditions are calculated.
(5) Based on MLB, the elastic strains and stresses of FRP beams are obtained. The factors of safety for material failure are obtained and compared to the panel strength properties obtained in Step (1).
(6) Following the explicit formulas of local and global buckling, the stability analysis of FRP shapes is implemented.
(7) For design and analysis of a deck-and-stringer bridge system, the stiffness properties and deformation of the stringers are obtained using MLB and Timoshenko's beam theory; while the material properties of orthotropic deck panels are obtained by performing the elastic equivalency analysis for cellular decks or using sandwich theory for honeycomb decks with facesheet properties obtained from Step (1) and core properties from Step (2).
(8) Based on limitations on serviceability for bridge systems (e.g., the deflection limit), the number of FRP stringers can be determined by the macro-flexibility analysis.
(9) Based on MLB, the maximum normal and shear stresses of the stringers and related factors of safety are calculated to meet the requirements for a safe design.
(10) A summary of design calculations and design sketches are used to complete the design.

Detailed design examples following the above design guidelines are provided in Davalos et al. (2006).

17.7 Conclusion

A systematic approach for material characterization, analysis, and design of all-fiber-reinforced polymer or plastic (FRP) composite structures is presented in this chapter. The suggested 'bottom-up' analysis concept (Fig.17.2) is applied throughout the procedure, from materials/microstructures, to macro components, to structural members, and finally to structural systems, thus providing a systematic analysis methodology for all-FRP composite structures. After a brief review of analysis of FRP composites in civil infrastructure, design procedures using several approaches are described (e.g., micro/macromechanics, Carpet plots, and homogenization theory) to obtain material properties, including constituent materials and ply properties, laminated panel engineering properties, core effective properties, and member stiffness properties. The mechanical behaviors of FRP shapes (e.g., bending and shear, deformation, local/global buckling, and material failure) are discussed in reasonable detail. An elastic equivalence analysis is used to obtain the apparent properties of cellular deck panels; while a first-order shear deformation macro-flexibility analysis is briefly introduced for composite deck-and-stringer systems, accounting for load distribution factors under various loading cases. The step-by-step design guidelines for FRP shapes described in this chapter are useful for practising engineers concerned with design of FRP composite structures. The systematic approach described in this chapter can be used efficiently to analyze and design FRP shapes and bridge systems and also develop new design concepts for all-composite structures.

17.8 References

Barbero, E.J., *Introduction to Composite Materials Design*. Taylor & Francis, Philadelphia, PA, 1999.

Barbero, E.J., Lopez-Anido, R., and Davalos, J.F., On the mechanics of thin-walled laminated composite beams. *Journal of Composite Materials*, 27(8): 806–829, 1993.

Barbero, E.J., Makkapati, S., and Tomblin, J.S., Experimental determination of the compressive strength of pultruded structural shapes. *Composites Science and Technology*, 59(13): 2047–2054, 1999.

Bleich, F., *Buckling Strength of Metal Structures*, McGraw-Hill, New York, 1952.

Brown, B., Experimental and analytical study of FRP deck-and-stringer short-span bridges. MSc thesis, West Virginia University, Morgantown, WV, 1998.

Chamis, C.C., *Simplified Composites Micromechanics Equations for Strength, Fracture Toughness, and Environmental Effects*. NASA TM-83696, 1984.

Chen, A. and Davalos, J.F., Buckling of honeycomb FRP core with partially restrained loaded edges. *Proceedings of Composites in Construction International Conference*, Cosenza, Italy, September 16–19, 2003a.

Chen, A. and Davalos, J.F., Bending strength of honeycomb FRP sandwich beams with sinusoidal core geometry. *Proceedings of 4th Canadian-International Composites Conference*, Ottawa, Canada, August 19–22, 2003b.

Chen, A. and Davalos, J.F., Bonding and aspect ratio effect on the behavior of honeycomb FRP sandwich cores. *Proceedings of 8th Pan American Congress of Applied Mechanics, PANCAM VIII*, January 5–9, 2004a.

Chen, A. and Davalos, J.F., Behavior of honeycomb FRP sandwich sinusoidal core panels with skin effect. *Proceedings of 9th ASCE Aerospace Division International Conference on Engineering, Construction and Operations in Challenging Environments*, Houston, TX, March 7–10, 2004b.

Chen, A. and Davalos, J.F., Development of facesheet for honeycomb FRP sandwich bridge deck panels. *Proceedings of 4th International Conference on Advanced Composite Materials in Bridges and Structures*, Calgary, Alberta, Canada, July 20–23, 2004c.

Chen, A. and Davalos, J.F., A solution including skin effect for stiffness and stress field of sandwich honeycomb core. *International Journal of Solids and Structures*, 42(9–10): 2711–2739, 2005.

Chen, A. and Davalos, J.F., Transverse shear including skin effect for composite sandwich with honeycomb sinusoidal core. *Journal of Engineering Mechanics, ASCE*, 133(3): 247–256, 2007.

Chen, A., Davalos, J.F., and Plunkett, J.D., Compression strength of honeycomb FRP core with sinusoidal geometry. *Proceedings of the American Society for Composites 17th Technical Conference*, West Lafayette IN, October 21–23, 2002.

Cusen, A.R. and Pama, R.P., *Bridge Deck Analysis*. John Wiley & Sons, New York 1975.

Davalos, J.F. and Chen, A., Buckling behavior of honeycomb FRP core with partially restrained loaded edges under out-of-plane compression. *Journal of Composite Materials*, 39(16): 1465–1485, 2005.

Davalos, J.F. and Qiao, P.Z., A computational approach for analysis and optimal design of FRP beams. *Computers and Structures*, 70(2): 169–183, 1999.

Davalos, J.F., Qiao, P., and Barbero, E.J., Multiobjective material architecture optimization of pultruded FRP I-beams. *Composite Structures*, 35: 271–281, 1996a.

Davalos, J.F., Salim, H.A., Qiao, P., Lopez-Anido, R., and Barbero, E.J., Analysis and design of pultruded FRP shapes under bending. *Composites, Part B: Engineering Journal*, 27(3–4): 295–305, 1996b.

Davalos, J.F., Qiao, P., Xu, X.F., Robinson, J., and Barth, K.E., Modeling and characterization of fiber-reinforced plastic honeycomb sandwich panels for highway bridge applications. *Composite Structures*, 52, 441–452, 2001.

Davalos, J.F., Qiao, P.Z., and Shan, L.Y., Advanced fiber reinforced polymer (FRP) structural composites for use in civil engineering. *In Advanced Civil Infrastructure Materials – Science, Mechanics and Applications*, edited by H.C. Wu. Woodhead Publishing, Cambridge, pp. 118–202, 2006.

Harris, J.S. and Barbero, E.J., Prediction of creep properties of laminated composites from matrix creep data. *Journal of Reinforced Plastics and Composites*, 17(4): 361–379, 1998.

Head, P.R., Advanced composites in civil engineering – a critical overview at this high interest, low use stage of development. *Proceedings of ACMBS*, M. El-Badry (ed.), Montreal, Quebec, Canada, pp. 3–15, 1996.

Jones, R.M., *Mechanics of Composite Materials*, Taylor & Francis, Philadelphia, PA, 1999.

Li, X.Z., Simplified analysis and design of fiber reinforced plastic structural shapes. Engineering Report for MSCE, The University of Akron, 2000.

Lopez-Anido, R., Analysis and design of orthotropic plates stiffened by laminated beams for bridge superstructures. PhD dissertation, Department of Civil and Environmental Engineering, West Virginia University, 1994.

Luciano, R. and Barbero, E.J., Formulas for the stiffness of composites with periodic microstructure. *International Journal of Solids and Structures*, 31(21): 2933–2944, 1994.

Madabhusi-Raman, P., and Davalos, J.F., Static shear correction factor for laminated rectangular beams. *Composites Part B: Engineering*, 27B, 285–293, 1996.

Noor, A., Burton, W.S., and Bert, C.W., Computational models for sandwich panels and shells. *Applied Mechanics Review*, 49(3): 155–199, 1996.

Pandey, M.D., Kabir, M.Z., and Sherbourne, A.N., Flexural-torsional stability of thin-walled composite I-section beams. *Composites Engineering*, 5(3): 321–342, 1995.

Plunkett, J.D., Fiber-reinforcement polymer honeycomb short span bridge for rapid installation. IDEA Project Report, November 1997.

Qiao, P.Z., Analysis and design optimization of fiber-reinforced plastic (FRP) structural beams. PhD dissertation, Department of Civil and Environmental Engineering, West Virginia University, 1997.

Qiao, P.Z. and Shan, L.Y., Explicit local buckling analysis and design of fiber-reinforced plastic composite shapes. *Composite Structures*, 70(4): 468–483, 2005.

Qiao, P.Z. and Wang, J.L., Mechanics of composite sinusoidal honeycomb cores. *Journal of Aerospace Engineering, ASCE*, 18(1): 42–50, 2005a.

Qiao, P.Z. and Wang, J.L., Transverse shear stiffness of composite honeycomb cores and efficiency of material. *Mechanics of Advanced Materials and Structures*, 12(2): 159–172, 2005b.

Qiao, P.Z. and Xu, X.F., Refined analysis of torsion and in-plane shear of honeycomb sandwich structures. *Journal of Sandwich Structures and Materials*, 7(4): 289–305, 2005.

Qiao, P.Z. and Zou, G.P., Local buckling of elastically restrained fiber-reinforced plastic plates and its application to box sections. *Journal of Engineering Mechanics*, 128(12): 1324–1330, 2002.

Qiao, P.Z., and Zou, G.P., Local buckling of composite fiber-reinforced plastic wide-flange sections. *Journal of Structural Engineering*, 129(1): 125–129, 2003.

Qiao, P., Davalos, J.F., and Barbero, E.J., Design optimization of fiber-reinforced plastic composite shapes. *Journal of Composite Materials*, 32(2): 177–196, 1998.

Qiao, P.Z., Davalos, J.F., Barbero, E.J., and Troutman, D., Equations facilitate composite designs. *Modern Plastics Magazine*, 76(11): 77–80, 1999.

Qiao, P.Z., Davalos, J.F., and Brown, B., A systematic analysis and design approach for single-span FRP deck/stringer bridges. *Composites, Part B: Engineering Journal*, 31: 593–609, 2000.

Qiao, P.Z., Davalos, J.F., and Wang, J.L., Local buckling of composite FRP shapes by discrete plate analysis. *Journal of Structural Engineering*, 127(3): 245–255, 2001.

Qiao, P.Z., Zou, G.P., and Davalos, J.F., Flexural-torsional buckling of fiber-reinforced plastic composite cantilever I-beams. *Composite Structures*, 60: 205–217, 2003.

Shan, L.Y., Explicit buckling analysis of fiber-reinforced plastic (FRP) composite structures. PhD dissertation, Washington State University, Pullman, WA, 2007.

Shan, L.Y. and Qiao, P.Z., Flexural-torsional buckling of fiber-reinforced plastic composite open channel beams. *Composite Structures*, 68(2): 211–224, 2005.

Tomblin, J.S., Compressive strength models for pultruded glass fiber reinforced composites. PhD dissertation, Department of Mechanical and Aerospace Engineering, West Virginia University, Morgantown, WV, 1994.

Troitsky, M.S., *Orthotropic Bridges, Theory and Design*. The James F. Lincoln ARC Welding Foundation, Cleveland, OH, 1987.

Tsai, S.W., *Composites Design*. Think Composites, Dayton, OH, 1988.

Vinson, J.R., *The Behavior of Sandwich Structures of Isotropic and Composite Materials*, Technomic Publishing, Lancaster, PA, 1999.

Wang, W.Q. and Davalos, J.F., Evaluation of facesheet-core delamination. *The 14th International Conference on Composite Materials* (Paper #1654), San Diego, California, July 14–18, 2003a.

Wang, W.Q. and Davalos, J.F., Numerical/experimental study of facesheet delamination for honeycomb FRP sandwich panels. *Proceedings of Composites in Constructions International Conference* (Paper #162), Cosenza, Italy, September 16–19, 2003b.

Wang, W.Q., Davalos, J.F., and Qiao, P., Study of facesheet-from-core delamination for honeycomb FRP sandwich panels. *Proceedings of ASC 17th Technical Conference* (Paper #080), West Lafayette, October 21–23, 2002.

Xu, X.F. and Qiao, P., Homogenized elastic properties of honeycomb sandwich with skin effect. *International Journal of Solids and Structure*, 39: 2153–2188, 2002.

Xu, X.F., Qiao, P., and Davalos, J.F., Transverse shear stiffness of composite honeycomb core with general configuration. *Journal of Engineering Mechanics*, 127(11): 1144–1151, 2001.

Index

A-scan, 131
ABAQUS, 373, 472
ABC techniques, 442
accessories, 354
ACI-318, 240
ACI 440.1R-06, 208, 212, 215–16
ACI318M-02, 209, 212
acoustic emission (AE), 129–30
acoustic-laser technique, 132
   detection of debonding in GFRP-wrapped concrete, 133
acrylonitrile–butadiene–styrene (ABS), 443
adhesive, 311–12
   bonding, 306
airbox, 296
alkalisation, 451, 454
Alside, 304
American Composites Manufacturers Association, 353
analytical modeling, 237–9
   deflection, 237–9
   strength, 239
analytical techniques
   fibre-reinforced polymer (FRP) composites, 470–2
      honeycomb FRP panel geometry, 471
ANSYS, 228, 232, 322–3
aramid fibre-reinforced polymer (AFRP), 353
ASTM 1583-04, 307
ASTM C192, 365
ASTM C393, 232, 238

ASTM C581, 460
ASTM C39-96, 365
ASTM C-393, 286
ASTM D4255, 364
ASTM D3039-76, 364
ASTM D3410-75, 364
ASTM D3762-98, 167
ASTM Standard D5364, 458
atmospheric pressure, 59
autoclave aerated concrete (AAC)
   hybrid fibre-reinforced polymer (FRP) design for structural applications, 226–44
      analytical modeling, 237–9
      comparing different panel designs, 233–7
      design graphs, 239–44
      performance issues, 227–9
   impact behaviour of panels for structural applications, 247–69
      analysing sandwich structures using energy balance model (EBM), 253–5
      analysis using energy balance model (EBM), 266, 268–9
      low velocity impact (LVI) and sandwich structures, 249–50
      low velocity impact (LVI) testing, 255–7
      materials, processing and methods of investigation, 229–33
      dimensions and reinforcement type of the tested panels, 232

Index

mechanical properties of plain AAC, 229
mechanical properties of SIKA carbon fibre composites, 230
mechanical properties of TYFO glass composites, 231
VART processed FRP-AAC specimens, 231
materials and processing, 250–3
  details of test specimens, 253
  mechanical properties of plain AAC, 251
  mechanical properties of SIKA carbon fibre composite, 252
  schematic diagram for CFRP/AAC sandwich panel, 252
results of impact testing, 258–66
  CFRP/AAC panels processed by VARTM technique, 264–6
  hand lay-up technique processed CFRP/AAC panels, 261–4
  plain AAC specimens, 258–60
automated fibre placement (AFP), 69–71
  advanced automated fibre placement machine, 69
  fibre waviness defect, 71

B-scan, 131
basalt FRP (BFRP), 219
bending tests, 472
biaxial FRP for flexural reinforcement (BFFS), 230–1
biaxial FRP for flexural reinforcement (BFFS) panel
  experimental results, 234–5
    load vs deflection graphs for UFFS and BFFS panels, 235
    specimen failure, 235
  finite element results, 237
    numerical vs experimental load deflection curves, 237
biocomposites
  matrices, 26–31
    biobased, 30–1

petrochemical-based, 26–9
plastics and their biodegradability, 27
performance, 36–43
  biological properties, 42–3
  mechanical properties, 36–40
  physical properties, 40–2
biodegradability, 24
biofibre reinforced polymer composites
  biocomposites matrices, 26–31
  biocomposites performance, 36–43
  disadvantages, 22–4
    biological resistance, 24
    dimensional instability, 23
    equilibrium moisture content of different natural fibres, 23
    fire resistance, 24
    moisture absorption, 23
    ultraviolet resistance, 24
  future trends, 43–5
  natural fibres modification, 24–6
  processing, 31–6
    fibre load on the odor concentration of abaca fibre–PP composites, 32
  reinforcing fibres, 19–22
  structural applications, 18–45
bond behaviour, 394–99
  fracture mechanics-based approaches, 398–9
  stress-based approaches, 395–8
    predicted distribution of shear and peeling stresses, 396
    various techniques to reduce local stress concentrations, 397
Bredts formula, 375
bridge columns
  compression loading, 333–8
    failure mode, 338
    specimen details, 333–4
    stress–strain response, 334–8
  impact loading, 338–43
    deflection vs time for tested columns, 342
    energy vs time for tested columns, 342
    impact test results summary, 340

instrumented drop weight low velocity impact test, 339
load vs displacement for tested columns, 341
load vs time for tested columns, 341
prefabricated wraps, 332–3
concept schematic, 333
bridge deck designs, 320–3
analysis and design procedures, 322–3
E-glass/PP woven tape composite, 322
bridge proposed by Aref and Parson, 330–1
cross-sections and thermoplastic E-glass vs PP bridges, 331
S-glass epoxy vs E-glass PP deck, 331
case studies, 323–9
design verification, 328–9
double-lane bridge deck system, 326–8
experimental vs analytical deflection, 329
plan, 326
single-lane bridge deck system, 323–5
comparison, 329–31
Lockheed-Martin bridge, 330
design criteria, 321–2
plan of single-lane bridge deck, 320
Bridge Design Specifications, 365
bridge superstructures
advantages of FRP composites, 350–1
bridge conditions in USA, 347–50
age of US bridges, 349
bridge maintenance priorities at state DOTs, 349
number of deficient bridges, 348
typical sources of bridge deterioration, 350
fibre-reinforced polymer (FRP) composites, 347–79
applications, 351–6
hybrid FRP concrete bridge, 356–78

bridges, 186–90
Aberfeldy footbridge, 190
asset bridge deck profile, 187
DuraSpan bridge deck profile, 188
FRP balustrade system, 189
FRP handrail system, 188
GFRP bridge enclosure, 189
Strongwell Composolite, 187
Building Code for Structural Concrete, 240
building exteriors, 181–6
composite floor support structure, 184
Duragrid composite grating system, 182
FRP wall panels used in a clean room, 181
GRP dome on a mosque, 185
GRP gatehouse, 186
GRP pillars and porch, 185
The Eyecatcher Building, 186
white GRP cladding of Mondial House, 181
building interiors, 181–6
composite floor support structure, 184
Duragrid composite grating system, 182
FRP wall panels used in a clean room, 181
GFRP chemically resistant doors, 184
GFRP door for house construction, 183
GRP mezzanine flooring and handrail system, 182
white GRP cladding of Mondial House, 181

C-glass, 9
C-scan, 131
carbon fibre-reinforced polymer (CFRP), 117, 179, 352
carbon fibres, 5, 7–9, 66
mechanical properties, 8
carbonisation, 8
Carpet plots, 480–3, 504

compressive strength, 482
FF shear strength, 483
laminate in-plane stiffness properties, 481
ratio plot for laminate in-plane stiffness properties, 482
cellulose fibres, 25
cement concrete board (CCB), 273
CFRP/AAC panels
  hand lay-up technique, 261–4
  vacuum assisted resin transfer moulding (VARTM), 264–6
charge-coupled device (CCD), 134
chemical weathering, 455–6
chimney liners, 425–6
  large diameter FRP chimney flue liner, 426
  view of stack liner installation, 426
chopped fibres *see* nonwoven fabrics
civil engineering
  advanced fibre-reinforced polymer (FRP) composites, 177–203
  building applications, 181–201
  future trends, 202–3
  materials use in construction, 178–80
classical lamination theory (CLT), 475, 477, 485
coal-fired plants, 423–9
  air pollution control devices, 425
cohesive zone model (CZM), 159–60, 162, 472
composite material, 410
  advanced processing techniques for structural applications, 54–72
    automated fibre placement, 69–71
    future trends, 71–2
    manual lay up, 54–5
    plate bonding, 55–6
    preforming, 56–7
    pultruded composites, 65–9
    vacuum assisted resin transfer moulding (VARTM), 57–65

composite structural insulated panels (CSIP), 274, 276–9, 288–98
  adhesives connections, 311–12
    results of single lap shear tests and specimen dimensions, 312
  connections, 311–15
    ultrasonic welding, 312–14
  connections details, 314–15
    typical cross section of a building with CSIP, 314
  expanded polystyrene foam (EPS) materials, 278–9
  manufacturing, 279, 307–11
    full-scale CSIP, 280
    pull off strengths of candidate adhesives, 308
    schematic of strength testing, 308
  manufacturing with film adhesives, 309–10
    film position, bottom facesheets and dry spots created on facesheet, 309
  manufacturing with TP spray adhesives, 310–11
    adhesive film vs spray systems time required to manufacture CSIP panels, 311
    hot-melt impregnated spray adhesive, 310
  proposed composite structural insulated sandwich panels, 276
  thermoplastics facesheets, 276–8
    manufacturing steps, 277
compounding, 32
compression
  members failure, 119–20
    behaviour of GFRP-wrapped concrete cylinders, 119
  compression loading, 333–8
    failure mode, 338
    specimen details, 333–4
      mechanical data for reinforcing materials, 334
    stress–strain response, 334–8
concrete, 416
  thickness, 362–3

concrete–adhesive (CA) interface, 152–3
constituent materials, 475
continuously reinforced concrete pavement (CRCP) test, 416–17
Cooling Technology Institute (CTI) Code STD 137, 458–9
cooling towers, 426–7
  Buckeye Power Inc. Cardinal Plant Unit 3 FGD Project, 428
core crushing, 282
corona treatment, 25
critical failure-based durability, 163–4

DATAQ recorder, 340
DDG-1000, 418–19
  stealth warship, 419
debond growth rate, 164–5
debonding modeling, 282–5
  critical wrinkling stress in the facesheet, 284–5
  interfacial tensile stress, 283–4
  types of stresses at compressive facesheet for CSIP due to debonding, 282
  Winkler foundation model, 283
decks, 353–4
deflections, 83–6, 286–7, 489
  CSIP floor, 286–7
  CSIP wall, 287
design graphs, 239–44
  comparison graphs, 243–4
    reinforced AAC vs CFRP/AAC floor panels, 243
    reinforced AAC vs FRP/AAC wall panels, 243
  design limits, 240
  floor panels, 240–2
    CFRP/AAC, 241
    GFRP/AAC, 241
  wall panels, 242
design guidelines, 503–4
design lane load, 321
design tandem, 321
design truck load, 321

digital shearography (DISH), 132, 134–5
  infrared thermographic imaging of CFRP-bonded concrete, 134
Direct ReInforcement Fabrication Technology (DRIFT), 318
disaster-resistant buildings
  innovative fibre-reinforced polymer (FRP) composites, 272–99
  designing CSIP for building applications under static loading, 279, 281–7
  innovative composite structural insulated panels (CSIP), 274, 276–9
  traditional and advanced panellised construction, 273–4
  flood effects on buildings, 275
  SIP panel illustration, 273
disbond, 281
displacement ductility factor, 211
distortion-induced fatigue cracking, 384
double-lane bridge deck system, 326–8
  section with top flat face and hat-sine rib, 326
  case 4 using glass/PP, 326
  top and bottom flat face with hat-sine rib, 327–8
  cross-sectional shape parameters, 327
  deflection in double-lane deck model, Plate X
  maximum deflection, 328
  shear stress in double-lane deck model, Plate XI
DRIFT process see hot-melt impregnation process
ductility, 118
durability
  fibre-reinforced polymer (FRP), 455–8
  cracking initiation, 457
  green composites, 452–3
Duralin steam treatment process, 451
Dyna Z-16, 307
Dynatup 8250, 256

E-glass, 9
efficiency of material (EOM), 472
energy balance model (EBM), 253–5, 266, 268–9
  data and parameters used to calculate energy absorbed, 268
  experimental vs predicted absorbed energy for FRP/AAC panels, 268
energy release rate (ERR), 156
environment-assisted subcritical debonding
  adhesive joint, 164–6
    schematic diagram, 165
  growth of FRP-concrete interface, 166–71
    fracture surfaces of wedge-driven tests, 170
    subcritical debonding of the epoxy-concrete interface, 168
    subcritical debonding of the epoxy-concrete interface comparison, 168
    wedge driving test specimen, 166
environmental engineering, 411
epoxy, 14
epoxy-concrete interface, 168–70
Euler buckling load, 490–2
  effective length of column for end-conditions, 491
  end-restraint coefficients for long column buckling, 491
Euler formula, 281
expanded polystyrene foam (EPS), 278–9
external post-tensioning, 384
extrusion, 33

facesheet/core debonding, 281–2
  upward and downward wrinkling, 282
fail-safe approach, 125–6
failure
  prevention strategies, 123–9
    behaviour of FRP-strengthened beams, 124
    FRP shear strengthening configurations, 127
    full-scale prestressed girder strengthened through bonding, 127
    NDT and SHM, 128–9
    redundant design against failure, 126–7
    safe-life vs fail-safe approaches, 125–6
  structural engineering applications, 116–23
    commercially available FRP systems, 117
    compression members, 119–20
    flexural members, 122–3
    load deflection behaviour of FRP strengthened beams, 123
    tensile stress–strain of selected unidirectional FRP systems, 118
    tension members, 120–2
failure mode, 83–6, 338
fatigue, 121, 457
  green composites, 452–3
fatigue tests, 430
fibre Bragg grating (FBG), 137
fibre failure, 479–80
fibre-reinforced polymer (FRP)
  advanced civil engineering applications, 177–203
  bridge superstructures, 347–79
    applications, 351–6
    hybrid FRP concrete bridge, 356–78
  building applications, 181–201
    bridges, 186–90
    epoxy soil nails, 199
    geotechnical applications, 197–99
    GRP sheet piling system components, 197
    GRP soil nail and netting, 198
    GRP soil nails and rock bolts, 198
    GRP soil nails and rock bolts installation, 199
    infrastructure, 190–5
    interiors and exteriors, 181–6
    pipes, 199–200
    railway infrastructure, 196–7

Index 515

conventional repair techniques and advantages, 383–6
  schematic of typical flexible web-gap detail, 385
design of all-composite structures, 469–505
  design guidelines, 503–4
  review on analysis, 470–2
  structural members, 485–502
  structural systems, 502–3
disaster-resistant buildings, 272–99
  designing CSIP for building applications under static loading, 279, 281–7
  innovative composite structural insulated panels (CSIP), 274, 276–9, 288–98
  traditional and advanced panellised construction, 273–4
environmental engineering applications, 410–62
  advantages and environmental benefits, 412–14
  chemical environmental applications, 414–17
  coal-fired plants, 423–9
  design codes and specifications, 458–61
  durability of composites, 455–8
  future trends, 461–2
  green composites, 447–55
  mining environments, 429–35
  modular building of environmental durability, 435–7
  recycling composites, 441–7
  sea-water environment, 418–23
  wraps, 437–41
failure in structural engineering applications, 116–23
failure modes and their prevention, 115–41
failure prevention strategies, 123–9
future trends, 140–1, 202–3
  Brisbane river walkway, 202
  GRP cassroom, 203
  Startlink modular construction system, 203

hybrid composites for structural applications, 205–22
  future trends, 221–2
  reinforced concrete beams internal reinforcement, 207–17
hybrid composites in bridge construction, 218–21
  arrangements of tendons and strands in various types of cables, 219
  FRP rebars and tendons used in some existing bridges, 218
materials use in construction, 178–80
  benefits, 178–80
  fabrication techniques, 180
non-destructive testing (NDT) and structural health monitoring (SHM), 129–40
strengthening steel structures, 382–405
  bond behaviour, 394–9
  case studies and field applications, 401–2
  flexural rehabilitation of steel and steel–concrete composite beams, 386–94
  future trends, 402–4
  repair of crack steel members, 399–400
  stabilising slender steel members, 400–1
systematic analysis and design methodology, 473–84
  lay-up of typical FRP box-section, 475
  macro components, 477–84
  material properties of constituents, 475
  materials and microstructures, 474–7
  schematic diagram, 473
types and arrangement, 3–16
  angle interlock weave, 16
  composites, 14–15
  fabrics, 10–14

Index

fibres, 5–10
  future trends, 15–16
  orthogonal weave structure, 15
fibre-reinforced polymer (FRP)-concrete interface
  durability, 163–71
  fracture analysis, 155–63
  interface stress analysis, 149–55
  rehabilitation of reinforced concrete structures, 148–71
fibre-reinforced polymer (FRP) wraps, 437–41
  strengthening of concrete structure, 437–8
    rehabilitation of T-shaped reinforced concrete, 438
  strengthening of timber structures, 438–40
    retrofitting of railroad bridges using FRP wraps, 439
  strengthening of waterway structures, 440–1
    aging infrastructure along US waterways, 440
fibre volume fraction, 476
fibres, 5–10
  chemical structure
    chemical composition of common natural fibres, 21
  properties, 21–2
    physico–mechanical properties of natural and synthetic fibres, 22
  source, 21
    commercial major fibre source, 20
  types, 20–1
film adhesives, 309–10
finite element
  BFFS panels, 237
    deflection contours, Plate V
    panel model with loading and boundary conditions, Plate IV
  UFFS panel, 235–7
    deflection contours, Plate III
    panel model with loading and boundary conditions, Plate II

finite element analysis, 322–3, 373
finite element/control volume (FE/CV), 63
fire resistance, 289–91
first-ply-failure, 479–80
5-harness satin (5-HS) weave, 13
Flame Seal FX-PL, 290–1
flax fibres, 448, 454
flexural behaviour, 386–93
  FRP strengthening effect on elastic stiffness, yield moment and ultimate strength, 392
  moment-curvature response of steel beams, 389
  moment-curvature response of steel–concrete composite beams, 390
  schematic of example beam studied in the parametric analysis, 388
  schematic of moment-curvature analysis of FRP strengthened steel composite, 387
flexural fibre-reinforced polymer wrapped beams, 86–94
  hand layup methods, 87
  shear flexural hand layup method, 92
  shear flexural VARTM method, 93–4
  theoretical and experimental results comparison, 89, 91
  VARTM method, 88
flexural properties, 37–8
flexural rehabilitation
  steel and steel–concrete composite beams, 386–94
    flexural behaviour, 386–93
    prestressed FRP strengthening, 393–4
flexural specific rigidity, 357–8
flood resistance, 291–6
  flood testing results, 293–6
    average failure load for four groups, 295
    average load-deflection curves for all groups, Plate VI
    average load-strain curves for all groups, Plate VII
    average results summary for four groups, 294

average stiffness for four groups, 295
debonding is the common failure mode, 294
schematic for CSIP test sample for flood testing, 293
specimens details, 292
test setup with test instrumentation ASTM C 393, 292
flow analysis network (FAN), 63
fracture analysis, 155–63
FRP rebar, 416–17
McKinleyville Bridge with reinforced concrete deck, 417
pavement with FRP rebar, 417
FRP-reinforced concrete (FRPRC), 207–8
analysis of beams, 208–9
design philosophy, 215–17
flexural strength and ductility improvement, 210–12
fusion bonding, 306–7
fuzzy-logic, 65

G–R model, 151
George Fischer beta-PP, 334
geosynthetics, 431–2
trailing dam and design concept, 432
GFRP I-beam, 413
glass fibre-reinforced polymer (GFRP), 117, 179
glass fibres, 9–10, 66
chemical structure, 10
mechanical properties, 10
global buckling, 279, 281
graphitisation, 8
green composites, 447–55
bioresins and natural fibres, 447–8
durability and fatigue, 452–3
hydrophilic nature of fibres, 451
incompatibility of fibre with resin, 451–2
pultrusion of natural composites, 448–51

mechanical properties and cost ratio of natural fibres, 448
tensile and flexural properties of NFRPs, 450
recent progress, 453–5

hand lay-up technique, 252
CFRP/AAC panels, 261–4
energy vs time and load vs time curve for LVI testing
H-1 specimen, 261
H-2 specimen, 261
H-3 specimen, 261
failure mode, 263
LVI data for hand layup process, 262
techniques limitations, 79–80
roller method, 79
spray method, 79
VARTM method of FRP application, 80
vs vacuum assisted resin transfer moulding (VARTM), 81–3, 94–7
CFRP wrapping scheme for flexural beam, 82
CFRP wrapping scheme for shear beam, 82
flexural FRP, 95
flexural FRP and experimental comparison, 95
flexural strengthened beam, 81
loading points, 82
shear flexural FRP, 96
shear flexural FRP and experimental comparison, 96
shear strengthened beam, 81
high airflow pressure loading actuator (HAPLA) system, 296
high velocity impact (HVI), 289
homogenisation theory, 471, 474, 484, 504
hot-melt impregnation process, 276–7
hybrid composites
fibre-reinforced polymer (FRP) for structural applications, 205–22
bridge construction, 218–21

future trends, 221–2
reinforced concrete beams internal reinforcement, 207–17
hybrid fibre-reinforced polymer (FRP)
autoclave aerated concrete (AAC) design for structural applications, 226–44
analytical modeling, 237–9
comparing different panel designs, 233–7
design graphs, 239–44
materials, processing and methods of investigation, 229–33
performance issues, 227–9
concrete bridge superstructure, 356–78
concrete structural system, 356–9
design concept of concrete bridge superstructure, 359–60
practical issues, 377–8
structural analysis, 373–7
design concept, 359–60
schematic diagram, 359
design features, 360–3
box section, 361
concrete layer thickness, 362–3
design philosophy and assumptions, 360–1
shear keys, 363
stacking sequences, 362
web inclination, 361–2
experimental study, 363–73
deformed shapes of top and bottom surfaces, 369
force-displacement in destructive flexural test, 372
force-displacement relationship at G-BOT-C, 368
GFRP materials properties, 365
instrumentation layout, 367
loading configurations, 366
longitudinal strain on exterior web at Section G, 370
materials, 364–5
stiffness degradation obtained from fatigue test, 371
test procedures, 366–8
test results, 368–73
test setup, 365–6
test specimen, 363–4
impact behaviour of panels for structural applications, 247–69
analysing sandwich structures using energy balance model (EBM), 253–5
analysis using energy balance model (EBM), 266, 268–9
impact testing results, 258–66
low velocity impact (LVI) and sandwich structures, 249–50
low velocity impact (LVI) testing, 255–7
materials and processing, 250–3

I-565 Highway bridge girder, 97–111
CFRP design calculation, 98, 100–1
1600.2 mm bulb tee girder sectional properties, 101
bending moments and shear forces for a typical beam, 102
bulb tee girder, 100
bulb tee girder sectional properties, 100
moment and shear, 103
number of CFRP layers calculation, 104
pre-cast AASHTO 1600.2 mm bulb tee girder sectional properties, 101
crack repair by injecting polyurethane foam, 99
initial, flexural and shear cracks, 99
schematic diagram, 98
VARTM, 101, 104–7
1,600.2 mm bulb tee girder longitudinal view, 105
bagging and de-bulking of layup, 109
bulb tee section with retrofitting strategy, 105
final view of reinforced section, 111
infusion progression and flow front throughout flange and web, 110

latex-based coating application, 111
 part after debagging and bottom flange detail, 110
 release film and distribution mesh placement, 109
 retrofitting scheme, 106
 sealant tape layout and fabric placement, 108
 surface preparation and pre-priming of vacuum bag sealant tape zone, 107
 VARTM cost evaluation for field implementation, 107–11
impact-echo (IE), 129–30
impact properties, 38–40
 damping index, 38
 notched Charpy strength comparison, 40
 tensile vs flexural modulus, 39
 tensile vs flexural strength, 39
impact resistance, 288–9
 high velocity impact (HVI), 289
  HVI on full-scale CSIP, 290
 low velocity impact (LVI), 288–9
  damage after LVI for OSB SIP and CSIP, 289
  typical impact vs time curves for CSIP and traditional SIP, 288
impregnation, 62
infrared thermography (IRT), 132, 134–5
 infrared thermographic imaging of CFRP-bonded concrete, 134
infrastructure, 190–5
 FRP utility poles, 192
 GRP absorbent noise barrier, 191
 GRP reflective noise barrier, 191
 GRP storage tank, 193
 GRP tank cover, 194
 heavy-duty trench covers, 195
 lightweight trench covers, 195
 motorway sign, 192
 sectional tank, 194
injection moulding, 33–4
 effect of filler type on mechanical strength performance of filled PP composites, 34

Instron Model 8250, 338
interface stress analysis, 149–55
 comparisons and verifications, 154–5
  interface stresses obtained by different methods, 155
 fundamental equations, 149–51
  free body diagram of the FRP strengthened RC beam, 150
  FRP-strengthened RC beam under concentrated load, 149
 three-parameter elastic foundation model, 153–4
 two-parameter elastic foundation model, 151–3
internal reinforcement, 207–17
 analysis of FRPRC beams, 208–9
 design philosophy of hybrid FRPRC beams, 215–17
  comparison of ρ_{bf} at design and testing stage, 216
  effectiveness of 135° hook in stirrup, 213–15
  load-displacement curves for over-reinforced FRPRC beam, 215
  properties and test results for the study of minimum reinforcement content, 215
 flexural strength and ductility improvement of FRPRC beams, 210–12
  load-displacement curves for pure and hybrid FRPRC over-reinforced beam, 210
 minimum flexural FRP reinforcement content, 212–13
  failure mode of FRPRC, 213
  FRPRC deflection profiles, 214
  load-displacement curves for FRPRC and SRC based on ACI code, 213

jute fibre-reinforced polypropylene, 44

kenaf fibres, 453–4
Kevlar, 5–7, 276
 chemical structure, 6
 properties, 7

Leadership in Energy and Environmental Design (LEED), 447
linear elastic fracture mechanics (LEFM), 156–8
  crack tip element of FRP-strengthened concrete beam, 156
  EERs vs crack lengths for a DCB specimen, 158
  phase angles vs crack lengths for a DCB specimen, 159
load, 83–6
Load and Resistance Factor Design (LRFD), 459–60
  adjustment factors for end-use conditions, 460
local buckling load, 493–6
  critical aspect ratios for FRP plates, 496
  geometry of orthotropic plate elements of FRP shapes, 494
  plates with various unloaded edge conditions, 495
  rotational restraint stiffness and stress resultant for FRP shapes, 497–8
LockHeed Martin bridge, 330
  cross-sections and comparison, 330
low velocity impact (LVI), 249–50, 288–9
  testing, 255–7
    jig or specimen holder, 257
    schematic diagram of Instron Model 8250 drop tower testing machine, 256

macro components, 477–84
  carpet plots for FRP panel properties, 480–3
  engineering properties of laminated panels, 477–8
    panel stiffness properties of FRP shapes, 478
    honeycomb cores and laminated facesheets, 483–4

equivalent material properties of sinusoidal core, 484
sinusoidal core element, 484
strength of pultruded PRF panels, 478–80
manual layup, 54–5
marker-and-cell (MAC), 63
material recycling, 442–3
mats, 60
mechanical fasteners, 306
mechanics of laminated beam (MLB) theory, 472, 485, 504
mercerisation, 25–6
Meyer's indentation law, 255
microcracks, 67
microwave non-destructive testing, 135–6
microwave scattering response of GFRP-bonded reinforced concrete, 136
modular building, 435–7
  FRP composite house, 436
  multi-purpose FRP building, 435
modular panelised construction
  thermoplastic composite structural insulated panels (CSIP), 302–15
    CSIP connections, 311–15
    CSIP manufacturing, 307–11
    precast panels joining in modular buildings, 305–7
    traditional structural insulated panel (SIP) construction, 304–5
modular track panels, 429–30
  cross-section of GFRP track panel, 430
moisture content, 42
mould-filling, 62, 64
Mylar, 253

natural composites, 413
  pultrusion, 448–51
natural fibre mat thermoplastic (NMT), 28
natural fibre reinforced composite, 447
natural fibre-reinforced structural insulated panels (NSIP), 44

natural fibres
  modification, 24–6
    chemical methods, 25–6
    physical methods, 25
Nexus, 474
Nomex, 249
non-destructive testing (NDT), 128–40, 141
  combined use of different methods, 139–40
  short term monitoring between a girder and supporting pier, 128
nonlinear fracture mechanics, 158–63
  flexural-shear crack induced debonding, 159
  interface stress distributions at different debonding stages, 162
  traction-separation law, 160
nonwoven fabrics, 12

ocean thermal energy conversion (OTEC), 422–3
  cold water pipe, 424
oriented strand board (OSB), 273
orthotropic plate analysis, 375

panelised systems, 226
para-aramid fibres, 5–7
pedestrian bridges, 353
periodic microstructure model, 476–7
petrochemical-based thermoplastic, 26–9
  CO_2 emissions, 29
  moisture absorption and electrical conductivity for HDPE-rice hull, 28
petrochemical-based thermosets, 29
physical weathering, 455
pipelines, 432–3
  testing of a 16-inch FRP pipe, 433
  uses of FRP vessels and pipelines, 433
pipes, 199–200
  GRP pipe production, 200
  large diameter GRP water pipe, 201
  pitch-based GRP pipe lining, 201

pitch, 7–8
plain AAC specimens, 258–60
  energy vs time and load vs time curve for LVI testing
    P-1 specimen, 259
    P-2 specimen, 259
    P-3 specimen, 260
  flexural failure and shear failure mode, 260
  LVI data, 258
plasma treatment, 25
plate–adhesive (PA) interface, 152–3
plate bonding, 55–6
Poisson's ratio, 281, 364
polyacrylontrile (PAN), 7–8
polycarbonate (PC), 443
polyethylene (PE), 26–7, 30
polyhydroxybutyrate-co-valerate (PHBV), 30
polylactic acid (PLA), 30
polypropylene (PP), 30
polyvinyl chloride (PVC), 313
preforming, 56–7, 420
prepreg, 4, 14, 70
pultruded composites, 65–9
  interlocking joints for pultruded building panels, 68
  pultrusion process, 65
pultrusion, 35, 65, 420–2
  4' GFRP composite sandwich panels with integral joint edges, 421
  full-scale panels under four point bend loading, 422

quasi-static impact response see low velocity impact (LVI)

railway infrastructure, 196–7
  GRP cable tray, 196
  station platform system, 196
  water catchpit for rail track drainage, 197
Rayleigh–Ritz method, 492
rayon, 7
recycling composites, 441–7
  material and thermal recycling, 442–3
  material research, 443–4

product development, 444–5
thermoplastics, 441–2
thermosets, 445–7
rehabilitation, 352–3
reinforced concrete cement (RCC), 273–4
reinforced concrete (RC)
   fibre-reinforced polymer (FRP)-concrete interface assessment, 148–71
      durability, 163–71
      fracture analysis, 155–63
      interface stress analysis, 149–55
reinforcement ratio, 209, 212
reinforcements, 55, 60
reinforcing bars, 353
reinforcing fibre, 178
reinforcing tendons, 353
repair welding, 384
resin, 178
resin transfer moulding (RTM), 34–5
   glass, hemp fibre mechanical properties, 35
resistive strain gages (RSG), 138
rock bolting, 433–5
   spirals in coal preparation process, 435
   underground mining, 434

safe-life approach, 125–6
sandwich panel, 228
sandwich structures, 249–50
scanning electron microscopy (SEM), 250
shape memory alloys (SMA), 402
shear fibre-reinforced polymer wrapped beams, 90–4
   shear flexural hand layup method, 92
   shear flexural VARTM method, 93–4
   theoretical and experimental results comparison, 91
shear flexural beam, 90
shear flexural hand layup beam, 91
shear flexural VARTM beam, 91
sheet piling, 422
   FRP handrail, ladder, platform, and walkway, 423

SHELL 99, 232–3
Shell 99 elements, 323
SIKA Carbon Fibre laminates, 229
SIKA WRAP HEX 103C, 229, 251, 333
SIKA WRAP HEX 113C, 229
SIKADUR HEX 300, 229, 251, 333
single-lane bridge deck system, 323–5
   section with top and bottom flat face with hat-sine rib, 325
      deflection in single-lane deck model, Plate VIII
      shear stress in single-lane deck model, Plate IX
   section with top flat face and hat-sine rib, 323–5
      cross sectional case parameters for cases 1 to 4 and 5 to 7, 324
      maximum deflection for single-lane deck models, 325
      parameters for case 4, 323
      parameters for case 5, 324
skin shearing, 249
smart damper, 220
SOLID 186, 232–3, 241
speckle pattern shearing interferometry (SPSI) *see* digital shearography (DISH)
stabilisation, 8
stabilising slender steel members, 400–1
Standard Test Method for Flexural Properties of Sandwich Constructions, 232
static tests, 430
steel reinforcement, 217
stitched fabrics, 14
strain, 83–6
strength, 285–6
   CSIP floor, 285
   CSIP wall, 285–6
strengthening steel structures
   case studies and field applications, 401–2
      Kentucky State Highway 32 using high-modulus CFRP materials, 403

fibre-reinforced polymer (FRP)
    composites, 382–405
  bond behaviour, 394–99
  conventional repair techniques
    and advantages, 383–6
  flexural rehabilitation of steel and
    steel–concrete composite beams,
    386–94
  future trends, 402–4
  repair of crack steel members,
    399–400
  stabilising slender steel members,
    400–1
stress intensity factors, 157
stress–strain response, 117, 334–8
  compression test results summary, 337
  failure comparison among CyPB6
    cylinders, 337
  stress vs strain
    cylinders, 334
    CyPB3 cylinders, 335
    CyPB6 cylinders, 335
stress wave methods, 129–31
  ultrasonic (A-scan) testing of FRP-
    bonded steel, 131
structural analysis, 373–7
  linear finite element analysis, 373–4
    finite element mesh, 373
  simple analysis method, 374–7
    bridge model for orthotropic plate
      analysis, 376
    deflection at midspan under the
      twice the tandem load, 377
    deformed shapes from different
      analyses, 378
    effective engineering properties of
      different laminates, 375
    representative units for
      orthotropic plate analysis, 377
structural applications
  impact behaviour of FRP/AAC
    panels, 247–69
    analysing sandwich structures
      using energy balance model
      (EBM), 253–5
    analysis using energy balance
      model (EBM), 266, 268–9

impact testing results, 258–66
low velocity impact (LVI) and
  sandwich structures, 249–50
low velocity impact (LVI) testing,
  255–7
materials and processing, 250–3
structural health monitoring (SHM),
  128–40
  combined use of different methods,
    139–40
  embedded sensor, 136–9
    fibre Bragg grating operation and
      transmission, 137
    FRP tendons in Stork Bridge,
      138
    short term monitoring between a
      girder and supporting pier, 128
structural insulated panel (SIP), 273
  economic analysis
    cost reductions in terms of
      construction time, 305
    cost reductions in terms of energy
      efficiency, 305
    cost reductions in terms of labour
      cost, 305
  traditional construction, 304–5
structural members, 485–502
  beam stiffness properties, 485–8
    bending and shear stiffness, 488
    global and local coordinator
      systems used in MLB, 486
  equivalent analysis of FRP decks,
    496, 499–502
    geometric notation, 499
  mechanical behaviours of FRP
    components, 488–96
    bending and shear deflections of
      Timoshenko beams, 489
    elastic deflections, 489–90
    elastic strains and stresses, 490
    Euler buckling, 490–2
    flexural-torsional buckling, 492–3
    local buckling, 493–6
    ultimate bending and shear failure,
      496
structural systems, 502–3
  deck-and-stringer bridge system, 502

Index

superstructure, 355–6
swelling, 42

tensile properties, 36–7
tension
  members failure, 120–2
    tensile fatigue failures, 122
thermal recycling, 442–3
thermoforming, 36, 319
thermoplastic composite structural insulated panels (CSIP)
  modular panelised construction, 302–15
    CSIP connections, 311–15
    CSIP manufacturing, 307–11
    precast panels joining in modular buildings, 305–7
    traditional structural insulated panel (SIP) construction, 304–5
thermoplastic composites
  bridge structures, 317–45
    bridge deck designs, 320–3
    comparing bridge deck designs, 329–31
    compression loading for bridge columns, 333–8
    design deck studies, 323–9
    impact loading for bridge columns, 338–43
    prefabricated wraps for bridge columns, 332–3
  manufacturing process, 318–19
    close-up view of woven E-glass/polypropylene fabric, 319
thermoplastic spray adhesives, 310–11
thermoplastics, 441–2
thermoplastics facesheets, 276–8
thermosets, 445–7
  compression moulding, 35
thermosetting resin, 55
three-dimensional preforms, 57
three-parameter elastic foundation model (3PEF), 153–4
  adhesive layer model, 153
through-thickness strain, 151
Timoshenko beam theory, 489, 504

Tinius–Olsen Universal Testing Machine, 333
Tonen Carbon Tow Sheet, 438
torsional rigidity, 501
traction-separation law, 161
Tsai–Hill theory, 322
Twaron, 5–7
two-dimensional preforms, 56–7
2D woven fabrics, 12–13
  5-HS weave plan, 13
  plain weave plan, 13
two-parameter elastic foundation (2PEF), 151–3
TYFO Glass Fibre, 229
TYFO SHE-51A Composite, 229

U-wrapped scheme, 239
ultimate displacement ratio, 211
ultrasonic pulse-echo, 131
ultrasonic testing (UT), 129–30
ultrasonic welding, 312–14
  healing of polymer-polymer interface, 313
  proposed CSIP ultrasonically welded to PVC extruded shape, 313
ultrasonics, 130
ultraviolet (UV) degradation, 6
ultraviolet (UV) light, 24
underground storage tanks (UST), 414–16
  first generation FRP tank, 414
Underwriters Laboratories (UL) 1316, 458
uniaxial FRP for flexural reinforcement (UFFS), 230–1
uniaxial FRP for flexural reinforcement (UFFS) panel
  experimental results, 233–4
    experimental load vs deflection curve, 234
    failure mode of specimen, 234
    failure specimen in shear span, 233
  finite element results, 235–7
    deflection of UFFS panel and FE analysis and experiment, 236
    experimental and FE load deflection curves, 236

unidirectional laminates, 11–12
  (0,90,0,90)$_s$ laminate composite, 12
UST Program, 415
utility poles, 427–9
  FRP utility poles, 429

vacuum-assisted resin transfer
    moulding (VARTM), 57–65, 180,
    229, 252, 420–2
  braiding, weaving, and knitting
    patterns, 61
  case study on the use of mould-filling
    simulation, Plate I
  CFRP/AAC panels, 264–6
    failure mode, 267
    LVI data for VARTM processed
      FRP/AAC specimens, 264
  energy vs time and load vs time
    curve for LVI testing
    V-1 specimen, 265
    V-2 specimen, 265
    V-3 specimen, 266
  external strengthening of structures,
    77–113
  flexural fibre-reinforced polymer
    wrapped beams, 86–90, 90–4
  future trends, 111–13
  hand layup techniques limitations,
    79–80
  I-565 Highway bridge girder,
    97–111
  load, strain, deflections and failure
    modes, 83–6
    control flexural, load vs deflection,
      failure at mid-span and cracking,
      84
    control shear, load vs deflection,
      shear failure at ends and
      cracking, 85
    theoretical and experimental
      results comparison, 85
  manufacturing steps, 60
  mould, 57

setup and processing, 59
shear fibre-reinforced polymer
    wrapped beams, 90–4
vs hand layup, 81–3, 94–7
  CFRP wrapping scheme for
    flexural beam, 82
  CFRP wrapping scheme for shear
    beam, 82
  flexural FRP, 95
  flexural FRP and experimental
    comparison, 95
  flexural strengthened beam, 81
  loading points, 82
  shear flexural FRP, 96
  shear flexural FRP and
    experimental comparison, 96
  shear strengthened beam, 81
vacuum bag, 58, 59–60
vacuum pressure, 59
vertical compression, 472
viscoelastic material, 220
volume of fluid (VOF), 63

water absorption, 41–2
  HDPE/biofibre composites with and
    without compatibilisers, 41
weathering, 24
windstorm resistance, 296–8
  capacity, 297
  failure mode, 296–7
    back view of CSIP panel under
      windstorm testing, 297
    general failure mode of CSIP,
      298
    schematic of failure consequence
      of CSIP under simulated
      loading, 298
Winkler hypothesis, 283
wood plastic composite (WPC), 19, 45
wood-rotting fungi, 43
wrinkling, 282

Young's modulus, 322, 365